Final Exam
Wed May 18
2 PM

INTRODUCTION TO TECHNICAL MATHEMATICS

LOREN RADFORD
Delaware County Community College

ANTHONY VAVRA
West Virginia Northern Community College

SHIRLEY RYCHLICKI
West Virginia Northern Community College

PRINDLE, WEBER & SCHMIDT
BOSTON

To Lorraine
L.E.R.

PWS PUBLISHERS
Prindle, Weber & Schmidt · Willard Grant Press · Duxbury Press
Statler Office Building · 20 Providence Street · Boston Massachusetts 02116

© Copyright 1983 by PWS Publishers

All rights reserved. No part of this book may be reproduced or transmitted in any form or by any means, electronic or mechanical, including photocopying, recording, or any information storage and retrieval system, without permission, in writing, from the publisher.

PWS Publishers is a division of Wadsworth, Inc.

Printed in the United States of America

83 84 85 86 87—10 9 8 7 6 5 4 3 2 1

Library of Congress Cataloging in Publication Data

Radford, L. E. (Loren E.)
  Introduction to technical mathematics.

  1. Engineering mathematics. I. Vavra, Anthony.
II. Rychlicki, Shirley. III. Title.
TA330.R25   1982   510′.2462   82-20419
ISBN 0-87150-339-5

ISBN 0-87150-339-5

Cover photograph by Bruce Iverson. © Copyright 1982 by Bruce Iverson. Used by permission of the photographer. All rights reserved.

Cover and text design by Deborah Schneider. Composition by Omegatype Typography, Inc. Artwork by Vantage Art. Text printed and bound by The Maple-Vail Book Manufacturing Group. Covers printed by The Lehigh Press.

PREFACE

This text has been developed to provide what the authors perceive as the essential elements of an intermediate level technical mathematics sequence. We use the term "intermediate level" to indicate a level of difficulty somewhere between courses that primarily emphasize arithmetic and algebraic skill development and those which are calculus-based. For this course, a comprehensive coverage of linear, quadratic, trigonometric, exponential, and logarithmic functions is needed. More importantly, the subject material must be related to the specific skills needed by technologists. Our colleagues in technical fields have been very helpful in suggesting topic areas for emphasis and in isolating skill deficiencies observed in their students.

There are four distinguishable levels within the text: Level 1 (Chapters 1–4) comprises an introduction to the skills of arithmetic, geometry, and algebraic expressions; Level 2 (Chapters 5–8) introduces basic algebra, trigonometry, and graphing; Level 3 (Chapters 9–12) provides a foundation in intermediate algebra; Level 4 (Chapters 13–15) concentrates on additional topics in trigonometry, graphing, and special functions. Those schools which require a preliminary course in the fundamental skills (Level 1) may wish to skip all or part of the material in the first four chapters.

We feel that the text is unique in several respects noted below.

1. Basic concepts of numbers, data handling, geometry, and algebra are treated early. The path through this material may be varied according to the prerequisite skills of the students. The presence of this material early in the text provides a valuable reference for students with specific deficiencies in their mathematical preparation.

2. We have chosen to emphasize measurement and data handling. Such material is often treated in basic physics or chemistry texts and receives little reinforcement in mathematics classes. The material on data analysis in the body of the text is supplemented by an appendix that has a more extensive treatment.

3. The use of the calculator is integrated throughout the text. Keystroke sequences for problem solutions on any calculator that uses algebraic logic are provided.

4. The chapter on linear functions and graphing is both practical and comprehensive. An alternative view of the mathematical function as a kind of processor is developed. This approach has worked well with our students and is useful in treating calculator functions. In graphing, emphasis is placed on producing a data table as the basis for the graph.

5. The text provides independent guidance to the student. Numerous worked examples illustrate the techniques of problem solution. Performance

objectives have been inserted in the text material. Chapter practice tests keyed to the performance objectives provide a quick measure of the student's mastery of the material. Practice test answers should be available to students for self-diagnosis or provided in a learning laboratory setting where they can be discussed with a tutor. We have followed the latter practice.

We would like to acknowledge the assistance of the following individuals in preparing this material for publication. Dolly Jaworski, Melanie Smith, and Margorie Klemm provided invaluable support in preparing the manuscript. We would also like to thank Deborah Schneider for her patient work as our production editor. Numerous colleagues and students provided suggestions and feedback on the material. We would like to particularly mention Don Adams, Richard Morrison, Frank Mulvaney, Reuben Aronovitz, Wilbur Wiley, Elliot Rothkopf, Jim Chetock, Connie Dale, Don Niemeier, Mark Goldstein, George R. Cash, Manatee Junior College, David Crystal, Rochester Institute of technology, and George W. Brewer, Odessa College.

CONTENTS

CHAPTER 1   **REVIEW OF NUMBER CONCEPTS AND OPERATIONS, 1**
     1.1   Whole Numbers, *2*
     1.2   Integers, *7*
     1.3   Rational Numbers, *13*
     1.4   Real Numbers, *27*
     1.5   Scientific Notation, *30*
     1.6   Rational Numbers Revisited—Percents, Ratios and Proportions, *39*
     1.7   Summary, *47*
     Chapter 1 Practice Test, *48*

CHAPTER 2   **MEASUREMENT AND DATA HANDLING, 51**
     2.1   Systems of Units, *52*
     2.2   Advantages of the SI System, *62*
     2.3   Operating on the Data—the Calculator, *68*
     2.4   Handling Significant Digits, *76*
     2.5   Summary, *83*
     Chapter 2 Practice Test, *84*

CHAPTER 3   **APPLIED GEOMETRY, 87**
     3.1   Lines, Angles and Polygons, *88*
     3.2   Measures Associated with Polygons, *97*
     3.3   Measures Associated with Circles, *111*
     3.4   Measures Associated with Solids, *117*
     3.5   Summary, *127*
     Chapter 3 Practice Test, *128*

CHAPTER 4   **ALGEBRAIC EXPRESSIONS, 131**
     4.1   Algebraic Terms and Expressions, *132*
     4.2   Evaluation of Algebraic Expressions, *135*
     4.3   The Definition of a Polynomial, *139*
     4.4   The Addition and Subtraction of Polynomials, *142*
     4.5   The Multiplication and Division of Algebraic Terms, *150*
     4.6   The Multiplication of Polynomials, *156*
     4.7   Summary, *163*
     Chapter 4 Practice Test, *164*

CHAPTER 5   **FACTORING AND ALGEBRAIC FRACTIONS, 167**
     5.1   Factoring Polynomials, *168*
     5.2   Factoring Special Products, *173*
     5.3   Definition of Algebraic Fractions, *178*
     5.4   Simplification of Algebraic Fractions, *179*
     5.5   Multiplication and Division of Algebraic Fractions, *183*

5.6 Addition and Subtraction of Algebraic Fractions, *186*
5.7 Summary, *193*
Chapter 5 Practice Test, *193*

CHAPTER 6

LINEAR EQUATIONS AND INEQUALITIES, 195

6.1 Introduction to Linear Equations in One Variable, *196*
6.2 Solving Linear Equations in One Variable, *199*
6.3 Applications of Linear Equations in One Variable, *207*
6.4 Introduction to Linear Inequalities in One Variable, *219*
6.5 Solving Linear Inequalities in One Variable, *223*
6.6 Summary, *229*
Chapter 6 Practice Test, *230*

CHAPTER 7

LINEAR FUNCTIONS AND GRAPHING, 233

7.1 Definition of a Function, *234*
7.2 The Linear Equation as a Function, *241*
7.3 The Data Table as a Basis for the Graph, *246*
7.4 The Rectangular Coordinate System, *249*
7.5 Constructing the Graph of a Linear Equation, *256*
7.6 Graphing a Line Using its Characteristics (Slope, Intercepts, Points), *259*
7.7 Finding the Equation of a Line from its Characteristics, *269*
7.8 Variation, *274*
7.9 Summary, *281*
Chapter 7 Practice Test, *283*

CHAPTER 8

BASIC TOPICS IN TRIGONOMETRY, 285

8.1 Constructing a Triangle, *286*
8.2 Angles Defined by Ratios, *289*
8.3 Solving a Right Triangle Using the Trigonometric Ratios, *299*
8.4 Solving a Right Triangle with the Aid of a Calculator, *302*
8.5 Solving Applied Problems Involving Right Triangles, *309*
8.6 Solving Oblique Triangles, *313*
8.7 Trigonometry in Surveying*, *321*
8.8 Summary, *326*
Chapter 8 Practice Test, *328*

CHAPTER 9

LINEAR SYSTEMS, 331

9.1 Solving Systems of Linear Equations Graphically, *332*
9.2 Solving Systems of Linear Equations Algebraically, *337*
9.3 Solving Systems of Linear Equations Using Determinants, *342*
9.4 Applications of Linear Equations in Two Variables, *348*
9.5 Solving Larger Systems of Linear Equations, *359*
9.6 Summary, *367*
Chapter 9 Practice Test, *367*

CHAPTER 10

RADICALS AND COMPLEX NUMBERS, 369

10.1 The Simplification of Radicals, *370*
10.2 The Addition and Subtraction of Radicals, *375*

CONTENTS

- 10.3 The Multiplication and Division of Radicals, *378*
- 10.4 The Definition of a Complex Number, *385*
- 10.5 The Addition and Subtraction of Complex Numbers, *387*
- 10.6 The Multiplication of Complex Numbers, *389*
- 10.7 The Division of Complex Numbers, *391*
- 10.8 Graphing a Complex Number, *393*
- 10.9 The *j*-Operator, *395*
- 10.10 Summary, *399*
  Chapter 10 Practice Test, *400*

CHAPTER 11 QUADRATIC EQUATIONS, 403

- 11.1 The Solution of a Quadratic Equation by Factoring, *404*
- 11.2 The Solution of a Quadratic Equation Using the Quadratic Formula, *410*
- 11.3 The Solution of a Quadratic Equation Using the Calculator, *415*
- 11.4 Summary, *418*
  Chapter 11 Practice Test, *419*

CHAPTER 12 GRAPHING QUADRATIC EQUATIONS AND SYSTEMS, 421

- 12.1 The Graph of a Quadratic Equation, *422*
- 12.2 Quadratic Systems, *437*
- 12.3 Applications of Quadratic Systems*, *444*
- 12.4 Summary, *458*
  Chapter 12 Practice Test, *459*

CHAPTER 13 ADDITIONAL TOPICS IN TRIGONOMETRY, 461

- 13.1 Angles: Measurement and Relation to Other Quantites, *462*
- 13.2 Trigonometric Functions (First Quadrant), *476*
- 13.3 Trigonometric Functions (All Quadrants), *483*
- 13.4 Vector Resolution, *497*
- 13.5 Vector Composition, *502*
- 13.6 Vector Addition, *508*
- 13.7 Statics Problems*, *513*
- 13.8 Summary, *515*
  Chapter 13 Practice Test, *517*

CHAPTER 14 GRAPHING THE TRIGONOMETRIC FUNCTIONS, 521

- 14.1 Trigonometric Function Values by Quadrant, *522*
- 14.2 Graphs of the Sine, Cosine, and Tangent, *527*
- 14.3 The Sine and Cosine with Amplitude and Phase, *536*
- 14.4 Sine Curves and Rotating Vector, *542*
- 14.5 Applications of Graphing (Electronics)*, *548*
- 14.6 Summary, *555*
  Chapter 14 Practice Test, *556*

CHAPTER 15   LOGARITHMS AND EXPONENTIALS,   559

15.1   The Exponential Function,   *560*
15.2   The Logarithmic Function,   *568*
15.3   The Laws of Logarithms,   *571*
15.4   Common Logarithms,   *574*
15.5   The Natural Logarithms,   *585*
15.6   Computations with Logarithms,   *590*
15.7   Applications of Logarithms and Exponentials*,   *593*
15.8   Summary,   *600*
       Chapter 15 Practice Test,   *601*

APPENDIX I   GRAPHING EXPERIMENTAL DATA,   A1

I.1   General Techniques of Displaying Data,   *A1*
I.2   Semi-logarithmic Plots,   *A5*
I.3   Pictorial Graphs,   *A13*
I.4   Vector Addition,   *A15*

APPENDIX II   DATA ANALYSIS,   A19

II.1   Linear Least Squares Fitting,   *A19*
II.2   Mean and Standard Deviation,   *A25*

APPENDIX III   DIMENSIONAL ANALYSIS,   A29

APPENDIX IV   TABLES,   A33

A   Trigonometric Functions,   *A34*
B   Logarithms to Base 10,   *A35*
C   Natural Logarithms,   *A37*
D   Powers of Base e,   *A39*

ANSWERS TO ODD-NUMBERED EXERCISES,   A41

*These application sections are indicated in text by a tab at the upper outside corner of the page.

CHAPTER 1

REVIEW OF NUMBER CONCEPTS AND OPERATIONS

1.1 Whole Numbers
1.2 Integers
1.3 Rational Numbers
1.4 Real Numbers
1.5 Scientific Notation
1.6 Rational Numbers Revisited—Percents, Ratios and Proportions
1.7 Summary

Since numbers are such an integral part of our lives, we tend to take them for granted. We forget that there was a time when the abstract concept of a number didn't even exist. When we learned to count, we had the names and symbols for the counting numbers at our disposal from the beginning. Collections of objects were used to illustrate the meaning of those names and symbols.

As we encountered measurement, we often found that the object being measured did not contain the unit of measurement a whole number of times. This necessitated the use of fractional parts of a unit. Going one step further, we also recognized that the measure of certain quantities could not be represented exactly, even by using fractional parts of units. So again new types of numbers arose. In this chapter we will consider various types of numbers and common operations on those numbers.

SECTION 1.1 WHOLE NUMBERS

The numbers with which we count, that is, the numbers 1, 2, 3, 4, etc., are called the **natural numbers.** We will denote the collection or set of natural numbers by N. If we include the number 0 along with the natural numbers, we get the set of **whole numbers,** which we will denote by W. Since N and W contain infinitely many numbers, we can't actually list every number. So we list enough numbers to establish the pattern of the numbers and put three dots (ellipses) to indicate that the pattern continues.

DEFINITION

NATURAL AND WHOLE NUMBERS

The **natural numbers** are the counting numbers 1, 2, 3, ...

The **whole numbers** are the counting numbers plus 0, that is, 0, 1, 2, 3, ...

We use a number line to represent the whole numbers graphically. A **number line** is constructed as follows: We draw a straight line, usually in a horizontal or vertical position although it can be in any position. We select an arbitrary point on the line, call this point the **origin** of the number line and assign it the number 0. We then decide on the distance that we will use for the length of one unit on the number line and the direction that will be the positive direction on the line. The number 1 is assigned to the point located one unit length in the positive direction from the origin; likewise, the number 2 is assigned to the point located two units length in the positive direction from the origin. Continuing on in this manner, each whole number is assigned to a point on the number line. Figure 1.1 shows two examples of number lines that exhibit the set of whole numbers. As the illustrations in Figure 1.1 show, horizontal number lines usually have

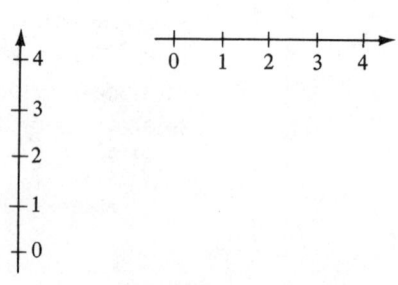

FIGURE 1.1

the positive direction to the right and vertical number lines usually have the positive direction upward. The arrow at the end of each number line indicates that the natural numbers continue indefinitely in the positive direction.

PERFORMANCE OBJECTIVE 1	*Locate 0, 5, 10, 15, 20, and 25 on a number line*
You should now be able to recognize whole numbers and locate them on a number line.	←——•——•——•——•——•——•——→ 　0　5　10　15　20　25

Next, we will briefly consider operations on whole numbers. We are all very familiar with addition, subtraction, multiplication, and division of whole numbers.

When whole numbers are added or multiplied, the resulting sum or product is always a whole number. We say that the whole numbers are **closed with respect to addition and multiplication.** That is,

If a and b are whole numbers, then $a + b$ and $a \cdot b$ are whole numbers.

(Note the use of a dot ($\cdot$) to indicate multiplication.)

The whole numbers are not closed with respect to subtraction because we cannot subtract a whole number from a smaller whole number and still get a whole number difference. For example, $2 - 5$ is not a whole number.

Likewise, the whole numbers are not closed with respect to division. We cannot divide a whole number by a larger whole number and get a whole number quotient. Even when we divide a whole number by a smaller whole number, we usually do not get a whole number quotient. We also cannot divide by 0. Division by 0 is not defined; it makes no sense. Some examples of the preceding comments are

$2 \div 3$ or $2/3$ or $\dfrac{2}{3}$ is not a whole number

$18 \div 5$ or $18/5$ or $\dfrac{18}{5}$ is not a whole number

$26 \div 0$ or $26/0$ is not defined

Division symbols $\div$, $/$, and the horizontal fraction bar will be used interchangeably.

We will have closure with respect to addition, multiplication, and subtraction when we consider a set of numbers called the **integers,** denoted by I. We will have closure with respect to addition, subtraction, multiplication, and division (except for division by 0) when we consider a set of numbers called the **rational numbers,** denoted by Q. Before considering integers and rational numbers, we will discuss some special properties of 0 and 1 and the prime factorization of natural numbers.

Note the following properties of the whole numbers 0 and 1:

1. When one of two addends is 0, the sum is the other addend. That is, $a + 0 = a$ for any number a. Because of this property, we call 0 the **additive identity.**

2. If we multiply any number by 1, we get that number as the product. That is $a \cdot 1 = a$ for any number a. Because of this property, we call 1 the **multiplicative identity.**

3. If at least one of the factors of a product is 0, then the product is 0.

4. If 0 is being divided by a nonzero number, the quotient is 0.

5. As we mentioned before, division by 0 is not defined. It is impossible.

Some examples that illustrate the properties of 0 and 1 are

1. $0 + 8 = 8$
2. $(4)(1) = 4$ (Note the use of parentheses to indicate multiplication).
3. $(3)(5)(0) = 0$
4. $\frac{0}{13} = 0$
5. $\frac{6}{0}$ is not defined.

PERFORMANCE OBJECTIVE 2 You should now be able to perform operations on whole numbers.	$(4)(0)(1) + \frac{0}{3} + 2$ $= 0 + 0 + 2$ $= 2$

A factorization of a natural number is a way of rewriting the number as a product of natural numbers. For example, 2×6, 3×4, 12×1, and $2 \times 2 \times 3$ are all the factorizations of the number 12.

WHOLE NUMBERS 1.1

DEFINITION

> **TWO CLASSIFICATIONS OF NATURAL NUMBERS**
>
> A **prime number** is not expressible as a product of factors other than 1 and itself.
>
> A **composite number** is expressible as a product of prime factors other than 1 and itself.

The prime numbers between 1 and 20 are 2, 3, 5, 7, 11, 13, 17, and 19.

The **prime factorization of a composite number** is the factorization of the number as a product of prime numbers. There is only one prime factorization of a number. Some examples of the prime factorization of a number are:

$$12 = 2 \cdot 2 \cdot 3$$
$$45 = 3 \cdot 3 \cdot 5$$
$$84 = 2 \cdot 2 \cdot 3 \cdot 7$$

To find the prime factorization of a number if it is not obvious, we can proceed in stages.

EXAMPLE Find the prime factorization of 1400.

SOLUTION We check to see if the prime number 2 is a factor of 1400. Since 1400 is an even number, 2 is a factor. We write

$$1400 = 2 \cdot 700$$

We now restrict our attention to 700. Since 700 is even, 2 is again a factor. So we have

$$1400 = 2 \cdot 700 = 2 \cdot 2 \cdot 350$$

Continuing in the same manner, we get

$$1400 = 2 \cdot 700 = 2 \cdot 2 \cdot 350 = 2 \cdot 2 \cdot 2 \cdot 175$$

Now 175 is not divisible by 2. So we check to see if it is divisible by the prime number 3. 175 is not divisible by 3 but it is divisible by the next prime number 5. Since $175 = 5 \cdot 35$ and $35 = 5 \cdot 7$, the factorization is completed as follows:

$$1400 = 2 \cdot 2 \cdot 2 \cdot 5 \cdot 35 = 2 \cdot 2 \cdot 2 \cdot 5 \cdot 5 \cdot 7$$

So the prime factorization of 1400 is $2 \cdot 2 \cdot 2 \cdot 5 \cdot 5 \cdot 7$.

PERFORMANCE OBJECTIVE 3	$174 = 2 \cdot 87$
You should now be able to prime factor natural numbers.	$= 2 \cdot 3 \cdot 29$ prime

EXERCISE SET 1.1

1. Answer true or false.
   a. Every natural number is also a whole number.
   b. N is closed with respect to addition and multiplication.
   c. $4 - 5$ is a whole number.
   d. $4 \div 0$ is a whole number.

2. Answer true or false.
   a. Every whole number is also a natural number.
   b. W is closed with respect to addition and multiplication.
   c. $10 - 20$ is a natural number.
   d. $0 \div 1$ is a natural number.

3. Perform the following operations using the properties of 0 and 1:

   a. $(2)(0)(5)$ b. $(1)(8)$ c. $\dfrac{0}{6} + 0$

4. Perform the following operations using the properties of 0 and 1:

   a. $(18)(25)(10)(0)$ b. $(12)(1)$ c. $0 + \dfrac{3-3}{1}$

5. Write the prime factorization of each of the following numbers. If the number is prime, say so.
   a. 54   b. 29   c. 180

6. Write the prime factorization of each of the following numbers. If the number is prime, say so.
   a. 78   b. 155   c. 59

SECTION 1.2 INTEGERS

In the last section we learned that W is not closed with respect to subtraction. If we are working with the set of whole numbers, we cannot subtract a number from a number smaller than itself. In order to be able to solve such subtraction problems, we need to introduce a set of numbers called the **negative integers.**

The natural numbers are often called **positive integers.** It is the case that for each natural number or positive integer a, there exists another number, denoted by $-a$ and read as "negative a," such that $a + (-a) = 0$. We call $-a$ **the additive inverse** of a or **the opposite** of a. For example, the additive inverse or opposite of 3 is -3 (negative 3) and we have that $3 + (-3) = 0$. The set of negative integers consists of the additive inverses of all the positive integers. The set of negative integers includes $-1, -2, -3, -4, \ldots$.

DEFINITION

> **INTEGERS**
>
> The set of **integers,** denoted by I, is the combination of the set of whole numbers and the set of negative integers, that is,
>
> $$\ldots -2, \ -1, \ 0, \ +1, \ +2, \ldots$$

We have already located the whole numbers on the number line. The negative integers are located on the opposite side of 0 from the positive integers. The point corresponding to the negative integer -1 is located the same distance from 0 as the positive integer 1 but in the opposite direction from 0. In general, for any positive integer a, the point corresponding to $-a$ lies the same distance from 0 as the point corresponding to a but in the opposite direction. Figure 1.2 shows examples of number lines that exhibit the set of integers. We recall that the numbers get larger as we move in the positive direction on the number line. We see that not only are the negative integers all less than zero, but also, when we compare negative integers, the one with the largest numerical value is actually the smallest number. For example, -3 is less than -1 and -40 is less than -26.

The negative integers have many practical applications. For example, if the temperature is 5 degrees below zero, we can write $-5°$ or if we have lost \$7 in a poker game, we can write $-\$7$ to represent the loss.

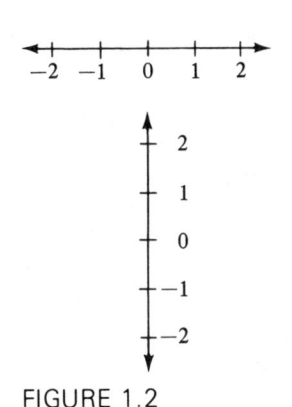

FIGURE 1.2

We have seen that a minus sign (−) denotes negative numbers. We can, although it is not necessary, put a plus sign (+) in front of positive numbers. If there is no sign before a nonzero number, it is understood that the number is positive.

DEFINITION

ABSOLUTE VALUE OF A NUMBER

The **absolute value** of a number is the distance of the number from the origin of the number line.

Since the distance from the origin to the point corresponding to −4 is 4 units, the absolute value of −4, denoted by |−4|, is 4. Similarly, the absolute value of 3, denoted by |3|, is 3; the absolute value of 0, denoted by |0|, is 0; the absolute value of −5, denoted by |−5|, is 5. Thus, the absolute value of a number is always greater than or equal to zero.

PERFORMANCE OBJECTIVE 4

You should now be able to recognize integers and locate them on a number line.

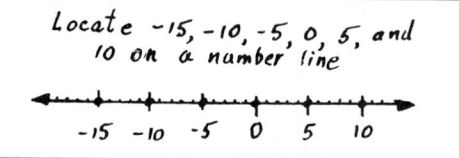

Locate −15, −10, −5, 0, 5, and 10 on a number line

[number line showing −15, −10, −5, 0, 5, 10]

We next discuss addition, subtraction, multiplication, and division problems that involve integers.

RULES FOR ADDING INTEGERS

1. To add two integers with the *same* sign, add the absolute values of the integers and give this sum the sign of the addends.

2. To add two integers with *different* signs, find the difference between the absolute values of the integers and give this difference the sign of the addend with the larger absolute value.

EXAMPLE   Add 2 + 9.

SOLUTION   2 + 9 = 11. Here both addends are positive. So we add the absolute values 2 and 9 and leave the sum positive.

EXAMPLE Add $-3 + (-5)$.

SOLUTION $-3 + (-5) = -8$. Here both addends are negative. So we add the absolute values 3 and 5 and make the sum negative.

EXAMPLE Add $6 + (-2)$.

SOLUTION $6 + (-2) = 4$. Here the addends have different signs. The absolute values of the addends are 6 and 2 and their difference is 4. We leave the 4 positive since the addend having the larger absolute value is the 6 and it is positive.

EXAMPLE Add $-5 + 3$.

SOLUTION $-5 + 3 = -2$. Again the addends have different signs. The difference of the absolute values 5 and 3 is 2. We give this difference a negative sign since the addend with the larger absolute value is negative.

To find the sum of more than two integers, we still add two integers at a time. We start by finding the sum of two of the integers; then we add a third integer to this sum; etc.

We can also use the number line to add integers. We add two integers as follows: We start at the origin and, just above the number line, draw an arrow from 0 to the first of the integers that we are attempting to add. We note the absolute value of the second integer; this absolute value is the number of units that we will move from the tip of the first arrow. If the second integer is positive, we start at the tip of the first arrow and move the necessary number of units in the positive direction; if the second integer is negative, we start at the tip of the first arrow and draw our second arrow in the negative direction. The number at the tip of the second arrow is the sum we are seeking. Some examples follow.

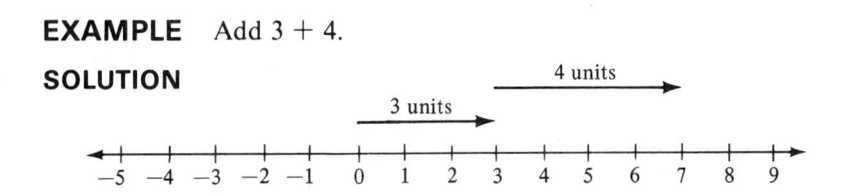

EXAMPLE Add $3 + 4$.

SOLUTION

The sum is 7.

EXAMPLE Add $8 + (-5)$.

SOLUTION

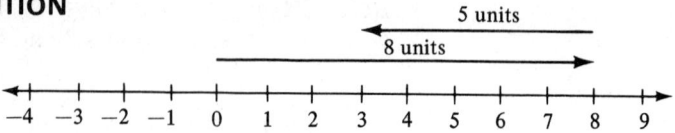

The sum is 3.

EXAMPLE Add $-7 + 4$.

SOLUTION

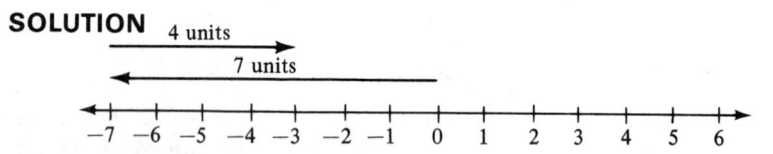

The sum is -3.

EXAMPLE Add $-5 + (-1)$.

SOLUTION

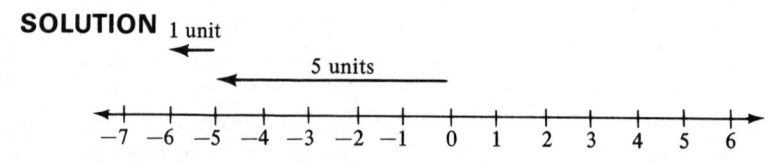

The sum is -6.

This procedure for adding two integers can also be used to add more than two integers.

EXAMPLE Add $6 + (-3) + (-5) + 1$.

SOLUTION

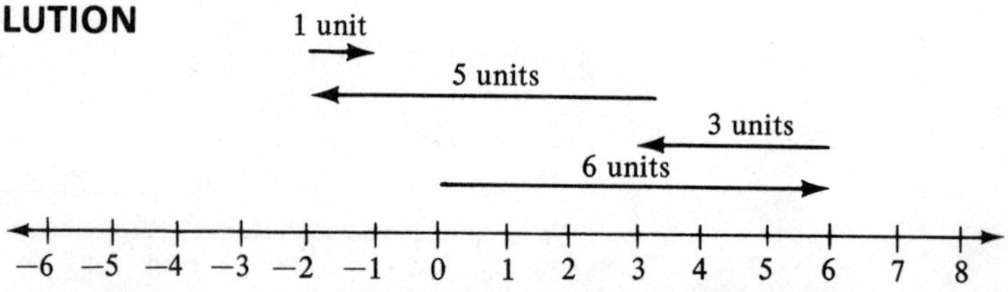

Procedure: We start at 0 and move to the first integer 6. To add -3, we move three units in the negative direction and get 3. To add -5, we move an additional five units in the negative direction and arrive at −2. Then to add 1, we move one unit in the positive direction. Our final answer is −1.

RULES FOR SUBTRACTING INTEGERS

Step 1. Change the sign of the subtrahend (the number being subtracted).

Step 2. Add using the rules for addition.

This procedure is illustrated in the following examples:

$$-4 - (-5) = -4 + 5 = 1$$
$$-11 - (-7) = -11 + 7 = -4$$
$$12 - 7 = 12 + (-7) = 5$$
$$-15 - 7 = -15 + (-7) = -22$$

PERFORMANCE OBJECTIVE 5	$-8 + (-2) + 3 = -10 + 3 = -7$
You should now be able to add and subtract integers.	$-18 - (-13) = -18 + 13 = -5$

RULES FOR MULTIPLYING OR DIVIDING INTEGERS

1. The absolute value of the product or quotient is obtained by multiplication or division of the absolute values of both integers.
2. The sign of the result is positive if both integers have the same sign.
3. The sign of the result is negative if the two integers have different signs.

Examples of multiplication and division of integers with like signs are

$$5(24) = 120, \quad (-7)(-3) = 21, \quad 18 \div 6 = 3, \quad -15 \div (-3) = 5$$

Examples of multiplication and division of integers with unlike signs are

$$3(-25) = -75, \quad -42 \div 7 = -6$$

The set of integers I is closed with respect to addition, subtraction, and multiplication, but it is not closed with respect to division. When we divide two integers, we do not always get an integer quotient. For example,

$$\frac{-9}{2} \text{ is not an integer}$$

In the next section, we will consider the set of rational numbers Q, which will be closed with respect to all four operations.

PERFORMANCE OBJECTIVE 6	
You should now be able to multiply and divide integers.	$\frac{(-9)(4)}{-6} = \frac{-36}{-6} = 6$

EXERCISE SET 1.2

1. Answer true or false. Every whole number is also an integer.

2. Answer true or false. Every integer is also a whole number.

3. Arrange the following integers in increasing order and locate them on a number line:

    $$6, \ -15, \ 0, \ -4, \ 10, \ -8$$

4. Arrange the following integers in increasing order and locate them on a number line:

    $$-9, \ 7, \ -2, \ 5, \ 0, \ -1, \ 3$$

5. a. Find the additive inverse of 12.
   b. Find the additive inverse of -7.
   c. What is the sum of a number and its additive inverse?

6. a. Find the additive inverse of 50.
   b. Find the additive inverse of -32.
   c. What number is its own additive inverse?

7. a. Find any integers a such that $|a| = 4$.
   b. Find any integers a such that $|a| = -2$.

RATIONAL NUMBERS 1.3

8. a. Find any integers a such that $|a| = 0$.
    b. Find any integers a such that $|a| = 6$.

9. a. Which of the following integers has the largest absolute value?

    $$-3, \quad 0, \quad -9, \quad 5$$

    b. Which of the integers in part (a) has the smallest absolute value?

10. a. Which of the following integers has the largest absolute value?

    $$-10, \quad 2, \quad -1, \quad 12$$

    b. Which has the smallest absolute value?

11. Use the number line to perform the following additions:
    a. $-2 + (-3)$
    b. $-5 + 7 + (-3)$

12. Use the number line to perform the following additions:
    a. $-8 + 6$
    b. $4 + (-11) + 8$

13. Perform the indicated operations.
    a. $18 + (-23)$
    b. $-1 + 9$
    c. $-5 - 5$
    d. $-9 - (-16)$
    e. $8(-11)$
    f. $(-4)(3)(-6)$
    g. $-76 \div (-19)$
    h. $\dfrac{-96}{6}$

14. Perform the indicated operations.
    a. $-34 + 27$
    b. $16 + (-15)$
    c. $-1 - (-1)$
    d. $23 - 32$
    e. $(-8)(-8)$
    f. $(-15)(-5)(-1)$
    g. $57 \div (-3)$
    h. $\dfrac{(-14)(5)}{-10}$

SECTION 1.3 RATIONAL NUMBERS

We will define rational numbers using two different forms: fraction form and decimal form.

DEFINITION

> **FRACTION FORM OF A RATIONAL NUMBER**
>
> A **rational number** is a number that can be written in the form
>
> $$\frac{a}{b}$$
>
> where a and b are integers and $b \neq 0$.
>
> *Note:* The symbol $\neq$ means "is not equal to."

The set of rational numbers is denoted by Q. Some examples of rational numbers in fraction form are

$$\frac{2}{3}, \quad \frac{1}{5}, \quad -\frac{3}{4}, \quad \text{and} \quad -\frac{5}{2}$$

Note that the negative fractions were written with the minus sign in front of the fraction. The fractions could also have been written with the minus sign in the numerator (the top half of the fraction) or in the denominator (the bottom half of the fraction). That is,

$$-\frac{a}{b} = \frac{-a}{b} = \frac{a}{-b}$$

For example,

$$-\frac{3}{4} = \frac{-3}{4} = \frac{3}{-4}$$

We see that an integer is also a rational number since any integer can be written as the integer over 1. For example,

$$2 = \frac{2}{1}, \quad -5 = \frac{-5}{1}, \quad \text{and} \quad 0 = \frac{0}{1}$$

So 2, −5, and 0 are rational numbers.

DEFINITION

> **DECIMAL FORM OF A RATIONAL NUMBER**
>
> A **rational number** is a number that can be written as a decimal that either terminates or repeats.

By **terminating decimal,** we mean one that eventually stops; it has only a finite number of nonzero decimal places. For example, 0.125, −0.5, 226.29, −17.4256, and 8.333 are terminating decimals. A

repeating decimal is one that continues on forever but that eventually begins to repeat the same digit or block of digits over and over without end. Examples of repeating decimals are 0.666 . . . , 2.08333 . . . , −0.8989 . . . , and −5.4123123 . . . . The dots in the preceding numbers indicate that the repeating digit or block of digits continues to repeat forever.

Each rational number is represented by a point on the number line. Figure 1.3 shows the points corresponding to the rational numbers

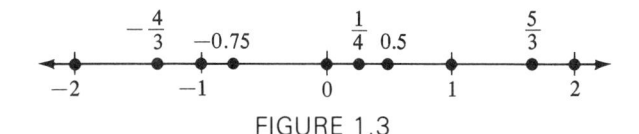

$$-2, \quad -\frac{4}{3}, \quad -1, \quad -0.75, \quad 0, \quad \frac{1}{4}, \quad 0.5, \quad 1, \quad \frac{5}{3}, \quad \text{and} \quad 2$$

FIGURE 1.3

For example, the point corresponding to the number 0.5 is located midway between the points corresponding to 0 and 1 and the point corresponding to $-\frac{4}{3}$ is $\frac{1}{3}$ of the way from −1 to −2.

Later in this section and again in Sections 1.5 and 1.6, we will work with rational numbers in decimal form. We will perform operations on decimals and convert fractions to decimals and decimals to fractions. We will also consider scientific notation, a special way of writing the decimal form of rational numbers. But first we will perform operations on rational numbers in fraction form.

PERFORMANCE OBJECTIVE 7	Locate $0, 1, -\frac{2}{3}, -1.5, 1.75,$ and -1 on a number line.
You should now be able to recognize rational numbers and locate them on a number line.	

DEFINITION	**FRACTION IN LOWEST TERMS**
	A fraction is in **lowest terms** if there are no common factors in the numerator and denominator.

We reduce a fraction to lowest terms by dividing both the numerator and denominator by the largest number that will divide into both (the **greatest common divisor** or **GCD**). If we cannot easily determine a number that will divide into both, it may help to write both the numerator and denominator as a product of prime numbers and proceed as in the following examples.

EXAMPLE Reduce $\frac{12}{18}$ to lowest terms.

SOLUTION **Method 1:** Six is the greatest common divisor of 12 and 18. We divide both 12 and 18 by 6 and get

$$\frac{12}{18} = \frac{2}{3}$$

Method 2: Both 12 and 18 are divisible by 2 since they are even numbers. If we divide 12 and 18 by 2, we get

$$\frac{12}{18} = \frac{6}{9}$$

Now 6 and 9 are both divisible by 3, so we get

$$\frac{6}{9} = \frac{2}{3}$$

Method 3: $\frac{12}{18} = \frac{2 \cdot 2 \cdot 3}{2 \cdot 3 \cdot 3} = \frac{2}{2} \cdot \frac{2}{3} \cdot \frac{3}{3} = 1 \cdot \frac{2}{3} \cdot 1 = \frac{2}{3}$

EXAMPLE Reduce $\frac{72}{120}$ to lowest terms by prime factorization of the numerator and denominator.

SOLUTION $\frac{72}{120} = \frac{2 \cdot 2 \cdot 2 \cdot 3 \cdot 3}{2 \cdot 2 \cdot 2 \cdot 3 \cdot 5} = \frac{2 \cdot 2 \cdot 2 \cdot 3}{2 \cdot 2 \cdot 2 \cdot 3} \cdot \frac{3}{5} = 1 \cdot \frac{3}{5} = \frac{3}{5}$

RULE FOR RAISING A FRACTION TO HIGHER TERMS

To raise a fraction to higher terms, multiply the numerator and denominator by the same number.

EXAMPLE Write $\frac{3}{4}$ as a fraction with denominator 36.

RATIONAL NUMBERS 1.3

SOLUTION Multiply both the numerator and denominator by 9 to get

$$\frac{3}{4} = \frac{3 \cdot 9}{4 \cdot 9} = \frac{27}{36}$$

EXAMPLE Write $\frac{-2}{5}$ as a fraction with denominator 20.

SOLUTION $\frac{-2}{5} = \frac{-2 \cdot 4}{5 \cdot 4} = \frac{-8}{20}$

We will often be raising fractions to higher terms when we add and subtract rational numbers.

PERFORMANCE OBJECTIVE 8	$\frac{12}{45} = \frac{2 \cdot 2 \cdot 3}{3 \cdot 3 \cdot 5} = \frac{3}{3} \cdot \frac{2 \cdot 2}{3 \cdot 5} = 1 \cdot \frac{4}{15} = \frac{4}{15}$
You should now be able to reduce a fraction to lowest terms and raise a fraction to higher terms.	$\frac{7}{13} = \frac{7 \cdot 5}{13 \cdot 5} = \frac{35}{65}$

RULES FOR ADDING FRACTIONS

1. If the fractions have the *same* denominator, add the numerators and put the sum over the denominator.

2. If the fractions have *different* denominators, first write the fractions with a common denominator, then add as in part 1.

EXAMPLE Add $\frac{2}{9} + \frac{5}{9}$.

SOLUTION In symbols, Rule 1 for adding fractions becomes

$$\frac{a}{c} + \frac{b}{c} = \frac{a + b}{c}$$

We find $\frac{2}{9} + \frac{5}{9} = \frac{2+5}{9} = \frac{7}{9}$.

EXAMPLE Add $\frac{1}{3} + \frac{2}{5}$.

SOLUTION In symbols, Rule 2 for adding fractions becomes

$$\frac{a}{b} + \frac{c}{d} = \frac{a \cdot d}{b \cdot d} + \frac{b \cdot c}{b \cdot d} = \frac{a \cdot d + b \cdot c}{b \cdot d}$$

We find $\frac{1}{3} + \frac{2}{5} = \frac{5}{15} + \frac{6}{15} = \frac{11}{15}$.

When we have to find a common denominator, we should try to find the **least common denominator (LCD)** so that we don't have to work with larger numbers than necessary.

RULES FOR FINDING THE LEAST COMMON DENOMINATOR (LCD)

Step 1. Write the prime factorization of each of the denominators.

Step 2. Write down each number that appears as a factor the greatest number of times it appears in any of the denominators.

Step 3. Multiply together all the factors listed in Step 2. This product is the LCD.

EXAMPLE Find the sum $\frac{3}{8} + \frac{5}{28}$.

SOLUTION First, we find the LCD:

$$8 = 2 \cdot 2 \cdot 2$$
$$28 = 2 \cdot 2 \cdot 7$$

The numbers that appear in the factorizations are 2 and 7. We use 2 as a factor three times since the greatest number of times 2 appears in one of the numbers is three. We use 7 as a factor once. So, the LCD = $2 \cdot 2 \cdot 2 \cdot 7 = 56$. Since

$$\frac{3}{8} = \frac{21}{56} \quad \text{and} \quad \frac{5}{28} = \frac{10}{56}$$

we have

$$\frac{3}{8} + \frac{5}{28} = \frac{21}{56} + \frac{10}{56} = \frac{21 + 10}{56} = \frac{31}{56}$$

If a negative fraction is involved in the addition of rational numbers, it is a good idea to write the minus sign in the numerator

RATIONAL NUMBERS 1.3   **19**

before adding. Handling the minus sign then becomes a matter of adding integers, which we learned to do in the last section.

EXAMPLE   Add $\frac{3}{10}$ and $-\frac{3}{5}$.

SOLUTION
$$\frac{3}{10} + \left(-\frac{3}{5}\right) = \frac{3}{10} + \left(\frac{-3}{5}\right)$$
$$= \frac{3}{10} + \left(\frac{-6}{10}\right)$$
$$= \frac{3 + (-6)}{10}$$
$$= \frac{-3}{10}$$

RULES FOR SUBTRACTING FRACTIONS

Step 1. Change the sign of the subtrahend.

Step 2. Add using the rules for addition.

Some examples are

$$\frac{2}{3} - \frac{1}{4} = \frac{2}{3} + \left(\frac{-1}{4}\right) = \frac{8}{12} + \left(\frac{-3}{12}\right) = \frac{8 + (-3)}{12} = \frac{5}{12}$$

$$\frac{5}{6} - \left(-\frac{1}{8}\right) = \frac{5}{6} + \frac{1}{8} = \frac{20}{24} + \frac{3}{24} = \frac{20 + 3}{24} = \frac{23}{24}$$

$$-\frac{8}{9} - \left(-\frac{3}{5}\right) = \frac{-8}{9} + \frac{3}{5} = \frac{-40}{45} + \frac{27}{45} = \frac{-40 + 27}{45} = \frac{-13}{45}$$

$$-3 - \frac{10}{7} = \frac{-3}{1} + \left(\frac{-10}{7}\right) = \frac{-21}{7} + \left(\frac{-10}{7}\right) = \frac{-21 + (-10)}{7} = \frac{-31}{7}$$

When we get an **improper fraction,** that is, a fraction in which the absolute value of the numerator is greater than that of the denominator, such as

$$\frac{-31}{7}$$

we may either leave the answer as the improper fraction in lowest terms or rewrite the answer as a mixed number as follows:

$$-4\tfrac{3}{7}$$

PERFORMANCE OBJECTIVE 9 You should now be able to add and subtract fractions.	$\frac{5}{12} + \frac{11}{54} = \frac{45}{108} + \frac{22}{108} = \frac{45+22}{108} = \frac{67}{108}$ $7 - \frac{5}{9} = 7 + \left(\frac{-5}{9}\right) = \frac{63}{9} + \left(\frac{-5}{9}\right)$ $\phantom{7 - \frac{5}{9}} = \frac{58}{9}$ or $6\frac{4}{9}$

RULES FOR MULTIPLYING FRACTIONS

Step 1. Multiply the numerators to find the numerator of the product.

Step 2. Multiply the denominators to find the denominator of the product.

Step 3. Reduce to lowest terms.

In symbols, the procedure for multiplication becomes

$$\frac{a}{b} \cdot \frac{c}{d} = \frac{ac}{bd}$$

The sign rules that we learned for multiplying integers hold for rational numbers as well. That is, when we multiply rational numbers with like signs, the product is positive; and when we multiply rational numbers with different signs, the product is negative. For example,

$$4 \cdot \frac{5}{9} = \frac{4}{1} \cdot \frac{5}{9} = \frac{20}{9}$$

$$-\frac{1}{4}\left(-\frac{3}{8}\right) = +\frac{3}{32}$$

$$\frac{7}{8}\left(-\frac{4}{5}\right) = -\frac{28}{40} = -\frac{7}{10}$$

Note that in the last example, we reduced our answer to lowest terms after we multiplied. We could have avoided the need to reduce our answer after multiplying by eliminating common factors in the numerator and denominator before multiplying. For example, we could have written

$$\frac{7}{8}\left(-\frac{4}{5}\right) = \frac{7}{2 \cdot \cancel{4}}\left(-\frac{\cancel{4} \cdot 1}{5}\right) = -\frac{7}{10}$$

RATIONAL NUMBERS 1.3

RULES FOR DIVIDING FRACTIONS

Step 1. Invert the divisor.

Step 2. Change the division sign to a multiplication sign.

Step 3. Perform the multiplication using the rules for multiplication.

In symbols, the procedure for division becomes

$$\frac{a}{b} \div \frac{c}{d} = \frac{a}{b} \cdot \frac{d}{c} = \frac{ad}{bc}$$

Some examples that use this procedure are

$$\frac{3}{4} \div \frac{1}{2} = \frac{3}{4} \cdot \frac{2}{1} = \frac{6}{4} = \frac{3}{2}$$

$$-\frac{7}{10} \div \frac{5}{3} = -\frac{7}{10} \cdot \frac{3}{5} = -\frac{21}{50}$$

$$-\frac{3}{2} \div (-4) = -\frac{3}{2} \div \left(-\frac{4}{1}\right) = -\frac{3}{2} \cdot \left(-\frac{1}{4}\right) = \frac{3}{8}$$

In discussing the multiplication and division of rational numbers, we should note a property that the set of rational numbers has that the set of integers did not have. For each rational number a except 0, there exists a rational number

$$\frac{1}{a}$$

such that $a \cdot \frac{1}{a} = 1$. We call $\frac{1}{a}$ the **multiplicative inverse** of a, or the **reciprocal** of a. For example,

$\frac{1}{2}$ is the multiplicative inverse of 2

$-\frac{4}{3}$ is the multiplicative inverse of $-\frac{3}{4}$

-5 is the multiplicative inverse of $-\frac{1}{5}$

PERFORMANCE OBJECTIVE 10	$\frac{7}{15} \cdot \frac{10}{17} = \frac{7}{3 \cdot \cancel{5}} \cdot \frac{2 \cdot \cancel{5}}{17} = \frac{14}{51}$
You should now be able to multiply and divide fractions.	$\frac{8}{13} \div \frac{4}{5} = \frac{8}{13} \cdot \frac{5}{4} = \frac{2 \cdot \cancel{4}}{13} \cdot \frac{5}{\cancel{4}} = \frac{10}{13}$

The sign rules we learned for performing operations on integers hold for rational numbers as well. So, to perform operations on rational numbers in decimal form, we only need to recall how to add, subtract, multiply, and divide positive decimal numbers. Consider the following examples.

EXAMPLE Add 7.387 + 9.546. **SOLUTION**
$$\begin{array}{r} 7.387 \\ +9.546 \\ \hline 16.933 \end{array}$$

EXAMPLE Subtract 12.4 − 7.932. **SOLUTION**
$$\begin{array}{r} 12.400 \\ -7.932 \\ \hline 4.468 \end{array}$$

EXAMPLE Multiply 5.4 × 7.68. **SOLUTION**
$$\begin{array}{r} 7.68 \\ \times 5.4 \\ \hline 3\;072 \\ 38\;40 \\ \hline 41.472 \end{array}$$

EXAMPLE Divide 47.4 ÷ 6.32. **SOLUTION**
$$\begin{array}{r} 7.5 \\ 6.32\overline{)47.40.0} \\ 44\;24 \\ \hline 3\;160 \\ 3\;160 \\ \hline 0 \end{array}$$

The preceding examples illustrate the following rules for performing operations on positive decimal numbers.

1. When we add or subtract, we line up the decimal points before we perform the addition or subtraction. If one number has fewer decimal places than the other, 0's can be used to create the same number of decimal places in both numbers. The process of adding additional 0's is equivalent to finding a common denominator.

2. When we multiply, the number of decimal places in the product is the sum of the number of decimal places in the factors.

3. When we divide, we move the decimal point in both the divisor and the dividend to the right, enough places to make the

RATIONAL NUMBERS 1.3

divisor a whole number. If necessary, 0's can be added at the end of the dividend. The decimal point in the quotient is placed directly above the decimal point in the dividend.

The following examples involve negative decimal numbers.

EXAMPLE Add -3.724 and -6.59.

SOLUTION $-3.724 + (-6.59) = -10.314$

Since both addends are negative, we add the absolute values 3.724 and 6.59 and make the sum negative.

EXAMPLE Add -7.385 and 6.142.

SOLUTION $-7.385 + 6.142 = -1.243$

The addends have different signs. We take the difference of the absolute values 7.385 and 6.142, then make the difference negative since the addend with the largest absolute value is negative.

EXAMPLE Subtract $-5.46 - (-9.57)$.

SOLUTION $-5.46 - (-9.57) = -5.46 + 9.57 = 4.11$

EXAMPLE Multiply -2.3 by -3.618.

SOLUTION $(-2.3)(-3.618) = 8.3214$

When we multiply or divide two numbers that have the same sign, the product or quotient is positive.

EXAMPLE Divide -8.75 by 0.125.

SOLUTION $-8.75 \div 0.125 = -70$

When we multiply or divide two numbers that have different signs, the product or quotient is negative.

PERFORMANCE OBJECTIVE 11	$-4 - (-2.58) = -4 + 2.58 = -1.42$
You should now be able to perform operations on rational numbers in decimal form.	$(-0.02)(-7.8) = 0.156$

We convert a rational number from fraction form to decimal form by dividing the numerator of the fraction by the denominator. The resulting decimal number will either terminate or repeat. Remember that we can use dots to indicate repeating blocks of digits.

EXAMPLE Convert $\frac{5}{8}$, $\frac{3}{7}$, $-\frac{21}{25}$, and $-\frac{19}{12}$ to decimal form.

SOLUTION

$$\frac{5}{8} = 5 \div 8 = 0.625, \quad \frac{3}{7} = 3 \div 7 = 0.428571428571\ldots$$

$$\frac{-21}{25} = -21 \div 25 = -0.84, \quad \frac{-19}{12} = -19 \div 12 = -1.58333\ldots$$

To convert a terminating decimal to fraction form, we must recall the names of the decimal places that we encounter as we move to the right of the decimal point. Moving from left to right the decimal places are, respectively, tenths, hundredths, thousandths, ten thousandths, etc. The number of decimal places involved tells us the denominator of the fraction. However, we should always be sure to reduce the fraction to lowest terms.

EXAMPLE Convert -0.8, 0.72, -0.001, and 0.0125 to fraction form.

SOLUTION

$$-0.8 = -\frac{8}{10} = -\frac{4}{5}$$

$$0.72 = \frac{72}{100} = \frac{18}{25}$$

$$-0.001 = \frac{-1}{1000}$$

$$0.0125 = \frac{125}{10,000} = \frac{1}{80}$$

We will not discuss the technique for converting repeating decimals to fractions. However, certain repeating decimals should be easily recognized. For example,

$$\frac{1}{3} = 0.333\ldots \quad \text{and} \quad \frac{2}{3} = 0.666\ldots$$

RATIONAL NUMBERS 1.3

PERFORMANCE OBJECTIVE 12	$\frac{2}{9} = 0.222\ldots$
You should now be able to convert rational numbers in fraction form to decimal form and vice versa.	$0.65 = \frac{65}{100} = \frac{13}{20}$

EXERCISE SET 1.3

1. Answer true or false. Every integer is also a rational number.

2. Answer true or false. Every rational number is also an integer.

3. Arrange the following rational numbers in increasing order and locate them on a number line:

   $$\frac{5}{6},\ -1.0,\ -\frac{7}{3},\ 3,\ \frac{11}{4},\ 1.5,\ -0.8$$

4. Arrange the following rational numbers in increasing order and locate them on a number line:

   $$-2,\ \frac{5}{3},\ -1.25,\ \frac{7}{2},\ 0,\ 2.75$$

5. Reduce the following fractions to lowest terms:
   a. $\frac{39}{91}$
   b. $-\frac{120}{126}$

6. Reduce the following fractions to lowest terms:
   a. $\frac{28}{70}$
   b. $-\frac{110}{198}$

7. a. Fill in the missing numerator: $\frac{-6}{7} = \frac{}{42}$.
   b. Fill in the missing denominator: $\frac{9}{16} = \frac{45}{}$.
   c. Write three fractions that are equal to $\frac{7}{8}$.

8. a. Fill in the missing numerator: $\frac{7}{9} = \frac{}{81}$.
   b. Fill in the missing denominator: $\frac{-4}{15} = \frac{-80}{}$.
   c. Write three fractions that are equal to $-\frac{5}{12}$.

9. a. Find the multiplicative inverse of $\frac{4}{3}$.
   b. Find the multiplicative inverse of -8.
   c. What is the product of a number and its multiplicative inverse?

10. a. Find the multiplicative inverse of $-\frac{5}{8}$.
    b. Find the multiplicative inverse of 10.
    c. Find a number that is its own multiplicative inverse. (There are two such numbers).

11. Perform the indicated operations. Give your answer in lowest terms.

    a. $-3 + \frac{8}{5}$
    b. $-\frac{2}{3} + \frac{5}{7}$
    c. $\frac{3}{8} - \frac{9}{14}$
    d. $-\frac{1}{2} - \left(-\frac{1}{2}\right)$
    e. $-\frac{3}{4}\left(-\frac{5}{6}\right)$
    f. $\frac{2}{5} \div \left(-\frac{24}{25}\right)$
    g. $-\frac{81}{2} \div (-27)$

12. Perform the indicated operations. Give your answer in lowest terms.

    a. $\frac{9}{4} + (-5)$
    b. $\frac{5}{8} - \frac{5}{12}$
    c. $-14\left(\frac{2}{7}\right)$
    d. $-\frac{7}{15} + \left(-\frac{7}{10}\right)$
    e. $0 - \left(-\frac{2}{3}\right)$
    f. $\frac{-24}{25} \div \left(-\frac{2}{5}\right)$
    g. $-\frac{18}{5} \div 72$

13. Perform the indicated operations.
    a. $-14.76 + 11.472$
    b. $0.569 - 4$
    c. $(10.32)(-4.188)$
    d. $(-63.788) \div (-8.62)$

14. Perform the indicated operations.
    a. $-3.891 + (-18.2)$
    b. $6 - 10.01$
    c. $(-0.125)(-0.64)$
    d. $-76.916 \div 6.7$

15. Convert the following rational numbers to decimal form:
    a. $\frac{19}{8}$
    b. $\frac{-2}{11}$
    c. $\frac{3}{16}$

16. Convert the following rational numbers to decimal form:
    a. $\frac{-83}{8}$
    b. $\frac{5}{12}$
    c. $\frac{3}{32}$

17. Convert the following rational numbers to fractions in lowest terms:

    a. 0.075   b. 0.0005   c. 0.666 . . .

18. Convert the following rational numbers to fractions in lowest terms:

    a. 0.008   b. 0.0018   c. 0.333 . . .

SECTION 1.4 REAL NUMBERS

We have discussed four sets of numbers so far in this chapter: N, W, I, and Q. We learned that natural numbers are also whole numbers, whole numbers are also integers and integers are, in turn, rational numbers. The question arises as to whether there is a bigger set of numbers that contains the rational numbers.

We have seen that each rational number corresponds to a point on the number line. We might then ask if there are other numbers that are not rational but are represented by points on the number line? The answer to both questions is "yes."

There do exist numbers that are not rational. Among the numbers that are not rational are the **irrational numbers.** When we combine the irrational numbers with the rational numbers, we form a bigger collection of numbers called the **set of real numbers.** The set of real numbers is denoted by R. The real numbers fill up the number line. That is, every real number corresponds to a point on the number line and every point on the number line corresponds to a real number.

But what type of numbers are the irrational numbers? Since an irrational number is not rational, it cannot be written in the form

$$\frac{a}{b}$$

where a and b are integers. Likewise, it cannot be written as a terminating or repeating decimal. If we were to write an irrational number in decimal form, the decimal expansion would continue forever and never begin to repeat the same block of digits. Some specific examples of irrational numbers are $\sqrt{2}$ (the square root of 2), $-\sqrt{3}$ (the negative square root of 3), and $\sqrt{6}$. It should be noted that not all square roots are examples of irrational numbers. This will be discussed further.

> **PERFORMANCE OBJECTIVE 13**
>
> You should now be able to explain the relationship between rational, irrational, and real numbers.

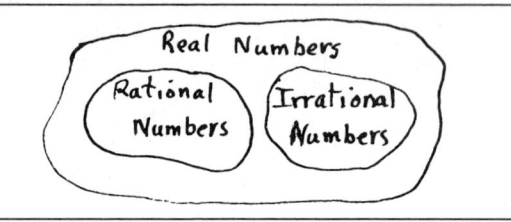

(Diagram: Real Numbers set containing Rational Numbers and Irrational Numbers.)

Since square roots provide us with a primary source of examples of irrational numbers and since square roots arise in many practical applications of mathematics, we will discuss them more thoroughly.

The **square root** of a given number is the positive number that, when multiplied by itself, yields the given number as the product. The symbol $\sqrt{}$, used to denote the square root of a number, is called a **radical sign.** The number under the radical sign is called the **radicand.** We say that the radicand is a **perfect square** if it can be written as some rational number times itself, that is, as the square of some rational number. That rational number is then the square root of the radicand. Thus the square root operation undoes what the squaring operation did, as the following examples illustrate.

$$\sqrt{16} = 4 \text{ since } (4)(4) = 16$$

$$\sqrt{0} = 0 \text{ since } (0)(0) = 0$$

$$\sqrt{\frac{4}{9}} = \frac{2}{3} \text{ since } \left(\frac{2}{3}\right)\left(\frac{2}{3}\right) = \frac{4}{9}$$

$$\sqrt{0.0001} = 0.01 \text{ since } (0.01)(0.01) = 0.0001$$

If a radicand is not a negative number and if it is not a perfect square, then its square root is an irrational number. For example

$$\sqrt{2}, \quad \sqrt{5}, \quad \sqrt{8}, \quad \sqrt{\frac{1}{3}}, \quad \text{and} \quad \sqrt{1.25}$$

are irrational numbers since

$$2, \quad 5, \quad 8, \quad \frac{1}{3}, \quad \text{and} \quad 1.25$$

are not perfect squares.

Note that the product of a real number times itself can never be negative even if the original number is negative. Thus it is not possible for us to get a real number answer as the square root of a negative number. In Chapter 10 we will learn that the square root of a negative number is an imaginary number or complex number. For now we will just say that it is nonreal. Thus,

$$\sqrt{-1}, \quad \sqrt{-2}, \quad \sqrt{-10}, \quad \sqrt{-\frac{1}{4}}, \quad \text{and} \quad \sqrt{-1.44}$$

are nonreal numbers since the radicands are negative.

There are several ways of evaluating square roots. Perhaps the easiest way is by means of a calculator. Other methods include square root tables, logarithms, and the square root algorithm, which is a step by step procedure that enables us to calculate square roots by hand computation. If the radicand is a perfect square, we can get an exact value for the square root. Otherwise we will obtain only an approximate value that has been rounded off to a certain number of decimal places. Except for the square roots of simple perfect squares, you will not need to evaluate square roots in this section. It will be sufficient if you can identify square roots as rational, irrational, or nonreal numbers.

PERFORMANCE OBJECTIVE 14	$-\sqrt{81} = -9$
You should now be able to evaluate simple square roots and identify others as irrational or nonreal numbers.	$\sqrt{18}$ is irrational. $\sqrt{-9}$ is nonreal.

EXERCISE SET 1.4

1. Identify the following numbers as rational, irrational, or real. Numbers will fall into more than one category.

    a. 2   b. $\sqrt{2}$ c. $-\frac{1}{2}$

    d. π e. 0   f. $-\sqrt{6}$

2. Identify the following numbers as rational, irrational or real. Numbers will fall into more than one category.

    a. -5 b. $-\sqrt{3}$ c. 4.285

    d. $-\pi$ e. 0   f. 3.14

3. Evaluate the following square roots of perfect squares:

   a. $\sqrt{64}$
   b. $\sqrt{100}$
   c. $\sqrt{\dfrac{1}{4}}$
   d. $\sqrt{\dfrac{9}{16}}$
   e. $\sqrt{0.01}$
   f. $\sqrt{0.49}$

4. Evaluate the following square roots of perfect squares:

   a. $\sqrt{36}$
   b. $\sqrt{144}$
   c. $\sqrt{\dfrac{1}{25}}$
   d. $\sqrt{\dfrac{81}{49}}$
   e. $\sqrt{0.09}$
   f. $\sqrt{1.21}$

5. Identify each of the following square roots as rational, irrational, or nonreal:

   a. $\sqrt{\dfrac{2}{3}}$
   b. $\sqrt{\dfrac{4}{25}}$
   c. $-\sqrt{1}$
   d. $\sqrt{11}$
   e. $\sqrt{-4}$
   f. $\sqrt{0.03}$

6. Identify each of the following square roots as rational, irrational, or nonreal:

   a. $-\sqrt{0.01}$
   b. $\sqrt{\dfrac{3}{4}}$
   c. $\sqrt{-16}$
   d. $-\sqrt{64}$
   e. $\sqrt{9.9}$
   f. $\sqrt{-1.2}$

SECTION 1.5 SCIENTIFIC NOTATION

Rational numbers are often written using a special notation called scientific notation. In order to understand scientific notation it is necessary to understand notation involving exponents and, in particular, power of ten notation.

A product that involves repeated factors of the same number can be written more concisely by using exponents. For example, we can

SCIENTIFIC NOTATION 1.5

write $2 \cdot 2 \cdot 2 \cdot 2 \cdot 2$ as 2^5. Two is called the **base** and five is called the **exponent.** We read 2^5 as "two to the fifth power" or "the fifth power of two" or just "two to the fifth." If we perform the indicated multiplications, we see that 2^5 is equal to 32.

When the exponent on a number is a 2 or a 3, we usually read the expression as "the number squared" or "the number cubed," respectively, instead of "the number to the second power" or "the number to the third power." A number on which no exponent appears is understood to have an exponent of one.

If an exponent is a natural number, then it indicates how many times the base should be used as a factor. We will have to view exponents that are not natural numbers differently. But for now we will restrict our attention to examples with natural number exponents. Consider the following examples:

$$4^4 = 4 \cdot 4 \cdot 4 \cdot 4 = 256$$

$$3^5 = 3 \cdot 3 \cdot 3 \cdot 3 \cdot 3 = 243$$

$$10^3 = 10 \cdot 10 \cdot 10 = 1000$$

$$\left(\frac{1}{2}\right)^6 = \frac{1}{2} \cdot \frac{1}{2} \cdot \frac{1}{2} \cdot \frac{1}{2} \cdot \frac{1}{2} \cdot \frac{1}{2} = \frac{1}{64}$$

$$\left(\frac{3}{4}\right)^2 = \frac{3}{4} \cdot \frac{3}{4} = \frac{9}{16}$$

$$(0.2)^7 = (0.2)(0.2)(0.2)(0.2)(0.2)(0.2)(0.2) = 0.0000128$$

In the previous examples, the bases are 4, 3, 10, $\frac{1}{2}$, $\frac{3}{4}$, and 0.2, respectively, and the exponents are 4, 5, 3, 6, 2, and 7, respectively.

In the evaluation of an expression involving an exponent, it is important to recognize that the base is just the factor that immediately precedes the exponent. If more than one factor or a negative sign is to be included as part of the base, there will be parentheses around the entire expression to indicate that it is the base. This is illustrated in the following examples:

$$-3^4 = -81 \quad \text{since the base is just 3}$$

$$(-3)^4 = 81 \quad \text{since the base is now } -3$$

$$2 \cdot 5^3 = 2 \cdot 125 = 250 \quad \text{since the base is just 5}$$

$$(2 \cdot 5)^3 = 10^3 = 1000 \quad \text{since } 2 \cdot 5 = 10 \text{ is now the base}$$

> **PERFORMANCE OBJECTIVE 15**
> You should now be able to interpret notation involving natural number exponents.
>
> $$-2^3 + \left(\tfrac{1}{2}\right)^2 = -8 + \tfrac{1}{4}$$
> $$= -\tfrac{32}{4} + \tfrac{1}{4} = -\tfrac{31}{4} \text{ or } -7\tfrac{3}{4}$$

In Chapter 4 we will discuss zero and negative integer exponents and in Chapter 10 we will cover fractional exponents. However, in this section it will be necessary to consider negative integer exponents on a base of 10. That is, we will consider integer powers of ten.

If a number can be written as a product of 10's or as 1 divided by a product of 10's, we say that the number is a **power of ten.** Some examples of powers of ten are

$$1000 = 10(10)(10), \quad \text{denoted by } 10^3$$

$$100 = 10(10), \quad \text{denoted by } 10^2$$

$$10, \quad \text{denoted by } 10^1$$

$$0.1 = \frac{1}{10}, \quad \text{denoted by } 10^{-1}$$

$$0.01 = \frac{1}{100} = \frac{1}{10(10)}, \quad \text{denoted by } 10^{-2}$$

$$0.001 = \frac{1}{1000} = \frac{1}{10(10)(10)}, \quad \text{denoted by } 10^{-3}$$

To write the ordinary form of the number denoted by 10^n, where n is an integer, we start with the number 1 and move the decimal point n places to the right if n is positive and n places to the left if n is negative. We show the movement of the decimal point in the following examples:

$$10^3 = 1.000 = 1000$$
$$10^2 = 1.00 = 100$$
$$10^{-1} = .1 = 0.1$$
$$10^{-3} = .001 = 0.001$$

To multiply a number by 10^n, where n is an integer, we move the decimal point in the number n places to the right if n is positive

SCIENTIFIC NOTATION   1.5

and n places to the left if n is negative. In the following examples, arrows are again used to show the movement of the decimal point.

$$3.458 \times 10^2 = 3.458 = 345.8$$

$$14.63 \times 10^3 = 14.630 = 14,630$$

$$-5.37 \times 10^{-1} = -5.37 = -0.537$$

$$27.291 \times 10^{-3} = 027.291 = 0.027291$$

| PERFORMANCE OBJECTIVE 16 | $175.32 \times 10^3 = 175,320$ |
| You should now be able to perform operations involving powers of ten. | $8.39 \times 10^{-3} = 0.00839$ |

DEFINITION

SCIENTIFIC NOTATION

A number is written in **scientific notation** if it is expressed as the product of a power of ten and a decimal number with one nonzero digit to the left of the decimal point.

Examples of numbers written in scientific notation are

$$3.97 \times 10^3, \quad 4.6 \times 10^{-2}, \quad -8.73 \times 10^1, \quad \text{and} \quad -3 \times 10^4$$

The method of writing a number in scientific notation will be presented in three cases.

CASE 1: THE NUMBER IS A WHOLE NUMBER.

The decimal point in a whole number is understood to follow the last digit. We must move it left until it follows the first (leftmost) digit. Then we multiply by the power of ten that corresponds to the number of places the decimal point has moved.

EXAMPLE   Convert 53,219 to scientific notation.

SOLUTION   Write 53,219. (add the decimal point).

Write 5.3219 (move the decimal point until it follows the first digit).

Write 5.3219×10^4 (multiply by 10^4 since the decimal point was moved four places)

CASE 2: THE NUMBER IS A DECIMAL NUMBER

In a decimal number the decimal point position is known. We simply move it left or right until it follows the leftmost nonzero digit. Then we multiply by the power of ten that corresponds to the number of places moved. Left moves require positive powers; right moves require negative powers.

EXAMPLE   Convert -45.37 to scientific notation.

SOLUTION   Write -4.537 (move the decimal point left one place)

Write -4.537×10^1 (multiply by the correct power of ten).

EXAMPLE   Convert 0.007146 to scientific notation.

SOLUTION   Write 7.146 (move the decimal point right three places).

Write 7.146×10^{-3} (multiply by the correct power of ten; note that it is negative since the decimal point moved right).

CASE 3: THE NUMBER INVOLVES POWER OF TEN NOTATION ALREADY

If the number is in power of ten notation with one nonzero digit left of the decimal point, then it is in scientific notation. If not, we deal with the number as in Case 1 or 2 (whichever applies) and change the power of ten by addition or subtraction of the number of places the decimal point has been moved.

EXAMPLE   Convert 55×10^3 to scientific notation.

SOLUTION   Write 5.5 (move the decimal point left one place).

Write 5.5×10^4 (add one to the power of ten).

EXAMPLE   Convert 45×10^{-3} to scientific notation.

SOLUTION   Write 4.5 (move the decimal point left one place).

Write 4.5×10^{-2} (add 1 to -3 to get -2).

EXAMPLE   Convert -0.00213×10^{-2} to scientific notation.

SOLUTION   Write −2.13 (move the decimal point right three places).
Write -2.13×10^{-5} (add −3 to −2 to get −5).

EXAMPLE   Convert 1170×10^3 to scientific notation.

SOLUTION   Write 1.170 (move the decimal point left three places).
Write 1.170×10^6 (add +3 to +3 to get +6).

PERFORMANCE OBJECTIVE 17	$-253.5 = -2.535 \times 10^2$
You should now be able to write a number in scientific notation.	(two left yields +2 as an exponent)

We will now perform operations on numbers written in scientific notation. In order to add or subtract two numbers that are written in scientific notation, we must have the same power of ten in both numbers. We then just add or subtract the numerical factors preceding the power of ten and multiply the sum or difference by that power of ten. Note that an additional step may be required to write the answer in either scientific notation or as an ordinary number.

EXAMPLE   Add $3.79 \times 10^3 + 6.43 \times 10^3$.

SOLUTION
$$\begin{array}{r} 3.79 \times 10^3 \\ +\,6.43 \times 10^3 \\ \hline 10.22 \times 10^3 \end{array}$$
or 1.022×10^4 or   10,220

EXAMPLE   Subtract $6.59 \times 10^{-2} - 4.138 \times 10^{-2}$.

SOLUTION
$$\begin{array}{r} 6.590 \times 10^{-2} \\ -\,4.138 \times 10^{-2} \\ \hline 2.452 \times 10^{-2} \end{array}$$
or   0.02452

EXAMPLE   Add $7.83 \times 10^4 + 5.9 \times 10^3$.

SOLUTION   Since the powers of ten are different, we must either write both numbers as ordinary numbers or we must convert one of the numbers so that it has the same power of ten as the other number. We will use both methods to perform the addition.

Method 1:

$$7.83 \times 10^4 = 78{,}300$$

$$5.9 \times 10^3 = 5900$$

$$\begin{array}{r} 78{,}300 \\ +5{,}900 \\ \hline 84{,}200 \end{array} \quad \text{or} \quad 8.42 \times 10^4$$

Method 2:

$$7.83 \times 10^4 = 78.3 \times 10^3$$

$$\begin{array}{r} 78.3 \times 10^3 \\ +5.9 \times 10^3 \\ \hline 84.2 \times 10^3 \end{array} \quad \text{or} \quad 8.42 \times 10^4$$

When multiplying powers of ten, we add the exponents to obtain the exponent on ten in the product. When dividing one power of ten by another power of ten, we subtract the exponent in the denominator from the exponent in the numerator to obtain the exponent on ten in the quotient. These rules use two of the laws of exponents that will be discussed in Chapter 4. However, the rules can be verified by writing out the values of powers of ten involved and then performing the operations. We use the rules to multiply and divide powers of ten in the following examples:

$$10^2 \cdot 10^3 = 10^5 = 100{,}000$$

$$10^{-4} \cdot 10^6 = 10^2 = 100$$

$$10^{-2} \cdot 10^{-1} = 10^{-3} = \frac{1}{10^3} = \frac{1}{1000}$$

$$\frac{10^8}{10^5} = 10^{8-5} = 10^3 = 1000$$

$$\frac{10^3}{10^4} = 10^{3-4} = 10^{-1} = \frac{1}{10}$$

$$\frac{10^{-1}}{10^3} = 10^{-1-3} = 10^{-4} = \frac{1}{10^4} = \frac{1}{10{,}000}$$

When multiplying or dividing numbers written in scientific notation, we do not need to have the same power of ten in both numbers. To perform the multiplication or division we first multiply or divide

SCIENTIFIC NOTATION 1.5

the numerical factors preceding the powers of ten, then multiply or divide the powers of ten, and finally multiply the products or quotients resulting from the previous two steps.

EXAMPLE Multiply $(4.6 \times 10^4) \times (2 \times 10^{-6})$.

SOLUTION $(4.6 \times 10^4) \times (2 \times 10^{-6}) = (4.6 \times 2) \times (10^4 \times 10^{-6})$
$= 9.2 \times 10^{-2}$ or 0.092

Note that in this example we have shown that

$$46{,}000 \times 0.000002 = 0.092$$

Had our original problem been written in this ordinary number form, it would still have been a good idea to first write the factors in scientific notation and then proceed with the multiplication. By doing this, there would be less chance of a decimal point error.

EXAMPLE Divide $(2.24 \times 10^2) \div (8.96 \times 10^{-3})$.

SOLUTION $\dfrac{2.24 \times 10^2}{8.96 \times 10^{-3}} = \dfrac{2.24}{8.96} \times \dfrac{10^2}{10^{-3}}$
$= 0.25 \times 10^5$
$= 2.5 \times 10^4$ or $25{,}000$

PERFORMANCE OBJECTIVE 18

You should now be able to perform operations on numbers in scientific notation.

$(4 \times 10^{-3}) + (8.1 \times 10^{-3}) = 12.1 \times 10^{-3}$
$= 1.21 \times 10^{-2}$ or 0.0121
$(5.36 \times 10^5) \times (3 \times 10^{-2}) = 16.08 \times 10^3$
$= 1.608 \times 10^4$ or $16{,}080$

EXERCISE SET 1.5

1. Evaluate the following expressions involving natural number exponents:

   a. 6^3 b. $(-2)^5$ c. $\left(\dfrac{2}{3}\right)^4$

   d. 0.5^2 e. -4^4 f. $3 \cdot 5^2$

2. Evaluate the following expressions involving natural number exponents:

   a. 5^4
   b. $(-1)^6$
   c. $\left(\dfrac{4}{5}\right)^3$
   d. 0.02^3
   e. -8^2
   f. $(-1 \cdot 3)^2$

3. Rewrite the following numbers using power of ten notation:
   a. 100,000
   b. 0.0001

4. Rewrite the following numbers using power of ten notation:
   a. 1,000,000
   b. 0.00001

5. Rewrite the following powers of ten as ordinary numbers:
   a. 10^{-5}
   b. 10^4

6. Rewrite the following powers of ten as ordinary numbers:
   a. 10^{-3}
   b. 10^5

7. Perform the following multiplications:
   a. 0.032×10^2
   b. -132.59×10^{-1}
   c. 20.125×10^{-3}
   d. -5.4×10^6

8. Perform the following multiplications:
   a. 7.0965×10^3
   b. -18.25×10^{-2}
   c. -0.08×10^4
   d. 0.45×10^5

9. Write the following numbers in scientific notation:
   a. 9563
   b. -49.5
   c. 0.0362
   d. 891×10^3
   e. 0.18×10^{-1}
   f. 76.02×10^{-4}

10. Write the following numbers in scientific notation:
    a. 325,000
    b. -678.2
    c. 0.82
    d. 64×10^3
    e. 0.0055×10^{-2}
    f. 2184×10^{-6}

11. Perform the following operations on numbers written in scientific notation. Write your answers in scientific notation and as ordinary numbers.
    a. $4.8 \times 10^{-3} + 6.7 \times 10^{-3}$
    b. $5.3 \times 10^4 - 5.82 \times 10^4$
    c. $7.63 \times 10^5 + 1.356 \times 10^6$
    d. $(3.75 \times 10^3) \times (8.3 \times 10^{-5})$
    e. $(1.397 \times 10) \times (6.4 \times 10^2)$
    f. $(2.54 \times 10^3) \div (5.08 \times 10)$
    g. $(3.152 \times 10^{-9}) \div (3.2 \times 10^{-7})$

12. Perform the following operations on numbers written in scientific notation. Write your answers in scientific notation and as ordinary numbers.
   a. $9.28 \times 10^{-4} + 7.84 \times 10^{-4}$
   b. $2.96 \times 10^3 - 5.75 \times 10^3$
   c. $6.1 \times 10^5 + 1.39 \times 10^6$
   d. $(7 \times 10^2) \times (4.943 \times 10^{-7})$
   e. $(9.85 \times 10^3) \times (8.3 \times 10^2)$
   f. $(6.729 \times 10^{-2}) \div (3 \times 10^3)$
   g. $(5.406 \times 10^4) \div (6.36 \times 10^6)$

SECTION 1.6

RATIONAL NUMBERS REVISITED—PERCENTS, RATIOS, AND PROPORTIONS

An understanding of rational numbers is important in working with many other mathematical concepts. Among them are percents, ratios, and proportions.

We have already discussed the conversion of fractions to decimals and vice versa. We will now consider the conversion of fractions and decimals to percents and of percents to fractions and decimals.

Percent, denoted by the symbol %, means "hundredth." So 1% is equivalent to 1/100 or 0.01. Likewise, 7% is equivalent to 7/100 or 0.07 and 63% is equivalent to 63/100 or 0.63. To convert a percent to a fraction or decimal, we divide the number of percent by 100. When converting to decimal form, we accomplish the division by moving the decimal point in the number two places to the left. When converting to fraction form, we put the number of percent over 100 and reduce the fraction to lowest terms. It is sometimes easier to first convert a percent to decimal form and then to use the decimal form to write the percent as a fraction. Examples of these conversions follow:

$$45\% = 0.45 = \frac{45}{100} = \frac{9}{20}$$

$$100\% = 1.00 = 1$$

$$12.5\% = 0.125 = \frac{125}{1000} = \frac{1}{8} \quad \left(\text{Note that } \frac{12.5}{100} = \frac{(12.5)(10)}{(100)(10)} = \frac{125}{1000} \right)$$

1 REVIEW OF NUMBER CONCEPTS AND OPERATIONS

$$175\% = 1.75 = 1\tfrac{75}{100} = 1\tfrac{3}{4}$$

$$\tfrac{3}{10}\% = 0.3\% = 0.003 = \tfrac{3}{1000} \quad \left(\text{Note that } \tfrac{\tfrac{3}{10}}{100} = \tfrac{3}{10} \cdot \tfrac{1}{100} = \tfrac{3}{1000}\right)$$

$$0.25\% = 0.0025 = \tfrac{25}{10000} = \tfrac{1}{400}$$

To perform the reverse operation, that is, to convert a fraction or decimal to a percent, we multiply by 100 and write the percent symbol. Multiplying by 100 is equivalent to moving the decimal point two places to the right. Some examples are

$$0.48 = 48\%$$

$$2.6 = 260\%$$

$$0.032 = 3.2\%$$

$$\tfrac{1}{2} = \tfrac{1}{2}(100\%) = 50\%$$

$$\tfrac{2}{3} = \tfrac{2}{3}(100\%) = \tfrac{200}{3}\% = 66\tfrac{2}{3}\%$$

$$\tfrac{3}{8} = \tfrac{3}{8}(100\%) = \tfrac{300}{8}\% = \tfrac{75}{2}\% = 37\tfrac{1}{2}\%$$

When working with a fraction whose decimal form terminates, it is sometimes easier to first convert the fraction to decimal form and then to write the decimal as a percent. For example,

$$\tfrac{3}{8} = 0.375 = 37.5\%, \quad \tfrac{7}{16} = 0.4375 = 43.75\%$$

RULES FOR CONVERSIONS INVOLVING PERCENTS

Divide by 100 to change a percent to a fraction or decimal.

Multiply by 100 to change a fraction or decimal to a percent.

PERFORMANCE OBJECTIVE 19

You should now be able to convert fractions and decimals to percents and vice versa.

$$\tfrac{7}{8} = 0.875 = 87.5\%$$

$$4.8\% = 0.048 = \tfrac{48}{1000} = \tfrac{6}{125}$$

PERCENTS, RATIOS, AND PROPORTIONS 1.6

The conversions that we have just discussed are helpful in three general types of problems. The first type is finding a certain percent of a given number. The second type is determining what percent one number is of another number. The third type is determining the number of which a given number is a certain percent. We will discuss each type and illustrate it with an example.

TYPE 1

To find a certain percent of a given number, we first convert the percent to a fraction or decimal and then multiply it times the given number.

EXAMPLE Find 55% of 430.

SOLUTION 55% = 0.55

0.55(430) = 236.5, so 55% of 430 is 236.5.

TYPE 2

To determine what percent one number is of another number, we divide the first number by the second number and convert the result to a percent.

EXAMPLE 27 is what percent of 72?

SOLUTION $\frac{27}{72} = 0.375 = 37.5\%$, so 27 is 37.5% of 72.

EXAMPLE If the price of newspaper increased from $.75 to $.90, what is the percent increase in price?

SOLUTION The increase in price is $.15. We need to determine what percent $.15 is of the original price of $.75.

$$\frac{\$.15}{\$.75} = 0.2 = 20\%$$

So the price increased 20%.

TYPE 3

To determine the number of which a given number is a certain percent, we convert the percent to a fraction or decimal and divide the result into the given number.

EXAMPLE 48 is 32% of what number?

SOLUTION 32% = 0.32

$$\frac{48}{0.32} = 150, \text{ so 48 is 32\% of 150.}$$

PERFORMANCE OBJECTIVE 20	
You should now be able to solve problems involving percents.	*72% of 96 ?* *(0.72)(96) = 69.12*

DEFINITION

RATIO

A **ratio** is a comparison of two quantities. We denote the ratio of *a* to *b* as

$$\frac{a}{b} \quad \text{or} \quad a:b$$

For example, the ratio of 3 to 5 is given by

$$\frac{3}{5} \quad \text{or} \quad 3:5$$

and the ratio of 4 to 30 is given by

$$\frac{2}{15} \quad \text{or} \quad 2:15 \left(\text{since } \frac{4}{30} = \frac{2}{15}\right)$$

The ratio of 6 pounds to 20 pounds is given by

$$\frac{6 \text{ lb}}{20 \text{ lb}} \quad \text{or} \quad 6 \text{ lb} : 20 \text{ lb}$$

which can be simplified to $\frac{3}{10}$ or $3:10$.

The ratio of 2 feet to 3 yards is given by

$$\frac{2 \text{ ft}}{3 \text{ yd}} \quad \text{or} \quad 2 \text{ ft} : 3 \text{ yd}$$

This ratio is complicated by the fact that the quantities have different units. However, we can simplify it as follows:

$$\frac{2 \text{ ft}}{3 \text{ yd}} = \frac{2 \text{ ft}}{9 \text{ ft}} = \frac{2}{9} \quad \text{or} \quad 2:9$$

PERCENTS, RATIOS, AND PROPORTIONS 1.6

As the previous examples illustrate, ratios, like fractions, should be given in lowest terms. Also, when a ratio involves like quantities, the units will cancel leaving a pure number without units.

DEFINITION

PROPORTION

A **proportion** is a statement that two ratios are equal.

An example of a true proportion is $\frac{2}{3} = \frac{12}{18}$. However, the statement $\frac{4}{5} = \frac{16}{25}$ is not a true proportion since $\frac{4}{5}$ is not, in fact, equal to $\frac{16}{25}$.

Consider the proportion $\frac{a}{b} = \frac{c}{d}$. The a, b, c, and d are called the **terms** of the proportion. The terms a and d are called the **extremes** of the proportion and the terms b and c are called the **means**. In a proportion the product of the extremes must equal the product of the means. That is, if

$$\frac{a}{b} = \frac{c}{d}$$

then $a \cdot d = b \cdot c$. We can use this fact to determine if two ratios are really equal. We can also use the relationship between the means and extremes to find an unknown quantity in a proportion.

EXAMPLE Find n in the proportion

$$\frac{n}{60} = \frac{5}{12}$$

SOLUTION Setting the product of the extremes equal to the product of the means, we get

$$12 \times n = 5 \times 60$$

That is, $12 \times n = 300$. To find n we must divide 300 by 12, the factor in front of n. Then

$$n = \frac{300}{12} = 25$$

EXAMPLE Find n in the proportion

$$\frac{4}{15} = \frac{n}{12}$$

SOLUTION Equating the product of the means and product of the extremes gives

$$15 \times n = 4 \times 12$$
$$= 48$$

Dividing 48 by the factor with n, that is, by 15, we get

$$n = \frac{48}{15} = 3\frac{1}{5} \text{ or } 3.2$$

We can interchange the means or the extremes of a proportion and still have a valid proposition. That is,

$$\text{if } \frac{a}{b} = \frac{c}{d}, \text{ then } \frac{a}{c} = \frac{b}{d}, \frac{d}{b} = \frac{c}{a}, \text{ and } \frac{d}{c} = \frac{b}{a}$$

PERFORMANCE OBJECTIVE 21	
You should now be able to solve a proportion for an unknown quantity.	$\frac{8}{5} = \frac{12}{n} \Rightarrow 8 \times n = 5 \times 12$ $8 \times n = 60$ $n = \frac{60}{8} = \frac{15}{2} = 7.5$

Proportions are useful for solving word problems in which comparisons are made between quantities. If we can use the information in the problem to set up a proportion in which three of the quantities are known, we can then solve for the unknown quantity. The following examples illustrate the use of proportions in solving word problems.

EXAMPLE A map is drawn to scale using 1 in. to represent 50 mi. If two cities are $3\frac{1}{4}$ in. apart on the map, what is the actual distance between the cities?

SOLUTION Since the map is drawn to scale, we know that the ratio of the map distances should equal the ratio of the actual distances. Letting n miles represent the actual distance between the two cities, we can write the proportion

PERCENTS, RATIOS, AND PROPORTIONS 1.6

$$\frac{1 \text{ in.}}{3.25 \text{ in.}} = \frac{50 \text{ mi}}{n \text{ mi}}$$

Since the units cancel, we have $\frac{1}{3.25} = \frac{50}{n}$. This proportion is solved as follows:

$$1 \times n = 50 \times 3.25$$
$$n = 162.5$$

So the actual distance between the cities is 162.5 mi.

Since the means and extremes of a proportion can be interchanged, we could also have used the following proportions to solve the problem:

$$\frac{1 \text{ in.}}{50 \text{ mi}} = \frac{3.25 \text{ in.}}{n \text{ mi}}$$

$$\frac{n \text{ mi}}{3.25 \text{ in.}} = \frac{50 \text{ mi}}{1 \text{ in.}}$$

However, we chose to use a proportion in which the ratios involved like quantities.

EXAMPLE After filling his tank with gas, a driver determined that he had traveled 319 mi on 22 gal of gas. At that rate, how many gallons of gas will be used to make a 580 mi trip?

SOLUTION We will let n gallons represent the amount of gas needed. Assuming the same rate of gas consumption, we can set up the proportion

$$\frac{n \text{ gal}}{22 \text{ gal}} = \frac{580 \text{ mi}}{319 \text{ mi}}$$

That is,

$$\frac{n}{22} = \frac{580}{319}$$

We solve the proportion as follows:

$$319 \times n = 22 \times 580$$
$$= 12{,}760$$
$$n = \frac{12{,}760}{319} = 40$$

The driver will use 40 gal of gas on a 580 mi trip.

PERFORMANCE OBJECTIVE 22	If the ratio of women to men at a college is 4:5 and if there are 3000 men enrolled, the number of women can be found by using the proportion: $\frac{4}{5} = \frac{n}{3000}$.
You should now be able to use proportions to solve word problems.	

EXERCISE SET 1.6

1. Convert the following percents to decimals:

   a. 5.81% b. 349% c. $\frac{3}{4}$%

2. Convert the following percents to decimals:

   a. 76.2% b. 1140% c. $\frac{7}{20}$%

3. Convert the following percents to fractions in lowest terms:
   a. 54% b. 9.3% c. 132%

4. Convert the following percents to fractions in lowest terms:
   a. 76% b. 6.4% c. 240%

5. Convert the following fractions or decimals to percents:

   a. 0.076 b. $\frac{1}{3}$ c. $\frac{13}{8}$ d. 0.0005

6. Convert the following fractions or decimals to percents:

   a. 2.3925 b. $\frac{2}{3}$ c. $\frac{15}{16}$ d. 0.002

7. a. Find 145% of 96. b. Find $1\frac{4}{5}$% of 180.
8. a. Find 69% of 348. b. Find 2.7% of 640.
9. a. 147.5 is what percent of 250?
    b. 85 is what percent of 34?
10. a. 152 is what percent of 475?
    b. 161 is what percent of 92?
11. a. 36 is 30% of what number?
    b. 105 is 175% of what number?

12. a. 418.6 is 80.5% of what number?

    b. 2457 is 325% of what number?

13. Write the ratio of 2 ft to 9 in. in simplest form.

14. Write the ratio of 20 oz to 2 lb in simplest form.

15. Solve the following proportions:

    a. $\dfrac{15}{9} = \dfrac{40}{n}$
    b. $\dfrac{0.5}{3.6} = \dfrac{n}{18}$

16. Solve the following proportions:

    a. $\dfrac{45}{m} = \dfrac{30}{12}$
    b. $\dfrac{m}{2.5} = \dfrac{0.24}{1.5}$

17. A 30 ft telephone pole casts a 24 ft shadow. A boy standing next to the pole casts a 4 ft shadow. How tall is the boy?

18. If $3\frac{1}{2}$ in. on a map represent a distance of 49 mi., how many miles are represented by 12 in.?

19. A car can go 448 mi on 16 gal of gas. How far will it go on 6.5 gal?

20. A car uses 22 gal of gas to go 374 mi. How many gallons will it take to go 595 mi?

SECTION 1.7 SUMMARY

In this chapter we have reviewed number concepts and operations on numbers. These concepts provide the groundwork on which we will build in upcoming chapters. The new concepts and definitions are summarized here.

Natural number: one of the numbers 1, 2, 3, 4, ...

Whole number: one of the numbers 0, 1, 2, 3, ...

Number line: a line on which we graphically represent numbers by points.

Origin of a number line: the point on a number line corresponding to the number 0.

Prime number: a natural number that is not expressible as a product of prime factors other than 1 and itself.

Composite number: a number that can be written as a product of prime factors other than 1 and itself.

Integer: one of the numbers . . . −3, −2, −1, 0, 1, 2, 3, ...

Absolute value of a number: the distance of the number from the origin of the number line; it is never negative.

Rational number: a number that can be written as a terminating or repeating decimal or in the form

$$\frac{a}{b}, \text{ where } a \text{ and } b \text{ are integers and } b \neq 0.$$

Irrational number: a number that is not rational.

Real number: either a rational or an irrational number.

Square root of a number: that positive number which, when multiplied by itself, yields the given number.

Radical sign: the symbol $\sqrt{}$ used to denote a root of a number.

Radicand: the expression under a radical sign.

Natural number exponent: a natural number, written in an elevated position to the right of a given number, to indicate the number of factors of the given number.

Base: the number that is raised to a power in an expression involving exponents.

Power of 10: an expression involving exponents in which the base is 10 and the exponent is an integer.

Scientific notation: a way of writing a number as the product of a power of ten and a decimal number with one (nonzero) digit to the left of the decimal point.

Ratio: a comparison of two quantities, denoted by $\frac{a}{b}$ or $a:b$.

Proportion: a statement that two ratios are equal.

CHAPTER 1 PRACTICE TEST

In Problems 1–9, consider the following numbers:

$$-\frac{1}{3}, \quad 5, \quad 0, \quad 1.5, \quad -4, \quad \sqrt{2}, \quad -\frac{4}{5}, \quad -3, \quad \frac{4}{5}, \quad -0.3$$

CHAPTER 1 PRACTICE TEST

List the numbers from above which are

1. (PO-1) Natural numbers
2. (PO-1) Whole numbers
3. (PO-4) Integers
4. (PO-7) Rational numbers
5. (PO-13) Irrational numbers
6. (PO-13) Real numbers

7. (PO-10) List the number for which a multiplicative inverse does not exist.

8. (PO-4) List the two numbers that are additive inverses of each other.

9. (PO-10) List the two numbers that are multiplicative inverses of each other.

10. (PO-1, 4, 7) Locate on a number line the points corresponding to the following numbers:

$$1.75, \quad -\frac{5}{3}, \quad 4, \quad 2.5, \quad 0, \quad -\frac{3}{4}, \quad -3$$

Write the prime factorization of each of the following numbers. If the number is prime, say so.

11. (PO-3) 130
12. (PO-3) 43
13. (PO-8) Reduce $\frac{168}{252}$ to lowest terms.
14. (PO-8) Write three fractions that are equal to $\frac{7}{10}$.

Perform the indicated operations. If a problem is undefined, say so. Give your answer in lowest terms.

15. (PO-2) $(4)(6)(0)$
16. (PO-6) $(1)(-1)$
17. (PO-2) $\frac{3}{0}$
18. (PO-4, 5) $|-5| + |0|$
19. (PO-5) $-32 + (-11)$
20. (PO-5) $14 - 22$
21. (PO-5) $-25 - (-72)$
22. (PO-6) $(-3)(5)(-1)(6)$
23. (PO-6) $0 \div (-12)$
24. (PO-6) $-\frac{5}{6} + \frac{1}{3}$
25. (PO-9) $\frac{8}{9} - \frac{5}{6}$
26. (PO-10) $\left(-\frac{6}{7}\right)\left(-\frac{4}{15}\right)$
27. (PO-10) $-\frac{3}{4} \div \frac{1}{12}$
28. (PO-10) $\frac{4}{5} \div (-8)$

29. (PO-11) $-34.51 + 19.645$ 30. (PO-11) $13 \div (-3.25)$

31. (PO-12) Convert $\frac{-9}{16}$ to decimal form.

32. (PO-12) Convert 0.015 to a fraction in lowest terms.

33. (PO-14) Evaluate $\sqrt{0.09}$.

34. (PO-14) Identify $\sqrt{-1}$ as a rational, irrational, or nonreal number.

35. (PO-15) Evaluate $(-6)^3$.

36. (PO-16) Perform the multiplication -1347.62×10^{-2}.

Write the following in scientific notation.

37. (PO-17) 6400   38. (PO-17) 0.745×10^{-3}

Perform the following operations on numbers written in scientific notation. Give your answers in scientific notation.

39. (PO-18) $6.7 \times 10^5 - 5.84 \times 10^5$

40. (PO-18) $(3.5 \times 10^{-2}) \times (2.25 \times 10^{-3})$

41. (PO-18) $(9.65 \times 10^7) \div (1.93 \times 10^4)$

42. (PO-19) Convert $\frac{24}{5}$ to a percent.

43. (PO-20) Find 38% of 1950.

44. (PO-21) Solve the following proportion for n.

$$\frac{6}{0.3} = \frac{8}{n}$$

45. (PO-22) If $2\frac{3}{4}$ in. on a map represent a distance of 110 mi, how many inches represent 290 mi?

CHAPTER 2

MEASUREMENT AND DATA HANDLING

2.1 Systems of Units
2.2 Advantages of the SI System
2.3 Operating on the Data—the Calculator
2.4 Handling Significant Digits
2.5 Summary

We may wonder why the process of conversion to the metric system of measurement in the United States is such a slow process. Surely it is not too difficult to convert numbers from one system of measurement to another. But the problem is not in converting numbers from one system to another. In converting to a new system we want to use established values that are convenient in the new system. Screw threads, for example, are somewhat different in the metric system. It would be foolish to consider that we had converted to metric by relabeling quantities with their metric values.

In this chapter we consider some of the practical problems associated with the measurement of real quantities. When we finish this chapter we will have developed some appreciation for the related activities of measurement, data handling on a calculator, and reporting of measured results.

SECTION 2.1 SYSTEMS OF UNITS

Consider our actions if we are asked to measure the width of a door opening. We would probably look for a yardstick or tape measure. Then we would compare the door opening with the measuring device as in Figure 2.1. Finally, we write down a number for the width of the door opening. Suppose our value for the width is $30\frac{1}{4}$ inches.

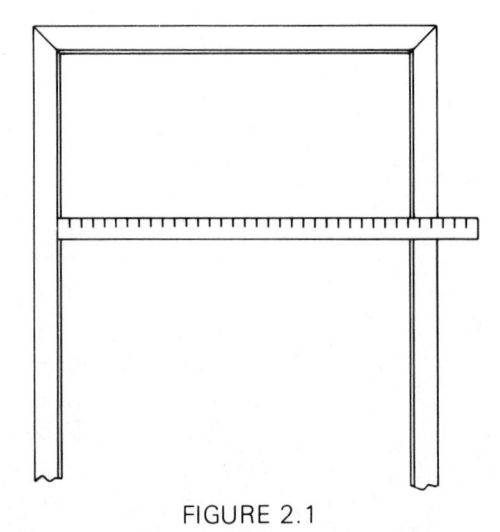

FIGURE 2.1

This process seems easy enough to understand. But we need to think carefully about each step in the process.

STEP 1: SELECTING A STANDARD UNIT OF MEASUREMENT

We selected a yardstick and recorded the measurement in inches. Perhaps this choice was a matter of convenience or maybe we wanted the measurement in inches because the lumber supplier records door dimensions in inches. If we are going to communicate the results of our measurement to others, we need to choose a standard that will be understood. It is for this reason that standard systems of units have been developed for use throughout the world. We will deal with two of these systems later in this chapter.

STEP 2: COMPARING THE QUANTITY BEING MEASURED WITH THE STANDARD

This involved placing the yardstick across the opening. We must use a valid method in making the comparison. For example, the yardstick should be horizontal. This step is a matter of technique and

SYSTEMS OF UNITS 2.1

varies from measurement to measurement. A machinist measuring the diameter of a rod to a thousandth of an inch would use a micrometer.

STEP 3: RECORDING AN APPROXIMATION TO THE QUANTITY BEING MEASURED

Here we wrote down a number and a unit. In dealing with measurements on real quantities, we must always include the units of measurement. Addition of a unit to the value indicates the size standard and the kind of measurement involved. For example, the inch and foot are both accepted units of length, but they are different in size. When we read $30\frac{1}{4}$ inches we know that a length was measured and the unit was the inch.

But what does the $30\frac{1}{4}$ mean? You might say that this means the door opening is exactly $30\frac{1}{4}$ inches wide. Let's take a closer look at the result of our comparison process in Figure 2.2.

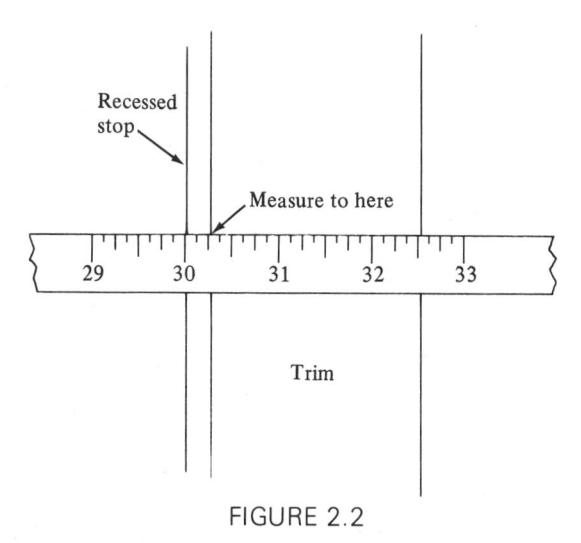

FIGURE 2.2

In Figure 2.2 it appears that the width is somewhat larger than $30\frac{1}{4}$ but not enough larger to record the result as $30\frac{3}{8}$. Thus, $30\frac{1}{4}$ is the nearest approximation to the length that can be made with the measuring device used. We could choose a measuring device with a finer scale and get a more precise indication of the length. But this more precise measurement would not necessarily be exact. This brings us to a fundamental property of numbers that result from measurement.

PROPERTY OF MEASURED VALUES

Results of measurements must be considered as approximate numbers.

This uncertainty is not a matter of concern in most practical situations. Those who need to deal with it are referred to Appendix II.

In the process of measurement, a standard unit is selected. Both an approximation to the measured quantity and the units of that quantity are reported.

EXAMPLE Record the time interval between 1:30 P.M. and 3:00 P.M. using an accepted system of units.

SOLUTION There is more than one answer to this problem depending on the size time unit chosen. We might report

$$\text{Time interval} = 1.5 \text{ hr}$$

$$\text{Time interval} = 90 \text{ min}$$

$$\text{Time interval} = 90 \times 60 = 5400 \text{ s}$$

In each case a number and a unit are reported.

EXAMPLE Record the weight of the object on the scale in Figure 2.3.

SOLUTION We estimate the reading to be 2.5. The scale is calibrated in pounds. The result of the measurement is then 2.5 lb.

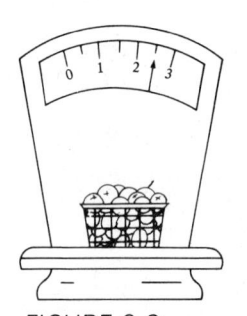

FIGURE 2.3

PERFORMANCE OBJECTIVE 1	
You should now be able to describe the steps in the measurement process.	1. Select a standard unit. 2. Make a comparison. 3. Record an approximation.

In this text we will use two systems of units. We will refer to these as the English and SI* systems. We list here several alternative names that are in common usage for these two systems.

English system	SI system*
also called:	also called:
British system	metric system**
Engineering system	mks system (for meter kilogram, second)
fps system (for foot, pound, second)	mksa system (for meter, kilogram, second, ampere)

*SI stands for Systeme Internationale.
**We could make a distinction between the SI system and the metric system but for the practical work the distinction is unnecessary.

SYSTEMS OF UNITS 2.1

In each system, units are defined for certain basic quantities.

DEFINITION

> **BASIC QUANTITY**
>
> A **basic quantity** is a physical quantity that is defined in terms of a standard without reference to other quantities.

DEFINITION

> **UNIT**
>
> A **unit** is a name given to the standard that represents a basic quantity.

We will consider only the five more common basic quantities. A total of seven basic quantities exist.

Quantity	English unit	SI unit
length	foot	meter
mass	slug (nb)*	kilogram
force	pound	newton (nb)*
time	second	second
current	ampere	ampere

*(nb) signifies nonbasic units.

The English system defines units for the basic quantities of length, force, time, and current while the SI system uses length, mass, time, and current as the basic quantities. Units for these quantities will frequently be designated by a one or two letter abbreviation.

English system	SI system
foot (ft)	meter (m)
slug (no abbrev.)	kilogram (kg)
second (s)	second (s)
pound (lb)	newton (N)
ampere (A)	ampere (A)

PERFORMANCE OBJECTIVE 2

You should now be able to list units for the basic quantities in the English and SI systems.

Some basic SI units are:
meter (m)
kilogram (kg)
second (s)

From these basic quantities we may build a set of derived quantities.

DEFINITION

> **DERIVED QUANTITY**
>
> A **derived quantity** is some combination of the basic quantities.

The quantity velocity is a derived quantity (as opposed to a basic quantity) since it is formed from a combination of the basic quantities length and time. The quantity **velocity** is defined as the quotient of length and time with length in the numerator. The units of derived quantities are some combination of the basic units.

EXAMPLE **Physical work** is defined as the product of force and distance. Find the units of work in the English system.

SOLUTION The unit of force is the pound. The unit of distance is the foot. Thus, the unit of work is

$$\text{pounds} \times \text{feet} = \text{pound-foot}$$

This unit is more commonly called the foot-pound (ft-lb).

EXAMPLE The quantity **kinetic energy** is defined as the product of the number $\frac{1}{2}$, the mass, and the velocity squared. Find the units of kinetic energy in the SI system.

SOLUTION One-half is a pure number and has no units. Mass has the units of kilograms (kg) and velocity has the units of meters per second (m/s) in the SI system. The units of kinetic energy are then

$$\text{kg} \times \left(\frac{\text{m}}{\text{s}}\right)^2 = \frac{\text{kg} - \text{m}^2}{\text{s}^2}$$

An important property of real quantities is their dimension. The dimension describes how the basic quantities enter into the definition of the quantity.

DEFINITION

> **DIMENSION OF A PHYSICAL QUANTITY**
>
> The **dimension of a physical quantity** is an expression of the combination of basic quantities that make up a physical quantity.

The dimension of a physical quantity is independent of the system of units in which it may be expressed.

EXAMPLE Find the dimension of velocity that is defined as the quotient of distance and time.

SOLUTION Since velocity is the quotient of distance and time we say that its dimension is

$$\frac{\text{length}}{\text{time}}$$

Note that no reference is made to a system of units.

EXAMPLE Find the dimension of the quotient of velocity and time (velocity in the numerator).

SOLUTION The dimension of velocity is length/time. Then the quotient will have the dimension

$$\frac{\text{length}}{\text{time}} \div \text{time} = \frac{\text{length}}{\text{time}^2}$$

The dimension of a quantity is a very important idea and is discussed in more detail in Appendix III.

PERFORMANCE OBJECTIVE 3	
You should now be able to determine the dimension of a physical quantity from its units.	*A speeding car travels more than 55 miles per hour. The dimension of speed is:* $\frac{length}{time}$

It is not our purpose to stress conversions between systems of units in this text. However, you should have some feeling for the relative size of the basic units. We offer the comparisons in Figure 2.4 as a review.

58   **2 MEASUREMENT AND DATA HANDLING**

Length: Foot   Meter
1 m = about 3.25 ft

Mass: Slug   kg
1 slug = approximately 15 kg

(a)

Force or weight (an object of one weight unit is shown on the same spring scale)

Pound scale   Newton

1 lb = about 4.5 N

(b)

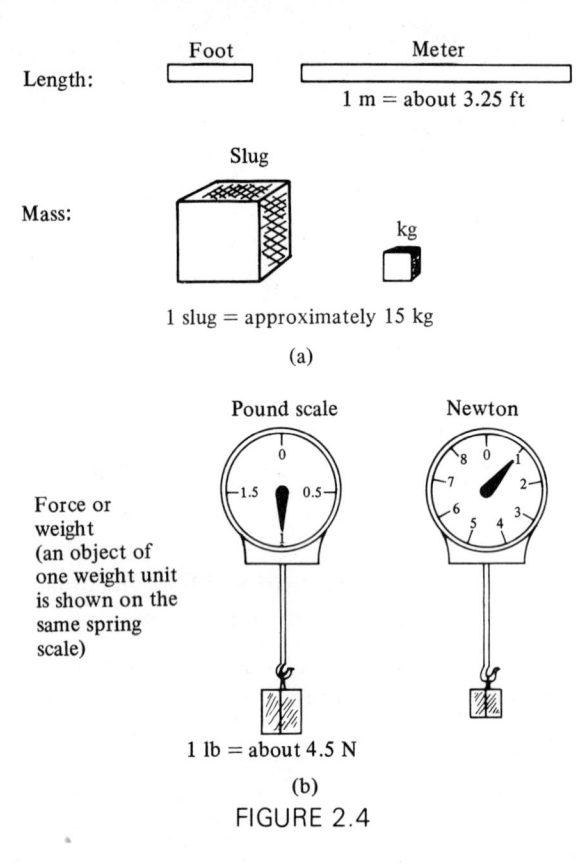

FIGURE 2.4

Comparisons of time and current units are not shown since the units are the same.

Since we are accustomed to packaging quantities in pounds in the English system and kilograms in the metric system, useful comparisons can be made between these units. See Figure 2.5.

Pound scale   Kilogram scale

Object weighing 1 lb   Same object with mass of 0.45 kg

(a)

Pound Scale   Kilogram Scale

Object weighing 2.2 lb   Same object with mass of 1 kg

(b)

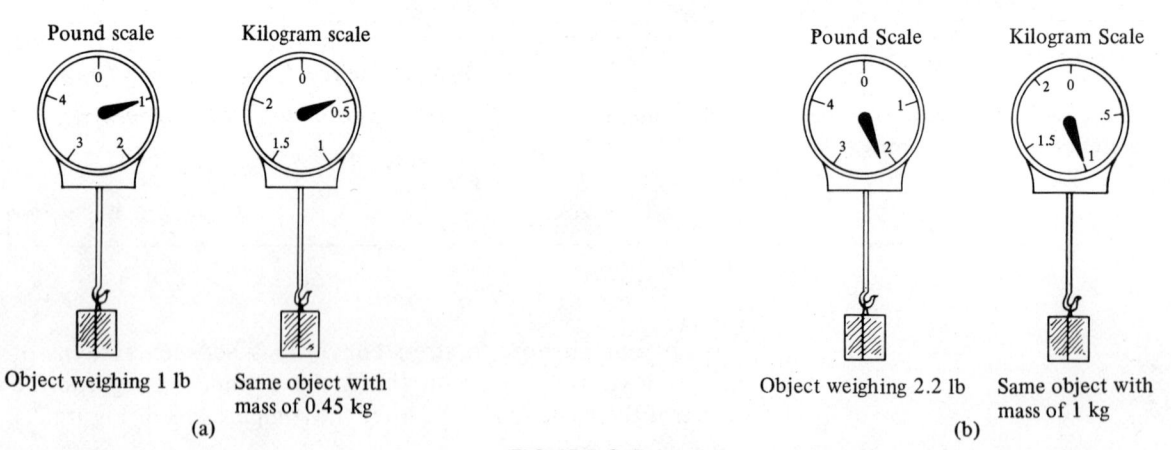

FIGURE 2.5

SYSTEMS OF UNITS 2.1

We emphasize that pounds and kilograms are not the same kinds of units so that the comparisons should not be stated as equalities.

To make exact conversions of values from one system of units to another, we generally look up a conversion factor in a table. Tables 2.1 and 2.2 provide factors for the basic units and for some of the derived units in each system. To make the conversion, multiply by the given factor.

TABLE 2.1 CONVERSION TABLE: ENGLISH TO SI

Value in English unit of:	× conversion factor	= value in SI unit of:
feet (ft)	0.3048 (m/ft)	meters (m)
pounds (lb)	4.45 (N/lb)	newtons (N)
seconds (s)	1	seconds (s)
slugs (lb-s^2/ft)	14.59 (kg/slug)	kilograms (kg)
velocity (ft/s)	0.3048 (m/ft)	(m/s)
work (ft-lb)	1.356 (J/(ft-lb))	joules (J)
power (ft-lb/s)	1.356	watts (W)
pounds (lb)	0.4536	kilograms (*)

(*) Equivalent size units only

TABLE 2.2 CONVERSION TABLE: SI TO ENGLISH

Value in SI unit of:	× conversion factor	= value in English unit of:
meters (m)	3.281 (ft/m)	feet (ft)
newtons (N)	0.2247 (lb/N)	pounds (lb)
seconds (s)	1	seconds (s)
kilograms (kg)	0.0685 (slug/kg)	slugs (lb-s^2/ft)
velocity (m/s)	3.281 (ft/m)	(ft/s)
work (J)	0.7376 (ft-lb/J)	(ft-lb)
power (W)	0.7375	(ft-lb/s)
kilograms (kg)	2.205	pounds (*)

* Equivalent size units only

PERFORMANCE OBJECTIVE 4

You should now be able to make approximate and exact conversions of quantities from one system to another.

1 mile = 5280 ft (0.3048) m/ft
= 1609.344 m
15 kg is equivalent to
15 kg (2.205) lb/kg = 33.075 lb

EXERCISE SET 2.1

1. Describe the steps necessary to measure the length of this page. Perform the measurement and report a result.

2. Describe the steps necessary to weigh a tomato in the produce market.

3. Write dimensions and units in the open blanks in the following table. Note that the quantities shown are both basic and derived.

Quantity	Dimension	English	SI
time			
acceleration		ft/s^2	
energy			kg-m^2/s^2
mass	mass		
momentum		slug-ft/s	

4. Write dimensions and units in the open blanks in the table.

Quantity	Dimension	English	SI
work		ft-lb	
inertia			kg-m^2
torque			kg-m^2/s^2
speed		ft/s	
power			kg-m^2/s^3

5. The mass density of a substance is defined as the mass per unit volume. Write the units of mass density in the SI system.

6. The weight density of a substance is defined as the weight per unit volume. Write the units of weight density in the English system. (Weight and force have the same dimensions).

7. Linear momentum is defined as the product of mass and velocity.
   a. Write the dimension of linear momentum.
   b. Write the units of linear momentum in the English system.

8. Angular momentum is defined as the product of linear momentum (defined in Problem 7) and a distance.
    a. Write the dimension of angular momentum.
    b. Write the units of angular momentum in the SI system.
9. Each of the following sentences refers to a physical quantity. Indicate whether the quantity is basic or derived and give its dimension.
    a. The car was traveling at 58 miles per hour (mph).
    b. The pressure in the cylinder was 700 pounds per square inch (psi).
    c. He jumped 75.8 in.
    d. The gasket had an area of 8.61 in.^{2}.
10. Each of the following sentences refers to a physical quantity. Indicate whether the quantity is basic or derived and give its dimension.
    a. The crank shaft weighed 89 lb.
    b. The piston displaced a volume of 51 in.^{3}.
    c. The inertia of the beam was 1500 kg-m^2.
    d. The metal had a density of 7.3 g/cm^3.
11. In this problem you are to use the approximate conversion factors in Figure 2.4 to estimate answers.
    a. What is the approximate kilogram mass of an 8 lb baby?
    b. Express the approximate weight of a 150 lb person in newtons.
    c. An I beam has a mass of 600 kg. What is its approximate mass in slugs and weight in pounds?
12. In this problem you are to use the approximate conversion factors in Figure 2.4 to estimate answers.
    a. A football player has a mass of 100 kg. What is his approximate mass in slugs and weight in pounds?
    b. Will a 4 m bar fit in a 14 ft rack?
    c. Should a 6000 kg truck cross a bridge with a 6 ton load limit?
13. Make exact conversions using Tables 2.1 and 2.2.
    a. 37.5 ft to SI
    b. 23 ft-lb to joules
    c. 950 W to the English system
    d. 55 lb to equivalent mass in SI
    e. 77 J to the English system

14. Make exact conversions using Tables 2.1 and 2.2.
    a. 147 m to English units
    b. 98 ft-lb/s to the SI system
    c. 45 kg to equivalent weight units in the English system
    d. 4.5 slugs to kilograms
    e. 23.1 ft-lb to the SI system

SECTION 2.2 ADVANTAGES OF THE SI SYSTEM

So far we see little reason to prefer one system of units over the other. Regardless of the system, we must define basic units and agree upon their values. However, we will find it difficult to do shop work that requires close tolerances if we have to work with a basic unit of length that is as long as a foot or meter. These units are just too big for such work. On the other hand, units such as inches or centimeters may be too small for planning interstate highways. We are faced with the problem of dividing the basic unit into smaller units or of producing larger units from the basic unit. We display some of these units for measurement of length only.

English Unit	Relation to Basic Unit	SI Unit	Relation to Basic Unit
mil	12,000 in 1 ft	millimeter	1000 in 1 m
inch	12 in 1 ft	centimeter	100 in 1 m
yard	3 times larger	decameter	10 times larger
mile	5280 times larger	kilometer	1000 times larger

We have picked only a few of the possible divisions. However, these few illustrate that the English system has no regular relationship between subdivisions. This would become more obvious if we discussed volumes in terms of pints, quarts, and gallons. Looking down the SI column we find that divisions of a unit differ only by powers of 10. Powers of ten were discussed in Chapter 1. We list some of them here.

$$10^{-3} = \frac{1}{1000} = 0.001 \qquad 10^1 = 10$$

ADVANTAGES OF THE SI SYSTEM 2.2

$$10^{-2} = \frac{1}{100} = 0.01 \qquad 10^2 = 100$$

$$10^{-1} = \frac{1}{10} = 0.1 \qquad 10^3 = 1000$$

Consider the advantages of this system.

FIRST ADVANTAGE OF THE SI (METRIC) SYSTEM

If we know the size of a liter (a measure of volume), then we know that a milliliter is 1/1000 of a liter. In the English system, we must memorize the fact that a quart is one-fourth of a gallon and a pint is one-half of a quart. Standard sizes for subunits and larger units are a major advantage of the SI system.

SECOND ADVANTAGE OF THE SI (METRIC) SYSTEM

Suppose we have a measured value of 0.00321 meter, which we want to convert to a smaller unit of millimeters. We can multiply by the appropriate power of 10. That power is

$$10^3 = 1000$$

since there are 1000 millimeters in 1 meter. Then 0.00321 meter = 0.00321 × 1000 = 3.21 millimeters.

On the other hand, consider the problem of converting 0.00519 feet to mils. We must multiply 0.00519 by 12,000. Where we could simply move the decimal point in the SI system, we must now perform a calculation.

There is also a logical system of naming the unit divisions in the SI system that holds for all quantities (lengths, masses, volumes, etc.). That system is as follows.

	Unit prefix	*Abbreviation*	*Multiplier*
large	giga	G	10^9
	mega	M	10^6
	kilo	k	10^3
	hecto	h	10^2
	deca	da	10^1
	deci	d	10^{-1}
	centi	c	10^{-2}
	milli	m	10^{-3}
	micro	μ	10^{-6}
	nano	n	10^{-9}
small	pico	p	10^{-12}

We can replace the power of ten multiplier in any number by the prefix or its letter abbreviation. That is, 10^9 = giga = G, 10^{-6} = micro = μ, etc. The prefix is used in pronouncing the unit. The abbreviation is normally used in writing the unit. Using these prefixes, we can write

1.5×10^3 m = 1.5 km   (we read this as 1.5 times ten to the third meters equals 1.5 kilometers.)

2.3×10^{-2} m = 2.3 cm

4.52×10^{-6} m = 4.52 μm

EXAMPLE   Replace the power of ten multiplier in the following expressions by the appropriate SI prefix:

$$1.53 \times 10^{-3} \text{ m}$$
$$3.84 \times 10^{-6} \text{ kg}$$
$$4.12 \times 10^{6} \text{ s}$$
$$9.23 \times 10^{-9} \text{ s}$$

SOLUTION 1.53×10^{-3} m = 1.53 mm

3.84×10^{-6} kg = 3.84 μkg = 3.84 mg (mg = milligram)

4.12×10^{6} s = 4.12 Ms

9.23×10^{-9} s = 9.23 ns

PERFORMANCE OBJECTIVE 5	
You should now be able to identify unit prefixes, multipliers, and their abbreviations in the SI system.	2.33×10^{-3} m = 2.33 mm 3.65×10^{3} m = 3.65 km 5.1×10^{-3} ℓ = 5.1 mℓ

Let us arrange the units in a manner that will suggest a method of conversion from one unit to another. We frequently see the units displayed in a staircase as in Figure 2.6.

ADVANTAGES OF THE SI SYSTEM 2.2

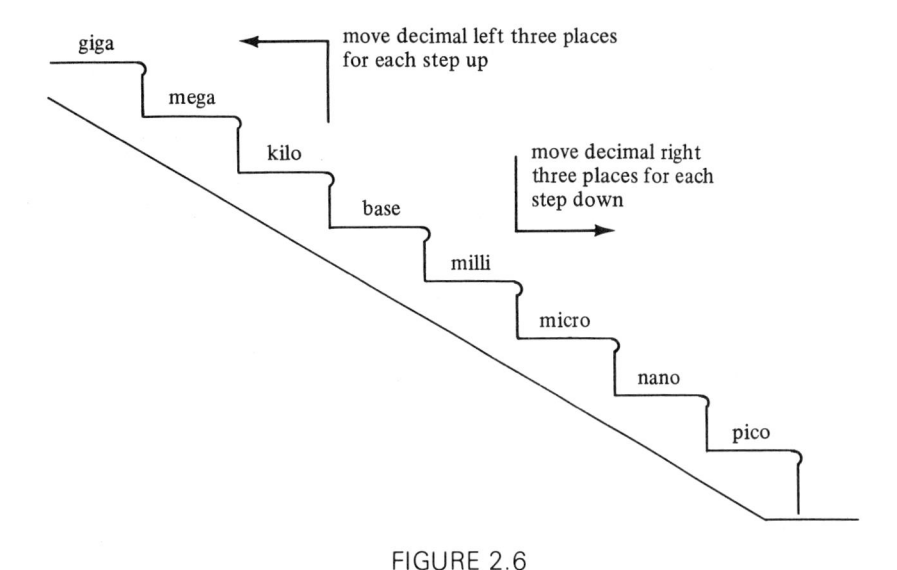

FIGURE 2.6

In this display we have shown only those steps which differ in size by a factor of 1000. The arrows and key at the right indicate the movement of the decimal point when we convert from one size unit to another. Consider the following examples, which illustrate the use of the staircase.

EXAMPLE Convert 1541 mm to meters.

SOLUTION 1541 mm = 1.541 m (move up one step; therefore the decimal point is moved left three places).

EXAMPLE Convert 2531 μm to mm.

SOLUTION 2531 μm = 2.531 mm (up one, left three)

EXAMPLE Convert 0.0159 mm to μm.

SOLUTION 0.0159 mm = 15.9 μm (down one, right three)

EXAMPLE Convert 0.000394 m to μm.

SOLUTION 0.000394 m = 394 μm (down two, right six)

PERFORMANCE OBJECTIVE 6	$263 \text{ cm} = 2.63 \text{ m}$
You should now be able to convert a given number from one size metric unit to another by shifting the decimal point.	$89 \text{ km} = 89{,}000{,}000 \text{ mm}$ $56 \mu m = 0.000056 \text{ m}$

THIRD ADVANTAGE OF THE SI (METRIC) SYSTEM

The SI system simplifies the recording of the precision of measurements. In Figure 2.7 we use two straight rules to measure the length of a block.

Centimeter rule

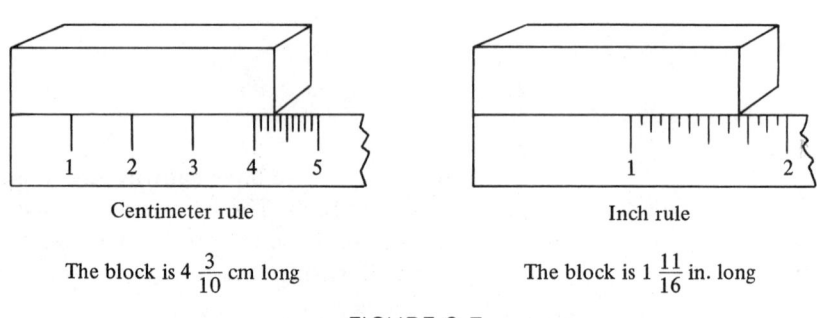

The block is $4\frac{3}{10}$ cm long

Inch rule

The block is $1\frac{11}{16}$ in. long

FIGURE 2.7

With the centimeter scale we get $4\frac{3}{10}$ centimeters, which is immediately expressed as 4.3 cm in the decimal system. On the inch scale we obtain $1\frac{11}{16}$ inches, which can only be converted to decimal form by dividing 11 by 16. The result of this division may not indicate the actual precision of the measurement. For example, we obtain 1.6875 inches, which might imply that we have made a measurement to the one ten-thousandths of an inch. Actually, we only measured to the nearest sixteenth of an inch.

ADVANTAGES OF THE SI SYSTEM

1. All quantities have the same unit subdivisions. Standard unit prefixes give the relative size.

2. Conversion between different size units only requires moving the decimal point.

3. Measurements can be written in the decimal system indicating both the size and the precision of the measurement.

EXERCISE SET 2.2

1. Convert the following numbers to the units indicated:
   a. 0.0000264 m to micro-meters (μm)
   b. 0.00381 ℓ to milliliters (mℓ)
   c. 50,000 tons to megatons
   d. 1512 megahertz (MHz) to gigahertz (GHz) (a hertz is a cycle per second with dimension of 1/time)
   e. 0.00055 microfarads (μF) to picofarads (pF)
   f. 22×10^{-9} F to picofarads (pF)
   g. 225,000 cm^2 to square meters (m^2) *

2. Convert the following numbers to the units indicated:
   a. 104 kilohertz (kHz) to hertz
   b. 2.2 pF to farads
   c. 7.5 μm to cm
   d. 0.86 ℓ to milliliters
   e. 1.4 kilocalories (kcal) to calories (cal)
   f. 2 m^3 to cubic centimeters (cm^3) *
   g. 1000 cm^2 to square meters *

3. Express the following numbers with a unit prefix rather than as a power of ten:
   a. 3.1×10^{-6} m
   b. 32×10^{-12} F
   c. 1.6×10^{3} m
   d. 42×10^{-3} ℓ

4. Express the following numbers with a unit prefix rather than as a power of ten:
   a. 55×10^{-6} A
   b. 1470×10^{6} Hz
   c. 51×10^{-2} m
   d. 3.1×10^{-9} coulombs (C)

5. Express the following numbers in power of ten notation by removing the metric prefix:
   a. 340 mm
   b. 5.5 MHz
   c. 45 picoseconds (ps)
   d. 0.32 cm

*Note that the quantity to be converted is raised to a power.

6. Express the following numbers in power of ten notation by removing the metric prefix:
   a. 5.5 μF
   b. 150 GHz
   c. 30 nanoseconds (ns)
   d. 1.75 km

SECTION 2.3 OPERATING ON THE DATA— THE CALCULATOR

We refer to the results of measurements as **data.** We are often required to operate on the data to obtain additional information about the system being studied. For example, suppose we have measured the voltage across a 14 picofarad capacitor to be 4.5 volts. We are asked to predict the amount of charge stored on the capacitor knowing that the charge on a capacitor equals the product of the capacitance and the voltage. Then

Charge = 0.000000000014 × 4.5 = 0.000000000063 Coulomb (C)

The preceding calculation is awkward as we have had to keep track of 12 decimal places. We could not use an electronic calculator to do the multiplication in the form shown since most calculators will not take more than seven or eight decimal places. Faced with this problem, we would probably multiply 14 × 4.5 and record the result as 63 pC. Computations with large and small numbers will continue to occur in our practical work. In this section we begin to develop methods of working with data using the electronic calculator. We recommend that you invest in a "scientific" calculator. As a minimum, such a calculator should have the following keys:

÷, ×, +, −	Arithmetic operations
+/−	
y^x	Exponentiation
sin, cos, tan	Trigonometric functions
$1/x$	the reciprocal
ln x or e^x	the natural log or power of e
INV	an inverse capability
x^2 or $\sqrt{x}$	square or square root
STO, RCL	Storage and recall of one number
EE	Entry of an exponent
=	Equals key
LOG	The log base 10 key

OPERATING ON THE DATA—THE CALCULATOR 2.3

In this section we will discuss the arithmetic operations and exponentiation. In later sections we will use the other capabilities of your calculator. We begin by showing a calculator face in Figure 2.8. Key placement may be different on your calculator, but the procedures for solving problems should be the same.

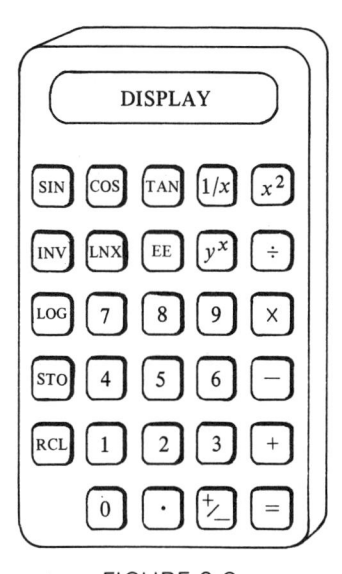

FIGURE 2.8

Your calculator will probably also have a value for pi (π) and a conversion from degrees to radians. In describing the following operations, we will use the word "Enter" to indicate that you press the correct numerical keys until the display shows the number being entered. "Press" will refer to using one of the keys other than a number or decimal point key. "Display" will indicate what is shown in the display of the calculator as a result of the "Enter" or "Press" steps. Consult your own calculator instruction book if it does not appear to follow the procedures indicated here.

EXAMPLE Add 15 + 3.2 on the calculator.

SOLUTION

Step	Enter	Press	Display
1	15		15.
2		+	15.
3	3.2		3.2
4		=	18.2

The answer is 18.2.

EXAMPLE Add 12.1 + 3.2 + 105.3.

SOLUTION

Step	Enter	Press	Display
1	12.1		12.1
2		[+]	12.1
3	3.2		3.2
4		[+]	15.3
5	105.3		105.3
6		[=]	120.6

The answer is 120.6.

EXAMPLE Evaluate 15.1 − 26.3 + 2.4.

SOLUTION

Step	Enter	Press	Display
1	15.1		15.1
2		[−]	15.1
3	26.3		26.3
4		[+]	−11.2
5	2.4		2.4
6		[=]	−8.8

The answer is −8.8.

In later chapters we will only show the enter and press steps for additions and subtractions. We would write the example above as

$$15.1 \; [-] \; 26.3 \; [+] \; 2.4 \; [=]$$

PERFORMANCE OBJECTIVE 7

You should now be able to use a calculator to perform a series of additions and/or subtractions.

16.21 [+] 3.85 [−] 8.02 [=]

Read 12.04 in the display.

OPERATING ON THE DATA—THE CALCULATOR 2.3

EXAMPLE Multiply 20.1 × 3.5.

SOLUTION

Step	Enter	Press	Display
1	20.1		20.1
2		×	20.1
3	3.5		3.5
4		=	70.35

The answer is 70.35.

EXAMPLE Find 3.612 ÷ 1.2.

SOLUTION

Step	Enter	Press	Display
1	3.612		3.612
2		÷	3.612
3	1.2		1.2
4		=	3.01

The answer is 3.01.

EXAMPLE Find 3.5 × 4.2 ÷ 1.4.

SOLUTION

Step	Enter	Press	Display
1	3.5		3.5
2		×	3.5
3	4.2		4.2
4		÷	14.7
5	1.4		1.4
6		=	10.5

The answer is 10.5.

2 MEASUREMENT AND DATA HANDLING

In later chapters we will only show the enter and press steps for multiplications and divisions. We would write the example above as

3.5 $\boxed{\times}$ 4.2 $\boxed{\div}$ 1.4 $\boxed{=}$

PERFORMANCE OBJECTIVE 8	
You should now be able to use a calculator to perform a series of multiplications and divisions.	*64.5* $\boxed{\div}$ *2.15* $\boxed{\times}$ *1.5* $\boxed{=}$ *Read 45 in the display.*

Now consider a problem such as the one discussed earlier.

EXAMPLE Multiply $0.000000000014 \times 4.5$.

SOLUTION If we try to follow the procedure used above we are unable to enter 0.000000000014 because the display cannot handle such a small number. However, we can perform the multiplication using exponential notation. Numbers that are very small or very large, in relation to 1, can be entered in exponential notation.

To enter a decimal number in exponential notation, we must express it as a product of a number and a power of ten. For example, 0.000000000014 can be written as 0.14×10^{-10}. To enter this number, complete the following steps.

Step	Enter	Press	Display
1	.14		0.14
2		$\boxed{EE}$	0.14 00
3	10		0.14 10
4		$\boxed{+/-}$	0.14 $-$ 10

We can now press any operation and complete a calculation involving this number. Let's complete the problem.

Step	Enter	Press	Display
5		$\boxed{\times}$	1.4 $-$ 11
6	4.5		4.5
7		$\boxed{=}$	6.3 $-$ 11

The answer is 6.3×10^{-11}.

OPERATING ON THE DATA—THE CALCULATOR 2.3

The number provided by the calculator in the last example has one digit left of the decimal point. This is a characteristic of scientific notation, a special form of exponential (power of ten) notation. Every number is expressed as the product of a value between 1 and 10 and a power of ten.

Let us consider how we would enter numerical quantities expressed as some multiple of a metric unit.

EXAMPLE Multiply 40 pF by 26 millivolts (mV).

SOLUTION In making the calculation, we should enter the prefixes indicating the unit size as exponents.

Step	Enter	Press	Display
1	40		40.
2		EE	40. 00
3	12		40. 12
4		+/−	40. − 12

At this stage we have 40 pF correctly displayed in power of ten notation. Note that it is not in scientific notation. We complete the calculation as follows:

5		×	4. − 11
6	26		26.
7		EE	26. 00
8	3		26. 03
9		+/−	26. − 03
10		=	1.04 − 12

The answer is 1.04×10^{-12} farad-volts (F-V) or 1.04×10^{-12} C.

EXAMPLE Perform the following division:

$$(-3.2 \times 10^{-19} \text{ C}) \div (6.68 \times 10^{-27} \text{ kg})$$

SOLUTION

Step	Enter	Press	Display	
1	3.2		3.2	
2		+/−	−3.2	(note this step)
3		EE	−3.2	00
4	19		−3.2	19
5		+/−	−3.2 −	19
6		÷	−3.2 −	19
7	6.68		6.68	
8		EE	6.68	00
9	27		6.68	27
10		+/−	6.68 −	27
11		=	−4.7904	07

The answer is -4.7904×10^7 C/kg.

In later chapters we will only show enter and press steps. The preceding problem would be written as

3.2 [+/−] [EE] 19 [+/−] [÷] 6.68 [EE] 27 [+/−] [=]

PERFORMANCE OBJECTIVE 9

You should now be able to enter numbers into the calculator in exponential notation (to include metric prefixes).

Enter 520 μm in basic units.

520 [EE] 6 [+/−] gives

520.−06 m

Before leaving this section, we offer some words of advice and caution about the use of the calculator. Remember that the calculator is like a tool. If it is to be of help, you must understand how it works and then you must learn to use it effectively. We suggest the following:

1. Practice correct entry of data. Review each entry in the display for accuracy. Without practice you may make more errors and work more slowly with a calculator than someone who uses pencil and paper.

2. Practice writing and interpreting the calculator's answer. This includes questioning whether the answer makes sense. If you attempt to multiply two large numbers together and obtain a small number as the answer, you have probably made an error in data entry.

3. Make as few data entries as possible. If you are adding or multiplying a series of numbers, do not reenter the result of each step. As we move through the text, we will suggest methods of using the storage capability and methods of arranging the problem to avoid unnecessary data entries.

EXERCISE SET 2.3

1. Perform the following operations on your calculator. Record your answers.
   a. 321 + 56.9 + 221.65
   b. 32.9 − 16.4 + 5.1
   c. 3.8 × 5 ÷ 2.5
   d. 32.5 ÷ 2 × 16.1

2. Perform the following operations on your calculator. Record your answers.
   a. 0.456 + 23.2 − 16.1
   b. 6.9 − 17.3 + 56.892
   c. 45.1 × 5.6 ÷ 2.8
   d. 2.5 ÷ 22.5 × 1.2

3. Enter the following numbers into your calculator using the $\boxed{\text{EE}}$ key. After each entry, press the $\boxed{=}$ key and record the value in the display.
   a. 0.00000015497
   b. 1,520,000,000
   c. 3.6×10^7
   d. -2.14×10^{-3}
   e. 1521×10^3
   f. -2136×10^{-7}
   g. 0.00159×10^{12}
   h. 0.0592×10^{-9}

4. Enter the following numbers into your calculator using the $\boxed{\text{EE}}$ key. After each entry, press the $\boxed{=}$ key and record the value in the display.
   a. 2,789,400,000
   b. 0.000000923
   c. 4.57×10^6
   d. -6.234×10^{-2}
   e. 3450×10^4
   f. -8273×10^{-9}
   g. 0.0008456×10^{14}
   h. -0.34×10^{-7}

5. Enter the following metric quantities using the prefix to determine the exponent. After each entry, press the [=] key and record the value in the display.
   a. 200 pF
   b. 96 GHz
   c. 3.15 μC
   d. 45 mA
   e. 65 mm
   f. 2.3 km

6. Enter the following metric quantities using the prefix to determine the exponent. After each entry, press the [=] key and record the value in the display.
   a. 45 ns
   b. 4.5 pF
   c. 731 cm
   d. 41.2 mℓ
   e. 56 mA
   f. 452 μm

7. Perform the following operations on your calculator. Write the answers in scientific notation.
   a. $55 \times 10^6 + 8.7 \times 10^7$
   b. $(1.67 \times 10^{-27})(2.9 \times 10^7)$ adjacent brackets indicate multiplication
   c. $(-1.6 \times 10^{-19})/(9.1 \times 10^{-31})$ "/" indicates division
   d. $(8.9 \times 10^{-12})(2.1 \times 10^{-2})/(1.5 \times 10^{-3})$
   e. $(15 \times 10^{-12})(55) + (30 \times 10^{-12})(55)$

8. Perform the following operations on your calculator. Write the answers in scientific notation.
   a. $74 \times 10^5 + 1.5 \times 10^6$
   b. $(3.34 \times 10^{-27})(4.5 \times 10^6)$ adjacent brackets indicate multiplication
   c. $(8.12 \times 10^{-19})/(9.1 \times 10^{-31})$ "/" indicates division
   d. $(7.1 \times 10^{-5})(21.2 \times 10^{-6})/(8.2 \times 10^{-5})$
   e. $(6.39 \times 10^{-9})(34) + (5.6 \times 10^{-9})(21)$

SECTION 2.4 HANDLING SIGNIFICANT DIGITS

When dealing with measurements, we must be concerned with the precision of the results. We noted that the SI system was particularly suited to recording the precision of data. Those who work with data have developed various methods of showing the precision of measured results. Sometimes a tolerance in the measurement is recorded.

HANDLING SIGNIFICANT DIGITS 2.4

For example, if we write 25.2 mm ± 0.05 mm, we know that the measurement is accurate to within 1/20 millimeter. That is, the actual value of the measured length is closer to 25.2 millimeters than to 25.1 millimeters or 25.3 millimeters. The digits 2, 5, and 2 all give significant information about the measurement. We are certain that each of these digits is correct although 25.2 may not be the precise value of the length. In this case, we would say that there are three significant digits in the measured result.

DEFINITION

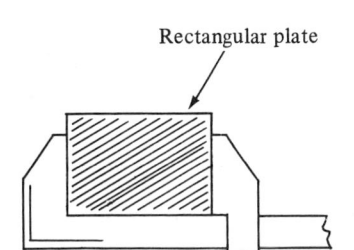

> **SIGNIFICANT DIGITS**
>
> **Significant digits** in a value are those digits which convey information about the known precision of that value.

In the example above, if we were to write 25.19 mm ± 0.005 mm, we would have a more precise measurement with four significant digits.

Those who do technical work in measurement would not write down (guess at) digits that were not indicated by the measuring device. But what if we use the measured result to do further calculations? Consider the following example.

We use a vernier to measure the dimensions of a rectangular plate to the nearest tenth of a millimeter.

We obtain 45.3 millimeters for the length and 25.7 millimeters for the width. When we multiply these two dimensions to obtain the plate area, we find 1164.21 square millimeters as the answer. This answer seems to imply that we are certain of the area to 1/100 square millimeter. However, we were uncertain of each dimension by 0.05 millimeters. We could draw a strip 0.05 millimeters wide that is the length of the sum of the dimensions to indicate the uncertain area. See Figure 2.10. This area is

$$(25.7 + 45.3) \times 0.05 = 3.55 \text{ mm}^2$$

FIGURE 2.9 — Rectangular plate

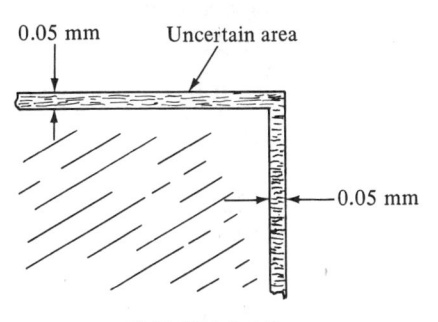

FIGURE 2.10

The uncertain area is over 3 square millimeters. It appears that we cannot be sure of the last three digits in the number 1164.21. We should then write it as 11$\underline{6}$0 square millimeters. The bar is placed under the last significant digit. In this case, we converted the non-significant digits to zero.

We give a procedure for fixing significant digits after multiplication or division.

FIXING SIGNIFICANT DIGITS AFTER MULTIPLICATION OR DIVISION

To fix significant digits after multiplication or division, retain as many significant digits in the result of a calculation as appeared in the least significant data used.

We illustrate this procedure with several examples.

EXAMPLE Multiply 2.45 × 5.1.

SOLUTION Multiplication yields 2.45 × 5.1 = 12.495, which has three more significant digits than the least significant piece of data entering the calculation (5.1). We write the result as 12.

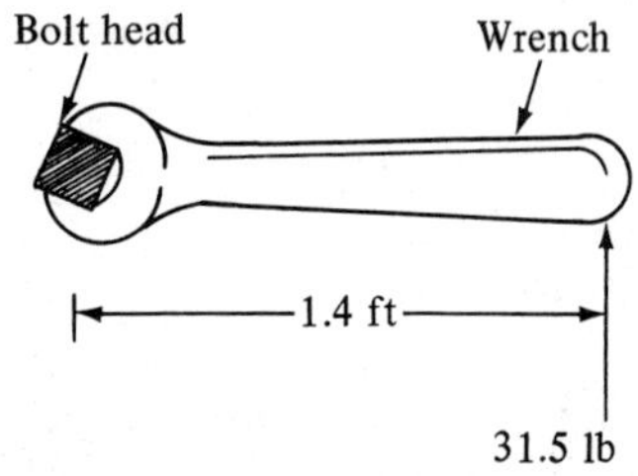

EXAMPLE Evaluate the torque on the bolt head shown in Figure 2.11.

SOLUTION Torque is obtained by multiplying the force by the distance of its point of application from the bolt head. That is, we multiply 31.5 lb by 1.4 ft. The calculator gives an answer of 44.1 to which we add the units foot-pounds.

Since distance had only two significant digits, we must reduce the answer to two significant digits. Thus, we report 44 ft-lb.

EXAMPLE Evaluate the speed of an object that moves 0.0122 m in 0.061 s.

SOLUTION Speed is obtained as the quotient of distance and time.

$$0.0122/0.061 = 0.2 \text{ m/s}$$

The answers appears to have only one significant digit while the least significant piece of data had two significant digits. We report 0.20 m/s as the answer.

We have added a zero to indicate that the answer is significant to the hundredth of a meter per second.

FIGURE 2.11

HANDLING SIGNIFICANT DIGITS 2.4

When we are adding or subtracting numbers, the procedure is somewhat different.

FIXING SIGNIFICANT DIGITS AFTER ADDITION OR SUBTRACTION

To fix significant digits after addition or subtraction, retain as significant the right-most column digit that is significant in all terms.

We illustrate this procedure with several examples.

EXAMPLE Add the numbers 5.234 and 0.5 and retain only significant digits in the answer.

SOLUTION

```
  5.234
+ 0.5
  -----
  5.734
```

Only the tenths place appears in both addends. As a result, we can only keep the tenths place in the sum. The answer is 5.7.

Note that, although 0.5 had only one significant digit, the answer has two significant digits. This illustrates the difference between the procedures for multiplication (division) and addition (subtraction).

EXAMPLE Find $345.92 + 5.21 - 23.0$.

SOLUTION $345.92 + 5.21 = 351.13$

$351.13 - 23.0 = 328.13$

Only the tenths place appears in all terms, so that the answer must be 328.1.

EXAMPLE Form the sum $2159 + 34.9 + 0.2349$

SOLUTION Addition yields 2194.1349, which must be reported as 2194 since the number 2159 has its right-most column in the one's position.

So far we have avoided some problems that arise in the reporting of significant digits. For example, if we multiply 55 by 900, we obtain 49,500, which should have only two significant digits. We cannot discard the right-most digits without altering the value of the number. We need some procedure for solving this and other problems. A procedure for indicating significant digits follows.

INDICATING SIGNIFICANT DIGITS

1. All digits reported are considered significant unless an underlining bar is used to mark the last significant digit.

2. If a number must carry more digits than are significant, mark the last significant digit by underlining.

3. Add trailing zeros to decimal numbers if they are significant.

4. Leading zeros in numbers are not significant.

5. Operations involving pure numbers do not change the number of significant digits.

The following examples illustrate the procedures:

Procedure 1 2.35 has three significant digits.
4000 has four significant digits.

Procedure 2 If 12,000 has four significant digits, write it as 12,0_00.

Procedure 3 If 3.25 should have four significant digits, write it as 3.250.

Procedure 4 0.00192 has three significant digits.

Procedure 5 $3 \times 2.5 = 7.5$
     ↑   ↑
  pure no. data

$6 \times 0.12 = 0.72$
     ↑   ↑
  pure no. data

$2 \times 0.125 = 0.250$
     ↑   ↑
  pure no. data

We can now see another advantage of using scientific notation. If a number is written in scientific notation, all digits in the number (excluding the power of ten) are significant. No underlining bars are necessary and leading zeros are not a problem. Thus, we simply look at the number and count its significant digits.

4.519×10^{-9} has four significant digits

3.1×10^{4} has two significant digits

9.10×10^{-31} has three significant digits

HANDLING SIGNIFICANT DIGITS 2.4

PERFORMANCE OBJECTIVE 10	
You should now be able to identify the correct number of significant digits in measured data or calculations.	*4.51 has three significant digits.* *0.013 has two significant digits.* *1.590 × 10³ has four significant digits.*

Consider a calculation in scientific notation. Let us determine the result of the following multiplication:

$$(9.10 \times 10^{-31})(3.7 \times 10^{11})$$

Setting this up on the calculator we find 3.367×10^{-19} as the answer. But this answer has four digits of which only two are significant. Note that the second factor has only two significant digits. How do we reduce this to two significant digits?

In the examples used so far we simply dropped insignificant digits or changed them to zeros. But we may have an uneasy feeling about changing 3.367 to 3.3. The number 3.367 is closer to 3.4 than to 3.3. We recommend rounding to the nearest digit when reducing a number to significant digits. The procedure to be used in rounding follows.

RULE FOR ROUNDING

Look at the first digit to be discarded. If it is in the range 0 to 4, simply discard the unwanted digits. If it is 5 to 9, increase by one the last digit retained and discard the unwanted digits.

EXAMPLE Reduce 4.561×10^{-9} to three significant digits.

SOLUTION The key digit is the fourth digit (the first discarded digit). That digit (1) is in the range 0 to 4, so we simply discard it and write

$$4.56 \times 10^{-9}$$

EXAMPLE Reduce 3.1876×10^4 to four significant digits.

SOLUTION The key digit is the fifth digit. That digit is in the range 5 to 9 so we change the last digit retained (7) to 8 and discard the 6. We write

$$3.188 \times 10^4$$

as the answer.

EXAMPLE Reduce 1.97×10^{-3} to two significant digits.

SOLUTION The answer is 2.0×10^{-3}. When we increased 9 by one, we had 10. The 1 had to be carried to the next place left giving 2. We retain the first zero after the decimal point because it is significant.

EXAMPLE Reduce 2157 to three significant digits.

SOLUTION The key digit is 7. We increase 5 by one and write 21_6_0. We need the underline to indicate the last significant place.

PERFORMANCE OBJECTIVE 11	
You should now be able to round a calculated value to the correct number of significant digits.	Round 22.597 to three significant digits. Result → 22.6

EXERCISE SET 2.4

1. Multiply or divide the following and report the result with the correct number of significant digits. Assume each value is a measured value.

    a. 2.5×4.376
    b. 12×250
    c. $(4.1 \times 10^4) \div (5.62 \times 10^{-7})$
    d. $(7.31 \times 10^{-6})(3.100 \times 10^{-2})$
    e. $0.0031 \div 4.295$

2. Multiply or divide the following and report the result with the correct number of significant digits. Assume each value is a measured value.

    a. 3.1×52.4
    b. 0.125×16
    c. $(6.10 \times 10^3) \div (3.321 \times 10^{-8})$
    d. $(8.56 \times 10^{-4})(3.5 \times 10^3)$
    e. $0.0067 \div 420$

3. Multiply the following and report the result with the correct number of significant digits. Assume the first value in each operation is a pure number.

    a. $14 \times 1.601 \times 10^{-19}$
    b. $4 \times 1.675 \times 10^{-27}$
    c. 8×0.0025
    d. 4×0.525
    e. $3 \times (4.0 \times 10^7)$

4. Multiply the following and report the result with the correct number of significant digits. Assume the first value in each operation is a pure number.

   a. $22 \times 2.512 \times 10^{-5}$
   b. $8 \times 1.25 \times 10^{7}$
   c. 16×0.550
   d. $5 \times (2.5 \times 10^{4})$
   e. $1.45 \times (7.1 \times 10^{-8})$

5. Round the following values to the indicated number of significant digits:

   a. 2.517×10^{-3} to three digits
   b. 2139 to three digits
   c. 9.172×10^{-31} to two digits
   d. 3.155×10^{7} to two digits
   e. 7.2196 to four digits

6. Round the following values to the indicated number of significant digits:

   a. 5.192×10^{-6} to three digits
   b. 4135 to three digits
   c. 1.567 to two digits
   d. 7.134×10^{12} to three digits
   e. 0.00569 to two digits

7. Write the values below with the indicated number of significant digits:

   a. 17,000 with four digits
   b. 4.559 with three digits
   c. 0.017 with four digits
   d. 6.518×10^{4} with three digits
   e. 33.99 with three digits

8. Write the values below with the indicated number of significant digits:

   a. 2563 with two digits
   b. 9.19 with two digits
   c. 0.0612 with four digits
   d. 44 with three digits
   e. 22.97 with three digits

SECTION 2.5 SUMMARY

In this chapter we have considered some practical problems that you will encounter not only in mathematics but in other technical work. You need to be able to deal with numbers that result from the measurement of real physical quantities. The new concepts and definitions are reviewed here.

The measurement process: a series of steps involving selection of a standard, comparison of the quantity to be evaluated with the standard, and recording a result.

SI system: a system of units in which the most commonly used basic quantities are mass (kilograms), length (meters), time (seconds), and current (amperes).

English system: a system of units in which the most commonly used basic quantities are force (pounds), length (feet), time (seconds), and current (amperes).

Basic quantity: a fundamental physical quantity that is defined in terms of an accepted standard.

Unit: a name given to the standard of a basic quantity.

Derived quantity: some combination of basic quantities. For example, velocity (length/time) is a derived quantity.

Dimension: an expression of the combination of basic quantities in some physical quantity. Dimension is independent of the system of units.

Unit prefix: a prefix that indicates a multiplier of a unit.

Significant digits: those digits in a number which give useful information about the precision of the number.

Rounding: a method for selecting the value of the last significant digit in a number when digits are to be discarded.

CHAPTER 2

PRACTICE TEST

1. (PO-1) You are asked to find the length of a small steel rod that appears to be about 2 in. long. Describe each step in the measurement process.

2. (PO-2, 3) Complete the following table:

SUMMARY 2.5

Quantity	Dimension	Units–English system	Units–SI system
mass	mass	lb-s^2/ft	
force			kg-m/s^2
length	length	ft	
pressure		lb/ft^2	
energy		ft-lb	
angular momentum		slug-ft^2/s	

3. (PO-4) Complete the following table. First enter an approximate value. Check your approximation with an exact conversion. Make entries in the boxes where a unit symbol is placed.

English value		SI value	
Approximate	Exact	Approximate	Exact
	7 ft	m	m
ft	ft		10 m
	5 lb	N	N
	3 lb	kg*	kg*
lb*	lb*		6 kg
slug	slug		30 kg
ft-lb	ft-lb		8 J

* = enter equivalent units

4. (PO-5) Express the following numbers with a unit prefix rather than as a power of ten:
   a. 4.5×10^{-4} F
   b. 55.9×10^{-3} A
   c. 1220×10^{6} Hz
   d. 125×10^{3} m

5. (PO-6) Make the following conversions:
   a. 22,000 pF to microfarads
   b. 0.062 ℓ to milliliters
   c. 990 MHz to gigahertz
   d. 420 m to kilometers

6. (PO-7, 8, 9, 10, 11) Perform the following operations with your electronic calculator. Carry the appropriate number of digits in the answer.

   a. $1521.0 + 685.43 - 501.29$

   b. $239 \times 436 \div 19$

   c. $(1.5 \times 10^{-9})(14.1 \times 10^8)/(3 \times 10^8)$

   d. $4(1.67 \times 10^{-27}) + 2(9.1 \times 10^{-31})$

   e. $14 \; \mu F \times 320 \; mV$

7. (PO-10, 11) Consider that each number below is the result of a measurement. Perform the operation indicated and report the answer to the correct number of significant digits.

   a. 5.31×6.5 b. $4\underline{4}0 \times 50$ c. $(3.5 \times 10^{-2})/(3.25)$

8. (PO-10, 11) Write the following values with the indicated number of significant digits. Round as necessary.

   a. 14.853 to three significant digits

   b. 12,425 to four significant digits

   c. 3.1 to three significant digits

   d. 0.0132 to two significant digits

CHAPTER 3

APPLIED GEOMETRY

3.1 Lines, Angles, and Polygons
3.2 Measures Associated with Polygons
3.3 Measures Associated with Circles
3.4 Measures Associated with Solids
3.5 Summary

Geometry is the study of shapes, their relationships, and their measurement. It is one of the most practical branches of mathematics. Since most occupations, in one way or another, deal with objects in the physical world around us, people need to be aware of geometric properties of objects and of techniques for measuring the objects. There are many ways in which an understanding of geometric principles can be useful in everyday life.

In this chapter we will study shapes from both plane and solid geometry. We will find the perimeter and area of plane figures and we will find the surface area and volume of solids.

SECTION 3.1 LINES, ANGLES, AND POLYGONS

A point is the most basic geometric figure. We represent it graphically by a dot and denote it by a capital letter such as *P*. A **line,** that is, a straight line, is a one-dimensional geometric figure consisting of points extending infinitely far in both directions. We often denote a line by a lower-case letter such as ℓ. However, we also denote a line by giving two points that lie on the line, for example $\overleftrightarrow{AB}$, as in Figure 3.1.

To denote just those points on the line between points *A* and *B*, and including *A* and *B*, we write $\overline{AB}$. $\overline{AB}$ is called a **line segment** and is depicted in Figure 3.2.

Consider the line with points *B*, *A*, and *C* as shown in Figure 3.3. The point *A* divides the line into parts. To represent the set of points containing *A* and all the points on the same side of *A* as *B*, we write $\overrightarrow{AB}$. Likewise, the set of points containing *A* and all the points on the same side of *A* as *C* is denoted by $\overrightarrow{AC}$. We call $\overrightarrow{AB}$ and $\overrightarrow{AC}$ **rays.** *A* is the **endpoint** of the rays.

An **angle** consists of a point and two rays extending from that point. The point is called the **vertex** of the angle and the two rays are called the **sides** of the angle. The angle is shown in Figure 3.4.

Two common ways of denoting angles are illustrated in Figure 3.5. In the first case we name the angle by giving the vertex preceded by the symbol $\angle$. In the second case we use the $\angle$ symbol followed by the vertex and a point on each side of the angle, with the vertex between the other two points.

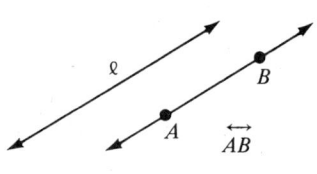

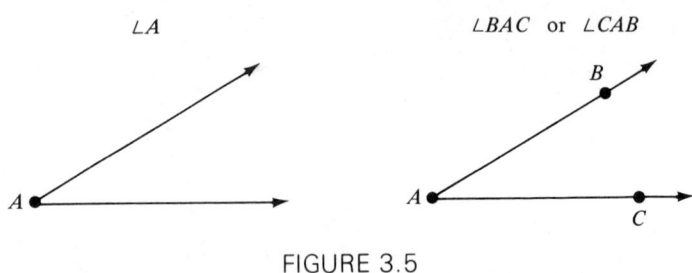

FIGURE 3.1

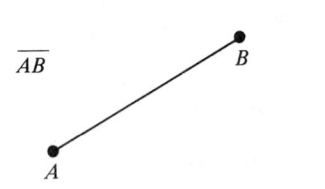

FIGURE 3.2

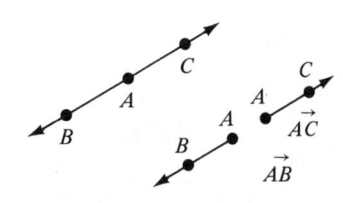

FIGURE 3.3

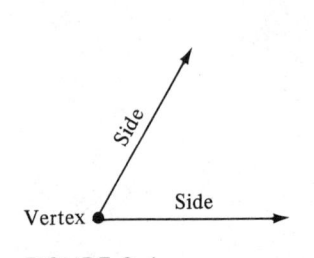

FIGURE 3.4

FIGURE 3.5

Two distinct lines, ℓ_1 and ℓ_2, will either intersect in a point or never intersect. If ℓ_1 and ℓ_2 never intersect we say that they are **parallel lines** and we write $\ell_1 \parallel \ell_2$. See Figure 3.6.

LINES, ANGLES, AND POLYGONS 3.1

Parallel lines

Intersecting lines

FIGURE 3.6

Perpendicular lines

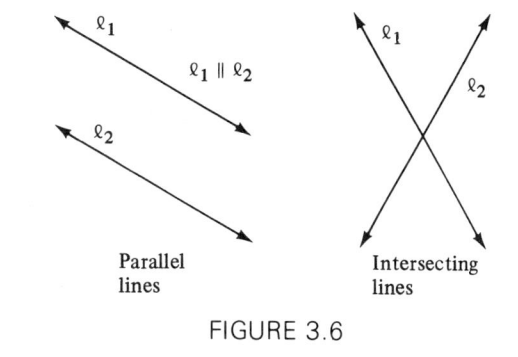

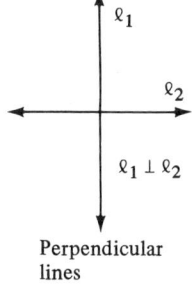

FIGURE 3.7

If ℓ_1 and ℓ_2 intersect in a point, then the point and the two lines form four angles. If the four angles are equal, that is, if they are the same size or have the same measure, we say that the lines are **perpendicular** and we write $\ell_1 \perp \ell_2$. A pair of perpendicular lines is shown in Figure 3.7.

We often characterize angles based on their size or measure. The angles formed when two perpendicular lines intersect are called **right angles.** The measure of a right angle is 90°. An angle whose measure is less than 90° is called an **acute angle.** An angle whose measure is greater than 90° is called an **obtuse angle.** Figure 3.8 illustrates these three types of angles.

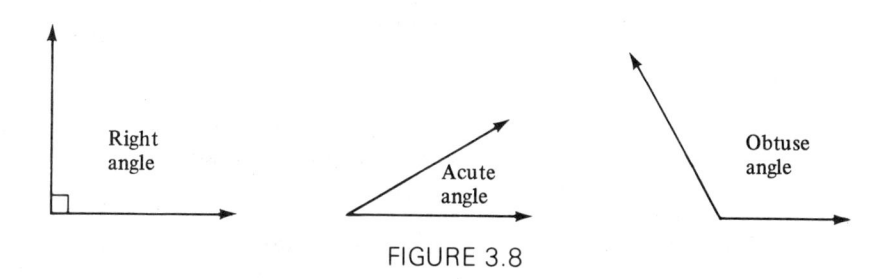

Right angle

Acute angle

Obtuse angle

FIGURE 3.8

NOTE: The symbol ⌐ used in denoting the right angle indicates that the sides are perpendicular.

We also have special names for two angles whose measures total to 90° or 180°. If the sum of the measures of two angles is 90°, we call the angles **complementary.** When the sum of the measures is 180°, we call the angles **supplementary.** See Figure 3.9.

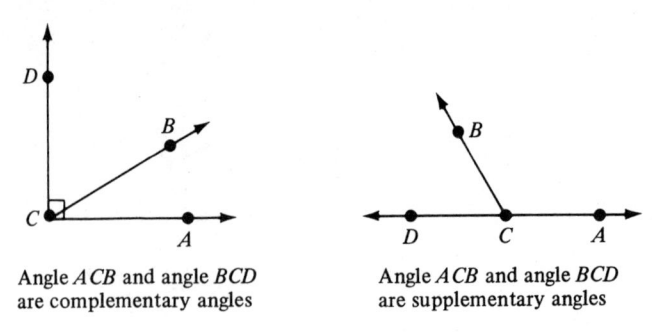

Angle ACB and angle BCD are complementary angles

Angle ACB and angle BCD are supplementary angles

FIGURE 3.9

PERFORMANCE OBJECTIVE 1

You should now be able to recognize the terminology and notation associated with lines and angles.

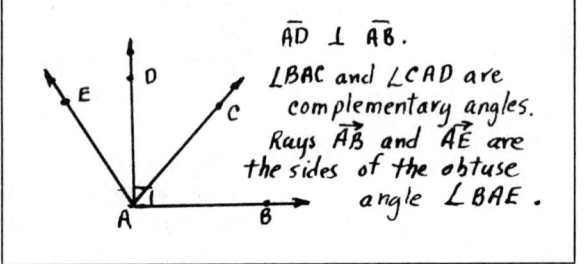

$\overline{AD} \perp \overline{AB}$.
$\angle BAC$ and $\angle CAD$ are complementary angles.
Rays $\overrightarrow{AB}$ and $\overrightarrow{AE}$ are the sides of the obtuse angle $\angle BAE$.

A two-dimensional geometric figure is called a **plane figure**. A plane figure that consists of line segments and angles and completely encloses a region is called a **polygon**.

We name polygons according to the number of sides (or angles) the polygon has. This is illustrated in the following table, which lists polygons with as many as eight sides:

Name of polygon	Number of sides (or angles)
triangle	3
quadrilateral	4
pentagon	5
hexagon	6
heptagon	7
octagon	8

Some examples of polygons are shown in Figure 3.10. We denote polygons by listing consecutive vertices of the angles that make up the polygons.

LINES, ANGLES, AND POLYGONS 3.1   **91**

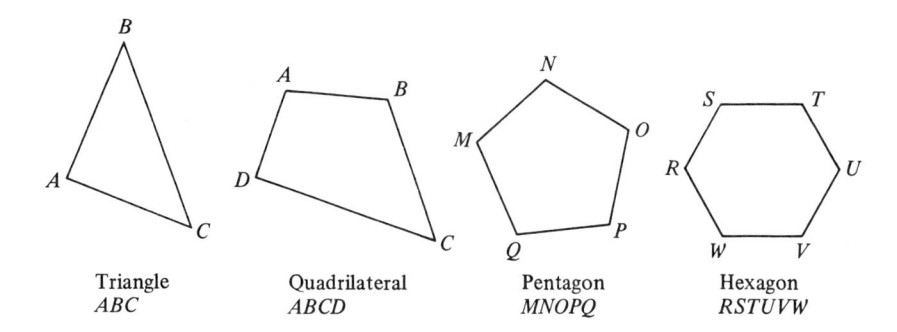

FIGURE 3.10

Triangle
ABC

Quadrilateral
ABCD

Pentagon
MNOPQ

Hexagon
RSTUVW

    A polygon with equal sides and equal angles is called a **regular polygon**. So a pentagon with equal sides and equal angles is a **regular pentagon** and a hexagon with equal sides and equal angles is a **regular hexagon**. We usually use special names for triangles and quadrilaterals with equal sides and angles. We call a regular triangle an **equilateral triangle** or an **equiangular triangle** and we call a regular quadrilateral a **square**.

    We also have specific names for some other triangles and quadrilaterals with special properties. Some of these are listed and identified as follows:

Name	Figure	Description
right triangle		a triangle with a right angle
isosceles triangle		a triangle with two equal sides
parallelogram		a quadrilateral with opposite sides parallel and equal
rectangle		a parallelogram with four right angles
trapezoid		a quadrilateral with one pair of opposite sides parallel

Right triangles are very important geometric figures and we will work with them frequently in trigonometry. The sides of a right triangle are given special names. The side opposite the right angle is the longest side of the triangle and is called the **hypotenuse.** The other two sides are called the **legs** of the right triangle. See Figure 3.11.

There is a special relationship between the legs and the hypotenuse of a right triangle. The relationship is specified by a result known as the **Pythagorean Theorem** which applies only to right triangles. The Pythagorean Theorem states that

$$a^2 + b^2 = c^2$$

where a and b are the lengths of the legs of the right triangle and c is the length of the hypotenuse. We will use this result later in the chapter.

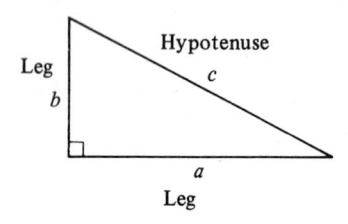

FIGURE 3.11

PERFORMANCE OBJECTIVE 2

You should now be able to identify and name polygons.

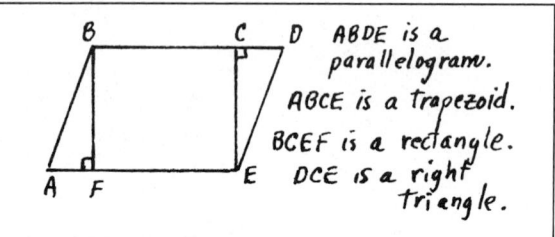

ABDE is a parallelogram.
ABCE is a trapezoid.
BCEF is a rectangle.
DCE is a right triangle.

We will now discuss two special relationships between polygons. They are congruency and similarity. **Congruent polygons** are polygons that have exactly the same size and shape. When discussing congruency and similarity, we use the concept of corresponding parts of polygons. We will illustrate corresponding angles and corresponding sides of polygons by referring to two congruent triangles in Figure 3.12.

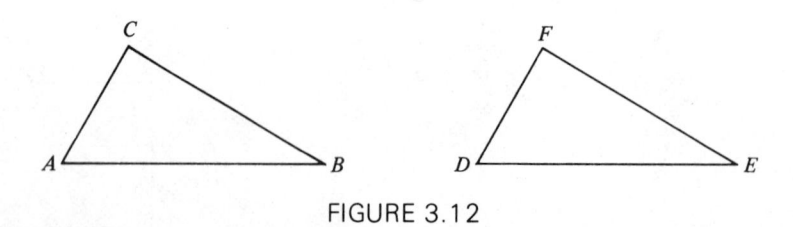

FIGURE 3.12

The corresponding angles of these congruent triangles are the angles whose measures are the same. These are the angles that would coincide if we were to superimpose one triangle on the other. The

LINES, ANGLES, AND POLYGONS 3.1

three pairs of corresponding angles are angles A and D, angles B and E, and angles C and F. The side in triangle DEF that corresponds to $\overline{AC}$ is $\overline{DF}$. We determine this by finding the side that is common to the two angles corresponding to angle A and angle C. Likewise $\overline{AB}$ corresponds to $\overline{DE}$ and $\overline{BC}$ corresponds to $\overline{EF}$. In congruent polygons, corresponding sides, as well as corresponding angles, are equal.

Similar polygons have the same shape but are not necessarily the same size. Corresponding angles of similar polygons are equal and corresponding sides are in proportion, that is, the ratios of the corresponding sides are equal. Consider the similar pentagons in Figure 3.13.

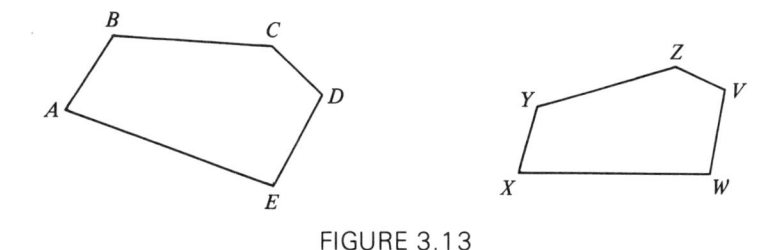

FIGURE 3.13

If pentagon $ABCDE$ is similar to pentagon $XYZVW$, then the corresponding angles are equal. That is,

$$\angle A = \angle X, \quad \angle B = \angle Y, \quad \angle C = \angle Z, \quad \angle D = \angle V, \quad \angle E = \angle W$$

Note that, when we name two similar polygons, we start with a pair of corresponding vertices and continue to name the other corresponding vertices using the same order for both polygons.

To say that the ratios of corresponding sides are equal means

$$\frac{AB}{XY} = \frac{BC}{YZ} = \frac{CD}{ZV} = \frac{DE}{VW} = \frac{EA}{WX}$$

where AB represents the length of side $\overline{AB}$, XY represents the length of $\overline{XY}$, etc. In the next section we will use these equal ratios to solve for unknown parts of similar polygons.

EXAMPLE Determine if the parallelograms in Figure 3.14 are similar.

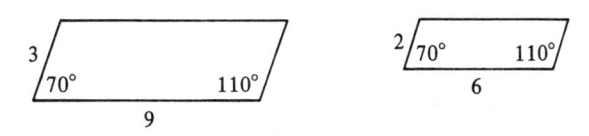

FIGURE 3.14

SOLUTION Yes, the parallelograms are similar. The corresponding angles are equal and the ratios of corresponding sides are the same.

$$\frac{3}{2} = \frac{9}{6}$$

Note that we only work with two ratios instead of four because the other two are the same as the first two.

EXAMPLE Consider a triangle with sides of length 3, 4, 5 and a triangle with sides of length 5, 6, 7. Is it possible for these triangles to be similar?

SOLUTION If the ratios of corresponding sides are equal, the possibility of the triangles being similar exists (since corresponding angles may be equal).

The shortest side of the first triangle will correspond to the shortest side of the second triangle and the longest sides of the triangles will correspond, etc. We then see that the ratios $\frac{3}{5}$, $\frac{4}{6}$, and $\frac{5}{7}$ are not equal. Therefore the triangles are not similar.

PERFORMANCE OBJECTIVE 3

You should now be able to recognize congruent and similar polygons.

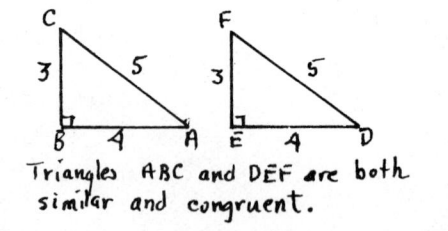

Triangles ABC and DEF are both similar and congruent.

EXERCISE SET 3.1

1. Name the following constructions:

   a.

   b.

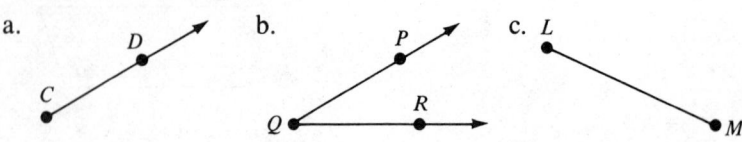

   c.

LINES, ANGLES, AND POLYGONS 3.1

2. Name the following constructions:
   a. 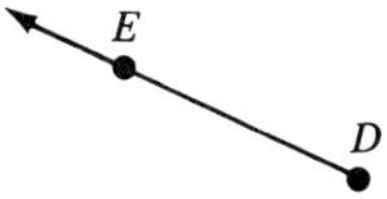
   b. 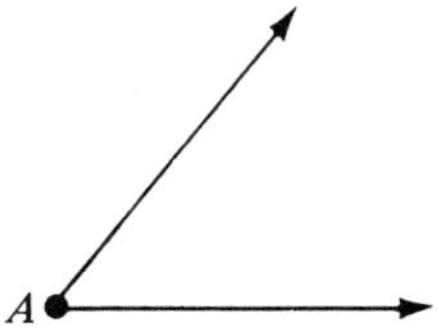
   c.

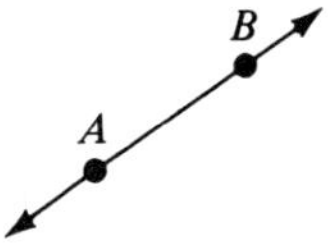

3. Classify the angles marked in the following figures as acute, right or obtuse:
   a. 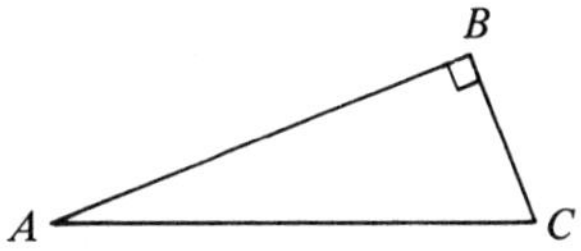
   b.

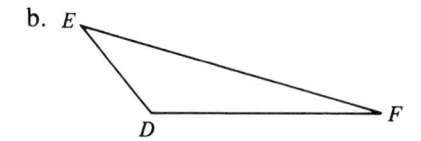

4. Classify the angles marked in the following figures as acute, right or obtuse:
   a. 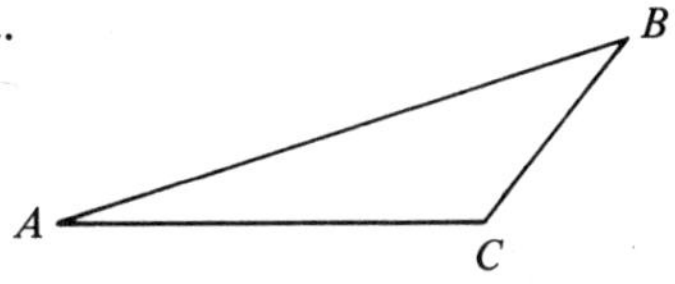
   b.

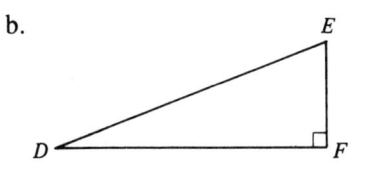

5. Identify the relationships between the following pairs of lines or angles:
   a. 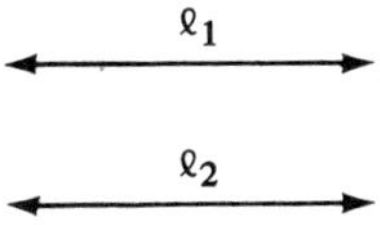
   b. 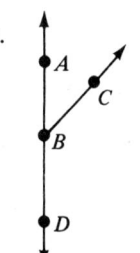
   c.

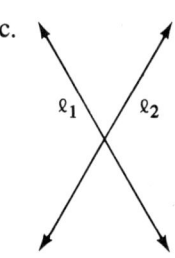

6. Identify the relationships between the following pairs of lines or angles:
   a. 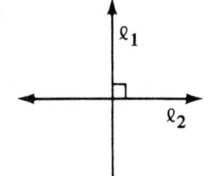
   b. 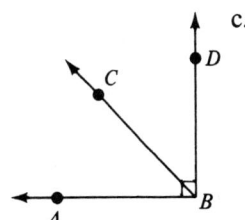
   c.

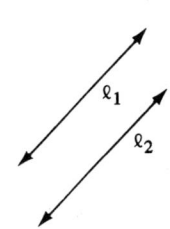

7. Name the polygons shown. Be as precise as possible.

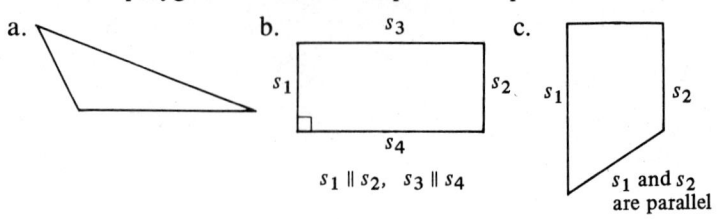

   a.   b. $s_1 \parallel s_2$, $s_3 \parallel s_4$ c. s_1 and s_2 are parallel

8. Name the polygons shown. Be as precise as possible.

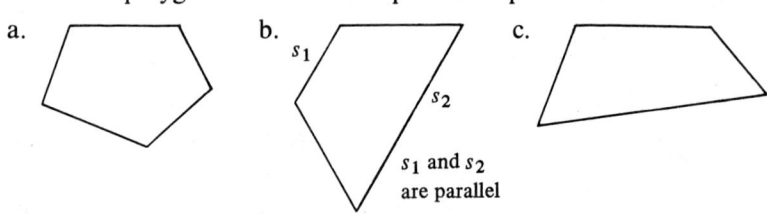

   a.   b. s_1 and s_2 are parallel   c.

9. Name the polygons shown. Be as precise as possible.

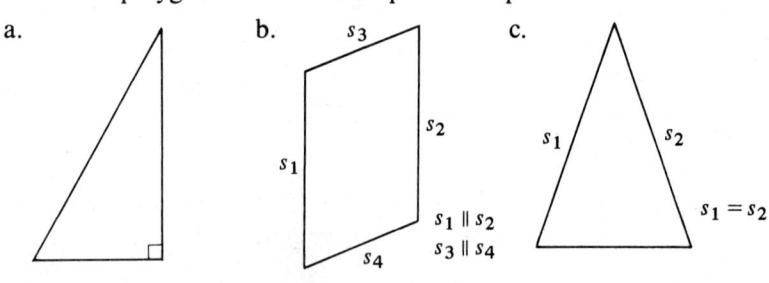

   a.   b. $s_1 \parallel s_2$, $s_3 \parallel s_4$ c. $s_1 = s_2$

10. Name the polygons shown. Be as precise as possible.

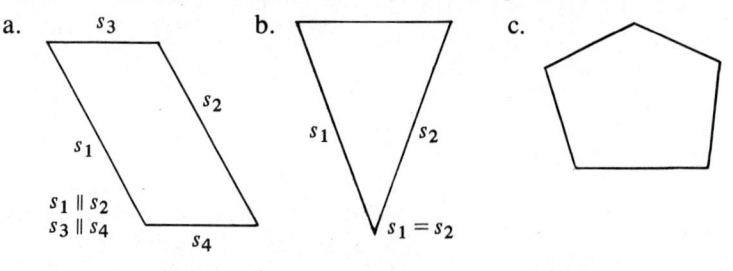

    a. $s_1 \parallel s_2$, $s_3 \parallel s_4$ b. $s_1 = s_2$ c.

11. Answer true or false.
    a. Congruent polygons are similar.
    b. A rectangle is a square.

12. Answer true or false.
    a. A regular polygon has equal sides.
    b. Similar polygons are congruent.

13. Assume that triangle ABC is congruent to triangle DEF (see Figure 3.15). Identify the corresponding angles and corresponding sides.

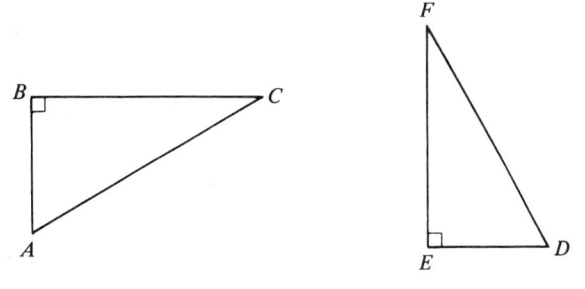

FIGURE 3.15

14. Assume that triangle RST is congruent to triangle XYZ (see Figure 3.16). Identify the corresponding angles and corresponding sides.

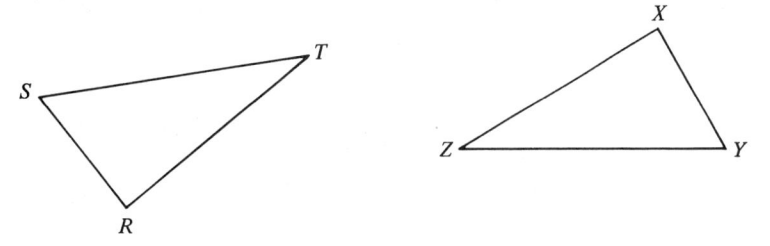

FIGURE 3.16

15. Consider a regular hexagon with a side 4 in. long and a regular hexagon with a side 5 in. long. Are these figures congruent? Are they similar?

16. Consider a regular pentagon with a side 5 cm long and a regular hexagon with a side 5 cm long. Are these figures congruent? Are they similar?

SECTION 3.2 MEASURES ASSOCIATED WITH POLYGONS

The **perimeter** of a polygon is the sum of the lengths of the sides of the polygon. If we want to find the perimeter of a regular polygon, we can multiply the length of one side by the number of sides. If we want to find the perimeter P of a rectangle, we can use the formula

$$P = 2w + 2l$$

where w is the width of the rectangle and l is the length. In some problems we may need to use given information to determine the lengths of sides whose lengths are not already specified. We can then proceed to find the perimeter. In other problems we may be given the perimeter and some additional information and we will have to find the lengths of the sides.

EXAMPLE Find the perimeter P of the polygon in Figure 3.17. All angles are right angles.

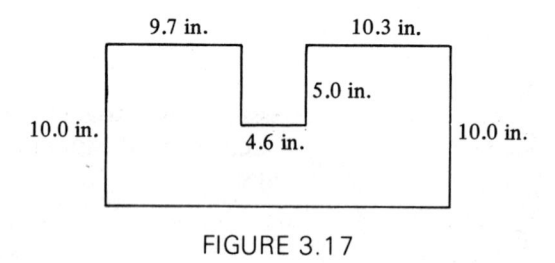

FIGURE 3.17

SOLUTION Figure 3.18 has the length of each side labeled.

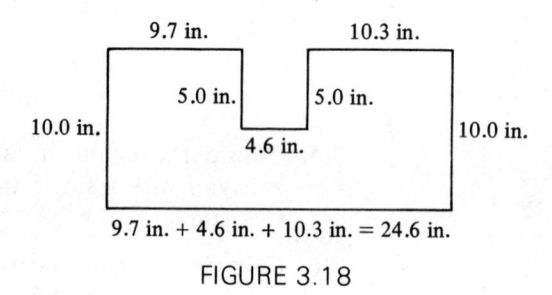

9.7 in. + 4.6 in. + 10.3 in. = 24.6 in.

FIGURE 3.18

$P = 10.0$ in. $+ 9.7$ in. $+ 5.0$ in. $+ 4.6$ in. $+ 5.0$ in. $+ 10.3$ in.
$+ 10.0$ in. $+ 24.6$ in.
$= 79.2$ in.

EXAMPLE Find the perimeter of a regular octagon with a side 30 cm long.

SOLUTION A regular octagon has eight equal sides. So

$$P = 8 \times 30 \text{ cm} = 240 \text{ cm}$$

MEASURES ASSOCIATED WITH POLYGONS 3.2

EXAMPLE Find the number of yards of fencing needed to enclose a rectangular swimming pool whose dimensions are 24 ft × 50 ft. Assume that the distance between the pool and the fence is to be 10 ft and that the fence is to be rectangular.

SOLUTION A sketch of the pool and fence is shown in Figure 3.19.

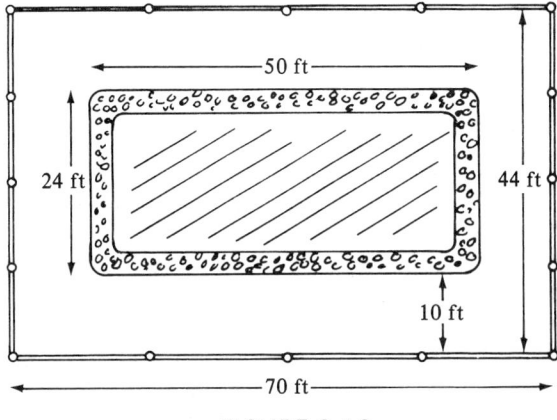

$$P = 2w + 2l$$
$$= 2(44 \text{ ft}) + 2(70 \text{ ft})$$
$$= 88 \text{ ft} + 140 \text{ ft}$$
$$= 228 \text{ ft}$$

FIGURE 3.19

Note that, although we have found the perimeter to be 228 ft, we have not completed the problem yet. We were asked to find the number of yards of fencing needed.

$$228 \text{ ft} = 228 \text{ ft} \left(\frac{1 \text{ yd}}{3 \text{ ft}}\right) = 76 \text{ yd}$$

EXAMPLE The perimeter of an isosceles triangle is 1 m. If the length of the unequal side is 40 cm, find the length of the equal sides.

SOLUTION Note that different units are used in the given measurements. We must either convert the perimeter to centimeters or the length of the given side to meters. We choose to convert the perimeter to centimeters as follows:

$$P = 1 \text{ m} = 100 \text{ cm}$$

A sketch may help to solve the problem. See Figure 3.20. We label the unknown sides with the same letter since they are equal.

The sum of the three sides is 100 cm. Since the third side is 40 cm, the sum of the two equal sides is 60 cm. Then each of the equal sides has length

$$x = \frac{60 \text{ cm}}{2} = 30 \text{ cm}$$

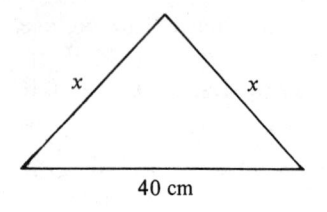

FIGURE 3.20

PERFORMANCE OBJECTIVE 4	
You should now be able to calculate the perimeter of polygons.	Since ABCD is a parallelogram, $P = 2(15) + 2(27)$ $= 30 + 54$ $= 84$

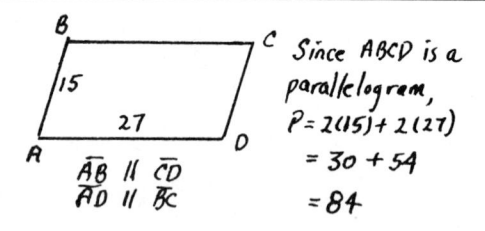

$\overline{AB} \parallel \overline{CD}$
$\overline{AD} \parallel \overline{BC}$

We need to be able to calculate the areas of regions enclosed by polygons. We will emphasize the formulas for areas of triangles and special quadrilaterals such as rectangles, parallelograms, and trapezoids. It will often be the case that the areas for other geometric figures can be found by dividing the regions into triangles and special quadrilaterals whose areas can easily be found.

For each geometric figure considered we will give a sketch of the figure, the area formula used to calculate its area, and examples that illustrate the use of the area formula. Recall that units involved in areas are square units.

The first formula is for the area of a triangle.

Name	Geometric figure	Area formula
triangle	(triangle with height h and base b)	$A = \frac{1}{2}bh$

Note that if the triangle is a right triangle, then the height is one of the legs and the base is the other leg.

EXAMPLE Find the area of a triangle with height 8.32 in. and base 11.5 in.

SOLUTION

$A = \frac{1}{2}bh$

$ = 0.5(11.5)(8.32)$

$ = 47.8 \text{ in.}^2$

Note that the answer is given to three significant digits.

EXAMPLE Find the area of a right triangle with perpendicular sides of length 6.0 cm and 8.0 cm.

SOLUTION

$A = 0.5(6.0)(8.0)$

$ = 24 \text{ cm}^2$

EXAMPLE Find the height of a triangle with base 5.0 ft and area 19 ft².

SOLUTION The area formula $A = 0.5bh$ can be rearranged to find the height h. Since $A = 0.5bh$, we can find h by dividing A by $0.5b$. That is,

$$h = \frac{A}{0.5b}$$

$$ = \frac{19}{0.5(5.0)}$$

$$ = 7.6 \text{ ft}$$

To evaluate the expression

$$\frac{19}{0.5(5.0)}$$

on a calculator, we may want to use the store $\boxed{\text{STO}}$ and recall $\boxed{\text{RCL}}$ keys. We would key in the problem as follows:

$$0.5 \; \boxed{\times} \; 5.0 \; \boxed{=} \; \boxed{\text{STO}} \; 19 \; \boxed{\div} \; \boxed{\text{RCL}} \; \boxed{=}$$

The calculator solution is also 7.6 ft. By using the $\boxed{\text{STO}}$ and $\boxed{\text{RCL}}$ keys we are able to perform the computation in the denominator first and then store the result until it is needed. We recall it from the calculator's memory as the divisor of 19.

3 APPLIED GEOMETRY

EXAMPLE Find the area of an isosceles triangle with base 6.0 cm and equal sides of length 5.0 cm.

SOLUTION A sketch of the triangle is given in Figure 3.21.
We need to find the height h of the triangle. The Pythagorean Theorem enables us to solve for h since we have a right triangle whose hypotenuse is 5.0 cm and whose legs are 3.0 cm and h.

We use the Pythagorean Theorem to find h as follows:

$$h^2 + (3.0)^2 = (5.0)^2$$
$$h^2 = (5.0)^2 - (3.0)^2$$
$$h = \sqrt{(5.0)^2 - (3.0)^2}$$
$$= \sqrt{25 - 9} = \sqrt{16}$$
$$= 4.0 \text{ cm}$$

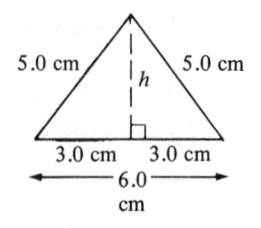

FIGURE 3.21

To solve this problem on the calculator follow the sequence

$$5 \boxed{x^2} \boxed{-} 3 \boxed{x^2} \boxed{=} \boxed{\text{INV}} \boxed{x^2}$$

which yields the result 4. So to take a square root on the calculator we key in $\boxed{\text{INV}}\ \boxed{x^2}$. This is because the square root is the inverse operation of squaring.

Now we will state the formula for the area of a rectangle.

Name	Geometric figure	Area formula
rectangle	(rectangle with Width (w) and Length (l))	$A = lw$

EXAMPLE Calculate the area of a template with dimensions as shown in Figure 3.22.

SOLUTION The template can be represented as two rectangles as shown in Figure 3.23.

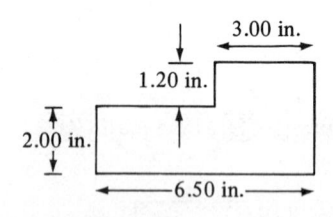

FIGURE 3.22

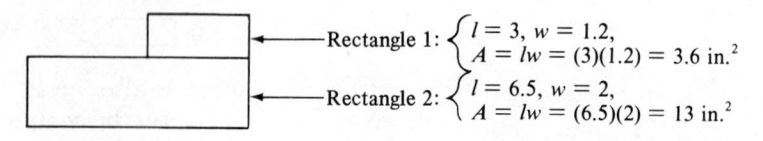

Rectangle 1: $\begin{cases} l = 3, w = 1.2, \\ A = lw = (3)(1.2) = 3.6 \text{ in.}^2 \end{cases}$

Rectangle 2: $\begin{cases} l = 6.5, w = 2, \\ A = lw = (6.5)(2) = 13 \text{ in.}^2 \end{cases}$

FIGURE 3.23

Thus the total area of the template is 16.6 in.^{2}.

The formula for the area of a parallelogram follows.

Name	Geometric figure	Area formula
Parallelogram	(parallelogram with height (h) and base (b))	$A = bh$

EXAMPLE A city lot is laid out in the shape of a parallelogram. The property runs 210 ft back from the street. If there are 26,000 ft^2 in the lot, what is the width of the property? Width here refers to the perpendicular distance from side to side.

SOLUTION We sketch the lot as shown in Figure 3.24. To solve for the width of the lot we need to rearrange the equation $A = bh$. The unknown in this problem is h. We divide both sides of the equation by b to obtain

$$\frac{A}{b} = \frac{bh}{b} = h$$

We substitute to find

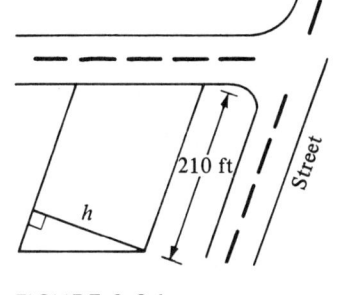

$$h = \frac{A}{b} = \frac{26,000}{210} = 123.81 \quad \text{or} \quad 124 \text{ ft}$$

FIGURE 3.24

Here is the formula for the area of a trapezoid.

Name	Geometric figure	Area formula
trapezoid	(trapezoid with base (b_1), height (h), base (b_2))	$A = \frac{1}{2}(b_1 + b_2)h$

EXAMPLE Calculate the area of a trapezoid with the following dimensions

$$b_1 = 14.5 \text{ in.}, \quad b_2 = 20.3 \text{ in.}, \quad h = 8.90 \text{ in.}$$

SOLUTION

$$A = \frac{1}{2}(b_1 + b_2)h = 0.5(14.5 + 20.3)(8.9) = 0.5(34.8)(8.9)$$
$$= 154.86 = 155 \text{ in.}^2$$

We illustrate a technique for solving this problem on the calculator. If we desire to do the multiplications from left to right as they are written, we will first need to evaluate $b_1 + b_2$ and store that value.

$$14.5 \boxed{+} 20.3 \boxed{=} \boxed{\text{STO}}$$

Then multiply

$$0.5 \boxed{\times} \boxed{\text{RCL}} \boxed{\times} 8.9 \boxed{=}$$

which yields 154.86 as before. Note the recall step by which we were able to enter the stored data value.

EXAMPLE A realtor claims to be selling a 1 acre lot. The lot has a trapezoidal shape as shown in Figure 3.25.

Dimension a is measured as 275 ft

Dimension b (the lot width) is 184 ft

Dimension c is 200 ft

If 1 acre = 43,560 ft^2, is the claim valid?

SOLUTION We recognize the base dimensions as $b_1 = 200$ and $b_2 = 275$. The height h is 184. The area is then

$$A = 0.5(200 + 275)(184) = 0.5(475)(184) = 43,700 \text{ ft}^2$$

The value slightly exceeds 43,560 ft^2, so the claim is valid.

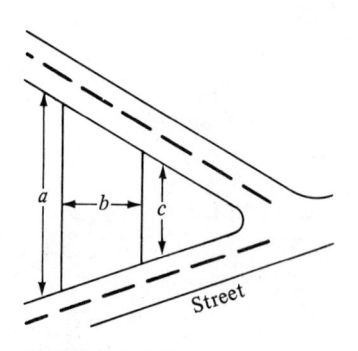

FIGURE 3.25

AREA FORMULAS FOR POLYGONS

triangle	$A = \frac{1}{2}bh$	parallelogram	$A = bh$
rectangle	$A = lw$	trapezoid	$A = \frac{1}{2}(b_1 + b_2)h$

PERFORMANCE OBJECTIVE 5	
You should now be able to calculate the area of polygons.	parallelogram $h = \sqrt{5^2 - 3^2}$ $= \sqrt{25 - 9}$ $= \sqrt{16} = 4$ $A = bh = (7)(4)$ $= 28$ square units

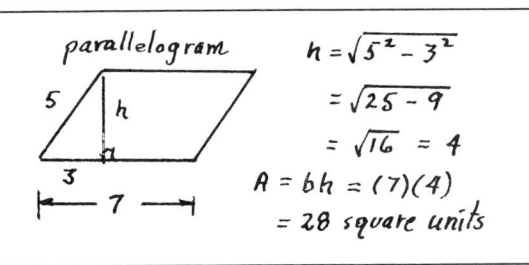

For any polygon with n sides, the sum of the measures of the angles is $(n - 2)180°$. Since the n angles in a regular polygon are equal, the measure of each angle in a regular polygon of n sides is

$$\frac{(n - 2)180°}{n}$$

EXAMPLE Find the sum of the measures of the angles in a triangle, a quadrilateral, and a pentagon.

SOLUTION

Figure	Number of sides	Sum of angles
triangle	3	$(3 - 2)180° = 1(180°)$ $= 180°$
quadrilateral	4	$(4 - 2)180° = 2(180°)$ $= 360°$
pentagon	5	$(5 - 2)180° = 3(180°)$ $= 540°$

EXAMPLE Find the measure of each angle in a regular octagon.

SOLUTION A regular octagon has eight equal angles. The measure of each angle is

$$\frac{(8 - 2)180°}{8} = \frac{6(180°)}{8} = 135°$$

PERFORMANCE OBJECTIVE 6	regular hexagon:
You should now be able to perform calculations involving the measures of the angles of polygons.	sum of angles = (6−2) 180° = 4(180°) = 720° measure of each angle = 720°/6 = 120°

We will now use the similarity of polygons to solve for unknown measures of the polygons. The measures given in the similarity problems should be interpreted as exact numbers.

EXAMPLE Assume that the quadrilaterals in Figure 3.26 are similar and that $\overline{AD}$ and $\overline{EH}$ are corresponding sides. Find the unknown sides of the quadrilaterals. They are labeled x, y, and z.

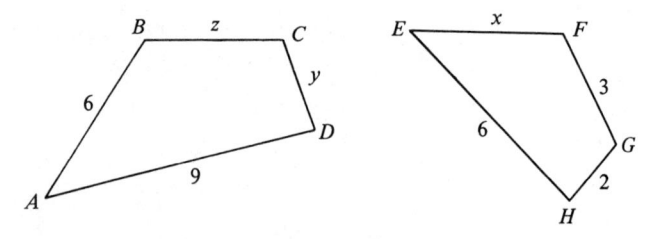

FIGURE 3.26

SOLUTION Since ratios of corresponding sides must be equal, we can set up the following proportions:

$$\frac{6}{x} = \frac{9}{6}, \quad \frac{y}{2} = \frac{9}{6}, \text{ and } \frac{z}{3} = \frac{9}{6}$$

Solving these proportions gives x, y, and z. Recall that we can solve proportions by setting the product of the means equal to the product of the extremes.

$$9x = 6(6) \qquad 6y = 2(9) \qquad 6z = 3(9)$$
$$9x = 36 \qquad 6y = 18 \qquad 6z = 27$$
$$x = 4 \qquad y = 3 \qquad z = \frac{27}{6} = 4.5$$

Consider the perimeters of the two quadrilaterals in the previous example. The perimeter of the first is

$$6 + 9 + 3 + 4.5 = 22.5$$

The perimeter of the second is

$$4 + 6 + 2 + 3 = 15$$

If we form the ratio of the first perimeter to the second perimeter, we obtain

$$\frac{22.5}{15}$$

which is equal to $\frac{9}{6}$, the ratio of corresponding sides.

We have the following general result.

If two polygons are similar, then the ratio of their perimeters is equal to the ratio of any two corresponding sides.

EXAMPLE Find the perimeter of the second triangle in Figure 3.27 if the given triangles are similar and $\overline{AB}$ and $\overline{DE}$ are corresponding sides.

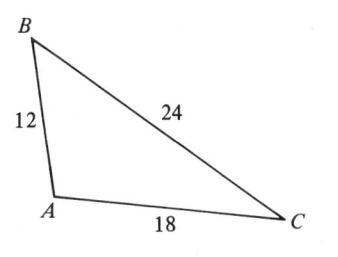

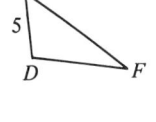

FIGURE 3.27

SOLUTION We could just use similarity to find sides $\overline{EF}$ and $\overline{DF}$ and then add the three sides to get the perimeter. However, we can eliminate a step if we use the fact that the ratio of the perimeters equals the ratio of the corresponding sides.

Let P represent the perimeter of triangle DEF. Note that the perimeter of triangle ABC is

$$12 + 24 + 18 = 54$$

We have the proportion

$$\frac{P}{54} = \frac{5}{12}$$

which we solve as follows:

$$12P = 5(54)$$
$$12P = 270$$
$$P = \frac{270}{12} = 22.5$$

So the perimeter of the second triangle is 22.5.

Another useful result about similar polygons is as follows:

The ratio of the areas of similar polygons is equal to the ratio of the squares of any two corresponding sides.

EXAMPLE Find the area of a regular hexagon with sides of length 5 in. if a similar regular hexagon has sides of length 8 in. and an area of 83.14 in.^{2}.

SOLUTION Let A represent the area being sought. Then

$$\frac{A}{83.14} = \frac{5^2}{8^2}$$

That is,

$$\frac{A}{83.14} = \frac{25}{64}$$

So

$$64A = 25(83.14)$$
$$64A = 2078.5$$
$$A = 32.48 \text{ in.}^2$$

EXAMPLE Two rooms have floors that are similar rectangles. The number of square yards of wall to wall carpeting needed for the smaller room is $\frac{9}{16}$ of that needed for the larger room. The dimensions of the smaller room are 9 ft × 12 ft. Find the dimensions of the larger room.

SOLUTION The ratio of the areas of the rooms is $\frac{9}{16}$. This ratio must equal the ratio of the squares of the room widths. If we call the width of the larger room x, we have the proportion

$$\frac{9}{16} = \frac{9^2}{x^2} \quad \text{or} \quad \frac{9}{16} = \frac{81}{x^2}$$

Then

$$9x^2 = 16(81)$$
$$9x^2 = 1296$$
$$x^2 = 144$$
$$x = \sqrt{144}$$
$$x = 12$$

Likewise if we call the length of the larger room y, we have the proportion

$$\frac{9}{16} = \frac{12^2}{y^2} \quad \text{or} \quad \frac{9}{16} = \frac{144}{y^2}$$

Then

$$9y^2 = 16(144)$$
$$9y^2 = 2304$$
$$y^2 = 256$$
$$y = \sqrt{256}$$
$$y = 16$$

So the dimensions of the larger room are 12 ft. × 16 ft.

PERFORMANCE OBJECTIVE 7

You should now be able to solve for unknown parts of polygons.

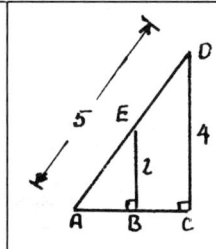

Triangles ABE and ACD are similar.

So $\frac{AE}{5} = \frac{2}{4}$

$AE = \frac{2(5)}{4} = 2.5$

EXERCISE SET 3.2

1. Calculate the perimeter of a regular pentagon of side 5.2 in.

2. Find the length of each side of a regular octagon with a perimeter of 14.4 cm.

3. A regular hexagonal bolt head is to be recessed into a metal plate. Each side of the hexagonal head is 0.25 in. long. What is the minimum diameter of a hole into which the bolt can be recessed?

   Hint: A hexagon is composed of 6 equilateral (equal-sided) triangles.

4. A square-headed bolt is to be recessed into a metal plate. Each side of the square head is 1 cm long. What is the minimum diameter of a hole into which the bolt can be recessed?

5. An ocean front restaurant in the shape of a rectangle is to have a deck on three sides as shown in Figure 3.28. How many square feet of deck must be built?

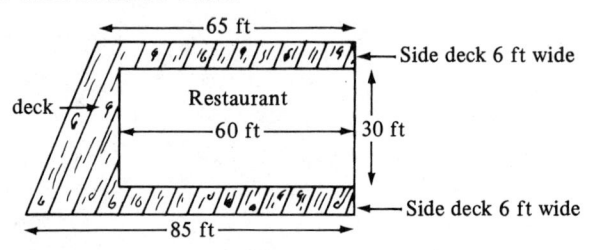

FIGURE 3.28

6. Using Figure 3.28, and assuming the restaurant is 40 ft long, how many square feet of deck must be built?

7. An enclosed rectangular yard (see Figure 3.29) is to be sodded except for a triangular area on one end and a 4 ft border on the other three sides. These edges are to be planted with shrubs and flowers. How many square feet of sod should be purchased?

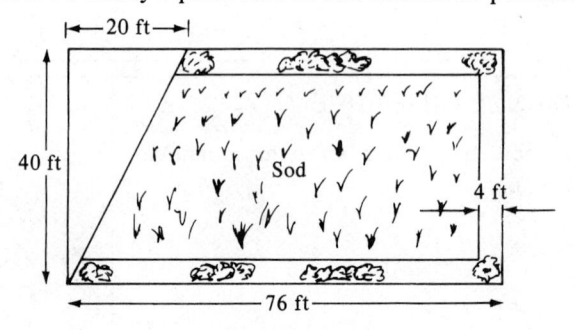

FIGURE 3.29

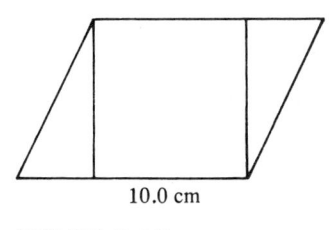

FIGURE 3.30

8. Consider the parallelogram in Figure 3.30. The shorter side of the parallelogram is 10.0 cm. The figure contained within the parallelogram is a square with area 64.0 cm^2. Find the area of the parallelogram.

9. A surveyor is setting out stakes for a foundation in the shape of a regular hexagon. What angle will he measure between adjacent sides?

10. Calculate the sixth angle of a hexagon when the other angles are 150°, 120°, 115°, 90°, and 105°.

11. Two similar triangles have corresponding sides of lengths 8 in. and 21 in. If the perimeter of the smaller triangle is 36 in., what is the perimeter of the larger triangle?

12. Two similar rectangular plots of land have corresponding widths of 4 m and 5 m. If the perimeter of the larger plot is 35 m, what are the lengths of the two rectangular plots?

13. Two square gardens are to be dug and the area of one is to be four times the area of the other. If the smaller garden is 10 ft on a side, what should be the dimensions of the larger garden? (Note that any two squares are similar.)

14. Two similar right triangles have corresponding legs 8 cm and 10 cm. If the other leg of the smaller triangle is 6 cm, find the area of the two triangles.

SECTION 3.3 MEASURES ASSOCIATED WITH CIRCLES

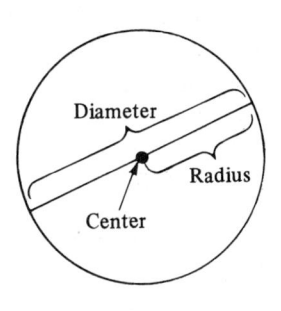

FIGURE 3.31

Other than a polygon, the most common plane figure that encloses a region is a circle. A **circle** consists of all the points that are a given distance from a fixed point. The fixed point is called the **center** of the circle. The distance between the center and the points on the circle is called the **radius** of the circle.

The line segment that connects two points on a circle and passes through the center is called a **diameter** of the circle. The length of the diameter of a circle is twice the radius. See Figure 3.31.

The line segment joining any two points on a circle is called a **chord** of the circle. We see that diameters are special chords. A line that touches a circle at exactly one point is called a **tangent** of the circle. Note that, except for the point at which it touches the circle,

112 3 APPLIED GEOMETRY

the tangent line must lie outside of the circle or it would, when extended, cross the circle in two points. See Figure 3.32.

A circle is said to be **inscribed** in a polygon if all the sides of the polygon are tangents of the circle. We also say that the polygon is **circumscribed** about the circle. Circle *O* is inscribed in quadrilateral *ABCD* in Figure 3.33. Or we could say that quadrilateral *ABCD* is circumscribed about circle *O*.

A circle is said to be circumscribed about a polygon if the polygon is contained within the circle and each vertex of the polygon lies on the circle, that is, if the sides of the polygon are chords of the circle. We also say that the polygon is inscribed in the circle. See Figure 3.34 in which circle *O* is circumscribed about quadrilateral *ABCD*.

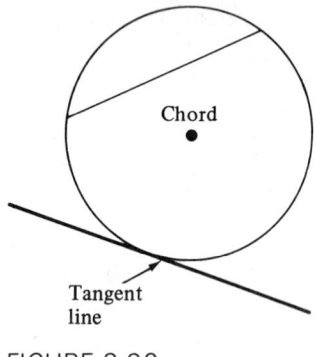

FIGURE 3.32

FIGURE 3.33

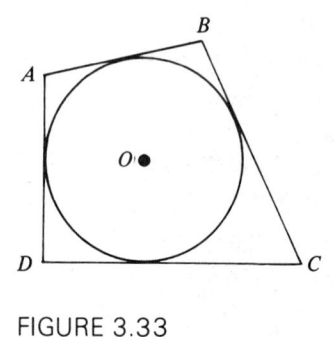

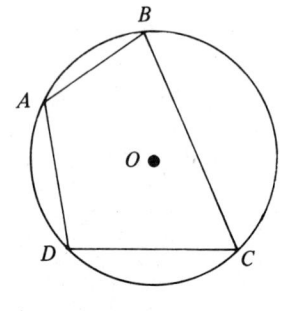

FIGURE 3.34

A circle may be circumscribed about or inscribed in any regular polygon.

PERFORMANCE OBJECTIVE 8	
You should now be able to recognize the terminology associated with a circle.	$\overline{OB}$ is a radius. $\overline{AD}$ is a diameter. $\overline{AB}$ is a chord. ABCDEF is an inscribed hexagon.

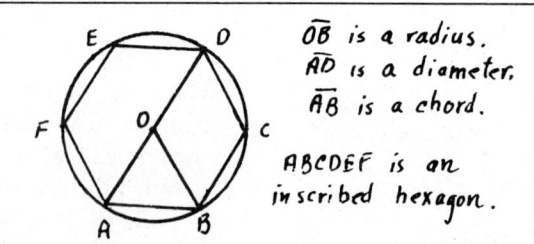

The term *perimeter* for polygons corresponds to the term *circumference* for circles. The **circumference of a circle** is the distance around the circle, that is, the length of the curved line that forms the circle. The formula for the circumference C of a circle is

$$C = 2\pi r$$

where r is the radius of the circle and $\pi \approx 3.14$ ($\approx$ means "approximately equal to"). Since the diameter d of a circle is equal to $2r$, the circumference formula can also be written as $C = \pi d$.

Note: The $\boxed{\pi}$ key on a calculator can be useful for performing calculations involving circles.

EXAMPLE Find the circumference of a circle whose radius is 0.50 in.

SOLUTION

$C = 2\pi r$
$ = 2(3.14)(0.50 \text{ in.})$
$ = 3.14 \text{ in.} \quad \text{or} \quad 3.1 \text{ in. (to two significant digits)}$

EXAMPLE A gardener has 40 ft of decorative wire fence. How large a circular flower garden should he design in order that it may be enclosed by the 40 ft of fence?

SOLUTION Since $C = \pi d$, the diameter of the circular garden is given by

$$d = \frac{C}{\pi} = \frac{40 \text{ ft}}{3.14} = 12.7 \text{ ft} = 12 \text{ ft}$$

This is a case when we would not want to round to 13 ft.

EXAMPLE Find the circumference of a circle that is inscribed in a regular hexagon with 4.0 in. sides.

SOLUTION To find the circumference of the circle, we first need to find the radius r. We can do this by using two facts about a circle inscribed in a regular hexagon:

1. The lines from the center of the circle to the vertices form equilateral triangles, that is, triangles with three equal sides.

2. A radius drawn to a point where the circle is tangent to a side is perpendicular to that side and divides the side into two equal segments.

A sketch that illustrates these facts is given in Figure 3.35.

We can restrict our attention to the right triangle in Figure 3.36 to solve for r.

By the Pythagorean Theorem, we have

$$r^2 + 2.0^2 = 4.0^2$$
$$r^2 + 4.0 = 16$$
$$r^2 = 12$$
$$r = \sqrt{12} = 3.5 \text{ in.} \quad \text{(to two significant digits)}$$

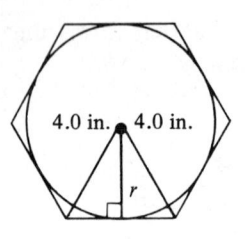

FIGURE 3.35

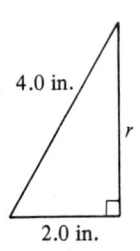

FIGURE 3.36

Then

$$C = 2\pi r$$
$$= 2(3.14)(3.5)$$
$$= 22 \text{ in.} \quad \text{(to two significant digits)}$$

PERFORMANCE OBJECTIVE 9

You should now be able to calculate the circumference of circles.

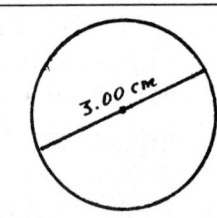

$$C = \pi d$$
$$= 3.14(3.00)$$
$$= 9.42 \text{ cm}$$

The **area of a circle** of radius r is given by the formula

$$A = \pi r^2$$

EXAMPLE Find the area of a circle whose diameter is 5.0 ft. Use $\pi = 3.14$.

SOLUTION The radius of the circle is half the diameter or 2.5 feet. So

$$A = \pi r^2$$
$$= 3.14(2.5)^2$$
$$= 20 \text{ ft}^2 \quad \text{(to two significant digits)}$$

MEASURES ASSOCIATED WITH CIRCLES 3.3

EXAMPLE A circular high-rise dormitory is to have 5000 ft² of space on each floor. What will be the radius of the building?

SOLUTION To solve this problem we use the equation $A = \pi r^2$. However, the unknown is r not A. We must rearrange this equation to solve for r. First divide by π to obtain

$$r^2 = \frac{A}{\pi}$$

Then take the square root of both sides to obtain

$$r = \sqrt{\frac{A}{\pi}} = \sqrt{\frac{5000}{\pi}} = 39.89 \text{ ft}$$

The calculator sequence for solving this equation is

$$5000 \boxed{\div} \boxed{\pi} \boxed{=} \boxed{\text{INV}} \boxed{x^2}$$

(*Note:* Only one data value needs to be entered if you have a $\boxed{\pi}$ key.)

EXAMPLE Find the area of a circle that is circumscribed about a square whose sides are 2.0 in.

SOLUTION The sketch in Figure 3.37 includes a right triangle that will enable us to solve for r. By the Pythagorean Theorem we have

$$r^2 = 1.0^2 + 1.0^2$$
$$r^2 = 1.0 + 1.0$$
$$r^2 = 2.0$$
$$r = \sqrt{2.0} = 1.4 \text{ in.}$$

Then

$$A = \pi r^2$$
$$= (3.14)(1.4)^2$$
$$= 6.2 \text{ in.}^2 \quad \text{(to two significant digits)}$$

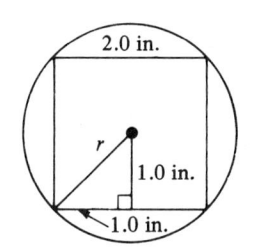

FIGURE 3.37

FORMULAS FOR CIRCLES

circumference $\quad C = 2\pi r$
area $\quad A = \pi r^2$

3 APPLIED GEOMETRY

PERFORMANCE OBJECTIVE 10

You should now be able to calculate the area of circles.

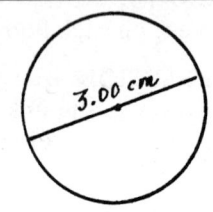

$A = \pi r^2$
$= 3.14 (1.5)^2$
$= 7.07$ sq. cm

(circle with diameter 3.00 cm)

EXERCISE SET 3.3

1. Solve for the circumference of the following circles and express your answer with the correct number of significant digits and with correct units.
    a. radius = 6.4 ft
    b. diameter = 10 mm
    c. radius = 5.1×10^3 m

2. Solve for the circumference of the following circles and express your answer with the correct number of significant digits and with correct units.
    a. radius = 12.8 cm
    b. diameter = 2.1 m
    c. radius = 1.25×10^2 in.

3. Calculate the area of the following circles. Express your answer with the correct number of significant digits and with appropriate units.
    a. radius = 5.62 in.
    b. radius = 1.3×10^{-3} m
    c. diameter = 30 ft

4. Calculate the area of the following circles. Express your answer with the correct number of significant digits and with appropriate units.
    a. radius = 18.9 cm
    b. radius = 0.76×10^3 m
    c. diameter = 24 in.

5. Calculate the radius of the following circles:
    a. A circle with area 5.2×10^{-4} m^2
    b. A circle with area 11,000 ft^2
    c. A circle with circumference 70 ft
    d. A circle with circumference 2.5×10^7 m

6. Calculate the radius of the following circles:
   a. A circle with area 48.6 cm^2
   b. A circle with area 33,000 ft^2
   c. A circle with circumference 26.0 cm
   d. A circle with circumference 8.67×10^4 m

7. Find the circumference of a circle inscribed in a square whose sides are 12 in.

8. Find the area of a circle inscribed in a square whose area is 49 cm^2.

9. Find the area of a circle circumscribed about a regular hexagon whose sides are 16.0 in. (Recall that the lines from the center of the circle to the vertices of the hexagon form equilateral triangles.)

10. Find the circumference of a circle circumscribed about a regular hexagon whose perimeter is 33.6 m.

SECTION 3.4   MEASURES ASSOCIATED WITH SOLIDS

A three-dimensional geometric figure is called a **solid.** Some specific solids that we will consider are prisms, circular cylinders, circular cones, and spheres.

A **prism** is a solid bounded by polygons, two of which are identical and parallel polygons, and the remainder of which are parallelograms. The identical and parallel polygons are called the **bases** of the prism and the parallelograms are called the **sides** of the prism. The height of the prism is the perpendicular distance between the parallel bases. If the sides of a prism are perpendicular to the bases, then the sides are rectangles and we call the prism a **right prism.** The height of a right prism is equal to the length of the rectangular sides. Some examples of prisms are sketched in Figure 3.38 where the identical and parallel polygons are shaded.

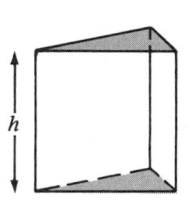

Triangular bases   Hexagonal bases   Rectangular bases

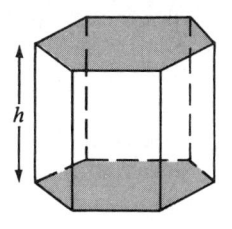

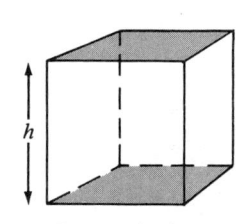

FIGURE 3.38

A **circular cylinder** (or just cylinder) is a solid consisting of two identical and parallel circles as bases and a curved surface joining the bases. If the curved surface is perpendicular to the bases, we call the cylinder a **right cylinder.** The height of a cylinder is the perpendicular distance between the bases. Sketches of cylinders are shown in Figure 3.39.

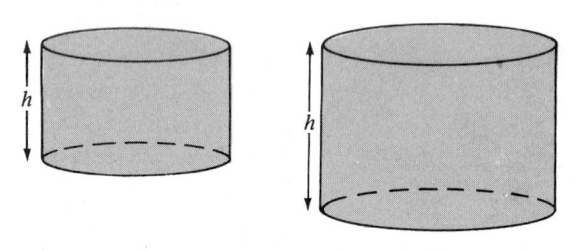

FIGURE 3.39

A **circular cone** (or just cone) is a solid consisting of a circular base and a curved surface that comes to a point called the **vertex** of the cone. The height of the cone is the perpendicular distance from the vertex to the base. If the perpendicular distance from the vertex to the base meets the base at the center of the circle, then the cone is a **right circular cone.** Sketches of cones are shown in Figure 3.40.

Circular bases

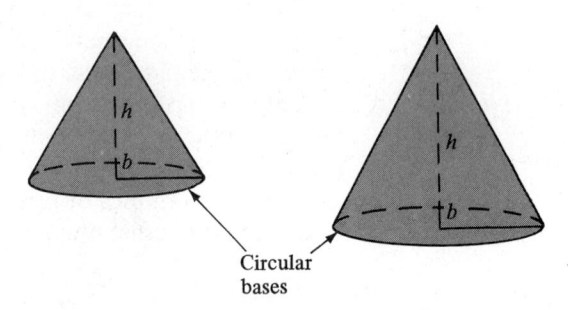

FIGURE 3.40

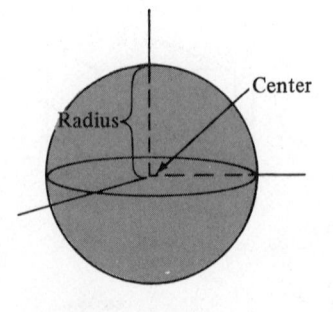

FIGURE 3.41

A **sphere** is a solid whose surface consists of all the points that are a given distance from a fixed point. The fixed point is called the **center** of the sphere and the distance between the surface and the center is called the **radius.** See Figure 3.41.

In calculating surface areas and volumes in the remainder of this section, we will restrict our attention to right prisms, right circular cylinders, right circular cones, and spheres.

MEASURES ASSOCIATED WITH SOLIDS 3.4

PERFORMANCE OBJECTIVE 11

You should now be able to identify various geometric solids.

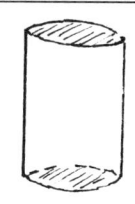

right circular cylinder with circular bases shaded

The **surface area of a right prism** is the sum of the areas of the bases and the sides. The units involved in surface area are square units.

EXAMPLE Find the surface area of the right prism in Figure 3.42.

SOLUTION Each base is a right triangle with legs 4.0 cm and 3.0 cm. We can find the hypotenuse of the right triangle by using the Pythagorean Theorem. We do this as follows:

$$c^2 = (4.0)^2 + (3.0)^2$$
$$= 16 + 9 = 25$$
$$c = \sqrt{25}$$
$$= 5.0$$

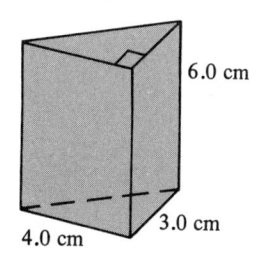

FIGURE 3.42

The area of each base is

$$A = 0.5(3.0)(4.0) = 6.0 \text{ cm}^2$$

The area of the three rectangular sides are

$$(4.0)(6.0) = 24 \text{ cm}^2$$
$$(3.0)(6.0) = 18 \text{ cm}^2$$
$$(5.0)(6.0) = 30 \text{ cm}^2$$

The surface area A_s of the prism is

$$2(6.0) + 24 + 18 + 30 = 12 + 24 + 18 + 30$$
$$= 84 \text{ cm}^2$$

The **surface area of a right circular cylinder** is given by the formula

$$A_s = 2A_b + C_b h$$

where A_b is the area of the circular base, C_b is the circumference of the base, and h is the height of the cylinder. See Figure 3.43. $A_b = \pi r^2$ and $C_b = 2\pi r$ since the base is a circle. A more useful expression for surface area may be

$$A_s = 2\pi r^2 + 2\pi rh$$

where r is the radius of the base and h is the height of the cylinder.

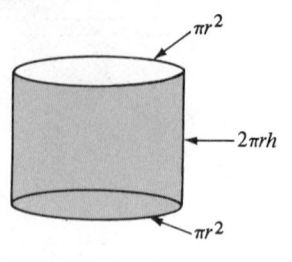

FIGURE 3.43

EXAMPLE Find the surface area of a right cylinder with radius 1.25 m and height 3.50 m.

SOLUTION

$$\begin{aligned} A_s &= 2\pi r^2 + 2\pi rh \\ &= 2(3.14)(1.25)^2 + 2(3.14)(1.25)(3.50) \\ &= 37.2875 \\ &= 37.3 \text{ m}^2 \quad \text{(to three significant digits)} \end{aligned}$$

Calculator solution:

2 $\boxed{\times}$ $\boxed{\pi}$ $\boxed{\times}$ 1.25 $\boxed{x^2}$ $\boxed{+}$ 2 $\boxed{\times}$ $\boxed{\pi}$ $\boxed{\times}$ 1.25 $\boxed{\times}$ 3.5 $\boxed{=}$

yields 37.306413, which is 37.3 m^2 to three significant digits.

The **surface area A_s of a sphere** is given by

$$A_s = \pi d^2$$

where d is the diameter of the sphere. Since $d = 2r$, we can substitute to obtain the equivalent formula

$$\begin{aligned} A_s &= \pi(2r)^2 \\ &= \pi(2r)(2r) \\ &= 4\pi r^2 \end{aligned}$$

where r is the radius of the sphere. See Figure 3.44.

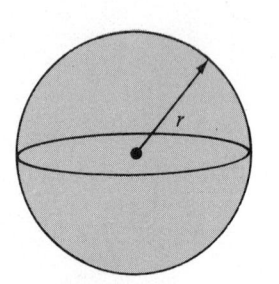

FIGURE 3.44

EXAMPLE The surface area of a spherical water tank is to be painted. The radius of the tank is 15 ft. How many gallons of paint should be purchased if each gallon will cover 300 ft^2?

SOLUTION

$A_s = 4\pi r^2 = 4(3.14)(15)^2$
$= 2826$ or $2\underline{8}00$ ft^2

To find the number of gallons of paint divide 2800 by 300.

$$\frac{2800}{300} = 9.3 \text{ gal}$$

We should buy 10 gal. Here is another example of when we violate the normal rules of rounding.

The **surface area of a right circular cone** is given by the formula

$$A_s = A_b + \frac{1}{2} C_b \sqrt{r^2 + h^2}$$

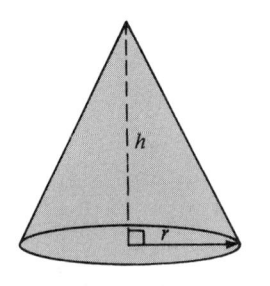

where A_b is the area of the circular base and C_b is the circumference of the base. See Figure 3.45.

Since $A_b = \pi r^2$ and $C_b = 2\pi r$, we can substitute these expressions into A_s to get

$$A_s = \pi r^2 + \frac{1}{2} \cdot 2\pi r \sqrt{r^2 + h^2}$$
$$A_s = \pi r^2 + \pi r \sqrt{r^2 + h^2}$$

FIGURE 3.45

This last formula is the more useful form for the surface area of a right circular cone.

EXAMPLE Find the surface area of a right circular cone with radius 12.5 cm and height 16 cm.

SOLUTION

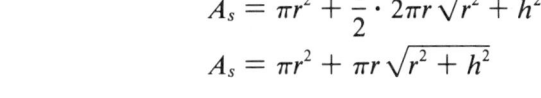

$A_s = \pi r^2 + \pi r \sqrt{r^2 + h^2}$
$= 3.14(12.5)^2 + 3.14(12.5)\sqrt{(12.5)^2 + (16)^2}$
$= 1287.5547$
$= 1\underline{3}00$ cm^2 (to two significant digits)

Calculator solution:

12.5 [x^2] [+] 16 [x^2] [=] [INV] [x^2] [×]
[π] [×] 12.5 [=] [+] [π] [×] 12.5 [x^2] [=]

yields 1288.2077, which is $1\underline{3}00$ cm^2 to two significant digits. Note that in the calculator solution we evaluated the square root first and then multiplied it by πr.

SURFACE AREA FORMULAS FOR SOLIDS

right circular cylinder $\quad A_s = 2\pi r^2 + 2\pi rh$

right circular cone $\quad A_s = \pi r^2 + \pi r \sqrt{r^2 + h^2}$

sphere $\quad A_s = 4\pi r^2$

PERFORMANCE OBJECTIVE 12

You should now be able to calculate the surface areas of specified solids.

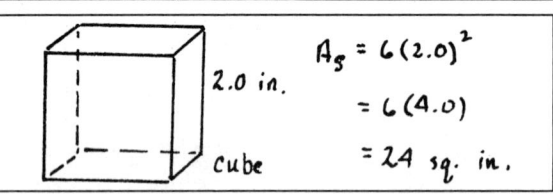

Cube, 2.0 in.

$A_s = 6(2.0)^2$
$= 6(4.0)$
$= 24$ sq. in.

The **volume V of a right prism** is the product of the area of the base A_b and the height h of the prism. That is,

$$V = A_b h$$

The units involved in volumes are cubic units.

EXAMPLE Find the volume of the right prism in Figure 3.42.

SOLUTION In a previous example we found the surface area of this prism. At that time we calculated the area of the base to be 6.0 cm². So

$$V = A_b h = (6.0)(6.0)$$
$$= 36 \text{ cm}^3$$

EXAMPLE Laundry trays are to be designed to hold 65 gal of water before overflowing. The trays are to have the typical shape of a right prism with a trapezoidal base. Determine how wide (h) the trays should be. See Figure 3.46.

MEASURES ASSOCIATED WITH SOLIDS 3.4

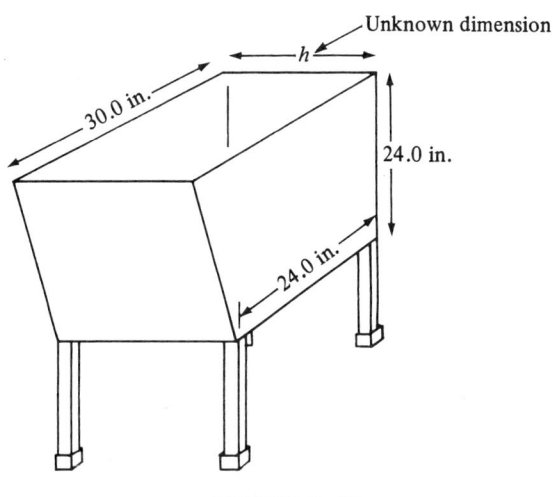

FIGURE 3.46

SOLUTION We first find the area of the base (the side of the tray in this case). Note that 1 gal = 231 in.^{3}.

$$A = \frac{1}{2}(b_1 + b_2)h$$
$$= (0.5)(24 + 30)(24) = 648 \text{ in.}^2$$

Next we find the required volume in cubic inches.

$$65 \text{ gal} = 65 \text{ gal}\left(\frac{231 \text{ in.}^3}{\text{gal}}\right) = 15{,}015 \text{ in.}^3$$

Since $V = A_b h$, we write

$$h = \frac{V}{A_b} = \frac{15{,}015}{648} = 23.171$$
$$= 23.2 \text{ in.} \quad \text{(to three significant digits)}$$

The **volume of a right circular cylinder** of height h is given by the formula

$$V = A_b h$$

where A_b is the area of the circular base or by the equivalent formula

$$V = \pi r^2 h$$

where r is the radius of the base.

EXAMPLE Calculate the number of gallons of water that can be held in a right circular cylinder that has a base of radius 5.0 ft and height of 12.0 ft (1 gal = 0.1337 ft^3).

SOLUTION

$$V = \pi r^2 h = (3.14)(5.0)^2(12.0) = 942 \text{ ft}^3$$

$$= 942 \text{ ft}^3 \left(\frac{1 \text{ gal}}{0.1337 \text{ ft}^3}\right) = 7045.6 \text{ or } 7\underline{0}00 \text{ gal}$$

EXAMPLE A cylindrical tank must fit upright in a room with an 8.0 ft ceiling. What radius should the tank have if it is to hold 2000 gal of water?

SOLUTION We know that the volume of a right circular cylinder is given by

$$V = \pi r^2 h$$

We need to rearrange this equation to find the unknown of interest in this problem. Divide both sides of this equation by πh.

$$\frac{V}{\pi h} = \frac{\pi r^2 h}{\pi h} = r^2$$

Then

$$r = \sqrt{\frac{V}{\pi h}} = \sqrt{\frac{(2000)(0.1337)}{(3.14)(8.0)}}$$

$$= 3.26 \text{ or } 3.3 \text{ ft} \quad \text{(to two significant digits)}$$

Equivalent formulas for the **volume of a right circular cone** of height h are

$$V = \frac{1}{3} A_b h$$

where A_b is the area of the circular base or

$$V = \frac{1}{3} \pi r^2 h$$

where *r* is the radius of the base.

EXAMPLE A conical pile of gravel is formed at the end of a conveyor belt. Calculate the volume of gravel in this pile when it is 8.0 ft high and has a diameter at the base of 13.0 ft. See Figure 3.47.

SOLUTION

$$V = \frac{1}{3}(3.14)(6.5)^2(8.0)$$
$$= 353.8 = 3\underline{5}0 \text{ ft}^3$$

FIGURE 3.47

The volume of a sphere of radius *r* is given by the formula

$$V = \frac{1}{3}A_s r$$

where A_s is the surface area of the sphere. Since $A_s = 4\pi r^2$, substituting $4\pi r^2$ for A_s yields the equivalent formula

$$V = \frac{4}{3}\pi r^3$$

EXAMPLE Calculate the number of gallons of water that can be contained in a spherical tank of inner radius 9.2 ft.

SOLUTION

$$V = \frac{4}{3}(3.14)(9.2)^3 = 3260.1 \text{ ft}^3$$
$$= 3260.1 \text{ ft}^3 \cdot \left(\frac{1 \text{ gal}}{0.1337 \text{ ft}^3}\right) = 24{,}384 \quad \text{or} \quad 2\underline{4}{,}000 \text{ gal}$$

VOLUME FORMULAS FOR SOLIDS

$$\text{right prism} \quad V = A_b h$$
$$\text{right circular cylinder} \quad V = \pi r^2 h$$
$$\text{right circular cone} \quad V = \frac{1}{3}\pi r^2 h$$
$$\text{sphere} \quad V = \frac{4}{3}\pi r^3$$

PERFORMANCE OBJECTIVE 13

You should now be able to calculate the volumes of specific solids.

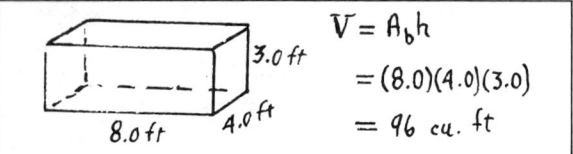

$V = A_b h$
$= (8.0)(4.0)(3.0)$
$= 96$ cu. ft

EXERCISE SET 3.4

1. A gas tank is to be designed as a rectangular based right prism. The base dimensions are 12.0 × 36.0 in. How deep should the tank be if it is to hold 16 gal?

2. The height of a square based right prism is 32.5 cm. The area of the square base of the prism is 144 cm^2. Find the surface area of the right prism.

3. A hot water heater cylinder is designed to be 3.1 ft high and have a radius of 8.5 in. Calculate the minimum amount of sheet metal needed to construct the tank assuming no waste.

4. A layer of gravel 5.0 cm deep is to be spread on a trapezoidal-shaped driveway. The length of the driveway (the height of the trapezoid) is 30.0 m. At its narrowest spot the driveway is 5.0 m wide and at its widest spot the width is 8.0 m. How many cubic meters of gravel are needed for the driveway?

5. Assuming the tank in Problem 3 is completely filled, how many gallons will it hold?

6. An empty cylindrical well is to be filled with dirt. Given that the well is 8.5 m deep and has a 3.0 m diameter, calculate the amount of dirt needed to fill the well.

7. Spherical balls of radius 4.21 cm are to be silver-plated. Given that it costs 12.7 cents to silver plate each square centimeter, calculate the cost of plating each ball.

8. If the spherical balls in Problem 7 are made of material whose density is 6.24 g/cm^3, what is the weight of each ball?

9. A piston moves 3.72 in. during the compression stroke in an engine. If the diameter of the cylinder is 2.58 in., what is the decrease in volume of the fuel mixture during compression?

10. A water tank in the shape of a hemisphere has a 5.8 ft diameter. How many gallons of water will the tank hold?

11. A sand dealer claims that a conical pile of sand has 1 yd^3 of sand in it. You measure the pile to be 4 ft high and 5.5 ft across at the base. Would you pay the dealer for a cubic yard or dispute his claim (1 yd^3 = 27 ft^3)?

12. A cardboard hat in the shape of a right circular cone is to be made for a Halloween costume. The hat is to be 25.0 cm high and have a base diameter of 22.0 cm. Calculate the minimum amount of cardboard needed to make the hat assuming no waste. (*Note:* Since the hat has an open base, the area of the circular base of the cone should not be included in the answer.)

SECTION 3.5 SUMMARY

In this chapter we have provided the necessary formulas for dealing with plane and solid figures. It is important to be able to recognize these figures and to select the correct formula from the problem description or sketch. At times it may be necessary to rearrange these formulas before solving. You should particularly note the units attached to areas and volumes.

We summarize the formulas for various measures of plane and solid figures.

Perimeter of a polygon: the sum of the lengths of the sides of the polygon.

Area formulas for polygons:

Polygon	Formula
triangle	$A = \frac{1}{2}bh$
rectangle	$A = lw$
parallelogram	$A = bh$
trapezoid	$A = \frac{1}{2}(b_1 + b_2)h$

Sum of the measures of the angles of a polygon with *n* sides:

$$(n - 2)180°$$

Measure of each angle in a regular polygon of n sides:

$$\frac{(n-2)180°}{n}$$

Circumference and area formulas for circles:

$$C = \pi d = 2\pi r$$

$$A = \pi r^2$$

Surface areas and volumes of solids:

Solid	Surface area	Volume
right prism	sum of the areas of the bases and sides	$V = A_b h$
right circular cylinder	$A_s = 2\pi r^2 + 2\pi r h$	$V = \pi r^2 h$
right circular cone	$A_s = \pi r^2 + \pi r \sqrt{r^2 + h^2}$	$V = \frac{1}{3}\pi r^2 h$
sphere	$A_s = \pi d^2 = 4\pi r^2$	$V = \frac{4}{3}\pi r^3$

CHAPTER 3 PRACTICE TEST

1. (PO-1, 2) Name the following constructions and figures:

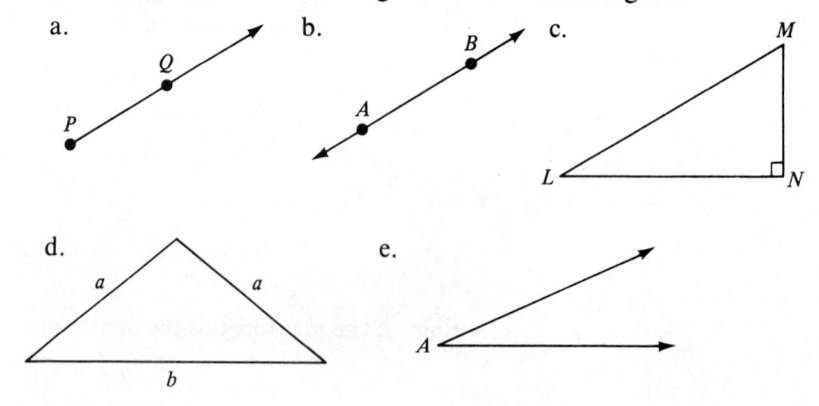

   a.

   b.

   c.

   d.

   e.

CHAPTER 3 PRACTICE TEST

2. (PO-3)

   a. Answer true or false.

   Any two equilateral triangles are similar.

   b. Answer true or false.

   If two similar polygons have a pair of corresponding sides equal, then the polygons are congruent.

   c. Consider a rectangle 3.2 in. wide and 9.6 in. long and a rectangle 4.8 in. wide and 6.4 in. long. Are these rectangles similar?

3. (PO-4) Calculate the perimeter of the following plane figures:

   a. A square of side 3.2 in.

   b. A regular hexagon of side 1.61 in.

4. (PO-5) Calculate the area of the plate in Figure 3.48. The unshaded area is a rectangular hole.

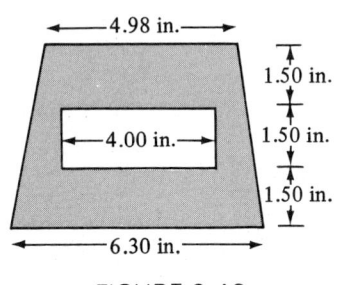

FIGURE 3.48

5. (PO-6) Calculate the fifth angle of a pentagon when the other angles are 100°, 100°, 120°, and 90°.

6. (PO-7) A right triangle whose sides are of length 6 cm, 8 cm, and 10 cm is similar to a right triangle whose perimeter is 18 cm.

   a. Find the hypotenuse of the second triangle.

   b. Find the area of the triangle whose perimeter is 18 cm.

7. (PO-8, 9, 10) Calculate the circumference and area of the following circles:

   a. $r = 1.3$ in.   b. $d = 7.5$ mm

   c. The circle inscribed in a square whose perimeter is 96 cm

8. (PO-11, 12) Calculate the surface area of the following solid figures:

   a. A right prism with a base in the shape of a trapezoid. The height of the prism is 7.82 in. The trapezoid dimensions are shown in Figure 3.49.

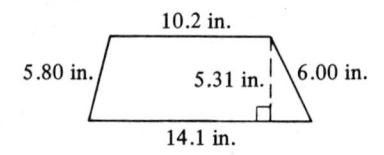

   ```
            10.2 in.
      5.80 in.   5.31 in.   6.00 in.
            14.1 in.
   ```

   FIGURE 3.49

   b. A right circular cylinder whose radius and height are both 7.00 in.

9. (PO-11, 13) Calculate the volume of the following solid figures:

   a. A right circular cone 9.5 ft high and base radius 15.2 ft

   b. A sphere of radius 7.12 in.

10. (PO-11, 12, 13) Calculate the values of the following:

    a. The radius of a sphere whose surface area is 18.1 cm^2.

    b. The base radius of a right circular cone of volume 22.5 ft^3 and 3.1 ft high.

CHAPTER 4

ALGEBRAIC EXPRESSIONS

4.1 Algebraic Terms and Expressions
4.2 Evaluation of Algebraic Expressions
4.3 The Definition of a Polynomial
4.4 The Addition and Subtraction of Polynomials
4.5 The Multiplication and Division of Algebraic Terms
4.6 The Multiplication of Polynomials
4.7 Summary

In arithmetic, the quantities we work with are specific numbers such as the rational numbers we considered in Chapter 1. We often call these specific numbers **constants.**

Algebra is a generalization of arithmetic. In algebra we work not only with constants but also with quantities that do not represent a specific value or whose value is not yet known. These quantities are usually represented by letters such as x, y, and z. We call these letters **literal numbers** or **variables.**

In this chapter we will discuss algebraic expressions and the rules for performing certain operations on two types of algebraic expressions, polynomials and rational algebraic terms.

SECTION 4.1 ALGEBRAIC TERMS AND EXPRESSIONS

We denote the addition, subtraction, and division of variables with the same symbols as we use for specific numbers, that is, with the symbols $+$, $-$, $\div$, or the fraction bar. However, this is not the case for multiplication. The symbol $\times$ is often used to denote the multiplication of specific numbers. Use of the $\times$ to represent multiplication of variables could lead to confusion since x is a commonly used variable. In Chapter 1 we introduced two other ways of indicating multiplication. They are often used with variables as well as with specific numbers. One method is the use of a raised dot. For example,

$$x \cdot y$$

means x times y, and

$$2 \cdot a \cdot b$$

means 2 times a times b. Another method is to use parentheses. For example,

$$5(6)$$

means 5 times 6, and

$$(4)(c)(d)$$

means 4 times c times d.

To indicate the multiplication of variables or a constant times a variable, it is not necessary to use any multiplication symbol at all. For example,

$$xyz$$

means x times y times z and

$$-3c$$

means -3 times c.

Note that we cannot omit the multiplication symbol when we multiply specific numbers. It is clear that 23 does not mean the same as $2 \cdot 3$ or $(2)(3)$ or 2×3.

Exponents are commonly used to indicate a product in which a variable or constant is used as a factor more than once. For example, using exponents

$$2xxxyyz$$

would be written

$$2x^3y^2z$$

since there are three factors of x and two factors of y. When no exponent is given on a number, such as on the 2 and on the z, it is understood that the exponent is 1 and that there is only one factor of that number in the product. We will discuss laws of exponents in a later section.

NOTATION FOR MULTIPLICATION

$x \cdot y$ multiplication symbol ($\cdot$) shown

$(x)(y)$ quantities in adjoining parentheses

xy multiplication implied—no separation of symbols

x^2y^3 exponents used for continued multiplication of the same quantity

---

PERFORMANCE OBJECTIVE 1	$3 \cdot 7 \cdot x \cdot x \cdot x \cdot y + (-2)(-4)(a)(a)$
You should now be able to understand the notation for operations with constants and variables.	$= 21 x^3 y + 8a^2$

Constants, variables, and combinations of constants and variables formed by taking products and quotients are called **algebraic terms**.

Algebraic terms, sums and differences of algebraic terms, and quotients of sums and differences of algebraic terms are called **algebraic expressions**.

Some examples of algebraic terms are

$$-3 \quad \text{(a constant)}$$

$$w \quad \text{(a variable)}$$

$$2y \quad \text{(a product of a constant and a variable)}$$

$$abc \quad \text{(a product of variables)}$$

$$\frac{x}{y} \quad \text{(a quotient of variables)}$$

$$\frac{4x^2y}{5z} \quad \text{(a quotient of products of constants and variables)}$$

Some examples of algebraic expressions are the preceding algebraic terms as well as the following combinations of algebraic terms:

$$xy - 1 \quad \text{(a difference of algebraic terms)}$$

$$x^3 + 2x^2 + 3 \quad \text{(a sum of algebraic terms)}$$

$$\frac{ab}{2c} - \frac{x}{y} + 6 \quad \text{(a sum and difference of algebraic terms)}$$

$$\frac{x + y}{y - 5} \quad \text{(a quotient of sums and differences of algebraic terms)}$$

When the variables in an algebraic expression appear only in sums, differences, and products, we call the algebraic expression a **polynomial**. Algebraic expressions that involve division by variables or by combinations of constants and variables are called **rational algebraic expressions**.

This chapter will emphasize the addition, subtraction, and multiplication of polynomials and the multiplication and division of rational algebraic terms. The more complex rational algebraic expressions will be considered in Chapter 5.

PERFORMANCE OBJECTIVE 2

You should now be able to recognize algebraic terms and algebraic expressions.

$$\underbrace{6x^2y^3}_{\text{algebraic terms}} - \underbrace{7y^2}_{} + \underbrace{5y^3}_{}$$

$$\underbrace{\hphantom{6x^2y^3 - 7y^2 + 5y^3}}_{\text{algebraic expression}}$$

EXERCISE SET 4.1

1. Identify the constants and the variables in each of the following expressions:

    a. $2ab - 3c$ b. $\dfrac{6x}{y}$

2. Identify the constants and the variables in each of the following expressions:

    a. $8x - 5y + 2$ b. $4a + \dfrac{b}{c}$

3. Rewrite the following products without any multiplication symbols:

    a. $3 \cdot 4 \cdot x \cdot y - 5 \cdot z$ b. $2(a)(c) + (-6)(-3)(b)$

4. Rewrite the following products without any multiplication symbols:
   a. $-2 \cdot 8 \cdot x \cdot z - y(-4)$
   b. $3 \cdot a \cdot x + (b)(y)(-2) - (-5)(-1)$

5. Use exponents to simplify the following products:
   a. $7aaaaa$
   b. $xxyyyzww$

6. Use exponents to simplify the following products:
   a. $-3xxxyyy$
   b. $2abbccccdd$

7. Determine whether the following are algebraic terms. Label each as either an algebraic term or an algebraic expression.
   a. $5(-y)x$
   b. $4a - 1$
   c. $\dfrac{x+2}{3x}$
   d. $\dfrac{-3x^2y}{5zw}$

8. Determine whether the following are algebraic terms. Label each as either an algebraic term or an algebraic expression.
   a. $8 - a$
   b. $x(-3y)$
   c. $\dfrac{a-b}{b+c}$
   d. $\dfrac{2r^2(-3s)}{t^2(-s)}$

SECTION 4.2 EVALUATION OF ALGEBRAIC EXPRESSIONS

To evaluate an expression, we substitute specific numbers in place of the variables and compute the value of the resulting expression. For example, if we evaluate $x + y$ for $x = 5$ and $y = -3$, we get

$$5 + (-3) = 2$$

We will sometimes be evaluating expressions that involve more than one operation. In such expressions, symbols of grouping are often used. Commonly used symbols of grouping are parentheses ( ), brackets [ ], and braces { }.

RULES FOR PERFORMING OPERATIONS WITH SYMBOLS OF GROUPING

1. Perform operations within symbols of grouping first.

2. If a pair of grouping symbols lies within another pair of grouping symbols, the operations within the innermost pair should be performed first.

EXAMPLE Evaluate $a(b + c)$ for $a = -3$, $b = 4$, and $c = 2$.

SOLUTION The parentheses indicate that we are to find the sum of b and c and then multiply by a. When we substitute the specific numbers in place of the variables and carry out the operations, we get

$$a(b + c) = -3(4 + 2) = -3(6) = -18$$

EXAMPLE Evaluate $a + [b \div (5 - c)]$ for $a = 5$, $b = -8$, and $c = 3$.

SOLUTION Substituting the specific numbers in place of variables, we get

$$5 + [-8 \div (5 - 3)]$$

We evaluate this expression as follows:

$$5 + [-8 \div (5 - 3)] = 5 + [-8 \div 2] = 5 + [-4] = 1$$

If the algebraic expression $a + [b \div (5 - c)]$ had been written in the form

$$a + \frac{b}{5 - c}$$

we would still have performed the same series of operations. In a sense, the fraction bar in

$$a + \frac{b}{5 - c}$$

is serving as the symbol of grouping. The bar indicates that we must first compute the denominator, $5 - c$, then divide it into the numerator, b, and finally add the quotient to a.

PERFORMANCE OBJECTIVE 3

You should now be able to evaluate algebraic expressions that involve symbols of grouping.

$3 \div \{8 + [6 \div (2 \cdot 3)]\} =$
$3 \div \{8 + [6 \div 6]\} =$
$3 \div \{8 + 1\} = 3 \div 9 = \frac{1}{3}$

Suppose we want to evaluate the algebraic expression $5 - 2y + 3x$ for $x = 4$ and $y = 7$. Substituting the specific numbers in place of variables gives $5 - 2(7) + 3(4)$. This expression involves three operations: addition, subtraction, and multiplication. So questions arise as to the order in which the operations should be performed. Do we subtract the 2 from the 5 before multiplying by 7? Do we add the 3 to what precedes it before multiplying by 4? The answer to both questions is no.

ORDER OF OPERATIONS IN ALGEBRAIC EXPRESSIONS

Step 1: Move through the expressions from left to right performing each multiplication and division as it appears. Leave additions and subtractions alone.

Step 2: Perform additions and subtractions after all multiplications and divisions are complete.

Recall that each subtraction step can be converted to addition by changing the sign of the subtrahend. So all that remains is addition and that can be performed in any order. We can rearrange the terms in a sum and perform a series of additions in any order we want because of the Commutative and Associative Laws of Addition. The **Commutative Law of Addition** states that

$$a + b = b + a$$

for any real numbers a and b. The **Associative Law of Addition** states that

$$(a + b) + c = a + (b + c)$$

for any real numbers a, b, and c.

EXAMPLE Simplify $5 - 2(7) + 3(4)$.

SOLUTION

$$\begin{aligned} 5 - 2(7) + 3(4) &= 5 - 14 + 12 \\ &= 5 + (-14) + 12 \\ &= -9 + 12 \quad [\text{or } 5 + (-2) \text{ or } 17 + (-14)] \\ &= 3 \end{aligned}$$

EXAMPLE Evaluate $x^2 - 3y \div z$ for $x = \frac{1}{2}$, $y = \frac{1}{4}$, and $z = \frac{3}{8}$.

SOLUTION

$$\begin{aligned} x^2 - 3y \div z &= \left(\frac{1}{2}\right)^2 - 3\left(\frac{1}{4}\right) \div \frac{3}{8} \\ &= \frac{1}{4} - \frac{3}{4} \div \frac{3}{8} = \frac{1}{4} - \frac{3}{4} \cdot \frac{8}{3} \\ &= \frac{1}{4} - 2 = \frac{1}{4} - \frac{8}{4} = -\frac{7}{4} \end{aligned}$$

The calculator may be of assistance in the evaluation of algebraic expressions particularly if the values to be substituted are whole numbers or can be expressed exactly as decimals. Some fractions such as $\frac{1}{3}$ and $\frac{5}{6}$ can never be expressed exactly on the calculator. But let us look at those cases in which the values can be expressed exactly. We redo the examples of this section using the calculator.

EXAMPLE Evaluate $a(b + c)$ for $a = -3$, $b = 4$, and $c = 2$.

SOLUTION We use the procedure outlined earlier. We perform the operations inside symbols of grouping first.

$$4 \boxed{+} 2 \boxed{=} \boxed{\times} 3 \boxed{+/-} \boxed{=} \text{ yields } -18$$

A key step is the press of the $\boxed{=}$ key after the 2 is entered. This totals the value inside the symbols of grouping. This is necessary since the calculator follows the order of operations described earlier in this chapter. Multiplications and divisions are performed before additions and subtractions.

EXAMPLE Evaluate $a + [b \div (5 - c)]$ for $a = 5$, $b = -8$, and $c = 3$.

SOLUTION We evaluate the expression within the innermost grouping symbols first and use the storage feature of the calculator.

$$5 \boxed{-} 3 \boxed{=} \boxed{\text{STO}} 8 \boxed{+/-} \boxed{\div} \boxed{\text{RCL}} \boxed{=} \boxed{+} 5 \boxed{=}$$

yields 1

EXAMPLE Evaluate $x^2 - 3y \div z$ for $x = \frac{1}{2}$, $y = \frac{1}{4}$, and $z = \frac{3}{8}$.

SOLUTION These three fractions can be expressed as exact decimals so the computation can be performed on the calculator.

$$\frac{1}{2} = 0.5, \quad \frac{1}{4} = 0.25, \quad \text{and} \quad \frac{3}{8} = 0.375$$

$$0.5 \boxed{\times} 0.5 \boxed{-} 3 \boxed{\times} 0.25 \boxed{\div} 0.375 \boxed{=} \text{ yields } -1.75$$

The answer is -1.75.

Note that without symbols of grouping we were able to move from left to right just as the expression is written. If we have an $\boxed{x^2}$ key, $0.5 \boxed{\times} 0.5$ can be replaced by $0.5 \boxed{x^2}$.

PERFORMANCE OBJECTIVE 4	$12 + 4 \cdot 9 - 24 \div 8 \cdot 15 =$
You should now be able to evaluate expressions that do not have symbols of grouping.	$12 + 36 - 3 \cdot 15 =$
	$12 + 36 - 45 = 3$

EXERCISE SET 4.2

Evaluate the following algebraic expressions for the given values of the variables:

1. $4a - 3b + 2 \quad$ for $a = 0$, $b = \dfrac{1}{3}$

2. $-2x^2 + 5xy - 3y \quad$ for $x = \dfrac{3}{2}$, $y = -1$

3. $\dfrac{8x^2 + 3y}{xy} \quad$ for $x = 3$, $y = -4$

4. $\dfrac{3x^2 + yw + 1}{x + 4} \quad$ for $x = -2$, $y = -5$, $w = 3$

5. $9 - 3[(a + b) - (c - d)] \quad$ for $a = 2$, $b = -3$, $c = 1$, $d = -2$

6. $a + b((b - a) - a] \quad$ for $a = 3$, $b = 8$

7. $ab \div c - d + 2c \quad$ for $a = 2$, $b = -6$, $c = 4$, $d = 5$

8. $y - x \div y^2 + 2(x + y) \quad$ for $x = 10$, $y = 2$

9. $3y^2 - x \div (4x - y) \quad$ for $x = 2$, $y = 4$

10. $\dfrac{4ab - (3 + b)}{a + b} \quad$ for $a = 0$, $b = -6$

SECTION 4.3 — THE DEFINITION OF A POLYNOMIAL

A polynomial is an algebraic expression consisting of sums, differences, and products of numbers and variables raised to whole number powers. Examples of polynomials are

$$x^2$$

$$x + 3$$

$$x^2 - 2x + 4$$

$$5x^2y - 3xy^2 + 26$$

$$\frac{1}{4}x^2y^3$$

As you can see from the examples, a polynomial may involve just one variable or more than one variable and a polynomial may consist of a single term or more than one term. In each term of a polynomial, the number, with the sign (+ or −) preceding the number, is called the **numerical coefficient** of the term. If no number appears with the variables in a term, the numerical coefficient of the term is understood to be a 1.

A POLYNOMIAL

The terminology associated with a polynomial includes the following:

Numerical coefficient of a term (−4)
$$\underbrace{3x^4}_{\text{A term of the polynomial}} - 4x^3 + 5\underset{\underset{\text{Whole number power of the variable (2)}}{\uparrow}}{x^2} + 9$$

The definition of a polynomial rules out expressions such as

$3x^{1/2} + 2$ (the power of x is not a whole number)

$\dfrac{1}{x} + x^2$ (the $\dfrac{1}{x}$ term involves division by the variable)

Polynomials that consist of one, two, or three terms are given special names.

DEFINITION

POLYNOMIALS OF ONE, TWO, OR THREE TERMS

A **monomial** has one term.

A **binomial** has two terms.

A **trinomial** has three terms.

EXAMPLE Write two polynomials of each type described in the definition.

SOLUTION

Monomials: $4y$, $\dfrac{1}{2}x^4$

Binomials: $x^2 - 4x$, $3y^4 + 16$

Trinomials: $x^2 + 2x + 1$, $x^2 + xy + y^2$

The degree of a polynomial depends on the exponents on the variables. We will start by learning to identify the degree of a monomial.

THE DEFINITION OF A POLYNOMIAL 4.3

DEFINITION

> **DEGREE OF A MONOMIAL**
>
> The **degree of a monomial** is the sum of the exponents on the variables in the monomial.

EXAMPLE Find the degree of the following monomials:

$$12x^3, \quad 9x^2y^4, \quad -6xy^3z^4$$

SOLUTION $12x^3$ has degree 3. We say that $12x^3$ is a third degree polynomial in x.

$9x^2y^4$ has degree 6. We say that $9x^2y^4$ is a sixth degree polynomial in x and y.

$-6xy^3z^4$ has degree 8. This is an eighth degree polynomial in x, y and z.

Note that when no number is written as an exponent on a variable the exponent is 1.

DEFINITION

> **DEGREE OF A POLYNOMIAL**
>
> The **degree of a polynomial** is the degree of the highest degree term in the polynomial.

EXAMPLE Find the degree of the following polynomials:

$$x^2y - 3xy^3 - y^3$$

$$13x^2y^3z - 2xy^2z + 3xy^6$$

SOLUTION In the polynomial $x^2y - 3xy^3 - y^3$, the degrees of the terms are 3, 4, and 3. Thus, the degree of the polynomial is 4.

In $13x^2y^3z - 2xy^2z + 3xy^6$, the degrees of the terms are 6, 4, and 7 so that 7 is the degree of the polynomial.

PERFORMANCE OBJECTIVE 5

You should now be able to identify the type and degree of a polynomial.

$$\underbrace{\overset{degree\,=\,5}{xy^4} - \overset{degree\,=\,6}{8x^2yz^3} + \overset{degree\,=\,4}{15xy^2z}}_{\text{6}^{th}\text{ degree polynomial in }x,y,\text{ and }z.}$$

EXERCISE SET 4.3

1. Identify the following polynomials by type:
   a. $-7ab^2 + 6a^2b$
   b. $3y^2 + 5y - 7$
   c. $7x^3y^4z$
   d. $9b - 12f + g$

2. Identify the following polynomials by type:
   a. $-4x^2yw$
   b. $8x^3 - 8$
   c. $x^4 + x^2 + x$
   d. $5xy^2z - 4x^2y^3z$

3. Identify the degree of the following polynomials:
   a. $3x^4 - 2x^3 - 4x^2 + x + 12$
   b. $3 + x - 5x^2 + 6x^3$
   c. $-2x^2yz^5$
   d. $9y^2z + 5x^3y^5z - 12x^2y^2$

4. Identify the degree of the following polynomials:
   a. $7 - 5x + x^3$
   b. $4a^2b^3c$
   c. $3x^3 + x^3y + 3y^2$
   d. $-r^2s^3t + 7st^2 + 8r^6t$

SECTION 4.4 THE ADDITION AND SUBTRACTION OF POLYNOMIALS

The key to adding and subtracting polynomials is the ability to recognize like terms.

DEFINITION

> **LIKE TERMS**
>
> Algebraic terms in which the variables and the exponents on the variables are exactly the same are called **like terms.**

Algebraic terms that are not like terms are called **unlike terms.** The numerical coefficients of the terms have no bearing on whether the terms are like or unlike terms. Examples of like and unlike terms follow:

x, $\frac{1}{2}x$, and $-3x$ are like terms.

$4x$ and $4xy$ are unlike terms since the variables are not identical.

THE ADDITION AND SUBTRACTION OF POLYNOMIALS 4.4

6*ab* and 2*ba* are like terms.

NOTE: *ab* is the same as *ba*. The fact that $ab = ba$ is called the **Commutative Law for Multiplication.**

$8x^2$ and $2x$ are unlike terms since the exponents on the variables are different.

$\frac{2}{3}a$ and $\frac{3}{4}b$ are unlike terms since the variables are not the same.

We can write the sum of like terms as a single term, but we cannot combine unlike terms into a single term. For example, we can write $4x + 5x$ as the single term $9x$, but we cannot write the sum $4x + 5y$ as a single term. We add the like terms $4x$ and $5x$ by adding their coefficients 4 and 5 to get 9 and keeping the variable x.

The principle that enables us to add like terms by adding their numerical coefficients is the **Distributive Law**. The Distributive Law is a very important principle in algebra and we will be using it often.

DEFINITION

DISTRIBUTIVE LAW

For any real numbers *a*, *b*, and *c*,

$a(b + c) = ab + ac$

NOTE: We also have $a(b - c) = ab - ac$.

We use the Distributive Law to add $4x$ and $5x$ as follows:

$$4x + 5x = (4 + 5)x \quad \text{(by the Distributive Law)}$$
$$= 9x \quad \text{(since } 4 + 5 = 9\text{)}$$

The following examples also show the use of the Distributive Law in the addition of like terms:

$$-6z + (-7z) = [-6 + (-7)]z = -13z$$

$$8w + (-15w) = [8 + (-15)]w = -7w$$

$$\frac{1}{2}ab + \frac{1}{3}ab = \left(\frac{1}{2} + \frac{1}{3}\right)ab = \left(\frac{3}{6} + \frac{2}{6}\right)ab = \frac{5}{6}ab$$

$$4x + (-4x) = [4 + (-4)]x = 0x = 0$$

NOTE: We call the terms $4x$ and $-4x$ **additive inverses** of each other since their sum is zero.

If we want to subtract like terms we proceed, as before, by changing the sign of the subtrahend and then adding. Examples of the subtraction of like terms are:

$$25x - (-7x) = 25x + 7x$$
$$= (25 + 7)x$$
$$= 32x$$

$$-6z - (-9z) = -6z + 9z$$
$$= (-6 + 9)z$$
$$= 3z$$

$$-\frac{3}{4}a - \frac{1}{8}a = \frac{-3}{4}a + \left(\frac{-1}{8}a\right)$$

$$= \left[\frac{-3}{4} + \left(\frac{-1}{8}\right)\right]a = \left[\frac{-6}{8} + \left(\frac{-1}{8}\right)\right]a$$

$$= \frac{-7}{8}a \quad \text{or} \quad -\frac{7}{8}a$$

PERFORMANCE OBJECTIVE 6 You should now be able to recognize like terms and combine them by addition and subtraction.	$\frac{3}{4}r^2y$ and $\frac{1}{3}r^2y$ are like terms. $\frac{3}{4}r^2y - \frac{1}{3}r^2y = \left(\frac{3}{4} - \frac{1}{3}\right)r^2y$ $= \left(\frac{9}{12} - \frac{4}{12}\right)r^2y = \frac{5}{12}r^2y$

To add two polynomials, we combine like terms in the polynomials. We can use either the horizontal method or the vertical method of addition. Each method involves identifying the like terms in the polynomials and adding each pair of like terms to obtain the desired sum.

To use the horizontal method of addition, we write the polynomials horizontally (across the page), group the like terms together, and add the pairs of like terms. To use the vertical method of addition, we write one polynomial under the other so that like terms are lined up in columns and then we add the terms in each column. The same answer is obtained regardless of which method is used. However, the vertical method is often easier to use since it enables

THE ADDITION AND SUBTRACTION OF POLYNOMIALS 4.4

us to put every term in a logical location with respect to all other terms and it emphasizes the similarity between the addition of polynomials and the addition of whole numbers in columns.

EXAMPLE Add the polynomials $3x^2 + 2x + 4$ and $4x^2 + 2x + 5$ using both methods.

SOLUTION

Horizontal method:

$$(3x^2 + 2x + 4) + (4x^2 + 2x + 5)$$
$$= (3x^2 + 4x^2) + (2x + 2x) + (4 + 5)$$
$$= (3 + 4)x^2 + (2 + 2)x + (4 + 5)$$
$$= 7x^2 + 4x + 9$$

Vertical method:

$$\begin{array}{r} 3x^2 + 2x + 4 \\ +\,4x^2 + 2x + 5 \\ \hline 7x^2 + 4x + 9 \end{array}$$

EXAMPLE Add the polynomials $(5x + 4y - 2xy)$ and $(7x - 6y + 3xy)$ using both methods.

SOLUTION

Horizontal method:

$$(5x + 4y - 2xy) + (7x - 6y + 3xy)$$
$$= 5x + 4y - 2xy + 7x - 6y + 3xy$$
$$= 5x + 4y + (-2xy) + 7x + (-6y) + 3xy$$
$$= [5x + 7x] + [4y + (-6y)] + [-2xy + 3xy]$$
$$= [5 + 7]x + [4 + (-6)]y + [-2 + 3]xy$$
$$= 12x + (-2y) + 1xy$$
$$= 12x - 2y + xy$$

Note that to avoid unnecessary signs in our answers, we converted the addition of the negative term $-2y$ to subtraction.

Vertical method:

$$\begin{array}{r} 5x + 4y - 2xy \\ +\,7x - 6y + 3xy \\ \hline 12x - 2y + xy \end{array}$$

You may find it easier to change the subtractions to additions before adding. This gives

$$\begin{array}{r} 5x + 4y + (-2xy) \\ +7x + (-6y) + 3xy \\ \hline 12x + (-2y) + xy \end{array} \quad \text{or} \quad 12x - 2y + xy$$

To subtract polynomials, we change the signs of all the terms in the subtrahend polynomial, and then add the resulting polynomials.

EXAMPLE $(5x + 2) - (4x - 3)$

SOLUTION

Horizontal method:

$$\begin{aligned} (5x + 2) - (4x - 3) &= (5x + 2) + (-4x + 3) \\ &= 5x + (-4x) + 2 + 3 \\ &= [5 + (-4)]x + 5 \\ &= x + 5 \end{aligned}$$

EXAMPLE Subtract $2x^2 + 4x + 3$ from $5x^2 + 3x + 4$.

SOLUTION

Vertical method:

$$\begin{array}{r} 5x^2 + 3x + 4 \\ -(2x^2 + 4x + 3) \\ \hline \end{array} \quad \begin{array}{c} \text{Change subtraction} \\ \text{to addition} \end{array} \rightarrow \quad \begin{array}{r} 5x^2 + 3x + 4 \\ +(-2x^2 - 4x - 3) \\ \hline 3x^2 - x + 1 \end{array}$$

EXAMPLE $(5x + 4y - 2xy) - (7x - 6y + 3xy)$

SOLUTION

Horizontal method:

$$\begin{aligned} (5x &+ 4y - 2xy) - (7x - 6y + 3xy) \\ &= (5x + 4y - 2xy) + (-7x + 6y - 3xy) \\ &= 5x + 4y - 2xy + (-7x) + 6y + (-3xy) \\ &= [5x + (-7x)] + [4y + 6y] + [-2xy + (-3xy)] \\ &= [5 + (-7)]x + [4 + 6]y + [-2 + (-3)]xy \\ &= -2x + 10y + (-5xy) \\ &= -2x + 10y - 5xy \end{aligned}$$

THE ADDITION AND SUBTRACTION OF POLYNOMIALS 4.4

We have not yet added or subtracted polynomials in which there is a term unlike all other terms. The following example illustrates this situation.

EXAMPLE $(3x^2 + x + 2) - (2x^4 + 5x^2 - x)$

SOLUTION

Vertical method:

$$\begin{array}{r} 3x^2 + x + 2 \\ -(2x^4 + 5x^2 - x) \end{array} \text{ Change subtraction to addition} \rightarrow \begin{array}{r} 3x^2 + x + 2 \\ +(-2x^4 - 5x^2 + x) \\ \hline -2x^4 - 2x^2 + 2x + 2 \end{array}$$

PERFORMANCE OBJECTIVE 7	$(11x - 6y + z) + (3x + 7y - 2z)$
You should now be able to add two polynomials or find their difference.	$= (11 + 3)x + (-6 + 7)y + (1 - 2)z$
	$= 14x + y - z$

Up to this point we have performed only one operation on polynomials in each problem. Let us now look at the procedure to be followed when more than one operation is involved in a given problem. A good approach for working a problem of this type is to work with only two polynomials at a time and to work in some orderly manner such as left to right.

EXAMPLE

$(4x^4 - 5x^3 + 2x^2 - x + 6) + (-3x^4 + 2x^3 + x^2 + 2x + 5)$
$- (5x^4 - 3x^3 - x^2 + 4x - 12)$

SOLUTION To work this problem we will add the first two polynomials and then subtract the third polynomial from this sum.

$$\begin{array}{r} 4x^4 - 5x^3 + 2x^2 - x + 6 \\ +(-3x^4 + 2x^3 + x^2 + 2x + 5) \\ \hline x^4 - 3x^3 + 3x^2 + x + 11 \\ -(5x^4 - 3x^3 - x^2 + 4x - 12) \end{array}$$

Changing subtraction to addition, this becomes

$$\begin{array}{r} x^4 - 3x^3 + 3x^2 + x + 11 \\ +(-5x^4 + 3x^3 + x^2 - 4x + 12) \\ \hline -4x^4 \phantom{{}+3x^3} + 4x^2 - 3x + 23 \end{array}$$

The final answer is $-4x^4 + 4x^2 - 3x + 23$.

Note that in the answer the x^3 term has been eliminated since $-3x^3 + 3x^3 = 0x^3 = 0$.

EXAMPLE Use the Distributive Law and the rules for adding and subtracting polynomials to simplify the following:

$$(5a - 2b + c) + 3(b - 2c) - 2(a + d)$$

SOLUTION Using the Distributive Law to multiply 3 times $b - 2c$ and 2 times $a + d$, we get

$$(5a - 2b + c) + (3b - 6c) - (2a + 2d)$$

We then proceed as follows:

$$(5a - 2b + c) + (3b - 6c) - (2a + 2d)$$
$$= 5a - 2b + c + 3b - 6c + (-2a - 2d)$$
$$= 5a - 2b + c + 3b - 6c + (-2a) + (-2d)$$
$$= 5a + (-2b) + c + 3b + (-6c) + (-2a) + (-2d)$$
$$= [5a + (-2a)] + [-2b + 3b] + [c + (-6c)] + (-2d)$$
$$= [5 + (-2)]a + [-2 + 3]b + [1 + (-6)]c + (-2d)$$
$$= 3a + b + (-5c) + (-2d)$$
$$= 3a + b - 5c - 2d$$

RULES FOR ADDING AND SUBTRACTING POLYNOMIALS

Step 1: Identify like terms in the polynomials.

Step 2: Combine the coefficients of these terms to find the coefficient of each term in the sum or difference polynomial.

PERFORMANCE OBJECTIVE 8	$(a - 4b + c) + (6a + 2b) - (3a - 2b)$
You should now be able to add or subtract more than two polynomials.	① $\begin{array}{r} a - 4b + c \\ + 6a + 2b \\ \hline 7a - 2b + c \end{array}$ ② $\begin{array}{r} 7a - 2b + c \\ -(3a - 2b) \\ \hline 4a + c \end{array}$

EXERCISE SET 4.4

1. Answer true or false.
   a. $\frac{3}{10}x$ and $\frac{4}{10}y$ are unlike terms.
   b. $-6abc$, bac, and $\frac{1}{3}cab$ are like terms.
   c. x^2y and xy^2 are like terms.
   d. $3x^2$ and $3x$ are unlike terms.

2. Answer true or false.
   a. $5xy$ and $5rs$ are like terms.
   b. $\frac{1}{2}x$ and $2y$ are unlike terms.
   c. $-ac$ and $\frac{3}{4}ca$ are like terms.
   d. x^2 and $-x$ are unlike terms.

3. Combine like terms to simplify the following:
   a. $-17ab - 13ab$
   b. $-\frac{1}{3}x + \frac{5}{6}x$
   c. $3y - 8 - 6y + 9$
   d. $-16 + 21x + (-11x) + 9$

4. Combine like terms to simplify the following:
   a. $-12x + \frac{9}{2}x$
   b. $\frac{3}{4}ab - \frac{3}{8}ba$
   c. $16x - 11 - 13x + 8$
   d. $-7y + (-32x) - y + 14x$

5. Add the polynomials.
   a. $12x + 3$
      $+ 9x - 5$
   b. $8x - 12y + 3xy$
      $+ x + 7y - 6xy$
   c. $16x^4 + 3x^2 + 14$
      $+ 2x^4 + x^2 + 11$
   d. $(2x - 8y + z) + (8y - 2x - z)$
   e. $(10x - 4x^3) + (4x^2 - 3x + 2)$

6. Add the polynomials.
   a. $-6x + 15y$
      $+ 5x - 7y$
   b. $18xy - 3y + 12$
      $+ xy - 8y - 16$
   c. $9x^4 + 5x^3 - x$
      $+ 3x^4 - 4x^3 + x$
   d. $(-y + 3x + 7z) + (x - y - 7z)$
   e. $(2x^4 - 9x^3 - 4x) + (3x + 8x^2)$

7. Subtract the polynomials.

   a. $2x^2 + 3x + 1$
      $-(5x^2 + 2x + 4)$

   b. $3x^4 + 12x^3$
      $-(8x^4 + 3x^3)$

   c. $3ax - 7by + 4cz$
      $-(5ax - 5by + 4cz)$

   d. $(3x^2 + 6x + 12) - (2x^2 + 5x + 16)$
   e. $(5x^2 + 5x + 5) - (6x^2 + 6x + 6)$

8. Subtract the polynomials.

   a. $10x + 13$
      $-(2x + 15)$

   b. $17x + 6xy - 9y$
      $-(8x + 3xy\quad)$

   c. $-4ac + 7bc + 12$
      $-(-4ac + 7bc - 13)$

   d. $(5x^2 - 8x - 1) - (x^2 - 4x - 1)$
   e. $(9 + 2x^3) - (5x^3 - x + 6x^2)$

Perform the indicated operations in Exercises 9–16.

9. $(3x^3 - 2x^2 + 5x - 4) + (5x^3 + 3x^2 - 2x + 5) - (6x^3 + 4x^2 - 5x - 4)$
10. $(14x^3 + 8x^2 - 3) - (x^3 - 10x^2 - 2) + (-9x^3 + x^2 - 5)$
11. $(5x^2 + 2x - 1) - (3x^2 - 4x + 2) + (4x^2 - 5x + 12)$
12. $(22x^3 - 16x) + (-15x^3 + 7x^2) - (-9x + 7)$
13. $(6x^3 + 2x) + (3x^3 + 4x^2 - 3) - (5x^3 - 3x - 6)$
14. $(4x^2 - 12xy + 9y^2) - (4x^2 - 9y^2)$
15. $(5x^2 - 4x + 8) - (3x^2 - 2x + 9) - (3x^3 + 2x^2 - 5x + 16)$
16. $(8x^3 - 5x^2 + 21) - (2x^3 - x^2) - (9x^3 + 6x^2 - x - 4)$

Simplify the expressions in Exercises 17 and 18.

17. $2(7a - 3b) - 4(b - 5c) + 3(-5a + 4b - 6c)$
18. $-3(5x + 2y - z) + 5(3x - z) - 2(4y - 1)$

SECTION 4.5 THE MULTIPLICATION AND DIVISION OF ALGEBRAIC TERMS

The multiplication of polynomials will require the use of the laws of exponents. So we will first discuss the laws of exponents and demonstrate their use in the multiplication and division of algebraic terms.

In this section, we will work with only integer exponents. Recall that x^n, where n is a natural number, means the product in which x is used as a factor n times. We call x the base and n the exponent. We still need to consider zero exponents and negative integer exponents. We handle these as follows:

DEFINITION

ZERO AND NEGATIVE INTEGER EXPONENTS

$a^0 = 1$ for $a \neq 0$

$a^{-n} = \dfrac{1}{a^n}$ where n is a natural number and $a \neq 0$

The following examples illustrate the definition:

$$3^0 = 1$$

$$a^{-6} = \frac{1}{a^6}$$

$$5^{-2} = \frac{1}{5^2} = \frac{1}{25}$$

$$(-2)^{-3} = \frac{1}{(-2)^3} = \frac{1}{-8} = -\frac{1}{8}$$

$$7^{-1} = \frac{1}{7^1} = \frac{1}{7}$$

$$\frac{1}{4^{-3}} = \frac{1}{\frac{1}{4^3}} = 1 \div \frac{1}{4^3} = 1 \cdot \frac{4^3}{1} = 4^3 = 64$$

LAWS OF EXPONENTS

1. $a^m \cdot a^n = a^{m+n}$
2. $(a^m)^n = a^{m \cdot n}$
3. $(ab)^m = a^m b^m$
4. $\dfrac{a^m}{a^n} = a^{m-n}$ for $a \neq 0$
5. $\left(\dfrac{a}{b}\right)^m = \dfrac{a^m}{b^m}$ for $b \neq 0$

Law 1 examples:

$$x^3 \cdot x^6 = x^{3+6} = x^9$$

$$2^3 \cdot 2^4 = 2^{3+4} = 2^7 = 128$$

Law 2 examples:

$$(x^3)^5 = x^{3 \cdot 5} = x^{15}$$

$$(2^3)^2 = 2^{3 \cdot 2} = 2^6 = 64$$

Law 3 examples:

$$(xy)^6 = x^6 y^6$$

$$(3c)^4 = 3^4 c^4 = 81 c^4$$

$$(-2x)^5 = (-2)^5 x^5 = -32 x^5$$

Law 4 examples:

$$\frac{b^8}{b^5} = b^{8-5} = b^3$$

$$\frac{x^4}{x^6} = x^{4-6} = x^{-2} = \frac{1}{x^2}$$

$$\frac{3^4}{3^4} = 3^{4-4} = 3^0 = 1$$

$$\frac{5^7}{5^{10}} = 5^{7-10} = 5^{-3} = \frac{1}{5^3} = \frac{1}{125}$$

Law 5 examples:

$$\left(\frac{x}{y}\right)^{12} = \frac{x^{12}}{y^{12}}$$

$$\left(\frac{3}{4}\right)^3 = \frac{3^3}{4^3} = \frac{27}{64}$$

$$\left(-\frac{2}{3}\right)^5 = \frac{(-2)^5}{3^5} = -\frac{32}{243}$$

The calculator can be used to evaluate expressions involving exponents. If fractions are involved we suggest you evaluate the numerator and denominator separately after obvious cancellations are made. The calculator is not of much help in reducing fractions to lowest terms. We evaluate some of the expressions worked earlier.

EXAMPLE Evaluate $\left(\frac{3}{4}\right)^3$.

SOLUTION Write $\left(\frac{3}{4}\right)^3 = \frac{3^3}{4^3}$.

$3 \boxed{y^x} 3 \boxed{=}$ yields 27

$4 \boxed{y^x} 3 \boxed{=}$ yields 64

The answer is $\frac{27}{64}$.

EXAMPLE Evaluate $\left(-\frac{2}{3}\right)^5$.

SOLUTION $\left(-\frac{2}{3}\right)^5 = \frac{(-2)^5}{3^5}$

If we try to perform the operation $(-2)^5$ using the $\boxed{y^x}$ key on the calculator we are likely to run into trouble. The reason is the method used to determine powers. We may understand this better after we study logarithms (Chapter 15). To form $(-2)^5$ we can do continued multiplications or form $(2)^5$ using $\boxed{y^x}$ and take its negative.

$2 \boxed{y^x} 5 \boxed{=}$ yields 32 (change this sign)

$3 \boxed{y^x} 5 \boxed{=}$ yields 243

The answer is $\frac{-32}{243}$.

PERFORMANCE OBJECTIVE 9	
You should now be able to use the laws of exponents to simplify expressions.	$\left(-\frac{3}{2}\right)^3 + \left(\frac{3}{4}\right)^0 = -\frac{27}{8} + 1$ $= -\frac{27}{8} + \frac{8}{8} = -\frac{19}{8}$

RULES FOR MULTIPLYING ALGEBRAIC TERMS

Step 1: Multiply the coefficients of the terms to form the coefficient of the product.

Step 2: Multiply the variables of the terms to form the variables of the product.

Step 3: Simplify using the laws of exponents.

If the algebraic terms are in rational form, that is, if they are fractions, then we multiply them just as we do rational numbers—we find the product of the numerators and the product of the denominators and reduce the answer to lowest terms. To reduce the answer to lowest terms we not only eliminate common factors of the specific numbers in the numerator and denominator, but we also eliminate common factors of the variables by using the law

$$\frac{a^m}{a^n} = a^{m-n}$$

Multiplication of algebraic terms usually involves rearrangement of the factors in order to perform simplifications. Rearrangement of the factors is permissible because of the Commutative and Associative Laws of Multiplication. In a previous section, we mentioned the Commutative Law of Multiplication, that is $ab = ba$ for any real numbers a and b. The **Associative Law of Multiplication** assures us that we can perform a series of multiplications in any order we want. That is,

$$(ab)c = a(bc)$$

for any real numbers, a, b, and c. The following are examples of the multiplication of algebraic terms:

$$(6x^5y)(-3x^3y^2) = 6(-3)x^5x^3yy^2 = -18x^8y^3$$

$$a^4b^3(-4a^3bc^5)^2 = a^4b^3(-4)^2(a^3)^2b^2(c^5)^2 = a^4b^3(16)a^6b^2c^{10}$$
$$= 16a^4a^6b^3b^2c^{10} = 16a^{10}b^5c^{10}$$

$$\frac{-x^2}{4y}\left(\frac{-5y^4}{x^2}\right) = \frac{-1(-5)x^2y^4}{4x^2y} = \frac{5}{4} \cdot \frac{x^2}{x^2} \cdot \frac{y^4}{y} = \frac{5}{4} \cdot 1 \cdot y^3 = \frac{5y^3}{4}$$

$$\frac{4a^6b^3}{5c^2} \cdot \frac{3a^2c^4d}{8b^5} = \frac{4 \cdot 3 \cdot a^6a^2b^3c^4d}{5 \cdot 8b^5c^2} = \frac{3a^8b^3c^4d}{10b^5c^2}$$
$$= \frac{3a^8d}{10} \cdot \frac{b^3}{b^5} \cdot \frac{c^4}{c^2} = \frac{3a^8d}{10} \cdot b^{-2} \cdot c^2$$
$$= \frac{3a^8c^2d}{10} \cdot \frac{1}{b^2} = \frac{3a^8c^2d}{10b^2}$$

We have already discussed how to reduce quotients of algebraic terms to lowest terms. Now we will consider how to divide a rational algebraic term by another rational algebraic term. We divide rational algebraic terms in the same way as we divide rational numbers.

THE MULTIPLICATION AND DIVISION OF ALGEBRAIC TERMS 4.5

RULES FOR DIVIDING RATIONAL ALGEBRAIC TERMS

Step 1: Invert the divisor.

Step 2: Change the division sign to multiplication.

Step 3: Perform the multiplication using the rules for multiplication.

The following examples illustrate this procedure:

$$\frac{-2x^8}{5y} \div \frac{3x^5}{10y} = \frac{-2x^8}{5y} \cdot \frac{10y}{3x^5} = \frac{-2 \cdot 10x^8 y}{3 \cdot 5x^5 y} = \frac{-4x^8}{3x^5} = \frac{-4x^3}{3}$$

$$-12xy \div \left(\frac{-4x^9}{y^4}\right) = \frac{-12xy}{1} \cdot \frac{y^4}{-4x^9} = \frac{-12xy^5}{-4x^9} = \frac{3y^5}{x^8}$$

PERFORMANCE OBJECTIVE 10

You should now be able to multiply and divide algebraic terms.

$$\frac{16a^2 b}{c^3} \div \left(-\frac{12b}{c^2}\right)$$

$$= \frac{16a^2 b}{c^3}\left(-\frac{c^2}{12b}\right) = -\frac{4a^2}{3c}$$

EXERCISE SET 4.5

1. Simplify the following expressions involving exponents. Where possible, give your answer as a single number.

   a. $\left(\frac{4}{5}\right)^0$ b. $(-6)^{-3}$ c. $\frac{1}{x^{-7}}$ d. $(-5x)^4$

   e. $3^3 \cdot 3^2$ f. $(y^4)^2$ g. $\frac{6^9}{6^7}$ h. $\left(\frac{a}{4}\right)^4$

2. Simplify the following expressions involving exponents. Where possible, give your answer as a single number.

   a. $(-8.2)^0$ b. 3^{-5} c. $6y^{-3}$ d. $(-4x)^3$

   e. $x^2 \cdot x^5$ f. $(2^3)^2$ g. $\frac{10^6}{10^8}$ h. $\frac{(-2x)^5}{3}$

3. Perform the following multiplications and divisions of algebraic terms. Give your answers in lowest terms.

   a. $(-5x^3y^3)(-3xy^2z)$

   b. $2^2a^2b^3c(-ab^2c^4)^3$

   c. $\dfrac{3a^5c}{10b^4} \cdot \dfrac{5a^3b}{9c}$

   d. $-\dfrac{w^3}{5} \div \dfrac{3}{5}$

   e. $\dfrac{24x^5}{5y^3} \div 16xy$

4. Perform the following multiplications and divisions of algebraic terms. Give your answers in lowest terms.

   a. $\left(8a^2b^3\right)\left(-\dfrac{3}{2}ac^3\right)$

   b. $(-4)^3xy^2w^4(-x^2yw^3)^2$

   c. $\left(-\dfrac{12x^2y^6}{x^7y^8}\right)\left(\dfrac{21x^3}{y^3}\right)$

   d. $-\dfrac{a^5}{8b} \div \left(\dfrac{5a^6}{-24b^3}\right)$

   e. $32x^2y \div \dfrac{12x^2}{5}$

SECTION 4.6   THE MULTIPLICATION OF POLYNOMIALS

To perform a multiplication in which the multiplier is a monomial and the multiplicand is a polynomial, we apply the Distributive Law and multiply every term of the multiplicand by the monomial. When we multiply the monomial times a given term, we multiply the numerical coefficients and add the exponents on the variables involved in the multiplication.

EXAMPLE   Perform the multiplication $5x(x^2 - 4)$.

SOLUTION   Using the Distributive Law, we write

$$5x(x^2) + 5x(-4)$$

and then complete the multiplications to obtain

$$5x^3 - 20x$$

We can perform the multiplication of polynomials by using either the vertical method or the horizontal method for multiplication.

THE MULTIPLICATION OF POLYNOMIALS 4.6

As was the case in the addition and subtraction of polynomials, the same answer is obtained regardless of which method is used. However, the vertical method is often easier to use and it emphasizes the similarity between the multiplication of polynomials and the multiplication of whole numbers.

EXAMPLE Perform the multiplication $3x(x + 5)$.

SOLUTION In this problem $3x$ is the multiplier and $x + 5$ is the multiplicand.

Horizontal method:

$3x(x + 5) = 3x(x) + 3x(5)$ (by applying the Distributive Law)
$= 3x^2 + 15x$ (by multiplying the numerical coefficients and adding the exponents on the variable)

Vertical method:

$$\begin{array}{r} x + 5 \\ 3x \\ \hline 3x^2 + 15x \end{array}$$

since $3x$ times 5 is $15x$ and $3x$ times x is $3x^2$

EXAMPLE Multiply $3x^2 - 2xy + 5$ by $2x^3y^2$.

SOLUTION

Vertical method:

$$\begin{array}{r} 3x^2 - 2xy + 5 \\ 2x^3y^2 \\ \hline 6x^5y^2 - 4x^4y^3 + 10x^3y^2 \end{array}$$

Horizontal method:

$$2x^3y^2(3x^2 - 2xy + 5) = 6x^5y^2 - 4x^4y^3 + 10x^3y^2$$

To save space in the writing of a multiplication problem, parentheses are usually placed around the polynomials to be multiplied and then the polynomials are written next to each other (in a horizontal format). However, this format should not discourage use of the vertical method for multiplication.

PERFORMANCE OBJECTIVE 11	$-6x^2y(2x - 3xy^3 + 7)$
You should now be able to multiply a polynomial by a monomial.	$= -12x^3y + 18x^3y^4 - 42x^2y$

To multiply two polynomials, neither of which is a monomial, we select one polynomial to be the multiplier and one to be the multiplicand. We then multiply every term of the multiplicand by every term of the multiplier, combine like terms, and write a nice compact product. The vertical method will be used to illustrate this procedure.

EXAMPLE Multiply $(4x + 6)$ by $(2x - 4)$.

SOLUTION

$$
\begin{array}{rl}
4x + 6 \rightarrow & \text{multiplicand} \\
\underline{2x - 4} \rightarrow & \text{multiplier} \\
-16x - 24 \rightarrow & -4(4x + 6) = -16x - 24 \\
\underline{8x^2 + 12x} \rightarrow & 2x(4x + 6) = 8x^2 + 12x \\
8x^2 - 4x - 24 \rightarrow & \text{(combining like terms)}
\end{array}
$$

Therefore, $(4x + 6)(2x - 4) = 8x^2 - 4x - 24$.

EXAMPLE Multiply $(3x^2y - 2x + 6y)$ by $(3xy - 8)$.

SOLUTION

$$
\begin{array}{rl}
3x^2y - 2x + 6y \rightarrow & \text{multiplicand} \\
\underline{3xy - 8} \rightarrow & \text{multiplier} \\
-24x^2y + 16x - 48y \rightarrow & -8(3x^2y - 2x + 6y) \\
\underline{9x^3y^2 - 6x^2y \qquad\qquad\qquad + 18xy^2} \rightarrow & 3xy(3x^2y - 2x + 6y) \\
9x^3y^2 - 30x^2y + 16x - 48y + 18xy^2 \rightarrow & \text{(combining like terms)}
\end{array}
$$

Therefore,

$$(3x^2y - 2x + 6y)(3xy - 8) = 9x^3y^2 - 30x^2y + 16x - 48y + 18xy^2$$

EXAMPLE Multiply $(4x^2 - 2x + 5)$ by $(-2x^2 - 3x + 4)$.

THE MULTIPLICATION OF POLYNOMIALS 4.6

SOLUTION

$$\begin{array}{r} 4x^2 - 2x + 5 \\ -2x^2 - 3x + 4 \\ \hline 16x^2 - 8x + 20 \\ -12x^3 + 6x^2 - 15x \\ -8x^4 + 4x^3 - 10x^2 \\ \hline -8x^4 - 8x^3 + 12x^2 - 23x + 20 \end{array}$$

- $16x^2 - 8x + 20 \rightarrow 4(4x^2 - 2x + 5)$
- $-12x^3 + 6x^2 - 15x \rightarrow -3x(4x^2 - 2x + 5)$
- $-8x^4 + 4x^3 - 10x^2 \rightarrow -2x^2(4x^2 - 2x + 5)$
- $-8x^4 - 8x^3 + 12x^2 - 23x + 20 \rightarrow$ (combining like terms)

Therefore,

$$(4x^2 - 2x + 5)(-2x^2 - 3x + 4) = -8x^4 - 8x^3 + 12x^2 - 23x + 20$$

RULES FOR MULTIPLYING POLYNOMIALS

Multiply every term of the multiplicand by every term of the multiplier. Combine like terms.

PERFORMANCE OBJECTIVE 12

You should now be able to multiply polynomials other than monomials.

$$\begin{array}{r} 2\gamma^2 - 3\gamma - 4 \\ 5\gamma + 1 \\ \hline 2\gamma^2 - 3\gamma - 4 \\ 10\gamma^3 - 15\gamma^2 - 20\gamma \\ \hline 10\gamma^3 - 13\gamma^2 - 23\gamma - 4 \end{array}$$

We will now investigate seven special products that will be very useful in subsequent work on factoring polynomials. The formulas for the seven products follow.

FORMULA

> **THE SQUARE OF A BINOMIAL IN WHICH THE TERMS ARE ADDED**
>
> $(a + b)^2 = a^2 + 2ab + b^2$

The product is a trinomial whose three terms can be found by squaring the first term in the binomial, squaring the second term in the binomial, and taking twice the product of the terms in the binomial.

$$\begin{array}{r} a + b \\ a + b \\ \hline ab + b^2 \\ a^2 + ab \\ \hline a^2 + 2ab + b^2 \end{array}$$

FORMULA

> **THE SQUARE OF A BINOMIAL IN WHICH THE TERMS ARE SUBTRACTED**
>
> $(a - b)^2 = a^2 - 2ab + b^2$

The product is a trinomial whose three terms can be found by squaring the first term in the binomial, squaring the second term in the binomial, and taking the negative of twice the product of the terms in the binomial.

$$\begin{array}{r} a - b \\ a - b \\ \hline -ab + b^2 \\ a^2 - ab \\ \hline a^2 - 2ab + b^2 \end{array}$$

FORMULA

> **THE CUBE OF A BINOMIAL IN WHICH THE TERMS ARE ADDED**
>
> $(a + b)^3 = a^3 + 3a^2b + 3ab^2 + b^3$

The four terms of the product can be found from the two terms in the binomial by cubing the first term in the binomial, cubing the second term in the binomial, taking three times the product of the second term and the square of the first term, and taking three times the product of the first term and the square of the second term.

$$\begin{array}{r} a + b \\ a + b \\ \hline a^2 + 2ab + b^2 \\ a + b \\ \hline a^2b + 2ab^2 + b^3 \\ a^3 + 2a^2b + ab^2 \\ \hline a^3 + 3a^2b + 3ab^2 + b^3 \end{array}$$

FORMULA

> **THE CUBE OF A BINOMIAL IN WHICH THE TERMS ARE SUBTRACTED**
>
> $(a - b)^3 = a^3 - 3a^2b + 3ab^2 - b^3$

The four terms of the product can be found from the two terms in the binomial by cubing the first term, cubing the second term, taking the negative of three times the product of the second term and the square of the first term, and taking three times the product of the first term and the square of the second term.

$$
\begin{array}{r}
a - b \\
a - b \\
\hline
a^2 - 2ab + b^2 \\
a - b \\
\hline
-a^2b + 2ab^2 - b^3 \\
a^3 - 2a^2b + ab^2 \\
\hline
a^3 - 3a^2b + 3ab^2 - b^3
\end{array}
$$

FORMULA

THE PRODUCT OF THE SUM AND DIFFERENCE OF THE SAME TWO TERMS

$(a + b)(a - b) = a^2 - b^2$

The product is found by squaring the first term of the binomials and subtracting the square of the second term of the binomials.

$$
\begin{array}{r}
a + b \\
a - b \\
\hline
-ab - b^2 \\
a^2 + ab \\
\hline
a^2 - b^2 \quad \text{or} \quad a^2 - b^2
\end{array}
$$

FORMULA

THE PRODUCT OF A BINOMIAL AND A TRINOMIAL THAT RESULTS IN A SUM OF CUBES

$(a + b)(a^2 - ab + b^2) = a^3 + b^3$

Note how the factors that yield a sum of cubes are related to each other. The binomial factor is a sum of terms and the three terms of the trinomial are obtained by squaring the first term of the binomial, squaring the second term of the binomial, and taking the negative of the product of the two terms.

$$
\begin{array}{r}
a^2 - ab + b^2 \\
a + b \\
\hline
a^2b - ab^2 + b^3 \\
a^3 - a^2b + ab^2 \\
\hline
a^3 + b^3
\end{array}
$$

FORMULA

> **THE PRODUCT OF A BINOMIAL AND A TRINOMIAL THAT RESULTS IN A DIFFERENCE OF CUBES**
>
> $(a - b)(a^2 + ab + b^2) = a^3 - b^3$

Note how the factors that yield a difference of cubes are related to each other. The binomial factor is a difference of terms and the three terms of the trinomial are obtained by squaring the first term of the binomial, squaring the second term of the binomial, and taking the product of the two terms in the binomial.

$$\begin{array}{r} a^2 + ab + b^2 \\ a - b \\ \hline -a^2b - ab^2 - b^3 \\ a^3 + a^2b + ab^2 \\ \hline a^3 - b^3 \end{array}$$

It is not necessary to memorize these seven special products. However, it is a good idea to become aware of the patterns in these multiplications.

PERFORMANCE OBJECTIVE 13

You should now be able to identify certain special products of polynomials.

$$(5x^2 - y)(5x^2 + y)$$
$$= 25x^4 - y^2$$

EXERCISE SET 4.6

Multiply the following:

1. $5x^2 + 3x + 4$ by $3x^3$
2. $4x^3 - 9x^2 - 3x + 2$ by $-5x^5$
3. $2x^5y^2 - 4z$ by $5xz^2$
4. $-8w^3x^8 + yx$ by $3x^2y$
5. $3x^3 - 2x^2 + 5x - 4$ by $-3x$
6. $16x^2 - 8xy + y^2$ by $-\frac{1}{2}y$
7. $4x^2 - 5x - 6$ by $-2x^4$
8. $-9xy^3z^2 + 21y^4z$ by $\frac{2}{3}x^3y^2z$
9. $(2x^3 - 3x + 4)(x^2y - 4y)$
10. $(9x^2 - 4y^2)(3x - 2y + 1)$

11. $(2x^2 - 4)(x^3 - 3x + 5)$ 12. $(2x^4 - 3x + 5)(x^2 - 6)$
13. $(5x^4 - 3x^2 + 5)(x^2 - 2)$ 14. $(-4x^2 + x - 2)(3x^2 + 8x - 9)$
15. $(2x^2 - 5x + 6)(3x^2 - 4x + 5)$ 16. $(6x^2y - 5xy^2)(x^2y + y^2)$
17. $(x - 3)(x^2 + 3x + 9)$ 18. $(2x + 5)(4x^2 - 10x + 25)$
19. $(x + 8)^2$ 20. $(9x - 7)^2$
21. $(x - 16y)(x + 16y)$ 22. $(12x + 5y)(12x - 5y)$
23. $(x - 4)^3$ 24. $(2x + 1)^3$

SECTION 4.7 SUMMARY

In this chapter, we have introduced the basic expressions with which we work in algebra. We have emphasized the evaluation of algebraic expressions, operations on polynomials, and the use of laws of exponents in multiplying and dividing algebraic terms.

Key terms and concepts introduced in this chapter are summarized here.

Constant: a specific number.

Variable (or literal number): a quantity that does not represent a specific value or whose value is not yet known.

Algebraic term: a constant, a variable, or a combination of constants and variables formed by taking products and quotients.

Numerical coefficient of a term: the constant factor associated with the term.

Algebraic expression: an algebraic term, a sum or difference of algebraic terms, or a quotient of sums and differences of algebraic terms.

Rational algebraic expression: an algebraic expression that involves division by variables or by combinations of constants and variables.

Evaluation of an expression: the process of substituting specific numbers in place of the variables and computing the value of the resulting expression.

Symbols of grouping: parentheses ( ), brackets [ ], or braces { } used to group together terms in an algebraic expression.

Order of operations: procedure for evaluating algebraic expressions when there are no symbols of grouping to indicate which operations are to be performed first.

Polynomial: an algebraic expression consisting of sums, differences, and products of numbers and variables raised to whole number powers.

Monomial: a polynomial with one term.

Binomial: a polynomial with two terms.

Trinomial: a polynomial with three terms.

Degree of a monomial: the sum of the exponents on the variables in the monomial.

Degree of a polynomial: the degree of the highest degree term in the polynomial.

Like terms: terms in which the variables and the exponents on the variables are exactly the same.

Distributive law: $a(b + c) = ab + ac$ for any real numbers a, b, and c.

Laws of exponents: rules for simplifying the products and quotients of terms involving exponents.

CHAPTER 4 PRACTICE TEST

1. (PO-1) Rewrite the following expression without any multiplication symbols and use exponents to simplify products wherever possible:

$$2 \cdot 5 \cdot a \cdot a \cdot b \cdot b \cdot b - (-3)(-4)(a)(c)(c)(d)$$

2. (PO-2) Identify the algebraic terms that make up the following algebraic expression:

$$\frac{2}{3}x^3 + 4y + 1$$

Evaluate the following algebraic expressions for the given values of the variables:

3. (PO-3) $x^2 - [-2y^3 - (4x - y)]$ for $x = 15$, $y = -5$

4. (PO-4) $a^3 - 3b^2 \div c + 3$ for $a = \frac{1}{2}$, $b = \frac{2}{3}$, $c = \frac{1}{3}$

CHAPTER 4 PRACTICE TEST

5. (PO-3, 4) $\dfrac{6(2a - b)^2}{c} + 4d \div (d - c)$ for $a = 4$, $b = 3$, $c = -5$, $d = -10$

6. (PO-5) A polynomial with two terms is called a _____.

7. (PO-5) $3x^3y^2 - 5x^2y^4$ is a _____ degree polynomial in x and y.

8. (PO-6) Answer true or false. rst, $\tfrac{3}{4}srt$, and $-2tsr$ are like terms.

Combine like terms to simplify the following:

9. (PO-6) $\dfrac{5}{6}ab + \dfrac{5}{12}ab - \dfrac{5}{18}ab$

10. (PO-7) $(11x - 5y + 6) + (5x - 8y + 4)$

11. (PO-7) $(3x^2 + 2xy - y^2) + (4x^2 - 6xy - 5y^2)$

12. (PO-7) $(3x^2 - 5x + 4) + (2x^3 + 5x^2 - 6)$

13. (PO-7) $(5x^3 - 3x^2 + x) - (2x^3 - 3x^2 - 5x)$

14. (PO-7) $(-7ab + 4ac - 2bc) - (3ab + 4ca - 3cb)$

15. (PO-7) $(5x^3 - 9x) - (-5x^2 - 6x + 2)$

16. (PO-8) $3(6a - 3b) - 5(4a - 4b + c) + 4(-2b + c)$

Simplify the following expressions involving exponents. Give your answer as a single number.

17. (PO-9) $(-3)^{-5}$

18. (PO-9) $\dfrac{5^8}{5^5} + 5^0$

Perform the following multiplications and divisions of algebraic terms. Give your answers in lowest terms.

19. (PO-10) $3^3 x^5 yz^4 (-x^2 yz^3)^4$

20. (PO-10) $\dfrac{-6a^3b^4}{35c^5} \cdot \dfrac{7a^2c^3}{12b^4d}$

21. (PO-10) $\left(-\dfrac{w^2}{4}\right)^3 \div \left(-\dfrac{w}{8}\right)$

22. (PO-11) Multiply $(2x^2y - 4xy + 16xy^2)$ by $5x^2y^3$.

23. (PO-12) Multiply $(4x^2 - 5x + 3)$ by $(2x - 4)$.

24. (PO-13) Multiply $(3x - 4)$ by $(3x + 4)$.

25. (PO-13) Multiply $(x - 5)(x^2 + 5x + 25)$.

CHAPTER 5

FACTORING AND ALGEBRAIC FRACTIONS

5.1 Factoring Polynomials
5.2 Factoring Special Products
5.3 Definition of Algebraic Fractions
5.4 Simplification of Algebraic Fractions
5.5 Multiplication and Division of Algebraic Fractions
5.6 Addition and Subtraction of Algebraic Fractions
5.7 Summary

In elementary mathematics we emphasize the ability to express a composite number as the product of prime factors. Those prime numbers whose product is x are called the prime factors of x. If $x = 30$, then the prime factors of x are 2, 3, and 5. The prime factorization of numbers is important in the reduction of numerical fractions to lowest terms. Prime factorization is also necessary to determine the least common denominator of numerical fractions to be combined.

The factorization of algebraic expressions is no less important. It allows us to express composite algebraic expressions as products of simpler prime algebraic factors. This process simplifies the addition and subtraction of algebraic fractions. Factoring may also lead us to the solution of algebraic equations.

SECTION 5.1 FACTORING POLYNOMIALS

The procedure known as **factoring polynomials** can best be viewed as the reverse procedure of multiplication of polynomials. When multiplying polynomials, we begin with two or more polynomials, multiply them and end up with a single product polynomial. In factoring polynomials, we begin with the product polynomial and try to find the original polynomials that were multiplied to yield the product.

We will consider three basic factoring techniques in this section. The techniques will enable us to

1. Factor any common number and/or variable from all terms in a polynomial.
2. Factor certain second degree polynomials in which the coefficient of the squared term is 1.
3. Factor certain second degree polynomials in which the coefficient of the squared term is an integer.

The Distributive Law is the principle that enables us to factor any common factors from all the terms in a polynomial. No matter what type of polynomial we are attempting to factor, the first step in the factorization process should be to remove any common factors from the terms.

EXAMPLE Factor any common factors from the polynomial

$$2x^2 + x$$

SOLUTION x is common to both terms so we write

$$2x^2 + x = x(2x + 1)$$

in factored form.

EXAMPLE Factor any common factors from the polynomial

$$3x^2 - 15x$$

SOLUTION Both 3 and x are common factors so we write

$$3x^2 - 15x = 3x(x - 5)$$

in factored form.

When we factor common factors from the terms of a polynomial, we can determine the remaining factor by dividing the terms of the original polynomial by the common factors removed.

PERFORMANCE OBJECTIVE 1	
You should now be able to factor any common number and/or variable from a polynomial.	$4y^3 - 2y^2 = 2y^2(2y - 1)$

Next we consider the technique used to factor a second degree polynomial of the form $x^2 + bx + c$ where b and c are integers. To factor the polynomial $x^2 + bx + c$, we must find two integers, say n_1 and n_2, whose product is c and whose sum is b. The polynomial is then factored as $(x + n_1)(x + n_2)$. The following examples illustrate this procedure.

EXAMPLE Find the factors of $x^2 + 7x + 10$.

SOLUTION We must find two integers whose product is 10 and whose sum is 7. The two integers will be 2 and 5. If you do not immediately find the desired pair of integers, list all the pairs of integers whose product is 10 and pick out the pair whose sum is 7. The pairs of integers to be considered would include

$$\begin{aligned} & 1, 10 \\ & -1, -10 \\ & 2, 5 \rightarrow \text{sum is 7} \\ & -2, -5 \end{aligned}$$

$x + 2$ and $x + 5$ are the factors of $x^2 + 7x + 10$ and we write

$$x^2 + 7x + 10 = (x + 2)(x + 5)$$

This factorization can be checked by multiplying $x + 5$ by $x + 2$ to verify that the product is $x^2 + 7x + 10$.

EXAMPLE Find the factors of $x^2 - 3x + 2$.

SOLUTION Two integers whose product is $+2$ and whose sum is -3 are -2 and -1. Therefore,

$$\begin{aligned} x^2 - 3x + 2 &= [x + (-2)][x + (-1)] \\ &= [x - 2][x - 1] \end{aligned}$$

EXAMPLE Find the factors of $x^2 - 4x - 12$.

SOLUTION Two integers that, when multiplied, give -12 and when added give -4 are -6 and 2. Therefore,

$$x^2 - 4x - 12 = (x - 6)(x + 2)$$

RULES FOR FACTORING $x^2 + bx + c$

Write $x^2 + bx + c = (x + n_1)(x + n_2)$ where $n_1 n_2 = c$ and $n_1 + n_2 = b$.

We should note that we cannot always find a pair of integers n_1 and n_2 that satisfy the two conditions needed to factor the polynomial. In such cases we will say that the polynomial is prime with respect to the integers.

PERFORMANCE OBJECTIVE 2	
You should now be able to factor certain second degree polynomials of the form $x^2 + bx + c$.	$x^2 - 5x + 6 = (x - 3)(x - 2)$

A second degree polynomial in which the coefficient of the squared term is any integer other than 1 will have the general form

$$ax^2 + bx + c$$

where $a \neq 1$. To factor the polynomial $ax^2 + bx + c$, we use the following procedure:

Step 1: Multiply the coefficient of x^2 and the constant; that is, multiply a and c.

Step 2: Find the integer factors of this product that when added will give the coefficient of x; that is, find the factors of ac whose sum is b. We will call these factors n_1 and n_2.

Step 3: Rewrite the polynomial using the new factors as coefficients of x; that is, substitute $n_1 x + n_2 x$ in place of bx to obtain the polynomial $ax^2 + n_1 x + n_2 x + c$.

Step 4: Group the four terms of the polynomial into pairs and factor each pair of terms by removing the common factors.

Step 5: Apply the Distributive Law and write the factors of the polynomial.

FACTORING POLYNOMIALS 5.1

The following examples illustrate this procedure. The five steps are labeled in each example.

EXAMPLE Factor $6x^2 + 13x - 28$.

SOLUTION

Step 1. $6(-28) = -168$

Step 2. Factors of -168 are listed as follows:

$+168$	-1	-168	$+1$
$+84$	-2	-84	$+2$
$+56$	-3	-56	$+3$
$+42$	-4	-42	$+4$
$+28$	-6	-28	$+6$
$+24$	-7	-24	$+7$
$+21$	-8	-21	$+8$
$+14$	-12	-14	$+12$

21 and -8 is the pair whose sum is $+13$.

Step 3. $6x^2 + 13x - 28 = 6x^2 - 8x + 21x - 28$

Step 4. $6x^2 - 8x + 21x - 28 = (6x^2 - 8x) + (21x - 28)$
$$= 2x(3x - 4) + 7(3x - 4)$$

Step 5. $2x(3x - 4) + 7(3x - 4) = (2x + 7)(3x - 4)$

Therefore, $(2x + 7)$ and $(3x - 4)$ are the factors of $6x^2 + 13x - 28$.

Note that the same answer would have been obtained if we had interchanged the $-8x$ and $21x$ in Step 3. We then would have had

$$6x^2 + 13x - 28 = 6x^2 + 21x - 8x - 28$$
$$= (6x^2 + 21x) - (8x + 28)$$
$$= 3x(2x + 7) - 4(2x + 7)$$
$$= (3x - 4)(2x + 7)$$

EXAMPLE Factor $4x^2 + 12x - 7$.

SOLUTION

Step 1. $4(-7) = -28$

Step 2. Factors of -28 are listed as follows:

$$\begin{array}{cccc} -28 & +1 & +28 & -1 \\ -14 & +2 & +14 & -2 \\ -7 & +4 & +7 & -4 \end{array}$$

$+14$ and -2 is the pair whose sum is $+12$.

Step 3. $4x^2 + 12x - 7 = 4x^2 - 2x + 14x - 7$

Step 4. $4x^2 - 2x + 14x - 7 = (4x^2 - 2x) + (14x - 7)$
$$= 2x(2x - 1) + 7(2x - 1)$$

Step 5. $2x(2x - 1) + 7(2x - 1) = (2x + 7)(2x - 1)$

Therefore, $(2x + 7)$ and $(2x - 1)$ are the factors of $4x^2 + 12x - 7$.

RULES FOR FACTORING $ax^2 + bx + c$

Write $ax^2 + bx + c = ax^2 + n_1x + n_2x + c$ where $n_1 n_2 = ac$ and $n_1 + n_2 = b$. Then complete a pairwise factorization of this new expression to identify common factors.

Again there is no guarantee that the numbers n_1 and n_2 can be found among the integers. In such cases we say that the second degree polynomial is prime with respect to the integers.

PERFORMANCE OBJECTIVE 3	$2x^2 + 7x - 15 = 2x^2 + 10x - 3x - 15$
You should now be able to factor certain second degree polynomials of the form $ax^2 + bx + c$.	$= 2x(x+5) - 3(x+5)$
	$= (2x - 3)(x + 5)$

EXERCISE SET 5.1

Factor the following polynomials if possible:

1. $-25x^2 + 10x$
2. $36x^2 - 12x$
3. $3x^2 - 3x + 9$
4. $-5x^2 + 5x + 25$
5. $7x^2 - 14x - 21$
6. $2x^2 + 6x - 20$
7. $x^2 + 12x + 20$
8. $x^2 + 10x + 21$
9. $x^2 - 7x - 18$
10. $x^2 + 2x - 24$
11. $x^2 - 8x + 16$
12. $x^2 - 12x + 36$

13. $6x^2 + 17x + 5$ 14. $4x^2 + 20x + 9$

15. $4x^2 - 2x - 12$ 16. $15x^2 + 25x - 60$

17. $3x^2 + x - 2$ 18. $4x^2 + 8x - 5$

19. $3x^2 + 17x - 9$ 20. $5x^2 + 19x - 6$

SECTION 5.2 FACTORING SPECIAL PRODUCTS

Special product is a name given to a polynomial that has a readily identified factorization. We discussed some special products in Chapter 4.

We consider first a factoring in which two identical factors occur. The factors are the sum of two monomials represented by the letters a and b. The factorization is then $(a + b)(a + b)$ and we call the original polynomial a **perfect square trinomial.**

DEFINITION

PERFECT SQUARE TRINOMIAL

A polynomial is a **perfect square trinomial** if it satisfies the following conditions:

1. Two of the terms are the squares of monomials.
2. The remaining term is twice the product of the monomials.

Typical perfect square trinomials thus have the form:

$$a^2 + 2ab + b^2$$

EXAMPLE Show that $x^2 + 2x + 1$ is a perfect square trinomial.

SOLUTION In $x^2 + 2x + 1$, the first term is the square of x, the third term is the square of 1 and the second term is twice the product of x and 1. Therefore, $x^2 + 2x + 1$ is a perfect square trinomial.

EXAMPLE Show that $4x^2 - 12x + 9$ is a perfect square trinomial.

SOLUTION The first term is the square of $2x$, the third term is the square of 3 and the second term is twice the product of $2x$ and 3 but with the wrong sign. However, we could consider the third term, 9, as the square of -3 so that the second term is twice the product of $2x$ and -3. Therefore, $4x^2 - 12x + 9$ is a perfect square trinomial.

EXAMPLE   Find the factors of $4x^2 - 12x + 9$.

SOLUTION   We have already determined that $4x^2 - 12x + 9$ is a perfect square trinomial. In the process of doing this we had to select a and b as $2x$ and -3. Then

$$4x^2 - 12x + 9 = (2x - 3)^2$$

Note that $(-2x + 3)^2$ is also a valid factorization.

Next we consider a factoring in which three identical factors occur. The factors are the sum of two monomials represented by the letters a and b. The factorization is then $(a + b)^3$ and we call the original polynomial a **perfect cube.**

DEFINITION

PERFECT CUBE FOUR TERM POLYNOMIAL

A polynomial with four terms is a perfect cube if it satisfies the following conditions:

1. Two of the terms are cubes of the monomials a and b, that is, a^3 and b^3.
2. One term is three times the product of the square of monomial a and monomial b, that is, $3a^2b$.
3. One term is three times the product of monomial a and the square of monomial b, that is, $3ab^2$.

A perfect cube polynomial would then have the form

$$a^3 + 3a^2b + 3ab^2 + b^3$$

EXAMPLE   Factor the polynomial $x^3 + 12x^2 + 48x + 64$.

SOLUTION   We note that the first term is the cube of x, the fourth term is the cube of 4, the second term is

$$3(x)^2(4) = 12x^2$$

and the third term is

$$3(x)(4)^2 = 3(x)(16) = 48x$$

The polynomial can be written as

$$(x)^3 + 3(x)^2(4) + 3(x)(4)^2 + (4)^3 = (x + 4)^3$$

FACTORING SPECIAL PRODUCTS 5.2

EXAMPLE Factor the polynomial $8x^3 + 36x^2 + 54x + 27$.

SOLUTION We note that the first term is the cube of $2x$, the fourth term is the cube of 3, the second term is

$$3(2x)^2(3) = 3(4x^2)(3) = 36x^2$$

and the third term is

$$3(2x)(3)^2 = 3(2x)(9) = 54x$$

The polynomial can be written as

$$(2x)^3 + 3(2x)^2(3) + 3(2x)(3)^2 + (3)^3 = (2x + 3)^3$$

EXAMPLE Factor the polynomial $125x^3 - 300x^2 + 240x - 64$.

SOLUTION We note that $125x^3$ is the cube of $5x$, -64 is the cube of -4,

$$-300x^2 = 3(5x)^2(-4)$$

and

$$240x = 3(5x)(-4)^2$$

The polynomial can be written as

$$(5x)^3 + 3(5x)^2(-4) + 3(5x)(-4)^2 + (-4)^3 = (5x - 4)^3$$

Next we look at a **difference of squares binomial** that factors into the form $(a + b)(a - b)$.

DEFINITION

DIFFERENCE OF SQUARES BINOMIAL

A polynomial is a **difference of squares binomial** if is satisfies the following conditions:

1. Both terms are perfect squares.
2. The terms have different signs.

A difference of squares binomial then has the form

$$a^2 - b^2$$

EXAMPLE Factor the polynomial $x^2 - 4$.

SOLUTION This binomial is the difference of squares since x^2 is the square of x and 4 is the square of 2. So the polynomial can be written as

$$x^2 - 4 = (x)^2 - (2)^2 = (x + 2)(x - 2)$$

EXAMPLE Factor the polynomial $25x^2 - 16$.

SOLUTION This binomial is also the difference of squares since $25x^2$ is the square of $5x$ and 16 is the square of 4. So the polynomial can be written as

$$25x^2 - 16 = (5x)^2 - (4)^2 = (5x + 4)(5x - 4)$$

The next polynomial considered is a **sum of cubes binomial,** which factors into the form $(a + b)(a^2 - ab + b^2)$.

DEFINITION

SUM OF CUBES BINOMIAL

A polynomial is a **sum of cubes binomial** if it satisfies the condition that both terms are perfect cubes.

The sum of cubes binomial then has the form

$$a^3 + b^3$$

The two factors are

1. The sum of the cube roots of the two terms, that is, $(a + b)$.
2. The sum of the squares of the cube roots of the two terms minus the product of the cube roots of the terms, that is, $(a^2 - ab + b^2)$. You should note that this factor is prime provided that a and b have no common factors.

EXAMPLE Factor the polynomial $x^3 + 27$.

SOLUTION This binomial is a sum of cubes since

$$x^3 + 27 = (x)^3 + (3)^3$$

Its factors are then

$$(x + 3)(x^2 - 3x + 9)$$

FACTORING SPECIAL PRODUCTS 5.2

EXAMPLE Factor the polynomial $64x^3 + 8$.

SOLUTION This binomial is a sum of cubes since

$$64x^3 + 8 = (4x)^3 + (2)^3$$

Its factors are then

$$(4x + 2)(16x^2 - 8x + 4)$$

EXAMPLE Factor the polynomial $27x^3 - 125$.

SOLUTION This binomial is a sum of cubes since

$$27x^3 - 125 = (3x)^3 + (-5)^3$$

Its factors are then

$$(3x - 5)(9x^2 + 15x + 25)$$

PERFORMANCE OBJECTIVE 4	$81x^2 - 36 = (9x - 6)(9x + 6)$
You should now be able to factor any special product.	$8x^3 + 27 = (2x + 3)(4x^2 - 6x + 9)$

EXERCISE SET 5.2

Factor the following polynomials if possible:

1. $x^3 + 125$
2. $x^3 + 512$
3. $x^2 - 36$
4. $x^2 - 49$
5. $x^2 - 6x + 9$
6. $x^2 - 8x + 16$
7. $x^3 + 3x^2 + 3x + 1$
8. $x^3 + 6x^2 + 12x + 8$
9. $x^3 - 18x^2 + 108x - 216$
10. $x^3 - 21x^2 + 147x - 343$
11. $x^2 + 18x + 81$
12. $x^2 + 20x + 100$
13. $x^3 - 64$
14. $x^3 - 729$
15. $x^2 + 49$
16. $x^2 + 64$
17. $x^3 + 343$
18. $x^3 + 216$
19. $x^2 + 10x + 25$
20. $x^2 + 24x + 144$

SECTION 5.3 DEFINITION OF ALGEBRAIC FRACTIONS

Recall that a fraction is any number that can be expressed in the form

$$\frac{a}{b}$$

where a and b are integers and $b \neq 0$. We define an algebraic fraction in a similar manner.

DEFINITION

> **ALGEBRAIC FRACTION**
>
> An expression of the form $\frac{P}{Q}$, where P and Q are any polynomials and $Q \neq 0$, is an **algebraic fraction**.

An algebraic fraction is said to be undefined for any values of the variable that make the denominator zero. Although algebraic fractions may involve more than one variable, most of the fractions we will work with in this chapter will involve a single variable. Examples of algebraic fractions are

$$\frac{x+3}{x^2+2x+1}$$

$$\frac{2x^3+y}{x^2-xy+y^2}$$

$$\frac{6}{x^2}+\frac{2}{x}$$

$$5x^4-x^3+2x-7$$

The first two examples are already in the form $\frac{P}{Q}$. Later in the chapter we will see how to write expressions such as the third example in the form $\frac{P}{Q}$ by finding a common denominator for the terms. The last example is a polynomial. A polynomial may be considered to be an algebraic fraction in which $Q = 1$.

PERFORMANCE OBJECTIVE 5

You should now be able to identify each algebraic expression as a polynomial or as an algebraic fraction.

$3y^3 - 1$ ... a polynomial

$\frac{5}{y} - \frac{2}{y^3}$ ... algebraic fraction

EXERCISE SET 5.3

Identify each algebraic expression as a polynomial or as an algebraic fraction.

1. a. $\dfrac{7x}{9 - 2x}$ b. $3x^2 - 2x + 6$

   c. $\dfrac{5}{x} - \dfrac{3}{x^2}$ d. $x^3 - \dfrac{2}{x^2} + 5x - 10$

2. a. $\dfrac{-5x}{3x + 4}$ b. $4x^2 - 3x + 12$

   c. $\dfrac{-2}{x^3} + \dfrac{7}{x}$ d. $3x^3 + \dfrac{5}{x^2} - 5x - 12$

Identify P and Q in each of the following algebraic fractions:

3. a. $\dfrac{7x^2 + 1}{3x}$ b. $3x^2 + 5x - 2$

   c. $\dfrac{1}{x^2 + 2}$

4. a. $\dfrac{3x^2 - 4}{5x}$ b. $4x^2 - 6x + 3$

   c. $\dfrac{12}{x^2 + 3x + 4}$

SECTION 5.4 SIMPLIFICATION OF ALGEBRAIC FRACTIONS

To **simplify** an algebraic fraction means to reduce the algebraic fraction to lowest terms. An algebraic fraction is said to be in **lowest terms** when the numerator and denominator do not contain a common factor. For instance,

$$\frac{15}{25x^2}$$

can be written as

$$\frac{3}{5x^2}$$

in lowest terms since 5 is a factor common to both 15 and $25x^2$. Remember also that an algebraic fraction such as

$$\frac{3}{5x^2}$$

is undefined for any value of x that makes the denominator zero. Therefore,

$$\frac{3}{5x^2}$$

is undefined for $x = 0$.

When simplifying algebraic fractions, we continually use the laws of exponents. Those laws as developed in Chapter 4 are reproduced here. Examples are given to illustrate each law.

1. $a^m \cdot a^n = a^{m+n}$

$$x^3 \cdot x^5 = x^{3+5} = x^8$$

$$(x-4)^2(x-4) = (x-4)^{2+1} = (x-4)^3$$

2. $(a^m)^n = a^{m \cdot n}$

$$(x^3)^5 = x^{3 \cdot 5} = x^{15}$$

$$[(x^2+3)^4]^2 = (x^2+3)^{4 \cdot 2} = (x^2+3)^8$$

3. $(ab)^m = a^m b^m$

$$(3x)^4 = 3^4 x^4 = 81x^4$$

$$[(2x+1)(x-2)]^3 = (2x+1)^3(x-2)^3$$

4. $\dfrac{a^m}{a^n} = a^{m-n}$ for $a \neq 0$

$$\frac{x^5}{x^5} = x^{5-5} = x^0 = 1$$

$$\frac{(x+y)^2}{x+y} = (x+y)^{2-1} = x+y$$

$$\frac{x^4}{x^7} = x^{4-7} = x^{-3} = \frac{1}{x^3}$$

If an exponent in a term is negative, we rewrite that term so that the exponent is positive.

5. $\left(\dfrac{a}{b}\right)^m = \dfrac{a^m}{b^m}$ for $b \neq 0$

$$\left(\dfrac{3}{x}\right)^3 = \dfrac{3^3}{x^3} = \dfrac{27}{x^3}$$

$$\left(\dfrac{x^2 + 3}{x + 4}\right)^4 = \dfrac{(x^2 + 3)^4}{(x + 4)^4}$$

PERFORMANCE OBJECTIVE 6

You should now be able to use the properties of exponents to simplify algebraic fractions.

$$\left[\dfrac{(2y-3)^2}{(2y-3)^3}\right]^4 = \dfrac{(2y-3)^8}{(2y-3)^{12}} = (2y-3)^{8-12}$$
$$= \dfrac{1}{(2y-3)^4}$$

Factoring is used extensively in simplifying algebraic fractions. Factoring the polynomials in the numerator and denominator allows us to identify any common factors. These factors are then removed by elimination. For example, to simplify

$$\dfrac{x^2 + 5x + 4}{x^2 + 3x - 4}$$

we factor the numerator and denominator and eliminate the common factor $x + 4$.

$$\dfrac{x^2 + 5x + 4}{x^2 + 3x - 4} = \dfrac{(x + 4)(x + 1)}{(x + 4)(x - 1)} = \dfrac{x + 1}{x - 1}$$

We now work four more examples to illustrate the method of simplification of algebraic fractions.

EXAMPLE Simplify $\dfrac{x + 2}{2 + x}$.

SOLUTION $\dfrac{x + 2}{2 + x} = \dfrac{x + 2}{x + 2} = 1$

since the $x + 2$ is a common factor in both the numerator and denominator.

5 FACTORING AND ALGEBRAIC FRACTIONS

EXAMPLE Simplify $\dfrac{-x + y}{x - y}$.

SOLUTION $\dfrac{-x + y}{x - y} = \dfrac{-1(x - y)}{x - y} = -1$

since factoring -1 from the terms in the numerator yields the common factor $x - y$ in both the numerator and denominator.

EXAMPLE Simplify $\dfrac{3x^2 - 3y^2}{x^3 + y^3}$.

SOLUTION $\dfrac{3x^2 - 3y^2}{x^3 + y^3} = \dfrac{3(x^2 - y^2)}{(x + y)(x^2 - xy + y^2)}$

$= \dfrac{3(x + y)(x - y)}{(x + y)(x^2 - xy + y^2)} = \dfrac{3(x - y)}{x^2 - xy + y^2}$

To factor the numerator, we first removed the factor 3 from both terms and then factored the difference of squares $x^2 - y^2$. To factor the denominator, we used the technique for factoring a sum of cubes. Factoring the numerator and denominator yielded the common factor $x + y$ which we eliminated.

EXAMPLE Simplify $\dfrac{25x^2 - 49}{25x^2 - 70x + 49}$.

SOLUTION $\dfrac{25x^2 - 49}{25x^2 - 70x + 49} = \dfrac{(5x - 7)(5x + 7)}{(5x - 7)(5x - 7)} = \dfrac{5x + 7}{5x - 7}$

The numerator is a difference of squares and the denominator is a perfect square trinomial. After factoring these polynomials and eliminating the common factor $5x - 7$, we could not simplify any further even though the numerator and denominator were very similar.

RULES FOR SIMPLIFYING ALGEBRAIC FRACTIONS

To simplify algebraic fractions, prime factor the numerator and denominator polynomials and eliminate any common factors.

PERFORMANCE OBJECTIVE 7	
You should now be able to simplify algebraic fractions by factoring.	$\dfrac{x^2 - 3x - 4}{x^2 + 2x + 1} = \dfrac{(x - 4)(x + 1)}{(x + 1)(x + 1)}$ $= \dfrac{x - 4}{x + 1}$

EXERCISE SET 5.4

Simplify each of the following algebraic fractions:

1. $\dfrac{x^3 + 27}{x^2 + 2x - 3}$

2. $\dfrac{x^3 + 125}{x^2 + 2x - 15}$

3. $\dfrac{3x - 6}{2x^2 - 7x + 6}$

4. $\dfrac{6x + 18}{4x^2 + 11x - 3}$

5. $\dfrac{-t - 1}{t + 1}$

6. $\dfrac{2x + 3}{-2x - 3}$

7. $\dfrac{x^2 - 4}{2x - 4}$

8. $\dfrac{x^2 - 16}{5x - 20}$

9. $\dfrac{-2x^2 - 18x - 28}{8 + x^3}$

10. $\dfrac{-5x^2 - 40x - 80}{64 + x^3}$

SECTION 5.5 MULTIPLICATION AND DIVISION OF ALGEBRAIC FRACTIONS

We multiply algebraic fractions in the same way as we multiply fractions. The numerators are multiplied to form the numerator of the product and the denominators are multiplied to form the denominator of the product. Since we want the product to be in lowest terms, it is best to eliminate common factors from the numerator and denominator before multiplying. In order to identify the common factors, it may be necessary to factor the numerators and denominators first. The following examples illustrate the multiplication of algebraic fractions. Final reduced products will be left in factored form.

EXAMPLE Perform the multiplication

$$\dfrac{x - 2}{x + 2} \cdot \dfrac{x + 2}{(x + 2)(x - 2)}.$$

SOLUTION

$$\dfrac{x - 2}{x + 2} \cdot \dfrac{x + 2}{(x + 2)(x - 2)} = \dfrac{(x - 2)(x + 2)}{(x + 2)(x + 2)(x - 2)} = \dfrac{1}{x + 2}$$

Observe that every polynomial was factored, so that we were able to eliminate common factors and then multiply.

5 FACTORING AND ALGEBRAIC FRACTIONS

EXAMPLE Perform the multiplication

$$\frac{x^2 - 14x + 49}{7x - x^2} \cdot \frac{5x - 35}{x^2 - 10x + 21}$$

SOLUTION We first factor the polynomials in order to identify common factors.

$$\frac{x^2 - 14x + 49}{7x - x^2} \cdot \frac{5x - 35}{x^2 - 10x + 21} = \frac{(x - 7)(x - 7)(5)(x - 7)}{x(7 - x)(x - 7)(x - 3)}$$

$$= \frac{(x - 7)(5)(x - 7)}{x(7 - x)(x - 3)}$$

After we factored, we eliminated immediately one factor of $x - 7$ from both the numerator and denominator. If we rewrite the $7 - x$ in the denominator as $(-1)(x - 7)$, we can then eliminate another factor of $x - 7$ in the numerator and denominator. We do this as follows:

$$\frac{(x - 7)(5)(x - 7)}{x(7 - x)(x - 3)} = \frac{(x - 7)(5)(x - 7)}{x(-1)(x - 7)(x - 3)} = \frac{5(x - 7)}{-x(x - 3)}$$

EXAMPLE Perform the multiplication

$$\frac{x^2 - x - 12}{x^3 - 2x^2 + 5x - 10} \cdot \frac{10 - 5x}{x^2 - 9}$$

SOLUTION The solution of this problem involves the same steps as used in the previous example. The factorization yields

$$x^2 - x - 12 = (x - 4)(x + 3)$$
$$10 - 5x = 5(2 - x)$$

$x^3 - 2x^2 + 5x - 10$ can be factored by pairs as follows:

$$x^3 - 2x^2 + 5x - 10 = (x^3 - 2x^2) + (5x - 10)$$
$$= x^2(x - 2) + 5(x - 2)$$
$$= (x^2 + 5)(x - 2)$$

$$x^2 - 9 = (x - 3)(x + 3)$$

Substitution of factors gives

$$\frac{(x-4)(x+3)(5)(2-x)}{(x^2+5)(x-2)(x-3)(x+3)} = \frac{(x-4)(5)(2-x)}{(x^2+5)(x-2)(x-3)}$$

$$= \frac{(x-4)(5)(-1)(x-2)}{(x^2+5)(x-2)(x-3)}$$

$$= \frac{-5(x-4)}{(x^2+5)(x-3)}$$

PERFORMANCE OBJECTIVE 8	
You should now be able to multiply algebraic fractions.	$\dfrac{y^2-9}{y^2-y-12} \cdot \dfrac{y^2-10y+24}{y^2-5y-6} =$ $\dfrac{(y+3)(y-3)}{(y-4)(y+3)} \cdot \dfrac{(y-6)(y-4)}{(y-6)(y+1)} = \dfrac{y-3}{y+1}$

When we divide algebraic fractions, we again return to fractions to serve as a guide for the division. When we divide fractions, the fraction that is the divisor is inverted and then we multiply as before. For example,

$$\frac{1}{2} \div \frac{1}{3} \quad \text{becomes} \quad \frac{1}{2} \cdot \frac{3}{1} = \frac{3}{2}$$

With algebraic fractions, we follow the same procedure.

EXAMPLE Perform the division $\dfrac{x^2-4}{x-3} \div \dfrac{x^2-3x+2}{x^2-9}$.

SOLUTION The problem becomes

$$\frac{x^2-4}{x-3} \cdot \frac{x^2-9}{x^2-3x+2} \quad \text{(divisor is inverted)}$$

Each polynomial is factored to give

$$\frac{x^2-4}{x-3} \cdot \frac{x^2-9}{x^2-3x+2} = \frac{(x-2)(x+2)}{x-3} \cdot \frac{(x-3)(x+3)}{(x-2)(x-1)}$$

$$= \frac{(x+2)(x+3)}{x-1} \quad \text{(common factors are eliminated)}$$

> **PERFORMANCE OBJECTIVE 9**
>
> You should now be able to divide algebraic fractions.
>
> $$\frac{x+4}{x^2-4} \div \frac{x^2+9x+20}{x^2-x-6}$$
> $$= \frac{x+4}{(x-2)(x+2)} \cdot \frac{(x-3)(x+2)}{(x+4)(x+5)} = \frac{x-3}{(x-2)(x+5)}$$

RULES FOR MULTIPLYING AND DIVIDING ALGEBRAIC FRACTIONS

Procedures for multiplying and dividing algebraic fractions are the same as for multiplying and dividing numerical fractions.

EXERCISE SET 5.5

Perform the operation that is indicated in the following exercises:

1. $\dfrac{2x}{x+4} \cdot \dfrac{x^2-16}{8x^2}$

2. $\dfrac{5x}{x-3} \cdot \dfrac{x^2-9}{20x^2}$

3. $\dfrac{x-5}{5-x} \cdot \dfrac{x+2}{2+x}$

4. $\dfrac{x-7}{7-x} \cdot \dfrac{x+8}{8+x}$

5. $\dfrac{x^2-3x+2}{x^2-1} \cdot \dfrac{x^2+4x+3}{4x-8}$

6. $\dfrac{x^2+11x+30}{x^2-2x-48} \cdot \dfrac{x^2-11x+24}{3x+15}$

7. $\dfrac{x^3-1}{x-1} \div \dfrac{x^3+1}{x+1}$

8. $\dfrac{x^3-216}{x-6} \div \dfrac{x^3+216}{x+6}$

9. $\dfrac{x^2+2x+1}{x^2+2x} \div \dfrac{x^2-x-2}{x^2-4}$

10. $\dfrac{x^2+4x+4}{x^2+4x} \div \dfrac{x^2-2x-8}{x^2-16}$

SECTION 5.6 ADDITION AND SUBTRACTION OF ALGEBRAIC FRACTIONS

When we add or subtract algebraic fractions, each fraction must have the same denominator before the numerators can be combined. If the algebraic fractions do not have the same denominator, it is necessary

ADDITION AND SUBTRACTION OF ALGEBRAIC FRACTIONS 5.6

to find the least common denominator (LCD). The method for finding the LCD was outlined for rational numbers in Chapter 1. That method is equally valid for algebraic fractions and it is repeated here.

RULES FOR FINDING THE LCD

Step 1: Write the prime factorization of each of the denominators.

Step 2: Write each number or expression that appears as a factor the greatest number of times it appears in any one of the denominators.

Step 3: Multiply all the factors listed in Step 2. This product is the least common denominator.

EXAMPLE   Find the LCD for the following algebraic fractions:

$$\frac{x+2}{3x-6} \text{ and } \frac{x+3}{2x-4}$$

SOLUTION

Step 1: $3x - 6 = 3(x - 2)$
$2x - 4 = 2(x - 2)$

Step 2: $2, x - 2,$ and 3 are the factors.

Step 3: LCD $= 2 \cdot 3(x - 2) = 6(x - 2)$

EXAMPLE   Find the LCD for the following algebraic fractions:

$$\frac{7x}{x^2 - 25} \text{ and } \frac{5x}{x^2 + 5x}$$

SOLUTION

Step 1: $x^2 - 25 = (x - 5)(x + 5)$
$x^2 + 5x = x(x + 5)$

Step 2: $x, x + 5,$ and $x - 5$ are the factors.

Step 3: LCD $= x(x + 5)(x - 5) = x(x^2 - 25)$

EXAMPLE   Find the LCD for the following algebraic fractions:

$$\frac{3}{x^2 + 10x + 25} \text{ and } \frac{5x}{3x^2 + 15x}$$

SOLUTION

Step 1: $x^2 + 10x + 25 = (x + 5)(x + 5)$

$3x^2 + 15x = 3x(x + 5)$

Step 2: 3, x, $x + 5$, and $x + 5$ are the factors. Note that $x + 5$ appears as a factor twice in one denominator and so it is written twice.

Step 3: LCD $= 3x(x + 5)(x + 5) = 3x(x + 5)^2$

PERFORMANCE OBJECTIVE 10	$\dfrac{x-3}{9x^2-12x+4} + \dfrac{x+4}{3x^2+7x-6}$
You should now be able to find the least common denominator for algebraic fractions.	$9x^2-12x+4 = (3x-2)(3x-2)$ $3x^2+7x-6 = (3x-2)(x+3)$ LCD $= (3x-2)^2(x+3)$

To add algebraic fractions that have the same denominator, we add the numerators and put the sum over the denominator. For example,

$$\frac{x^2 - 3}{x^2 - 4} + \frac{4x + 7}{x^2 - 4} = \frac{x^2 - 3 + 4x + 7}{x^2 - 4} = \frac{x^2 + 4x + 4}{x^2 - 4}$$

$$= \frac{(x + 2)(x + 2)}{(x + 2)(x - 2)} = \frac{x + 2}{x - 2}$$

Answers should always be reduced to lowest terms as in the preceding example.

To add algebraic fractions with different denominators, we find the least common denominator for the fractions, write each fraction in an equivalent form that has the LCD as its denominator, and then add the fractions as previously discussed.

RULES FOR ADDING ALGEBRAIC FRACTIONS WITH DIFFERENT DENOMINATORS

Step 1: Find the LCD.

Step 2: Write each algebraic fraction in an equivalent form that has the LCD as its denominator.

Step 3: Add the numerators and place over the common denominator.

Step 2 in the procedure will require some explanation. To write a fraction in an equivalent form having the LCD as its denominator, we must multiply both numerator and denominator by those factors

ADDITION AND SUBTRACTION OF ALGEBRAIC FRACTIONS 5.6

in the LCD which are missing in the fraction's denominator. For example, to place the fraction

$$\frac{5}{x-3}$$

in an equivalent form when the LCD is $2(x-3)(x+5)$, we would multiply both numerator and denominator by $2(x+5)$ since those two factors are in the LCD but not in the fraction's denominator. Then

$$\frac{5}{(x-3)} = \frac{5(2)(x+5)}{(x-3)(2)(x+5)}$$

To illustrate the procedure for adding algebraic fractions with different denominators, we will complete the examples for Performance Objective 10.

EXAMPLE Perform the addition $\dfrac{x+2}{3x-6} + \dfrac{x+3}{2x-4}$.

SOLUTION The LCD has been identified as $6(x-2)$. The first denominator ($3x-6$) must be multiplied by 2 to convert it to the LCD. So we must also multiply the first numerator by 2. The second denominator ($2x-4$) must be multiplied by 3 to convert it to the LCD. So we must also multiply the second numerator by 3. Thus, the problem becomes

$$\frac{2(x+2)}{6(x-2)} + \frac{3(x+3)}{6(x-2)} = \frac{2x+4+3x+9}{6(x-2)} = \frac{5x+13}{6(x-2)}$$

This answer cannot be reduced any further.

EXAMPLE Perform the addition

$$\frac{7x}{x^2-25} + \frac{5x}{x^2+5x}$$

That is,

$$\frac{7x}{(x+5)(x-5)} + \frac{5x}{x(x+5)}$$

SOLUTION The LCD has been identified as

$$x(x+5)(x-5) \quad \text{or} \quad x(x^2-25)$$

We must multiply the first numerator and denominator by x and the second numerator and denominator by $x - 5$. Thus, the problem becomes

$$\frac{(7x)(x)}{x(x + 5)(x - 5)} + \frac{5x(x - 5)}{x(x + 5)(x - 5)}$$
$$= \frac{(7x)(x) + 5x(x - 5)}{x(x + 5)(x - 5)} = \frac{7x^2 + 5x^2 - 25x}{x(x + 5)(x - 5)}$$
$$= \frac{12x^2 - 25x}{x(x + 5)(x - 5)} = \frac{x(12x - 25)}{x(x + 5)(x - 5)}$$

Reducing to lowest terms gives

$$\frac{12x - 25}{x^2 - 25} \quad \text{(The } x \text{ has been eliminated from the numerator and the denominator.)}$$

EXAMPLE Perform the addition

$$\frac{3}{x^2 + 10x + 25} + \frac{5x}{3x^2 + 15x}$$

That is,

$$\frac{3}{(x + 5)(x + 5)} + \frac{5x}{3x(x + 5)}$$

SOLUTION The LCD has been identified as $3x(x + 5)^2$. We must multiply the first numerator and denominator by $3x$ and the second numerator and denominator by $x + 5$. Thus the problem becomes

$$\frac{3(3x)}{3x(x + 5)^2} + \frac{5x(x + 5)}{3x(x + 5)^2} = \frac{9x + 5x^2 + 25x}{3x(x + 5)^2}$$
$$= \frac{5x^2 + 34x}{3x(x + 5)^2} = \frac{x(5x + 34)}{3x(x + 5)^2}$$
$$= \frac{5x + 34}{3(x + 5)^2}$$

PERFORMANCE OBJECTIVE 11	$\dfrac{x-3}{9x^2-12x+4} + \dfrac{x+4}{3x^2+7x-6}$
You should now be able to add algebraic fractions.	$= \dfrac{(x-3)(x+3)}{(3x-2)^2(x+3)} + \dfrac{(x+4)(3x-2)}{(3x-2)^2(x+3)} = \dfrac{4x^2+10x-17}{(3x-2)^2(x+3)}$

ADDITION AND SUBTRACTION OF ALGEBRAIC FRACTIONS 5.6

To subtract algebraic fractions, we change the sign of every term in the numerator of the subtrahend and then add the fractions by following the addition procedure.

EXAMPLE $\dfrac{7}{3x} - \dfrac{2}{3x}$

SOLUTION $\dfrac{7}{3x} - \dfrac{2}{3x} = \dfrac{7}{3x} + \dfrac{-2}{3x} = \dfrac{5}{3x}$

EXAMPLE $\dfrac{2x}{x+2} - \dfrac{3}{x+2}$

SOLUTION $\dfrac{2x}{x+2} - \dfrac{3}{x+2} = \dfrac{2x}{x+2} - \dfrac{-3}{x+2} = \dfrac{2x-3}{x+2}$

EXAMPLE $\dfrac{x+2}{3x-6} - \dfrac{x+3}{2x-4}$

SOLUTION $\dfrac{x+2}{3x-6} - \dfrac{x+3}{2x-4} = \dfrac{x+2}{3x-6} + \dfrac{-x-3}{2x-4}$

$= \dfrac{x+2}{3(x-2)} + \dfrac{-x-3}{2(x-2)}$

$= \dfrac{(x+2)(2)}{3(x-2)(2)} + \dfrac{(-x-3)(3)}{2(x-2)(3)}$

$= \dfrac{2x+4}{6(x-2)} + \dfrac{-3x-9}{6(x-2)} = \dfrac{-x-5}{6(x-2)}$

EXAMPLE $\dfrac{7x}{x^2-25} - \dfrac{5x}{x^2+5x}$

SOLUTION

$\dfrac{7x}{x^2-25} - \dfrac{5x}{x^2+5x} = \dfrac{7x}{x^2-25} + \dfrac{-5x}{x^2+5x}$

$= \dfrac{7x}{(x-5)(x+5)} + \dfrac{-5x}{x(x+5)}$

$= \dfrac{7x(x)}{(x-5)(x+5)(x)} + \dfrac{-5x(x-5)}{(x)(x+5)(x-5)}$

$= \dfrac{7x^2}{(x-5)(x+5)(x)} + \dfrac{-5x^2+25x}{(x)(x-5)(x+5)}$

$= \dfrac{2x^2+25x}{(x)(x-5)(x+5)} = \dfrac{x(2x+25)}{(x)(x-5)(x+5)}$

$= \dfrac{2x+25}{(x-5)(x+5)}$

PERFORMANCE OBJECTIVE 12

You should now be able to subtract algebraic fractions.

$$\frac{x-3}{9x^2-12x+4} - \frac{x+4}{3x^2+7x-6}$$

$$= \frac{(x-3)(x+3)}{(3x-2)^2(x+3)} - \frac{(x+4)(3x-2)}{(3x-2)^2(x+3)} = \frac{-2x^2-10x-1}{(3x-2)^2(x+3)}$$

RULES FOR ADDING AND SUBTRACTING ALGEBRAIC FRACTIONS

Procedures for adding and subtracting algebraic fractions are the same as for adding and subtracting numerical fractions. These are

Step 1: Find the LCD of all expressions to be combined.

Step 2: Put each numerator over the LCD. (Multiply by the missing LCD factors in the denominator of that expression.)

Step 3: Combine the numerators over the LCD and simplify.

EXERCISE SET 5.6

Perform the indicated operations in the following exercises:

1. $\dfrac{x+1}{x-1} + \dfrac{x-1}{x+1}$

2. $\dfrac{x-4}{x+4} + \dfrac{x+4}{x-4}$

3. $\dfrac{x+2}{x^2+2x-15} + \dfrac{x-2}{x^2-25}$

4. $\dfrac{x+6}{x^2+x-20} + \dfrac{x-6}{x^2-16}$

5. $\dfrac{x-1}{x^2+7x+12} + \dfrac{x+2}{x^2+x-6}$

6. $\dfrac{x-5}{x^2-9x+14} + \dfrac{x+4}{x^2-6x-7}$

7. $\dfrac{x+5}{x^2+5x+6} - \dfrac{x+3}{x^2-x-6}$

8. $\dfrac{x+2}{x^2-2x-15} - \dfrac{x+1}{x^2+9x+18}$

9. $\dfrac{x+3}{x^2+7x+10} - \dfrac{x-4}{x^2-x-6}$

10. $\dfrac{x-7}{x^2+5x+4} - \dfrac{x+3}{x^2-3x-4}$

11. $\dfrac{3x}{3x^2-5x-12} - \dfrac{5x}{9x^2-16} + \dfrac{8x}{3x^2+13x+12}$

12. $\dfrac{2x}{2x^2+3x-20} + \dfrac{5}{4x^2-25} - \dfrac{3x}{6x^2-x-35}$

SECTION 5.7 SUMMARY

This chapter has emphasized the procedures for addition, subtraction, multiplication, division, and simplification of algebraic fractions. Techniques of factoring, using the laws of exponents, and finding the LCD are required in these operations. These techniques were developed in this and in previous chapters. Their use with algebraic fractions should serve to develop your skills in performing these important algebraic operations. We highlight the new concepts and definitions.

Factoring a polynomial: expressing the polynomial as a product of prime polynomials.

Special product: a polynomial that has a known factorization. Types of special products are:

  perfect square trinomial: factors into the square of a sum or difference of two monomials.

  perfect cube polynomial: factors into the cube of a sum or difference of two monomials.

  difference of squares binomial: factors into the product of the sum and difference of two monomials.

  sum of cubes binomial: factors into the product of a binomial and a trinomial.

Algebraic fraction: a quotient of two polynomials.

CHAPTER 5 PRACTICE TEST

1. (PO-2) Factor $x^2 - 5x - 6$.
2. (PO-4) Factor $x^3 + 343$.
3. (PO-3) Factor $3x^2 - 14x + 8$.
4. (PO-4) Factor $9x^4 - 4$.
5. (PO-1) Factor $6x^3 - 3x^2 - 12x$.
6. (PO-6) Simplify $\dfrac{5x^3y^2}{3x^4y}$.
7. (PO-7) Simplify $\dfrac{x^2 + x - 12}{x^2 + 5x + 4}$.

8. (PO-7) Simplify $\dfrac{x^3 - 125}{x^2 + x - 30}$.

9. (PO-7) Simplify $\dfrac{-x^2 - 2x + 15}{x^2 + 10x + 25}$.

10. (PO-8) $\dfrac{x^2 - 7x + 12}{x^2 - 8x - 20} \cdot \dfrac{x^2 - 15x + 50}{x^2 - 8x + 15}$

11. (PO-8) $\dfrac{2x^2 - 7x - 15}{2x^2 - 15x + 18} \cdot \dfrac{x^2 - 2x - 24}{x^2 - x - 20}$

12. (PO-8) $\dfrac{x^4 - 16}{x^3 - 8} \cdot \dfrac{x^2 + 2x + 4}{x^2 + 2x}$

13. (PO-8) $\dfrac{x^2 + 10x + 25}{2x^2 + 5x - 25} \cdot \dfrac{4x^2 - 25}{25 + 5x - 2x^2}$

14. (PO-9) $\dfrac{x^2 - 3x - 10}{x^2 - 6x + 8} \div \dfrac{x^2 - 2x - 35}{x^2 + 12x + 32}$

15. (PO-9) $\dfrac{x^2 - 2x - 15}{6 - x - x^2} \div \dfrac{x^2 + x - 30}{x^2 + 4x - 12}$

16. (PO-9) $\dfrac{x^3 + 1}{x - 1} \div \dfrac{x^2 - x + 1}{x^2 - 2x + 1}$

17. (PO-9) $\dfrac{6x^2 + 7x - 20}{3x^2 - 10x + 8} \div \dfrac{2x + 5}{2 - x}$

18. (PO-10, 11) $\dfrac{x - 5}{10x - 5} + \dfrac{x + 4}{12x - 6}$

19. (PO-10, 11) $\dfrac{x + 5}{x^2 - 2x - 3} + \dfrac{x - 4}{x^2 + 4x + 3}$

20. (PO-10, 11) $\dfrac{x - 7}{x^2 + 3x - 10} + \dfrac{x - 2}{x^2 - 2x - 35}$

21. (PO-10, 11) $\dfrac{x + 6}{x^2 + 11x + 24} + \dfrac{4 - x}{x^2 + 6x - 16}$

22. (PO-12) $\dfrac{-12}{5x} - \dfrac{-16}{5x}$

23. (PO-12) $\dfrac{2x + 1}{12x - 6} - \dfrac{5x}{3(4x - 2)}$

24. (PO-10, 12) $\dfrac{x - 2}{x^2 + 3x - 4} - \dfrac{x + 5}{x^2 + x - 12}$

25. (PO-10, 12) $\dfrac{2 - 5x}{x^2 + 10x + 25} - \dfrac{8 - 3x}{x^2 - 25}$

CHAPTER 6

LINEAR EQUATIONS AND INEQUALITIES

6.1 Introduction to Linear Equations in One Variable
6.2 Solving Linear Equations in One Variable
6.3 Applications of Linear Equations in One Variable
6.4 Introduction to Linear Inequalities in One Variable
6.5 Solving Linear Inequalities in One Variable
6.6 Summary

Algebraic expressions were defined and evaluated in Chapter 4. The value of such expressions changed as different values were assumed for the variables. The algebraic equation in one variable is somewhat more complex. It is like a message to be decoded. The appearance of the equals sign implies that only certain values of the variable will cause the expression to evaluate correctly. We will be concerned with finding those values.

In this chapter we consider a special class of algebraic equations—the linear equations in one variable. In a recent survey of engineers and supervisors in technical occupations, we found the ability to solve linear equations high on the list of mathematical skills necessary for a technician.

SECTION 6.1 INTRODUCTION TO LINEAR EQUATIONS IN ONE VARIABLE

An equation is a mathematical statement of equality between two expressions. An example of an equation that involves only specific numbers is $3 + 7 = 10$. An example of an equation that involves both a variable and specific numbers is $x + 4 = 6$.

DEFINITION

LINEAR EQUATION IN ONE VARIABLE

A **linear equation in one variable** is an equation that can be written in the form $ax + b = c$, where a, b, and c are real numbers and $a \neq 0$.

In the first three sections of this chapter, we will work with linear equations in one variable. These are often called **first degree equations in one variable.**

The reason for the use of the phrase, in one variable, is clear if we remind ourselves that a, b, and c are specific numbers. The equations are called first degree equations because the variable appears only to the first power, that is, with an exponent of 1.

Some examples of linear equations in one variable are

$$3x + 7 = 13$$

$$y - 8 = 17$$

$$6z = -21$$

$$4x + 9 = 36 - 5x$$

$$2x + 3 - 6x = 14 - 7x - 5$$

We will see in the next section how we can write equations such as the last two in the form $ax + b = c$.

The following equations are not linear equations in one variable and cannot be put in the form $ax + b = c$:

$x^2 = 4$ (the variable has an exponent of 2)

$4x^3 - 2x = 1$ (in one of the terms the variable has an exponent of 3)

$5x - 6y = 2$ (there are two variables in the equation)

PERFORMANCE OBJECTIVE 1	
You should now be able to recognize linear equations in one variable.	linear equations in one variable → $-\frac{2}{3}x = 8$, $y = x - 1$, $2x + 4 = 0$, $3x^2 + 2 = 6$

If we substitute 2 in place of x in the equation $x + 4 = 6$, we get the true statement $2 + 4 = 6$. If, however, we substitute any other number in place of x, we will not get a true statement. We call 2 the solution or root of the equation $x + 4 = 6$.

To solve an equation means to find the value or values of the variable that yield a true statement. Some equations have no solution; others have one, two, or more solutions. Linear equations in one variable always have a single solution. In the next section, we will learn how to find that unique solution. For now, we will use substitution to determine if a certain value is the solution to a given equation.

EXAMPLE Determine if -3 is the solution of the equation

$$7y + 16 = 3y + 4$$

SOLUTION We substitute -3 in place of y and simplify each side of the equation.

$$7(-3) + 16 = 3(-3) + 4$$
$$-21 + 16 = -9 + 4$$
$$-5 = -5 \quad \text{(true statement)}$$

Yes, -3 is the solution.

EXAMPLE Determine if 8 is the solution of the equation

$$\frac{1}{2}x - 3 = x - 6$$

SOLUTION Substituting and simplifying, we get

$$\frac{1}{2}(8) - 3 = 8 - 6$$
$$4 - 3 = 8 - 6$$
$$1 = 2 \quad \text{(false statement)}$$

No, 8 is not the solution.

EXAMPLE Determine if $-\frac{1}{2}$ is the solution of the equation

$$5x + 13 + 3x = 11 - 4x - 4$$

SOLUTION

$$5\left(-\frac{1}{2}\right) + 13 + 3\left(-\frac{1}{2}\right) = 11 - 4\left(-\frac{1}{2}\right) - 4$$

$$-\frac{5}{2} + 13 + \left(-\frac{3}{2}\right) = 11 - \left(-\frac{4}{2}\right) - 4$$

$$13 + \left(-\frac{8}{2}\right) = 11 + \frac{4}{2} + (-4)$$

$$13 + (-4) = 11 + 2 + (-4)$$

$$9 = 9 \quad \text{(true statement)}$$

Yes, $-\frac{1}{2}$ is the solution.

PERFORMANCE OBJECTIVE 2	$4y - 3 = 9 - 2y$
You should now be able to use substitution to determine if a number is the solution to a linear equation in one variable.	$4(-1) - 3 = 9 - 2(-1)$ $-4 - 3 = 9 + 2$ $-7 = 11$ (false statement) So -1 is not a solution.

PROPERTIES OF A LINEAR EQUATION IN ONE VARIABLE

The equation

1. Can be written in the form $ax + b = c$.
2. Has a unique solution.

EXERCISE SET 6.1

1. Determine whether or not each of the following equations is a linear equation in one variable:

    a. $8 = \frac{2}{3}x - 4$ b. $x^2 + 4x = 5$

    c. $7 - 2x + 4 = x - 9 + 2x$ d. $3x = 2y + 6$

2. Determine whether or not each of the following equations is a linear equation in one variable:
   a. $-5x + 4 = 10y$
   b. $\frac{3}{4}x = 2x - 5$
   c. $-3x = 0$
   d. $x^2 - 8 = 1$

3. For each of the following equations, determine if the given value of the variable is the solution:
   a. $4x + 1 - 5x = x - 1,\quad x = -1$
   b. $14y - 3 = 11y + 2,\quad y = \frac{5}{3}$
   c. $\frac{7}{8}w + \frac{1}{3} = \frac{5}{6}w + \frac{3}{4},\quad w = 10$

4. For each of the following equations, determine if the given value of the variable is the solution:
   a. $6 - 2x + 3 = -5x,\quad x = -3$
   b. $\frac{2}{3}z = \frac{4}{9}z + 2,\quad z = -9$
   c. $8y + 5 = 20 - 12y,\quad y = \frac{3}{4}$

SECTION 6.2 SOLVING LINEAR EQUATIONS IN ONE VARIABLE

In solving linear equations, we want to arrive at a statement $x =$ something. That is, we want to isolate the variable on one side of the equation. To do this, we will be making constant use of two basic principles—the Addition Principle and the Multiplication Principle for equations.

ADDITION PRINCIPLE FOR EQUATIONS

Adding the same expression to (or subtracting the same expression from) both sides of an equation produces an equation with the same solution as the original equation.

We use the Addition Principle when we want to eliminate an entire term from one side of an equation. Since the sum of a term and its additive inverse is always 0, we eliminate the term by adding its additive inverse to both sides of the equation.

EXAMPLE Solve the equation $x + 9 = 15$.

SOLUTION To solve this equation, we want to eliminate the 9 from the left side of the equation so that we isolate x on the left side. We can do this by adding -9, the additive inverse of 9, to both sides of the equation. Note that we cannot add -9 to just the left side of the equation. The Addition Principle requires that whatever is added to one side of an equation be added to the other side as well. If we add -9 to both sides of this equation, we get

$$x + 9 + (-9) = 15 + (-9)$$
$$x + 0 = 6$$
$$x = 6$$

EXAMPLE Solve the equation $x - 3 = 18$ (or $x + (-3) = 18$).

SOLUTION To solve this equation for x, we must eliminate -3 from the left side of the equation. We will add 3, the additive inverse of -3, to both sides of the equation to get

$$x - 3 + 3 = 18 + 3$$
$$x + 0 = 21$$
$$x = 21$$

NOTE: It is usually the case that the $x + 0$ step is omitted.

PERFORMANCE OBJECTIVE 3	
You should now be able to solve equations by applying the Addition Principle.	$x - 7 = -10$ $x - 7 + 7 = -10 + 7$ $x = -3$

MULTIPLICATION PRINCIPLE FOR EQUATIONS

Multiplying (or dividing) both sides of an equation by the same nonzero number produces an equation with the same solution as the original equation.

We use the Multiplication Principle when we want to eliminate from one side of an equation a number that is a factor of one of the terms. Since the product of a number and its multiplicative inverse is always 1, we eliminate the factor by multiplying both sides of the equation by its multiplicative inverse.

SOLVING LINEAR EQUATIONS IN ONE VARIABLE 6.2

NOTE: Multiplying by the multiplicative inverse of a given number is the same as dividing by the given number.

EXAMPLE Solve the equation $\frac{3}{4}x = 21$.

SOLUTION To solve the equation, we want to eliminate the factor $\frac{3}{4}$ from the left side of the equation so that we isolate x on the left side. We can do this by multiplying both sides of the equation by $\frac{4}{3}$, the multiplicative inverse of $\frac{3}{4}$. Multiplying both sides by $\frac{4}{3}$ yields

$$\frac{4}{3}\left(\frac{3}{4}x\right) = \frac{4}{3}(21)$$

Simplifying, we get $1x = 28$ or $x = 28$.

EXAMPLE Solve $3x = 126$.

SOLUTION Dividing both sides of this equation by 3 will eliminate the factor of 3 from the left side of the equation to yield

$$\frac{3x}{3} = \frac{126}{3}$$
$$x = 42$$

Since dividing by 3 is equivalent to multiplying by $\frac{1}{3}$, we could have multiplied both sides of the equation by $\frac{1}{3}$ to get the same result.

PERFORMANCE OBJECTIVE 4	
You should now be able to solve equations by applying the Multiplication Principle.	$-\frac{2}{3}x = 24$ $\left(-\frac{3}{2}\right)\left(-\frac{2}{3}x\right) = \left(-\frac{3}{2}\right)(24)$ $x = -36$

So far we have applied only one of the two basic principles in each equation we have solved. We will often have to apply both principles to solve a given equation. The following examples illustrate this.

EXAMPLE Solve the equation $3x + 5 = 17$ for x.

SOLUTION We see that in forming the left side of this equation, x was multiplied by 3 and then 5 was added to the product. In order to solve for x, we must reverse the steps used to form the expression $3x + 5$, remembering to perform the necessary operations on both sides of the equation. To eliminate the 5 from the left side of the equation, we add a -5 to both sides getting

$$3x + 5 + (-5) = 17 + (-5)$$
$$3x = 12$$

Now to eliminate the factor of 3 on the left side of the equation $3x = 12$, we divide both sides by 3 to get

$$\frac{3x}{3} = \frac{12}{3}$$
$$x = 4$$

EXAMPLE Solve the equation $\frac{1}{4}w - 9 = -11$ for w.

SOLUTION If we add 9 to both sides of this equation, we get

$$\frac{1}{4}w - 9 + 9 = -11 + 9$$
$$\frac{1}{4}w = -2$$

We now multiply both sides of the equation by 4 to get

$$4\left(\frac{1}{4}w\right) = 4(-2)$$
$$w = -8$$

PERFORMANCE OBJECTIVE 5	
You should now be able to solve equations by applying both the Addition Principle and the Multiplication Principle.	$3x + 15 = -12$ addition of -15 yields $3x = -27$ division by 3 yields $x = -9$

The equations we have solved up to this point have been in the form $ax + b = c$, where a, b, and c were specific numbers. In some equations, terms involving the variable may appear on both sides of the equation. In such cases, we must get all terms involving the variable onto one side of the equation and combine them. We then complete the solution of the equation by proceeding as in previous examples.

EXAMPLE Solve the equation $7y + 16 = 3y + 4$ for y.

SOLUTION We want to get both terms involving y onto one side of the equation so that we can combine them. We do this by adding $-3y$ to both sides of the equation. The complete solution of the equation is as follows:

$$7y + 16 = 3y + 4$$
$$7y + (-3y) + 16 = 3y + (-3y) + 4$$
$$[7 + (-3)]y + 16 = 0 + 4$$
$$4y + 16 = 4$$
$$4y + 16 + (-16) = 4 + (-16)$$
$$4y = -12$$
$$\frac{4y}{4} = \frac{-12}{4}$$
$$y = -3$$

EXAMPLE Solve the equation $5x + 13 + 3x = 11 - 4x - 4$ for x.

SOLUTION Note that on each side of the equation there are like terms that can be combined. We can simplify our solution by first combining them. Since $5x + 3x = 8x$ and $11 - 4 = 7$, the original equation becomes $8x + 13 = 7 - 4x$. The complete solution is

$$5x + 13 + 3x = 11 - 4x - 4$$
$$8x + 13 = 7 - 4x$$
$$8x + 4x + 13 = 7 - 4x + 4x$$
$$12x + 13 = 7$$
$$12x + 13 + (-13) = 7 + (-13)$$
$$12x = -6$$
$$\frac{12x}{12} = \frac{-6}{12}$$
$$x = -\frac{1}{2}$$

PERFORMANCE OBJECTIVE 6	
You should now be able to solve equations that have the variable on both sides of the equation.	$8x - 9 = 4 + 7x - 1$ addition of $-7x$ yields $x - 9 = 4 - 1$ addition of $+9$ yields $x = 9 + 4 - 1 = 12$

So far the equations in our examples have not involved many fractions. If an equation seems complicated due to fractions, we can simplify the solution by first clearing it of fractions. We do this by multiplying both sides of the equation by a common denominator, preferably the least common denominator. We can also clear an equation of decimals by multiplying both sides of the equation by the appropriate power of 10 (that is, 10, 100, 1000, etc.) The following examples illustrate these techniques.

EXAMPLE Solve $\frac{3}{4}x + \frac{2}{3} = \frac{1}{2}x + \frac{5}{6}$ for x.

SOLUTION We can clear this equation of fractions by multiplying both sides of the equation by the least common denominator 12. The complete solution is

$$\frac{3}{4}x + \frac{2}{3} = \frac{1}{2}x + \frac{5}{6}$$

$$12\left(\frac{3}{4}x\right) + 12\left(\frac{2}{3}\right) = 12\left(\frac{1}{2}x\right) + 12\left(\frac{5}{6}\right)$$

$$9x + 8 = 6x + 10$$

$$9x + (-6x) + 8 = 6x + (-6x) + 10$$

$$3x + 8 = 10$$

$$3x + 8 + (-8) = 10 + (-8)$$

$$3x = 2$$

$$\frac{3x}{3} = \frac{2}{3}$$

$$x = \frac{2}{3}$$

EXAMPLE Solve $2.5x - 3.7 = 0.15x + 1$ for x.

SOLUTION To clear this equation of decimals, we multiply both sides of the equation by 100 to get

$$250x - 370 = 15x + 100$$

We complete the solution as follows:

$$250x + (-15x) - 370 = 15x + (-15x) + 100$$

$$235x - 370 = 100$$

$$235x - 370 + 370 = 100 + 370$$

$$235x = 470$$

$$\frac{235x}{235} = \frac{470}{235}$$

$$x = 2$$

SOLVING LINEAR EQUATIONS IN ONE VARIABLE 6.2   205

PERFORMANCE OBJECTIVE 7	$\frac{2}{3}x - \frac{5}{6} = \frac{2}{9}$
You should now be able to solve equations that involve fraction or decimal coefficients.	multiplication by 18 yields $12x - 15 = 4$ $12x = 19 \quad x = \frac{19}{12}$

Some equations have symbols of grouping on one or both sides of the equation. In order to solve such an equation, we must first eliminate the grouping symbols.

Recall that we learned how to remove symbols of grouping preceded by a minus sign when we discussed the subtraction of polynomials in Chapter 4. If there is a number to be multiplied times the grouped expression, we use the Distributive Law to carry out the multiplication. After removing the symbols of grouping, we proceed to solve the equation using our previously discussed techniques.

EXAMPLE Solve $4(x - 6) = x - 3(2x + 1)$.

SOLUTION Using the Distributive Law to carry out the multiplications, we get

$$4x - 24 = x - (6x + 3)$$

Next, we eliminate the parentheses.

$$4x - 24 = x + (-6x - 3)$$
$$4x - 24 = x + (-6x) + (-3)$$

The solution is then completed as follows:

$$4x + (-24) = -5x + (-3)$$
$$4x + 5x + (-24) = -5x + 5x + (-3)$$
$$9x + (-24) = -3$$
$$9x + (-24) + 24 = -3 + 24$$
$$9x = 21$$
$$\frac{9x}{9} = \frac{21}{9}$$
$$x = \frac{7}{3}$$

EXAMPLE Solve $3x + 2(7 - x) = 6 - 5(2x + 4)$.

SOLUTION We use the Distributive Law to eliminate the parentheses and complete the solution as follows:

$$3x + 2(7 - x) = 6 - 5(2x + 4)$$
$$3x + (14 - 2x) = 6 - (10x + 20)$$
$$3x + 14 - 2x = 6 + (-10x - 20)$$
$$3x + (-2x) + 14 = 6 + (-10x) + (-20)$$
$$x + 14 = 6 + (-20) + (-10x)$$
$$x + 10x + 14 = -14 + (-10x) + 10x$$
$$11x + 14 = -14$$
$$11x + 14 + (-14) = -14 + (-14)$$
$$11x = -28$$
$$\frac{11x}{11} = \frac{-28}{11}$$
$$x = \frac{-28}{11}$$

Remember that you can always check your solution to an equation by substituting your answer in the original equation and evaluating the resulting expressions to see if you get a true statement.

PERFORMANCE OBJECTIVE 8	$4 - 2(3y + 14) = 0$
You should now be able to solve equations that involve symbols of grouping.	$4 - (6y + 28) = 0$
	$4 - 6y - 28 = 0$ (symbols removed)
	Solve in normal manner.

EXERCISE SET 6.2

Solve each of the following linear equations for x:

1. $x + 5 = -9$
2. $x - 15 = -6$
3. $\frac{-3}{4}x = -12$
4. $-\frac{5}{6}x = \frac{2}{3}$

5. $9x + 3 = 3$
6. $54 - 6x = 0$
7. $2 - 6x = 20$
8. $\frac{1}{3}x - 4 = -7$
9. $x - 15 = 1 - x$
10. $4x - 1 = 3 - 4x$
11. $2x - x - 3 = 13 + 7x - 4$
12. $10x + 3x - 20 = x - 24$
13. $\frac{3}{4}x + \frac{5}{4} - \frac{2}{3}x + \frac{1}{3} = 2$
14. $\frac{7}{10}x - 3 + \frac{2}{5}x = \frac{3}{5}x + 6$
15. $1.2x + 1 = 1.6$
16. $6.25x - 10.5 = x$
17. $5(3x - 4) - (4 + 7x) = 0$
18. $4(3x + 8) = 8$
19. $3(4x - 5) = 11 - 4(2x - 1)$
20. $6 - (6x - 4) = 2(8 - 5x) + 4$

SECTION 6.3 — APPLICATIONS OF LINEAR EQUATIONS IN ONE VARIABLE

The key step in solving practical application problems is to translate problems stated in words into mathematical equations.

The general procedure for solving application problems using linear equations in one variable is outlined here.

RULES FOR SOLVING APPLIED PROBLEMS (LINEAR EQUATIONS IN ONE VARIABLE)

Step 1. Read the problem as many times as it takes to understand it thoroughly.

Step 2. Identify the quantity that is being sought and represent it by some variable, say x. If more than one quantity is to be found, write expressions for each quantity in terms of the same variable. In such cases, we usually let x represent the smallest value.

Step 3. Write an equation that must be satisfied by the variable. Be sure that the information used to formulate the equation has not already been used in expressing one of the unknown quantities.

Step 4. Solve the equation for the variable and state the answer to the problem, making sure to give a value for each of the quantities being sought.

Step 5. Check the answer by making sure it satisfies the conditions in the stated word problem.

The five steps in this procedure will be explicitly displayed in certain examples. You will find that you can often set up the solution in more than one way. However, the answers will always be the same.

EXAMPLE The difference of two numbers is 5. If three times the smaller number is seven more than the larger number, find the two numbers.

SOLUTION

Step 1. Have you read the problem carefully? Do you understand it?

Step 2. We are trying to find two numbers. So we need to represent each of the numbers in terms of the variable. Let $x = $ the smaller number. Since the difference of the two numbers is 5, we can represent the larger number by adding 5 to the smaller number. That is,

$$\text{Let } x + 5 = \text{the larger number}$$

Step 3. We make use of the rest of the information in the problem to write a linear equation. We are told that three times the smaller number is seven more than the larger number. We denote three times the smaller number by $3x$. We need to show that $3x$ is seven more than $x + 5$. If we add 7 to $x + 5$ we will get

$$(x + 5) + 7 = 3x$$

Step 4. We solve this equation for x.

$$x + 5 + 7 + (-3x) = 3x + (-3x)$$
$$-2x + 12 = 0$$
$$-2x + 12 + (-12) = 0 + (-12)$$
$$-2x = -12$$
$$x = 6$$

So, we know that the smaller number is 6. Then, the larger number is $x + 5 = 6 + 5 = 11$.

Step 5. We check to make sure that our answers 6 and 11 satisfy the conditions of the problem. The difference between 6 and 11 is 5. Three times 6 is 18 and 18 is seven more than 11. Thus, our solution is correct.

EXAMPLE Three-halves of a number exceeds five-sixths of the number by four-thirds. Find the number.

SOLUTION

Step 1. Reread the problem if you do not understand it.

Step 2. Let $x =$ the number.

Step 3. We write three halves of the number as $\frac{3}{2}x$ and five-sixths of the number as $\frac{5}{6}x$. To show that $\frac{3}{2}x$ exceeds $\frac{5}{6}x$ by $\frac{4}{3}$, we subtract $\frac{5}{6}x$ from $\frac{3}{2}x$ and get the difference $\frac{4}{3}$. That is,

$$\frac{3}{2}x - \frac{5}{6}x = \frac{4}{3}$$

Step 4. We solve this equation as follows:

$$6\left(\frac{3}{2}x\right) - 6\left(\frac{5}{6}x\right) = 6\left(\frac{4}{3}\right)$$
$$9x - 5x = 8$$
$$4x = 8$$
$$x = 2$$

So, the number is 2.

Step 5. Check.

Three-halves of 2 is $\left(\frac{3}{2}\right) \cdot 2 = 3$

Five-sixths of 2 is $\left(\frac{5}{6}\right) \cdot 2 = \frac{5}{3}$

Three exceeds $\frac{5}{3}$ by $\frac{4}{3}$ since

$$3 - \frac{5}{3} = \frac{9}{3} - \frac{5}{3} = \frac{4}{3}$$

PERFORMANCE OBJECTIVE 9	Let $x =$ smaller number.
You should now be able to solve number problems using linear equations in one variable.	Let $x + 7 =$ larger number. $2x = (x + 7) + 1$

To solve word problems that involve geometric figures such as rectangles and triangles, we will use certain facts that we learned in Chapter 3 on applied geometry. Among these facts are the following:

1. The perimeter P of a rectangle is equal to twice its width w plus twice its length l. That is,

$$P = 2w + 2l$$

2. The sum of the measures of the angles in a triangle is 180°.

When we solve geometric word problems, it sometimes helps to draw a sketch of the geometric figure.

EXAMPLE The length of a rectangle is 6 ft less than three times its width. The perimeter of the rectangle is 140 ft. Find the dimensions of the rectangle.

SOLUTION

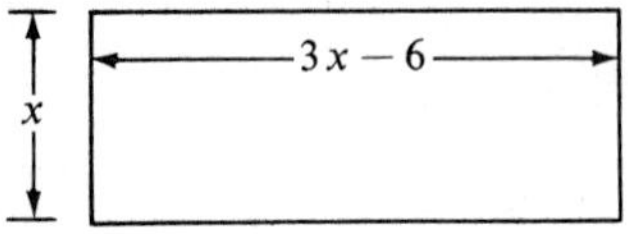

Step 1. If the problem is not clear to you, read it again before proceeding.

Step 2. Let $x =$ the width of the rectangle in feet. Then $3x - 6 =$ the length of the rectangle in feet. (See Figure 6.1)

Figure 6.1

Step 3. Since the perimeter is 140 ft, we have

$$2x + 2(3x - 6) = 140$$

Step 4.
$$2x + 6x - 12 = 140$$
$$8x - 12 = 140$$
$$8x = 152$$
$$x = 19$$

$$3x - 6 = 3(19) - 6 = 57 - 6 = 51$$

So, the dimensions of the rectangle are 19 ft by 51 ft.

Step 5. $P = 2(19) + 2(51) = 38 + 102 = 140$
Also $3(19) - 6 = 51$

EXAMPLE The second angle of a triangle is 6° more than the first angle. The third angle is 14° less than twice the second angle. Find the three angles.

SOLUTION Refer to Figure 6.2.

Let $x =$ the number of degrees in the first angle.

Then $x + 6 =$ the number of degrees in the second angle.

APPLICATIONS OF LINEAR EQUATIONS IN ONE VARIABLE 6.3

$2(x + 6) - 14 =$ the number of degrees in the third angle.

Since the sum of the three angles is 180°, we have

$$x + (x + 6) + 2(x + 6) - 14 = 180$$
$$x + x + 6 + 2x + 12 - 14 = 180$$
$$4x + 4 = 180$$
$$4x = 176$$
$$x = 44$$

$$x + 6 = 44 + 6 = 50$$

$$2(x + 6) - 14 = 2(50) - 14 = 100 - 14 = 86$$

So, the three angles are: 44°, 50°, and 86°. You should verify that these angles satisfy the conditions of the problem and add to 180°.

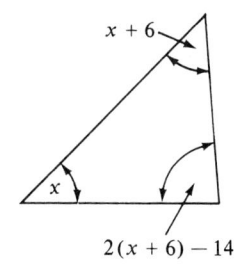

Figure 6.2

PERFORMANCE OBJECTIVE 10	
You should now be able to solve geometric problems using linear equations in one variable.	$y + y + 3y + 5 = 180$

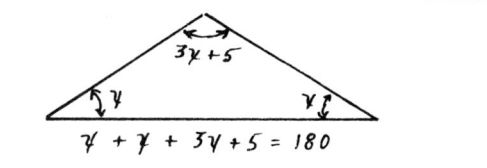

When we combine two or more components together to form a mixture, we will make use of the following facts:

1. The amount of a substance in the final mixture is equal to the sum of all the amounts of that substance in the components that made up the mixture.
2. The value of the final mixture is equal to the sum of the values of the components in the mixture.

EXAMPLE A chemist wants to mix an 18% iodine solution and a 6% iodine solution to make 60 oz of the 10% iodine solution. How much of each solution must he use to make the mixture?

SOLUTION

Step 1. Do you understand that we are mixing strong and weak solutions to produce a solution of intermediate strength?

Step 2. Let $x =$ the number of ounces of the 18% solution.

Then $60 - x =$ the number of ounces of the 6% iodine solution.

Step 3.

The amount of pure iodine in the 18% solution is 18% of x, that is, $0.18x$ oz.

The amount of pure iodine in the 6% solution is 6% of $60 - x$, that is, $0.06(60 - x)$ oz.

The amount of pure iodine in the 10% mixture is 10% of 60, that is, $0.10(60) = 6$ oz.

Since the amount of iodine in the mixture is the sum of the amounts of iodine in the components, we have the equation

$$0.18x + 0.06(60 - x) = 6$$

Step 4. Multiplying both sides of the equation by 100 and solving gives

$$18x + 6(60 - x) = 600$$
$$18x + 360 - 6x = 600$$
$$12x + 360 = 600$$
$$12x = 240$$
$$x = 20$$

$60 - x = 60 - 20 = 40$. So, the chemist should use 20 oz of the 18% solution and 40 oz of the 6% solution.

Step 5. $20(0.18) + 40(0.06) = 3.6 + 2.4 = 6.0 = 60(0.10)$

It may be helpful to summarize the information in a mixture problem by setting up a table. The table would include information about each of the components as well as the final mixture. The information given would include the value of each unit of substance, the amount of each substance, and the total value of each substance. The table can be set up as follows:

Component	*Unit value*	*Amount*	*Total value*
component 1			
component 2			
mixture			

APPLICATIONS OF LINEAR EQUATIONS IN ONE VARIABLE 6.3

The table format can be used in the previous example as follows:

Component	Unit value	Amount	Total value
18% solution	18% = 0.18	x	$0.18x$
6% solution	6% = 0.06	$60 - x$	$0.06(60 - x)$
mixture	10% = 0.10	60	$0.1(60) = 6$

Note that the unit value refers to the ounces of iodine per ounce of solution, the amount refers to the number of ounces of each iodine solution, and the total value is the number of ounces of pure iodine in each solution. Since the sum of the total values of the component solutions must equal the total value of the mixture, we get the equation

$$0.18x + 0.06(60 - x) = 6$$

which has already been solved.

EXAMPLE How much pure alcohol must be added to 64 oz of a 70% alcohol solution to strengthen it to a 90% alcohol solution?

SOLUTION Let $x =$ the number of ounces of pure alcohol to be added.

Component	Unit value	Amount	Total value
pure alcohol	100% = 1	x	$1x$
70% solution	70% = 0.7	64	$0.7(64) = 44.8$
mixture	90% = 0.9	$x + 64$	$0.9(x + 64)$

Note that the final mixture will consist of $(x + 64)$ oz of solution. So the amount of alcohol in the final mixture will be 90% of $x + 64$, that is, $0.9(x + 64) = (0.9x + 57.6)$ oz.

The amount of alcohol in the original 70% solution was $0.7(64) = 44.8$ oz. When we add x oz of pure alcohol to the 44.8 oz of pure alcohol in the 70% solution, we get the $(0.9x + 57.6)$ oz of pure alcohol in the final mixture. That is, we have the equation

$$x + 44.8 = 0.9x + 57.6$$

Multiplying both sides of this equation by 10 and solving, we get

$$10x + 448 = 9x + 576$$
$$x + 448 = 576$$
$$x = 128$$

So, 128 oz of pure alcohol must be added.

EXAMPLE A candy store mixes candy worth $1.50 a pound with candy worth $2.25 a pound. If the mixture is to weigh 50 lb and sell for $1.95 a pound, how many pounds of each kind of candy should be used to make the mixture?

SOLUTION It must be assumed in this problem that the candy store will make the same amount of money by selling the mixture as it would have made by selling the two kinds separately. That is, the value of the mixture is the sum of the values of the components.

Let $x =$ the number of pounds of the $1.50 candy.

Then $50 - x =$ the number of pounds of the $2.25 candy.

The value of the $1.50 candy is $1.50x$.

The value of the $2.25 candy is $2.25(50 - x)$.

The value of the mixture is $1.95(50) = 97.50.

The following table summarizes this information:

Component	Unit value	Amount	Total value
$1.50 candy	$1.50	x	$1.50x$
$2.25 candy	$2.25	$50 - x$	$2.25(50 - x)$
mixture	$1.95	50	$1.95(50) = 97.50

So, we have the equation

$$1.50x + 2.25(50 - x) = 97.50$$

which we solve as follows:

$$1.5x + 112.5 - 2.25x = 97.5$$
$$150x + 11{,}250 - 225x = 9750$$
$$11{,}250 - 75x = 9750$$
$$-75x = -1500$$
$$x = 20$$

APPLICATIONS OF LINEAR EQUATIONS IN ONE VARIABLE   6.3   215

$50 - x = 50 - 20 = 30$. So, the mixture should contain 20 lb of the $1.50 candy and 30 lb of the $2.25 candy.

PERFORMANCE OBJECTIVE 11

You should now be able to solve mixture problems using linear equations in one variable.

Component	Unit Value	Amount (ml)	Total Value
40% acid solution	0.40	x	$0.40x$
Water	0	$104 - x$	0
Mixture	0.25	104	$0.25(104)$

$$0.40x + 0 = 0.25(104)$$

We will consider motion problems in which the speed or rate of travel is constant. Such problems are called uniform motion problems.

The basic relationship used in solving motion problems is that the distance traveled equals the rate times the time. That is, $D = R \cdot T$. The rate of travel and the time must involve the same time unit.

We can use tables to summarize the information in a motion problem just as we did with mixture problems. The information given in the table would be interpreted as follows:

Distance takes the place of "total value"

Rate or $\dfrac{\text{distance}}{\text{time}}$ takes the place of "unit value"

Time takes the place of "amount"

The table can be set up as follows:

Object	Rate	Time	Distance
object 1			
object 2			

EXAMPLE   Two cars that are 230 mi apart are moving toward each other. If their speeds differ by 5 mph and if they will meet in 2 hr, what is the speed of each car?

SOLUTION   We need to represent the speed of each car in terms of the variable x.

Let $x =$ the speed in miles per hour of the slower car.

Then $x + 5 =$ speed in miles per hour of the faster car.

We know that each car will travel for 2 hr. So, $2x$ represents the distance traveled by the slower car and $2(x + 5)$ represents the distance traveled by the faster car.

Object	Rate (mph)	Time (hr)	Distance (mi)
slow car	x	2	$2x$
fast car	$x + 5$	2	$2(x + 5)$

Since the two cars cover a total distance of 230 mi, we have

$$2x + 2(x + 5) = 230$$
$$2x + 2x + 10 = 230$$
$$4x + 10 = 230$$
$$4x = 220$$
$$x = 55$$
$$x + 5 = 60$$

The speeds of the cars are 55 mph and 60 mph.

EXAMPLE Two cars are traveling on the same road in the same direction. One car is 30 mi ahead of the other. The car in front is going 40 mph and the second car is going 50 mph. How long will it take the second car to overtake the first car?

SOLUTION Let $x =$ the number of hours for the second car to overtake the first car. Then the following table gives the pertinent information in the problem:

Object	Rate (mph)	Time (hr)	Distance (mi)
slow car	40	x	$40x$
fast car	50	x	$50x$

Since the second car travels 30 mi more than the first car in order to overtake it, we have

$$50x = 40x + 30$$
$$10x = 30$$
$$x = 3$$

The solution is 3 hr.

PERFORMANCE OBJECTIVE 12

You should now be able to solve motion problems using linear equations in one variable.

Object	Rate (mph)	Time (hrs)	Distance (miles)
speeding car	x	$\frac{1}{5}$	$\frac{1}{5}x$
police car	$x+10$	$\frac{1}{5}$	$\frac{1}{5}(x+10)$

$$\frac{1}{5}(x+10) = \frac{1}{5}x + 2$$

EXERCISE SET 6.3

Solve the following word problems using linear equations in one variable:

1. The sum of two numbers is 35. Twice the smaller number exceeds the larger number by 4. Find the two numbers.

2. Find two numbers such that their sum is 72 and their difference is 14.

3. The difference of two numbers is 35. Find the two numbers if the larger is two more than four times the smaller.

4. The larger of two numbers is three times the smaller number. Find the two numbers if five times their sum is 100.

5. Find three numbers such that the second number is three more than the first, the third number is five more than the second, and twice the sum of the smaller two is nine less than three times the largest.

6. Find three numbers such that the smallest is one-third of the largest, the middle value is four more than the smallest, and the sum of the three numbers is four times the middle number.

7. The first side of a triangle is one-half the length of the second side, and the third side is 20 ft less than the second side. The perimeter of the triangle is 180 ft. Find the three sides.

8. The second side of a triangle is 15 cm less than three times the shortest side. The longest side is 3 cm more than the second side. Find the three sides if the perimeter of the triangle is four times the difference of the longest and shortest sides.

9. The length of a rectangle is four times its width. The perimeter of the rectangle is 150 ft. Find the dimensions of the rectangle.

10. Two sides of a rectangle differ by 3.5 cm. Find the dimensions of the rectangle if its perimeter is 67 cm.

11. The third angle of an isosceles triangle is 16° less than the sum of the two equal angles. Find the three angles of the triangle.

12. One acute angle of a right triangle is 12° more than twice the other acute angle. Find the acute angles of the right triangle.

13. How many ounces of a 16% alcohol solution must be added to 36 oz of a 12% alcohol solution to make a 15% alcohol solution?

14. How many liters of distilled water must be added to 15 ℓ of a 72% antifreeze solution to make a 50% antifreeze solution?

15. If an alloy containing 25% copper is mixed with a 40% copper alloy to get 300 lb of 30% alloy, how much of each must be used?

16. If an 85% acid solution is to be mixed with a 40% acid solution to get 150 mℓ of a 58% solution, how many milliliters of each solution should be used to make the mixture?

17. A store mixes nuts worth $2.75 a pound with nuts worth $1.75 a pound. If the mixture is to weigh 30 lb and sell for $2.35 a pound, how many pounds of each kind of nut should be used to make the mixture?

18. A grocer plans to mix his most expensive ($3.50/lb) and least expensive ($2.25/lb) coffee beans to get a blend of intermediate cost. If the mixture is to weigh 50 lb and sell for $3.00 a pound, how many pounds of each kind of bean should be used to make the mixture?

19. The distance between City A and City B is 60 mi. A car traveling 50 mph leaves City A for City B at the same time that another car traveling 40 mph leaves City B for City A. How long will it take for the two cars to meet?

20. Two cars start out from the same point and travel in opposite directions. The speeds of the cars are 48 mph and 57 mph. How long will it take for the two cars to get 378 mi apart?

21. Two cars start out from the same point and travel in opposite directions. The speed of one car is 15 mph faster than the speed of the other car. After 4 hr, the two cars are 380 mi apart. Find the speed of each car.

22. Two cars that are 600 mi apart are moving toward each other. If their speeds differ by 6 mph and if the cars are 123 mi apart after $4\frac{1}{2}$ hr, what is the speed of each car?

23. Mary lives in the country and works in the city. In going to work she drives her car from her home to the outskirts of the city and then takes a bus the rest of the way. The distance between Mary's home and her job is 24 mi and it takes her 36 min ($\frac{3}{5}$ hr) to get to work. If Mary drives her car 45 mph and if the bus travels at 30 mph, how much time does she spend in her car and how much time does she spend on the bus?

24. Consider Problem 23 again. Due to icy road conditions on a winter day, the car and bus speeds were greatly reduced and Mary spent 20 min traveling on the bus and 30 min in her car. If the bus speed was half of the car speed, what were the two speeds?

SECTION 6.4 INTRODUCTION TO LINEAR INEQUALITIES IN ONE VARIABLE

An **inequality** is a mathematical statement that one expression is related to another in one of the following ways: less than, less than or equal to, greater than, or greater than or equal to.

Before we can work with inequalities, we need to discuss the mathematical symbols that represent the preceding four relationships. The inequality symbols and their meanings are

$<$ less than
$\leq$ less than or equal to
$>$ greater than
$\geq$ greater than or equal to

Examples of inequalities that involve only specific numbers are

$6 < 8$ (read as "6 is less than 8")
$13 \leq 20$ (read as "13 is less than or equal to 20")
$10 > 9$ (read as "10 is greater than 9")
$7 \geq 7$ (read as "7 is greater than or equal to 7")

Note that while it is incorrect to say $13 = 20$, we can say $13 \leq 20$. This is because one of the two relationships, namely $<$, does hold. Likewise, while it is incorrect to say $7 > 7$, we can write $7 \geq 7$. Again, this is because one of the two relationships, namely $=$, does hold.

When an inequality involves only specific numbers, we can readily decide whether or not it is a true statement. However, when an inequality involves both a variable and specific numbers, we will find that the truth of the inequality depends on the value of the variable.

In this section and the next section we will work with linear inequalities in one variable. The discussion will parallel the presentation used in the previous sections on linear equations in one variable.

DEFINITION

> **LINEAR INEQUALITY IN ONE VARIABLE**
>
> A **linear inequality in one variable** is an inequality that can be written in the form
>
> $$ax + b < c \quad (\text{or} \leq c, > c, \geq c)$$
>
> where a, b, and c are real numbers and $a \neq 0$.

Some examples of linear inequalities in one variable are

$$3x + 7 < 13$$

$$y - 8 > 17$$

$$6z \leq -21$$

$$4x + 9 \geq 36 - 5x$$

$$2x + 3 - 6x < 14 - 7x - 5$$

We will see in the next section how we can write the last two inequalities in the form $ax + b \geq c$ and $ax + b < c$, respectively.

The following inequalities are not linear inequalities in one variable.

$$x^2 \geq 4 \quad \text{(the variable has an exponent of 2)}$$

$$4x^3 - 2x < 1 \quad \text{(the variable in one of the terms has an exponent of 3)}$$

$$5x - 6y \leq 2 \quad \text{(there are two variables in the inequality)}$$

PERFORMANCE OBJECTIVE 13

You should now be able to recognize linear inequalities in one variable.

linear inequalities in one variable
- $-\frac{2}{3}x < 8$
- $y > x + 1$
- $2x + 9 \geq 0$
- $3x^2 + 2 \leq 6$

Consider the linear inequality $x + 4 \leq 6$. If we substitute 2 in place of x in the inequality, we get the true inequality $2 + 4 \leq 6$. Likewise, if we substitute 1, $\frac{1}{4}$, 0 or $-\frac{5}{2}$ in place of x, we obtain the true inequalities

$$1 + 4 \leq 6 \quad \text{(i.e. } 5 \leq 6\text{)}$$

INTRODUCTION TO LINEAR INEQUALITIES IN ONE VARIABLE 6.4

$$\frac{1}{4} + 4 \leq 6 \quad (\text{i.e. } 4\frac{1}{4} \leq 6)$$

$$0 + 4 \leq 6 \quad (\text{i.e. } 4 \leq 6)$$

$$-\frac{5}{2} + 4 \leq 6 \quad (\text{i.e. } \frac{3}{2} \leq 6)$$

In fact every number less than or equal to 2 yields a true inequality when substituted into $x + 4 \leq 6$. The numbers greater than 2 yield false inequalities. We call the numbers that are less than or equal to 2 solutions of the inequality $x + 4 \leq 6$. We see that the inequality has infinitely many solutions.

Linear inequalities in one variable always have infinitely many solutions. In the next section, we will learn how to find those solutions. For now, we will use substitution to determine if a certain value is a solution to a given inequality.

EXAMPLE Determine if -5 is a solution of the inequality

$$7y + 18 \leq 3y + 4$$

SOLUTION We substitute -5 in place of y and simplify each side of the inequality.

$$7(-5) + 18 \leq 3(-5) + 4$$
$$-35 + 16 \leq -15 + 4$$
$$-19 \leq -11 \quad \text{(true statement)}$$

Yes, -5 is one of the infinitely many solutions.

EXAMPLE Determine if 9 is a solution of the inequality

$$\frac{1}{2}x - 3 > x - 6$$

SOLUTION Substituting and simplifying, we get

$$\frac{1}{2}(9) - 3 > 9 - 6$$

$$\frac{9}{2} - 3 > 3$$

$$\frac{3}{2} > 3 \quad \text{(false statement)}$$

No, 9 is not one of the solutions.

EXAMPLE Determine if $-\frac{1}{2}$ is a solution of the inequality

$$5x + 13 + 3x < 11 - 4x - 4$$

SOLUTION In Section 6.1 we worked with the same expressions in an equation. We evaluated both sides for $x = -\frac{1}{2}$ and obtained a value of 9 for each expression. So, now we have the inequality

$$9 < 9$$

This is a false statement, so $-\frac{1}{2}$ is not a solution of the inequality. (Note that $-\frac{1}{2}$ will be a solution if we replace $<$ by $\leq$ in the inequality.)

PERFORMANCE OBJECTIVE 14	
You should now be able to use substitution to determine if a number is a solution to a linear inequality in one variable.	$4x - 3 \leq 9 - 2x$ $4(-1) - 3 \leq 9 - 2(-1)$ $-4 - 3 \leq 9 + 2$ $-7 \leq 11$ (true statement) So -1 is a solution.

PROPERTIES OF A LINEAR INEQUALITY IN ONE VARIABLE

1. The inequality can be written in the form $ax + b < c$ (or $\leq c$, $> c$, $\geq c$).
2. The inequality has infinitely many solutions.

EXERCISE SET 6.4

1. Determine whether or not each of the following is a linear inequality in one variable:

   a. $8 \leq \frac{2}{3}x - 4$ b. $x^2 + 4x > 5$

   c. $2x + 3 = 7x$ d. $7 - 2x + 4 < x - 9 + 2x$

   e. $3x - 2y \geq 6$

2. Determine whether or not each of the following is a linear inequality in one variable:

   a. $-5x + 4 \geq 10y$ b. $\frac{3}{4}x \leq 2x - 5$

   c. $7x + 3 = -4$ d. $-3x > 0$

   e. $x^2 - 8 < 1$

3. For each of the following inequalities, determine if the given value of the variable is a solution:

   a. $4x + 1 - 5x < x - 1$, $x = -\dfrac{1}{2}$

   b. $14y - 3 \geq 11y + 2$, $y = 0$

   c. $\dfrac{7}{8}w - \dfrac{1}{3} < \dfrac{5}{6}w + \dfrac{3}{4}$, $w = 24$

4. For each of the following inequalities, determine if the given value of the variable is a solution:

   a. $6 - 2x + 3 < -5x$, $x = 2$

   b. $\dfrac{2}{3}z \leq \dfrac{4}{9}z + 2$, $z = 36$

   c. $8y + 5 > 20 - 12y$, $y = \dfrac{3}{4}$

SECTION 6.5   SOLVING LINEAR INEQUALITIES IN ONE VARIABLE

In solving linear inequalities, we want to arrive at one of the following statements: $x <$ some number, $x \leq$ some number, $x >$ some number, or $x \geq$ some number. That is, we want to isolate the variable on one side of the inequality, just as we did with equations. To do this, we will use the Addition Principle and the Multiplication Principle for inequalities.

ADDITION PRINCIPLE FOR INEQUALITIES

Adding the same expression to (or subtracting the same expression from) both sides of an inequality produces an inequality with the same solutions as the original inequality.

We recognize that the Addition Principle for inequalities is the same as the Addition Principle for equations. We use the Addition Principle when we want to eliminate an entire term from one side of an inequality.

Since there are infinitely many solutions to a linear inequality, we will graphically represent the solutions by locating them on a number line.

EXAMPLE Solve the inequality $x + 9 \leq 15$.

SOLUTION We add -9 to both sides of the inequality to get

$$x + 9 + (-9) \leq 15 + (-9)$$
$$x \leq 6$$

A graph of these solutions on a number line is shown in Figure 6.3.

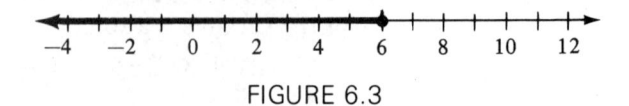

FIGURE 6.3

Note that every number less than or equal to 6 has been marked. The arrow at the end of the shaded region indicates that the solutions continue on forever in the negative direction.

EXAMPLE Solve the inequality $x - 3 > 18$.

SOLUTION Adding 3 to both sides of the inequality gives

$$x - 3 + 3 > 18 + 3$$
$$x > 21$$

We graph these x's as shown in Figure 6.4.

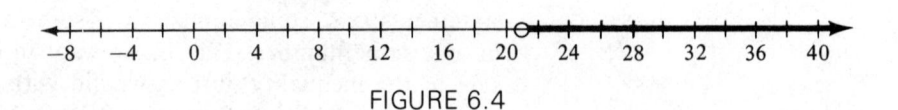

FIGURE 6.4

Note the use of an open circle at the number 21. This indicates that 21 is not a solution of the inequality. Only the numbers larger than 21 are solutions.

PERFORMANCE OBJECTIVE 15	$x - 7 \geq -10$
You should now be able to solve inequalities by applying the Addition Principle.	$x - 7 + 7 \geq -10 + 7$
	$x \geq -3$

MULTIPLICATION PRINCIPLE FOR INEQUALITIES

1. If we multiply or divide both sides of an inequality by the same positive number, we obtain an inequality with the same solutions as the original inequality.

2. If we multiply or divide both sides of an inequality by the same negative number, we must also change the sense or direction of the inequality sign in order to obtain an inequality with the same solutions as the original inequality.

Note that the Multiplication Principle for inequalities is not the same as the Multiplication Principle for equations when we are multiplying or dividing by a negative number. In this situation, we say that we must change the sense or direction of the inequality. This means that we replace $<$ by $>$, $\leq$ by $\geq$, $>$ by $<$, and $\geq$ by $\leq$. As was the case with equations, we use the Multiplication Principle when we want to eliminate from one side of an inequality a number that is a factor of one of the terms. The following examples illustrate the use of the Multiplication Principle. We again graph the solutions on a number line.

EXAMPLE Solve the inequality $\frac{3}{4}x \leq -21$.

SOLUTION To eliminate the factor $\frac{3}{4}$ from the left side of the inequality, we multiply both sides by $\frac{4}{3}$. Since we are multiplying by a positive number, we make no other changes in the inequality.

$$\left(\frac{4}{3}\right)\left(\frac{3}{4}x\right) \leq \frac{4}{3}(-21)$$

$$x \leq -28$$

See Figure 6.5.

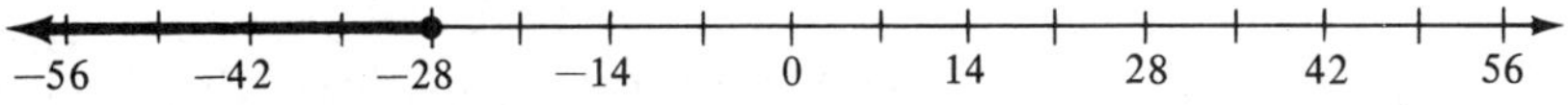

FIGURE 6.5

EXAMPLE Solve $-3x < 126$.

SOLUTION We divide both sides of the inequality by the negative number, -3, remembering to change the sense of the inequality.

$$\frac{-3x}{-3} > \frac{126}{-3}$$

$$x > -42$$

See Figure 6.6.

226 **6 LINEAR EQUATIONS AND INEQUALITIES**

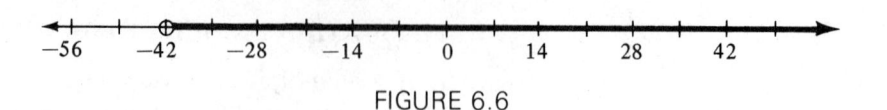

FIGURE 6.6

Note the open circle at -42 since -42 is not one of the solutions.

EXAMPLE Solve $-\frac{1}{2}x > -2$.

SOLUTION Multiplying by -2 and changing the sense of the inequality, we get

$$(-2)\left(-\frac{1}{2}x\right) < (-2)(-2)$$
$$x < 4$$

See Figure 6.7.

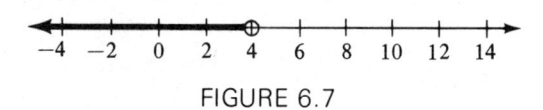

FIGURE 6.7

PERFORMANCE OBJECTIVE 16	$-\frac{2}{3}x > 24$
You should now be able to solve inequalities by applying the Multiplication Principle.	$\left(-\frac{3}{2}\right)\left(-\frac{2}{3}x\right) < \left(-\frac{3}{2}\right)(24)$
	$x < -36$

So far we have had to apply only one of the two basic principles in each inequality we have solved. However, as was the case with equations, we will usually have to apply both principles to solve a given inequality. The following examples include problems involving the variable on both sides of the inequality, fraction or decimal coefficients, and symbols of grouping.

EXAMPLE Solve the inequality $-3x + 7 \leq 17$ for x.

SOLUTION We first add -7 to both sides of the inequality.

$$-3x + 7 + (-7) \leq 17 + (-7)$$
$$-3x \leq 10$$

SOLVING LINEAR INEQUALITIES IN ONE VARIABLE 6.5

We then divide both sides by -3 and change the sense of the inequality.

$$\frac{-3x}{-3} \geq \frac{10}{-3}$$

$$x \geq \frac{-10}{3}$$

A graph of the solutions is given in Figure 6.8.

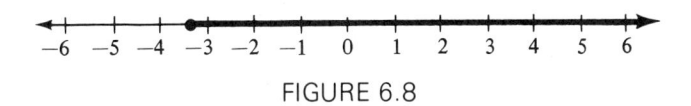

FIGURE 6.8

EXAMPLE Solve $\frac{3}{4}x + \frac{2}{3} > \frac{1}{2}x + \frac{5}{6}$ for x.

SOLUTION Multiplying by the least common denominator (12) clears the inequality of fractions.

$$12\left(\frac{3}{4}x\right) + 12\left(\frac{2}{3}\right) > 12\left(\frac{1}{2}x\right) + 12\left(\frac{5}{6}\right)$$

$$9x + 8 > 6x + 10$$

$$9x + (-6x) + 8 > 6x + (-6x) + 10$$

$$3x + 8 > 10$$

$$3x + 8 + (-8) > 10 + (-8)$$

$$3x > 2$$

$$x > \frac{2}{3}$$

See Figure 6.9

FIGURE 6.9

EXAMPLE Solve $x - 3(2x + 1) \geq 4(x - 6)$.

SOLUTION Using the Distributive Law to carry out the multiplications, we get

$$x - (6x + 3) \geq 4x - 24$$

We eliminate the parentheses and complete the solution as follows:

$$x - 6x - 3 \geq 4x - 24$$
$$-5x - 3 \geq 4x - 24$$
$$-5x + (-4x) - 3 \geq 4x + (-4x) - 24$$
$$-9x - 3 \geq -24$$
$$-9x - 3 + 3 \geq -24 + 3$$
$$-9x \geq -21$$
$$\frac{-9x}{-9} \leq \frac{-21}{-9}$$
$$x \leq \frac{7}{3}$$

See Figure 6.10.

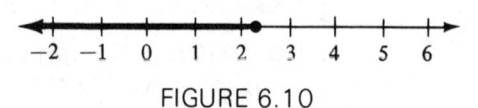

FIGURE 6.10

Recall that we can check a specific solution to see that we get a true inequality when we substitute it in place of the variable. However, we obviously cannot check each and every one of the infinite number of solutions. Nevertheless, checking one or more solutions can help identify certain types of errors, such as an error in the sense of the inequality.

PERFORMANCE OBJECTIVE 17

You should now be able to solve inequalities by applying the Addition and Multiplication Principles.

$$3y + 15 > -12$$
$$3y + 15 + (-15) > -12 + (-15)$$
$$3y > -27$$
$$y > -9$$

EXERCISE SET 6.5

Solve each of the following linear inequalities for x. Graph the solutions on a number line.

1. $x + 5 \leq -9$
2. $x - 15 \leq -6$

3. $\dfrac{-3}{4}x > -12$ 4. $-\dfrac{5}{6}x > \dfrac{2}{3}$

5. $9x + 3 < 3$ 6. $54 - 6x < 0$

7. $2 - 6x \leq 20$ 8. $\dfrac{1}{3}x - 4 \leq -7$

9. $x - 15 \geq 1 - x$ 10. $4x - 1 \leq 3 - 4x$

11. $2x - x - 3 < 13 + 7x - 4$ 12. $10x + 3x - 20 > x - 24$

13. $\dfrac{3}{4}x + \dfrac{5}{4} - \dfrac{2}{3}x + \dfrac{1}{3} > 2$ 14. $\dfrac{7}{10}x - 3 + \dfrac{2}{5}x < \dfrac{3}{5}x + 6$

15. $1.2x + 1 \leq 1.6$ 16. $6.25x - 10.5 \geq x$

17. $5(3x - 4) - (4 + 7x) \geq 0$ 18. $4(3x + 8) \leq 8$

19. $3(4x - 5) < 11 - 4(2x - 1)$

20. $6 - (6x - 4) > 2(8 - 5x) + 4$

SECTION 6.6 SUMMARY

In this chapter we have emphasized the solution of linear equations and inequalities in one variable. We have also solved application problems using linear equations in one variable.

The new concepts and definitions from this chapter are summarized.

Linear equation in one variable: an equation that can be written in the form $ax + b = c$, where a, b, c are real numbers and $a \neq 0$.

Addition principle for equations: adding or subtracting the same term on both sides of an equation doesn't change the solutions.

Multiplication principle for equations: multiplying or dividing both sides of an equation by the same nonzero number doesn't change the solutions.

Linear inequality in one variable: an inequality that can be written in the form $ax + b < c$ (or $\leq c$, $> c$, $\geq c$) where a, b, c are real numbers and $a \neq 0$.

Addition principle for inequalities: adding or subtracting the same term on both sides of an inequality doesn't change the solutions.

Multiplication principle for inequalities:

1. Multiplying or dividing both sides of an inequality by the same positive number doesn't change the solutions.

2. Multiplying or dividing both sides of an inequality by the same negative number requires that we change the direction of the inequality in order to obtain the same solutions.

CHAPTER 6 PRACTICE TEST

(PO-1, 13) Identify each of the following equations and inequalities as a linear equation in one variable, a linear inequality in one variable, or neither:

1. $\frac{1}{3}x^2 + 2x = 1$ 2. $5x - \frac{1}{2} \leq 0$ 3. $y > 3x + 4$

(PO-2, 14) For each of the following equations and inequalities, determine if the given value of the variable is a solution:

4. $3x + 2 - 5x = 2x - 1, \quad x = \frac{3}{4}$

5. $\frac{1}{3}x - \frac{1}{2}x < 2, \quad x = 12$

6. $6x - 2 \geq 4x + 1, \quad x = -1$

Solve each of the following linear equations for x:

7. (PO-3, 4, 5, 6) $x + 33 = 1 - x$

8. (PO-6) $7x + 5 - 2x + 6 = 2 - 4x$

9. (PO-7) $\frac{3}{20}x - 1 = \frac{4}{5}$

10. (PO-7) $3.2x - 1.75 = 2.55x + 1.5$

11. (PO-8) $-2(x - 5) + 10 - (2 - 5x) = 0$

Solve each of the following linear inequalities for x. Graph the solutions on a number line.

12. (PO-15) $x - 10 > -26$

13. (PO-16) $-27x < -81$
14. (PO-17) $3 + 14x - 5 \geq 11x - 2$
15. (PO-17) $\frac{3}{2}x + \frac{1}{6} < \frac{2}{3}x - \frac{2}{3}$
16. (PO-17) $8x - 3(x - 5) \geq 7x - 1$

Solve the following word problems using linear equations in one variable:

17. (PO-9) One-half of a number is eight more than one-third of the number. Find the number.

18. (PO-10) Two sides of a triangle are the same length. The third side is 6 in. less than the sum of the equal sides. If the perimeter of the triangle is 46 in., find the three sides.

19. (PO-11) How many gallons of 80% acid solution must be added to 15 gallons of 6% acid solution to make a 20% acid solution?

20. (PO-12) A car speeding at 65 mph is 3 mi. beyond a police car when the police car starts in pursuit at 80 mph. How long does it take the police car to overtake the speeding car?

CHAPTER 7

LINEAR FUNCTIONS AND GRAPHING

7.1 Definition of a Function
7.2 The Linear Equation as a Function
7.3 The Data Table as a Basis for the Graph
7.4 The Rectangular Coordinate System
7.5 Constructing the Graph of a Linear Equation
7.6 Graphing a Line Using its Characteristics—(Slope, Intercepts, Points)
7.7 Finding the Equation of a Line from its Characteristics
7.8 Variation
7.9 Summary

The concept of a function is fundamental to an understanding of the interdependence of two or more quantities. For example, the restoring force of a spring is considered to be a function of the stretching of the spring. The further we stretch the spring, the larger the restoring force.

The dependence expressed in a function can often be visualized by the use of a graph. In this chapter we will develop the characteristics of the graph of a linear function. This will lead us into a discussion of more complex dependencies under the section on variation. In a later chapter we will consider quadratic functions in some detail.

SECTION 7.1 DEFINITION OF A FUNCTION

In Chapter 6 we discussed linear equations in one variable. However, we are often confronted with equations that involve two or more variables. In such equations, if we arbitrarily assign values to all but one variable, the remaining variable is determined by the equation. Those variables whose values are assigned arbitrarily are called **independent variables.** The remaining variable is called the **dependent variable** since its value depends on the values of the independent variables. We give a name to the set of values that the independent variable can have. We call this set of values the **domain.** The set of values the dependent variable can have is called the **range.** When we have an equation with two variables, we may want to emphasize the pairing of independent variable values (domain values) with dependent variable values (range values). We can do this by writing the corresponding values as ordered pairs having the form

(independent variable value, dependent variable value)

or

(domain value, range value)

EXAMPLE In the equation $y = 2x$, find the range values that correspond to the domain values of $x = -3, -2, -1, 0, 1, 2, 3$ and write as ordered pairs.

SOLUTION The two variables involved in the equation are x and y. If we assign a specific value to x, the equation determines a corresponding value for y. The independent variable is x and y is the dependent variable.

When $x = -3$, $y = 2(-3) = -6$
When $x = -2$, $y = 2(-2) = -4$
When $x = -1$, $y = 2(-1) = -2$
When $x = 0$, $y = 2(0) = 0$
When $x = 1$, $y = 2(1) = 2$
When $x = 2$, $y = 2(2) = 4$
When $x = 3$, $y = 2(3) = 6$

DEFINITION OF A FUNCTION 7.1

The domain values are −3, −2, −1, 0, 1, 2, 3, and the corresponding range values are −6, −4, −2, 0, 2, 4, 6. So we write the following ordered pairs:

$$(-3, -6), \quad (-2, -4), \quad (-1, -2), \quad (0, 0), \quad (1, 2), \quad (2, 4), \quad (3, 6)$$

EXAMPLE In the equation $y = x - 3$, find the range values corresponding to the domain values of $x = 1, 2, 3, 4, 5$ and write as ordered pairs.

SOLUTION If we assign a specific value to x, the equation requires us to determine a corresponding value for y by subtracting 3 from x. The independent variable is x and y is the dependent variable.

When $x = 1$, $y = 1 - 3 = -2$
When $x = 2$, $y = 2 - 3 = -1$
When $x = 3$, $y = 3 - 3 = 0$
When $x = 4$, $y = 4 - 3 = 1$
When $x = 5$, $y = 5 - 3 = 2$

The domain values are 1, 2, 3, 4, 5 and the range values are −2, −1, 0, 1, 2. So we write the following ordered pairs:

$$(1, -2), \quad (2, -1), \quad (3, 0), \quad (4, 1), \quad (5, 2)$$

PERFORMANCE OBJECTIVE 1	Equation: $y = x^2 + 2$
You should now be able to write ordered pairs given an equation and specific domain values for the independent variable.	domain range 1 3 → (1, 3) 3 11 → (3, 11) 5 27 → (5, 27)

In each of the previous examples we determined a set of ordered pairs of numbers. Any set of ordered pairs is called a **relation**. If the set of ordered pairs also satisfies a special property, then the relation is called a **function**. The special property is that no two ordered pairs can have the same x value and different y values.

DEFINITION

FUNCTION

A **function** is a set of ordered pairs (x, y) such that each domain value x is associated with exactly one range value y.

EXAMPLE Determine whether the set of ordered pairs $(-3, -6)$, $(-2, -4)$, $(-1, -2)$, $(0, 0)$, $(1, 2)$, $(2, 4)$, $(3, 6)$ is a function.

SOLUTION The set of ordered pairs is a function since no two ordered pairs have the same x value and different y values.

EXAMPLE Determine whether the set of ordered pairs $(1, -2)$, $(2, -1)$, $(3, 0)$, $(4, 1)$, $(5, 2)$ is a function.

SOLUTION The set of ordered pairs is a function since no two ordered pairs have the same x value and different y values.

EXAMPLE Determine whether the set of ordered pairs $(-3, 9)$, $(-2, 4)$, $(-1, 1)$, $(0, 0)$, $(1, 1)$, $(2, 4)$, $(3, 9)$ is a function.

SOLUTION The set of ordered pairs is a function since no two ordered pairs have the same x value and different y values. Note that there are ordered pairs with the same y value and different x values. This function was generated from the equation

$$y = x^2$$

by squaring the domain values $-3, -2, -1, 0, 1, 2, 3$.

EXAMPLE Determine whether the set of ordered pairs $(3, 8)$, $(2, 0)$, $(3, 0)$, $(-2, 2)$, $(3, -2)$ is a function.

SOLUTION The set of ordered pairs is not a function since the ordered pairs $(3, 8)$, $(3, 0)$, $(3, -2)$ have the same x value and different y values.

PERFORMANCE OBJECTIVE 2	
You should now be able to determine if a set of ordered pairs is a function.	$(-3, 2)$ $(-1, 4)$ $(1, 8)$ $(-1, 5)$ } set fails the function test

---

In previous examples, functions were generated by using specific values of x in the equations

$$y = 2x$$

$$y = x - 3$$

$$y = x^2$$

DEFINITION OF A FUNCTION 7.1

Although functions are actually sets of ordered pairs, the equations used to generate the ordered pairs are often called the functions.

For example, we can require evaluation of the function $y = 2x$ for domain values of $x = -3, -2, -1, 0, 1, 2, 3$. To emphasize that the value of the dependent variable is determined by evaluating an expression involving x, a notation called **functional notation** is commonly used. In this notation a letter such as f is used as the name of the function and we write $f(x)$ to represent the range value associated with a specific domain value x. Then $f(x)$ can be written in place of y in the equation $y = 2x$. This yields $f(x) = 2x$.

EXAMPLE Evaluate the function $f(x) = 2x$ for domain values of $x = -3, -2, -1, 0, 1, 2, 3$ and write the ordered pairs.

SOLUTION

Function evaluation	Ordered pair
$f(-3) = 2(-3) = -6$	$(-3, -6)$
$f(-2) = 2(-2) = -4$	$(-2, -4)$
$f(-1) = 2(-1) = -2$	$(-1, -2)$
$f(0) = 2(0) = 0$	$(0, 0)$
$f(1) = 2(1) = 2$	$(1, 2)$
$f(2) = 2(2) = 4$	$(2, 4)$
$f(3) = 2(3) = 6$	$(3, 6)$

EXAMPLE Evaluate the function $g(x) = x^2$ for domain values of $x = -3, -2, -1, 0, 1, 2, 3$ and write the ordered pairs.

SOLUTION

Function evaluation	Ordered pair
$g(-3) = (-3)^2 = 9$	$(-3, 9)$
$g(-2) = (-2)^2 = 4$	$(-2, 4)$
$g(-1) = (-1)^2 = 1$	$(-1, 1)$
$g(0) = (0)^2 = 0$	$(0, 0)$
$g(1) = (1)^2 = 1$	$(1, 1)$
$g(2) = (2)^2 = 4$	$(2, 4)$
$g(3) = (3)^2 = 9$	$(3, 9)$

7 LINEAR FUNCTIONS AND GRAPHING

PERFORMANCE OBJECTIVE 3	
You should now be able to use functional notation to write ordered pairs $(x, f(x))$.	Function: $f(x) = x^2 - 4$ Ordered pairs: $(0, -4), (1, 3), (2, 0)$

A function, whether expressed as a set of ordered pairs or an equation, can be thought of as a processor since it associates quantities in the domain with quantities in the range. It processes a value of the independent variable x to produce a value of the dependent variable y. The processor role of a function is diagrammed in Figure 7.1.

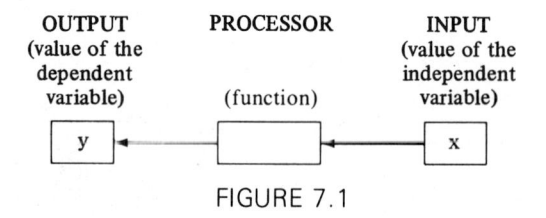

```
OUTPUT           PROCESSOR          INPUT
(value of the                       (value of the
 dependent                           independent
 variable)        (function)          variable)

   [y] ←──────── [        ] ←──────── [x]
```

FIGURE 7.1

EXAMPLE Show the function $y = x - 3$ as a processor of $x = 1, 2, 3, 4, 5$.

SOLUTION The function processes 1, 2, 3, 4, 5 as shown in Figure 7.2.

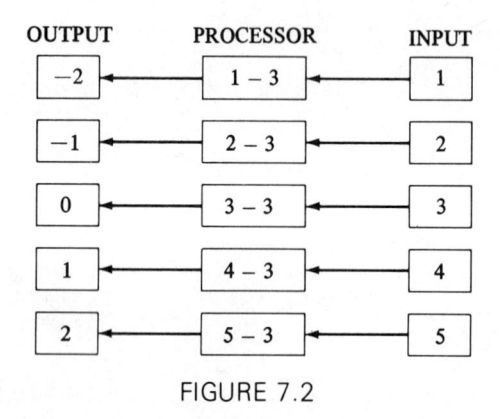

```
OUTPUT       PROCESSOR      INPUT
 [-2]  ←───   [1 - 3]  ←───  [1]
 [-1]  ←───   [2 - 3]  ←───  [2]
 [ 0]  ←───   [3 - 3]  ←───  [3]
 [ 1]  ←───   [4 - 3]  ←───  [4]
 [ 2]  ←───   [5 - 3]  ←───  [5]
```

FIGURE 7.2

The processing action of the function on 1, 2, 3, 4, 5 can be represented by the ordered pairs $(1, -2), (2, -1), (3, 0), (4, 1), (5, 2)$.

DEFINITION OF A FUNCTION 7.1

EXAMPLE Show the function $f(x) = x^2$ as a processor of $x = -3, -2, -1, 0, 1, 2, 3$.

SOLUTION The function processes $-3, -2, -1, 0, 1, 2, 3$ as shown in Figure 7.3.

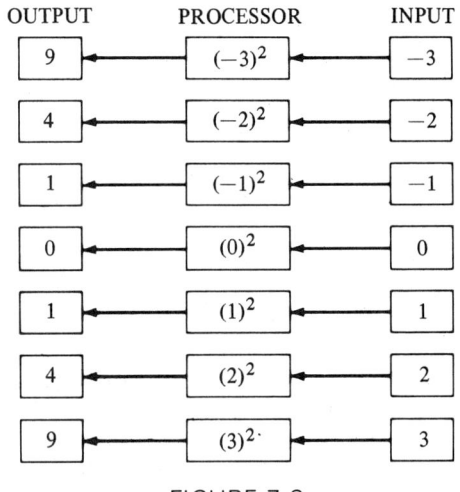

OUTPUT	PROCESSOR	INPUT
9	$(-3)^2$	-3
4	$(-2)^2$	-2
1	$(-1)^2$	-1
0	$(0)^2$	0
1	$(1)^2$	1
4	$(2)^2$	2
9	$(3)^2$	3

FIGURE 7.3

EXAMPLE Show that the set of ordered pairs $(3, 8), (2, 0), (3, 0), (-2, 2), (3, -2)$ is not a function.

SOLUTION The processor is not a function because it associates the domain value three with several different range values. (See Figure 7.4.)

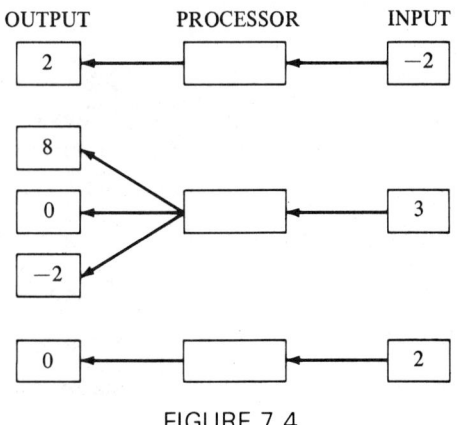

OUTPUT	PROCESSOR	INPUT
2		-2
8, 0, -2		3
0		2

FIGURE 7.4

EXERCISE SET 7.1

1. For the given domain values, find the corresponding range values and write as ordered pairs.
   a. $y = -4x$, $x = -2, -1, 0, 1, 2$
   b. $y = x + 5$, $x = -5, -3, 0, 3, 5$
   c. $y = x^3$, $x = -2, -1, 0, 1, 2$

2. For the given domain values, find the corresponding range values and write as ordered pairs.
   a. $y = -x$, $x = -2, -1, 0, 1, 2$
   b. $y = x + 2$, $x = -5, -3, 0, 3, 5$
   c. $y = x^4$, $x = -2, -1, 0, 1, 2$

3. Determine whether each set of ordered pairs is a function.
   a. $(-2, 8), (-1, 4), (0, 0), (1, -4), (2, -8)$
   b. $(-5, 0), (-3, 2), (0, 5), (3, 8), (5, 10)$
   c. $(-2, -8), (-1, -1), (0, 0), (1, 1), (2, 8)$
   d. $(-2, -3), (-1, -1), (0, 0), (-1, -3), (-2, -1)$

4. Determine whether each set of ordered pairs is a function.
   a. $(-2, 2), (-1, 1), (0, 0), (1, -1), (2, -2)$
   b. $(-5, -3), (-3, -1), (0, 2), (3, 5), (5, 7)$
   c. $(-2, 16), (-1, 1), (0, 0), (1, 1), (2, 16)$
   d. $(-3, -2), (-1, -1), (0, 0), (-3, -1), (-1, -2)$

5. Find the value of $f(x)$ for $x = -4, -2, 0, 1, 3$ in each of the following equations:
   a. $f(x) = 3x + 4$
   b. $f(x) = x^2 - 5$
   c. $f(x) = 5$
   d. $f(x) = x + 2$

6. Find the value of $f(x)$ for $x = -4, -2, 0, 1, 3$ in each of the following equations:
   a. $f(x) = -5x + 2$
   b. $f(x) = x^2 + 3$
   c. $f(x) = -4$
   d. $f(x) = x - 7$

7. Write the ordered pairs for Problem 5.

8. Write the ordered pairs for Problem 6.

9. Are the equations in Problem 5 functions?

10. Are the equations in Problem 6 functions?

SECTION 7.2 THE LINEAR EQUATION AS A FUNCTION

In Chapter 6 we discussed linear equations in one variable and their solutions. In this section we do the same for linear equations in two variables.

DEFINITION

LINEAR EQUATION IN TWO VARIABLES

A **linear equation in two variables** is one that can be written in the form $ax + by = c$, where a, b, and c are real numbers and either $a \neq 0$ or $b \neq 0$.

The variables are x and y, and we call the equation linear because the x and y appear to the first power. Some examples of linear equations in two variables are

$$3x + 2y = 6 \quad \text{and} \quad 4x - 7y = 14$$

If $b = 0$ in $ax + by = c$, we get the equation $ax = c$ and if $a = 0$, we get the equation $by = c$. So equations such as $x = 4$ and $5y = 8$, although they involve only one variable, can be thought of as linear equations in two variables.

Equations such as $3x^2 + y = 1$ and $x^3 + 4y^2 = 5$ are not linear equations in two variables because one or both variables appear with an exponent other than one.

PERFORMANCE OBJECTIVE 4

You should now be able to recognize linear equations in two variables.

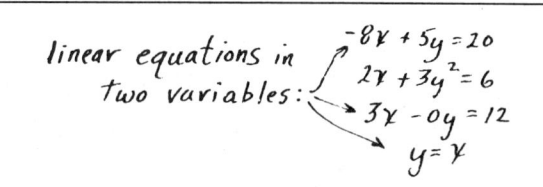

linear equations in two variables:
$-8x + 5y = 20$
$2x + 3y^2 = 6$
$3x - 0y = 12$
$y = x$

To solve a linear equation in the single variable x, we had to find the value for x that yielded a true statement. To find a solution for a linear equation in two variables, x and y, we must find a pair of values, one for x and one for y, that yield a true statement when substituted into the equation. We denote the pair of values by the ordered pair (x, y).

Unlike linear equations in one variable, which have only one solution, linear equations in two variables have infinitely many solutions. Although we cannot possibly list all the solutions, we do need to be able to find some specific ordered-pair solutions.

RULES FOR FINDING A SOLUTION OF $ax + by = c$, $a \neq 0$, $b \neq 0$

Substitute any real number for x and solve for y, or substitute any real number for y and solve for x.

EXAMPLE Find an ordered pair that satisfies the equation $3x + 2y = 6$.

SOLUTION Let $x = 0$, to obtain

$$3(0) + 2y = 6$$
$$0 + 2y = 6$$
$$y = 3$$

So $(0, 3)$ is a solution. Other solutions are $(2, 0)$, $(1, \frac{3}{2})$, $(-4, 9)$, $(\frac{4}{3}, 1)$.

RULES FOR FINDING A SOLUTION OF $ax + by = c$, $a \neq 0$, $b = 0$

x can only take on the value $\frac{c}{a}$.
y can be any real number.

EXAMPLE Find an ordered pair that satisfies the equation $2x + 0y = 8$ (that is, $2x = 8$).

SOLUTION Since $2x = 8$, then $x = 4$. The variable y is not restricted in any way. So y can be any real number. Some solutions are

$$(4, 0), \quad (4, 2), \quad (4, -1), \quad \left(4, \frac{1}{3}\right), \quad \left(4, -\frac{3}{2}\right)$$

RULES FOR FINDING A SOLUTION OF $ax + by = c$, $a = 0$, $b \neq 0$

x can take on any value.
y can only take on the value $\frac{c}{b}$.

EXAMPLE Find an ordered pair that satisfies the equation $0x + 4y = 3$ (that is, $4y = 3$).

SOLUTION If $4y = 3$, then $y = \frac{3}{4}$ and x can be any real number. Some solutions are

$$\left(0, \frac{3}{4}\right), \quad \left(6, \frac{3}{4}\right), \quad \left(-5, \frac{3}{4}\right), \quad \left(\frac{3}{4}, \frac{3}{4}\right)$$

PERFORMANCE OBJECTIVE 5	$\begin{aligned} -6x + 4y &= 12 \\ -6(1) + 4y &= 12 \\ 4y &= 18 \\ y &= \frac{9}{2} \end{aligned}$ $\left. \right\}$ $\left(1, \frac{9}{2}\right)$ is a solution.
You should now be able to determine solutions to linear equations in two variables.	

We can use the Addition and Multiplication Rules for equations to solve $ax + by = c$ for y provided that $b \neq 0$. We do this as follows:

$$ax + by = c$$
$$ax - ax + by = -ax + c$$
$$by = -ax + c$$
$$\frac{by}{b} = \frac{-ax}{b} + \frac{c}{b}$$
$$y = -\frac{a}{b}x + \frac{c}{b}$$

With the equation in this form it becomes easier to determine ordered pair solutions to the equation. We can take any value for x, substitute it into

$$y = -\frac{a}{b}x + \frac{c}{b}$$

and obtain the corresponding value of y.

Every equation of the form $ax + by = c$, $b \neq 0$, satisfies the special property of a function, because each domain value x is associated with the unique y value

$$-\frac{a}{b}x + \frac{c}{b}$$

We call such a function a **linear function**. However, an equation of the form $ax = c$, where $b = 0$, does not represent a function. The x value $\left(\frac{c}{a}\right)$ can be associated with infinitely many values of y since there are no restrictions on y.

EXAMPLE Solve $2y - x = 0$ for y, if possible, to determine how the function processes the following values of the independent variable: $-2, -1, 0, 1, 2$. Write the result of the processing as ordered pairs.

SOLUTION Solving for y yields

$$2y - x = 0$$
$$2y = x$$
$$y = \frac{1}{2}x$$

$(-2, -1)$, $\left(-1, -\frac{1}{2}\right)$, $(0, 0)$, $\left(1, \frac{1}{2}\right)$, $(2, 1)$

EXAMPLE Solve $3y + 5x = 9$ for y, if possible, to determine how the function processes the following values of the independent variable: $-6, -3, 0, 3, 6$. Write the result of the processing as ordered pairs.

SOLUTION Solving for y yields

$$3y + 5x = 9$$
$$3y = -5x + 9$$
$$y = -\frac{5}{3}x + 3$$

$(-6, 13)$, $(-3, 8)$, $(0, 3)$, $(3, -2)$, $(6, -7)$

EXAMPLE Solve $2x = 5$ for y, if possible, to determine how the function processes the following values of the independent variable: $-3, -1, 0, 1, 3$. Write the result of the processing as ordered pairs.

SOLUTION Since the equation is of the form $ax = c$, where $b = 0$, it does not represent a function. The equation cannot be solved for y, and x cannot take on the values $-3, -1, 0, 1, 3$. The variable x must have the value $\frac{5}{2}$. The x-value $\left(\frac{5}{2}\right)$ can be associated with infinitely many values of y.

PERFORMANCE OBJECTIVE 6	
You should now be able to rearrange a linear equation in two variables into the form $y = f(x)$ and find ordered pairs satisfying the equation.	$4x - 2y = 3$ $-2y = -4x + 3$ $y = 2x - \frac{3}{2}$ when $x = -1$, $y = -\frac{7}{2}$ $\left. \right\}$ $\left(-1, -\frac{7}{2}\right)$ is an ordered pair.

EXERCISE SET 7.2

1. Determine whether or not each of the following equations is a linear equation in two variables:
   a. $y = 5x + 1$ b. $2x + y^2 = 3$ c. $4x + 0y = 2$

2. Determine whether or not each of the following equations is a linear equation in two variables:
   a. $y = x^2$ b. $-3y + 4 = 2x$ c. $0x - y = -5$

3. Determine three solutions to each of the following linear equations in two variables:
   a. $5x - 2y = 20$
   b. $y = -3x + 2$
   c. $\frac{1}{4}y + \frac{1}{2}x = 3$
   d. $0x + 2y = 6$

4. Determine three solutions to each of the following linear equations in two variables:
   a. $-4x + 3y = -12$
   b. $y = -6x + 5$
   c. $\frac{2}{5}y + \frac{3}{10}x = 0$
   d. $4x + 0y = 8$

Solve the following linear equations for y, if possible, to determine how the function processes the given values of the independent variable. Write the result of the processing as ordered pairs. If the equation does not represent a function, so indicate.

5. $3x + y = 8$; $-2, 0, 1, \frac{8}{3}$

6. $2y + 4x = 0$; $-3, \frac{1}{2}, 0, 2$

7. $5x - 4y = 8$; $-4, 0, \frac{8}{5}, 4$

8. $5x + 0y = 10$; $-1, 0, 2, 5$

9. $\frac{1}{2}x + \frac{3}{4}y = \frac{3}{2}$; $-3, 0, \frac{3}{2}, 3$

10. $\frac{1}{3}x + \frac{5}{6}y = \frac{4}{3}$; $-2, -1, 0, 1$

SECTION 7.3

THE DATA TABLE AS A BASIS FOR THE GRAPH

Graphs originate with a mathematical relationship or with observed data. Prior to constructing a graph we will most frequently organize the data into a table. The graph is then a more visual presentation of the data in the table. We emphasize the process in the diagram given in Figure 7.5.

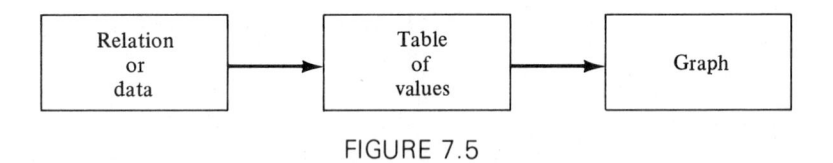

Relation or data → Table of values → Graph

FIGURE 7.5

In this chapter we will consider only mathematical relations. Graphing experimental data is considered in Appendix I.

Production of a table is generally an intermediate step in displaying the organized data. However, sometimes the table is the final step. To construct a table from a mathematical relation we must first identify the variable or variables that are to be processed. These are the independent variables. Then we must determine the permitted values of the independent variable or decide on the independent variable values that are of interest. Having done this, we compute the dependent value corresponding to each selected value of the independent variable. The result is the table.

In Section 7.2 we saw that a linear equation in two variables could be viewed as a linear function by letting x be the independent variable and y the dependent variable. In doing this we created ordered pairs of numbers. We can now construct a table representing these ordered pairs.

The table is constructed by labeling the first row of numbers as x and the second row of numbers as y. The ordered pairs are put into the table by arranging x-values from smallest to largest. The following table includes the ordered pairs generated from the equation $y = x - 3$ in Section 7.1.

(1, −2)　(2, −1)　(3, 0)　(4, 1)　(5, 2)
 ↓　　　 ↓　　　 ↓　　　↓　　　↓

x	1	2	3	4	5
y	−2	−1	0	1	2

THE DATA TABLE AS A BASIS FOR THE GRAPH 7.3

EXAMPLE Construct a table representing the ordered pairs

$(-2, 0)$, $(-1, 3)$, $(0, 6)$, and $(1, 9)$

SOLUTION

x	-2	-1	0	1
y	0	3	6	9

EXAMPLE Construct a table representing the ordered pairs

$(-1, 2)$, $(0, 5)$, $(3, 10)$, $(1, 8)$, and $(-2, 2)$

SOLUTION

x	-2	-1	0	1	3
y	2	2	5	8	10

EXAMPLE The resistance of a circuit element (in ohms) is predicted to depend on temperature in the following manner:

$$R = 150.0 + 0.1T + 0.002T^2$$

where T is measured in degrees Celsius. Make a table showing the value of R as the temperature ranges from 0° Celsius to 80° Celsius in 10 degree steps.

SOLUTION The independent variable is temperature. The values of interest range from 0 to 80 in intervals of 10. The dependent variable is resistance. We evaluate R for each value of T by substitution.

When $T = 0$, $R = 150.0 + 0.1(0) + 0.002(0)^2 = 150.0$

When $T = 10$, $R = 150.0 + 0.1(10) + 0.002(10)^2 = 151.2$
etc.

Next we construct the table.

TABLE RESISTANCE VERSUS TEMPERATURE
(assuming $R = 150.0 + 0.1T + 0.002T^2$)

T	0	10	20	30	40	50	60	70	80
R	150.0	151.2	152.8	154.8	157.2	160.0	163.2	166.8	170.8

This table gives us a better idea of what happens to the resistance as the temperature rises.

RULES FOR CONSTRUCTING A TABLE

Step 1. Identify the independent and dependent variables.

Step 2. Select values for the independent variable(s).

Step 3. Compute the dependent variable values.

Step 4. Construct and label the table.

PERFORMANCE OBJECTIVE 7	$s = 32t - 16t^2$
You should now be able to develop a data table from a mathematical relation.	<table><tr><td>t</td><td>1</td><td>2</td><td>3</td><td>4</td></tr><tr><td>s</td><td>16</td><td>0</td><td>-48</td><td>-128</td></tr></table>

EXERCISE SET 7.3

1. Construct a table using the following ordered pairs:
   a. $(-1, -1), (0, 1), (1, 3), (2, 5), (3, 7)$
   b. $(0, 0), (-2, -2), (1, 1), (-3, -3), (5, 5)$

2. Construct a table using the following ordered pairs:
   a. $\left(-1, \frac{1}{2}\right), \left(0, \frac{1}{2}\right), \left(1, \frac{1}{2}\right), \left(2, \frac{1}{2}\right), \left(3, \frac{1}{2}\right)$
   b. $(-1, -1), (2, 5), (-3, -5), (-2, -3), (0, 1)$

3. The restoring force exerted by a stretched spring is known to be given by $F = 8x$ N where x is the extension from the natural length of the spring in meters. Develop a table of values of F for values of x from 0 to 0.4 m in steps of 0.05 m.

4. The circumference of a circle is given by $C = 2\pi r$. If we let $\pi = 3.14$, develop a table of values of C for values of r from 0.5 to 2 m in steps of 0.5 m.

5. The position of a projectile fired vertically is given by
$$y = 300t - 16t^2$$

where y is the position above ground in feet and t is the elapsed time in seconds. Construct a table displaying y and t values as t ranges from 0 to 13 sec.

6. The position of a projectile fired vertically is given by

$$y = 20.0 + 49.0t - 4.9t^2$$

where y is the position above ground in meters and t is the elapsed time in seconds. Construct a table displaying y and t values as t ranges from 0 to 10 sec.

7. Construct a data table displaying the values of v (the dependent variable) for values of x (the independent variable) from $x = 0$ to 12 in integer steps. The relation between v and x is given by

$$v = 30 - 5x$$

8. Construct a data table displaying the values of w (the dependent variable) for values of z (the independent variable) from $z = 0$ to 6 in steps of 0.5. The relation between w and z is given by

$$w = 15 - 3z$$

SECTION 7.4 THE RECTANGULAR COORDINATE SYSTEM

By defining a coordinate system we are able to represent values of independent and dependent variables graphically. In this section we will develop a general two dimensional coordinate system that is based on the use of number lines for both axes. This general system is called the **rectangular coordinate system.** Such a system will be useful in graphing mathematical relations where both the independent and dependent variables represent numbers.

We construct a rectangular coordinate system by intersecting a horizontal number line and a vertical number line at their origins (zero positions). We call the point of intersection the **origin** of the coordinate system. The horizontal number line is called the ***x*-axis** and has positive direction to the right; the vertical number line is called the ***y*-axis** and has positive direction upward. The four regions formed by the two number lines are called **quadrants.** These are shown and labeled in Figure 7.6.

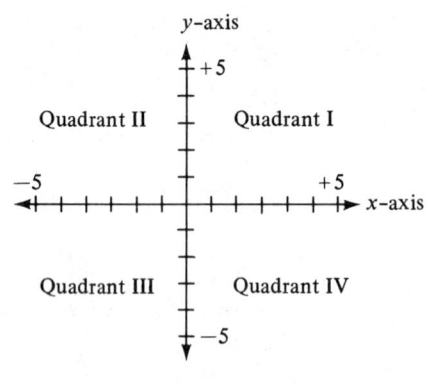

FIGURE 7.6

DEFINITION

RECTANGULAR COORDINATE SYSTEM

A **rectangular coordinate system** is formed by the intersection of two perpendicular number lines. The point of intersection of the number lines, which occurs at their respective zero points, is called the origin of coordinates. The number lines divide the plane into four equal regions called quadrants.

We construct the rectangular coordinate system in order to graph corresponding values of independent and dependent variables. Each pair of values of the independent and dependent variables corresponds to a point in the coordinate system. We call the numbers associated with a point the **x-coordinate** and the **y-coordinate** of the point. The x-coordinate is the independent variable value while the y-coordinate is the corresponding dependent variable value.

DEFINITION

COORDINATES OF A POINT

The **coordinates** of a point are an ordered pair of numbers giving, in turn, the direction and distance of a point along the x-axis and its direction and distance along the y-axis from the origin.

We locate the point that corresponds to the coordinates as follows:

Step 1. Start at the origin and move left or right along the x-axis until the desired x-coordinate is reached.

Step 2. Move along a vertical line (up or down) from this x-value until a point directly across from the desired y-coordinate is reached.

EXAMPLE Locate the point corresponding to (−1, 4).

SOLUTION Starting at the origin, we move one unit left. From this point we move four units up and place a dot to represent the ordered pair. See Figure 7.7.

EXAMPLE Locate the point corresponding to (5, −2).

SOLUTION We move five units to the right starting at the origin. From this point we move two units down and place a dot. See Figure 7.7.

EXAMPLE Locate the point corresponding to (0, −3).

SOLUTION We move zero units along the x-axis. We move three units down and place a dot. Notice that this dot will appear on the y-axis. See Figure 7.7.

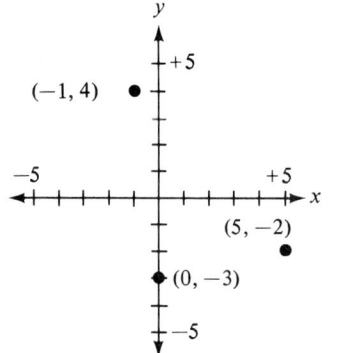

FIGURE 7.7

EXAMPLE Identify the coordinates of points A, B, and C in Figure 7.8.

SOLUTION Point A appears to be located four units to the right of the origin and two units above. Its coordinates are (4, 2).

Point B is located two units to the left of the origin and zero units above or below. Its coordinates are (−2, 0). You should notice that points that have one coordinate equal to zero lie on one of the axes. The converse of this statement is also true.

Point C is three units left of the origin and four units below so its coordinates are (−3, −4).

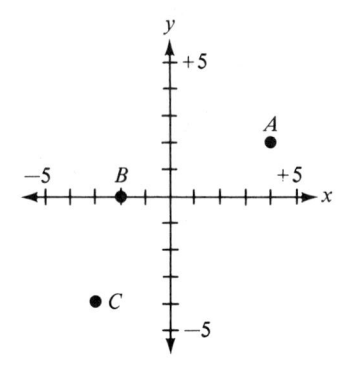

FIGURE 7.8

PERFORMANCE OBJECTIVE 8

You should now be able to plot and interpret ordered pairs of numbers as points in a rectangular coordinate system.

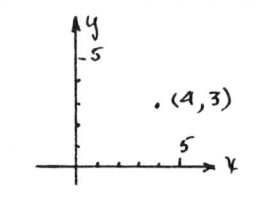

As noted earlier, the rectangular coordinate system is of particular value in displaying mathematical relationships. We display such relationships by forming point plots or line graphs. A point plot consists

7 LINEAR FUNCTIONS AND GRAPHING

of points in a rectangular coordinate system that represent ordered pairs of numbers satisfying the relationship.

EXAMPLE Construct a point plot of the mathematical relation

$$y = -5 + x^2$$

as x varies from -3 to 4.

SOLUTION We first construct a table of ordered pairs that satisfy the relation. We choose values of x one unit apart.

x	-3	-2	-1	0	1	2	3	4
y	4	-1	-4	-5	-4	-1	4	11

This table is obtained by substitution into the equation as follows: When $x = 4$,

$$y = -5 + (4)^2 = -5 + 16 = 11$$

Next we plot each ordered pair on an x-y coordinate system as shown in Figure 7.9. Figure 7.9 is a point plot of the relation $y = -5 + x^2$. Eight representative points have been shown.

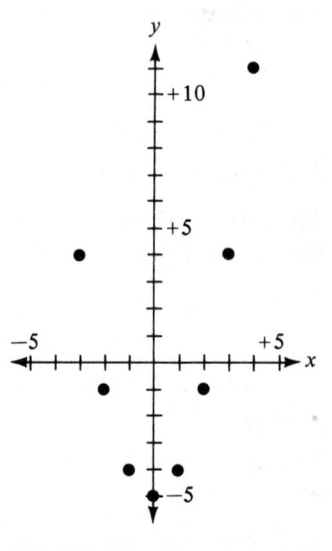

FIGURE 7.9

Mathematical relations are more clearly portrayed if we can construct a line graph rather than just a set of points. A line graph is formed by connecting representative points by a smooth curve. To illustrate this, consider the examples below.

EXAMPLE The y distance of an object from the origin is given by the equation $y = 2x + 1$. Construct a line graph that represents this relation. Let x range from 0 to 5.

SOLUTION We first construct a table of representative points.

x	0	1	2	3	4	5
y	1	3	5	7	9	11

THE RECTANGULAR COORDINATE SYSTEM   7.4   253

Next we plot the points on a rectangular coordinate system.

To construct a line graph we connect these points with a smooth curve as shown in Figure 7.10.

EXAMPLE   Construct a line graph representing the relation

$$y = 3 + \frac{5}{x}$$

Let x range from 1 to 6.

SOLUTION   We construct a table using the given relation.

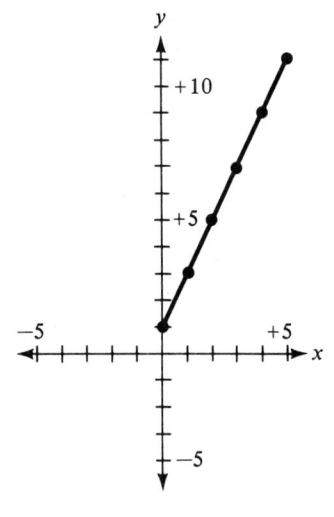

FIGURE 7.10

x	1	2	3	4	5	6
y	8	$5\frac{1}{2}$	$4\frac{2}{3}$	$4\frac{1}{4}$	4	$3\frac{5}{6}$

Then we plot representative points given in the table as shown in Figure 7.11.

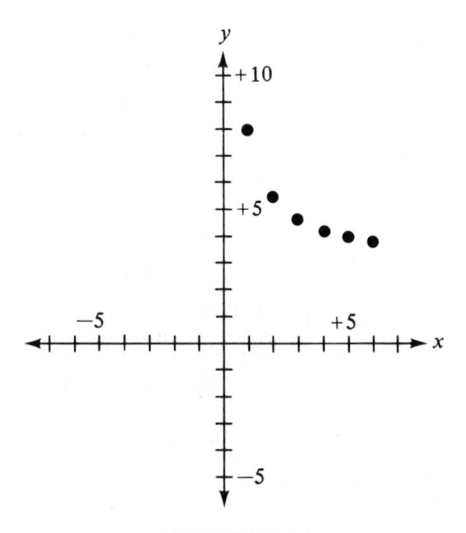

FIGURE 7.11

Finally we connect these points with a smooth curve. (See Figure 7.12.)

7 LINEAR FUNCTIONS AND GRAPHING

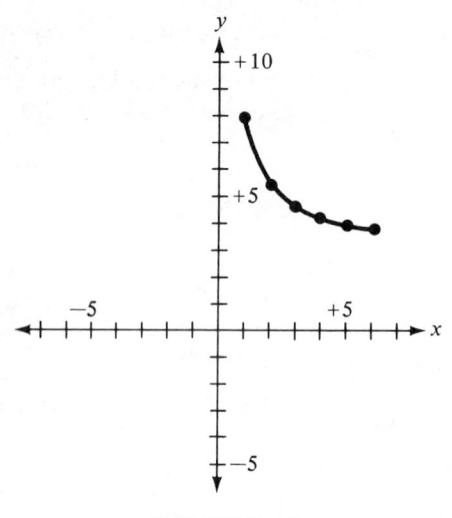

FIGURE 7.12

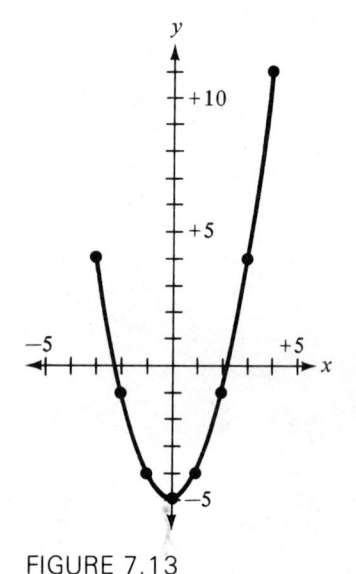

EXAMPLE Convert the point plot of the relation $y = -5 + x^2$ (in Figure 7.9) to a line graph.

SOLUTION The line graph is displayed in Figure 7.13. The points have been connected by a smooth curve to form the line graph.

FIGURE 7.13

In summary, we see that the rectangular coordinate system provides a method of representing ordered pairs of numbers graphically. We call such a representation a point plot. We will encounter point plots when graphing data sets. This coordinate system can also be used to display mathematical relations as line graphs. The line graph begins with a point plot. The line graph is used to indicate that the mathematical relation exists over all values of the independent variable —not just those representative values which have been plotted.

PERFORMANCE OBJECTIVE 9

You should now be able to construct a line graph to represent a mathematical relation.

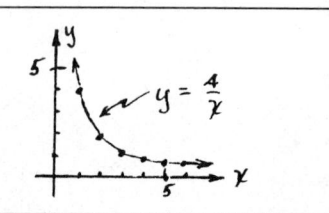

EXERCISE SET 7.4

1. Plot the following points on a rectangular coordinate system:
   a. (4, −5)   b. (−3, −7)   c. (−5, 2)
   d. (2, 6)   e. (0, 5)   f. (4, 0)
   g. (−5, −3) h. (0, 0)

2. Plot the following points on a rectangular coordinate system:
   a. (2, −3)   b. (−4, −5)   c. (−6, 3)
   d. (4, 8)   e. (0, 8)   f. (2, 0)
   g. (−2, −5) h. (0, −4)

3. Write the coordinates of the points displayed in Figure 7.14 (to the nearest integer value).

4. Write the coordinates of the points displayed in Figure 7.15 (to the nearest integer value).

5. Construct a point plot to represent the following mathematical relations:
   a. $y = -5 + 2x$ b. $y = -x + 6$ c. $y = x^2 - 5$
   Let x range from −3 to 3 in integral steps in each case.

6. Construct a point plot to represent the following mathematical relations:
   a. $y = -3 + 3x$ b. $y = -x + 3$ c. $y = x^2 - 4$
   Let x range from −3 to 3 in integral steps in each case.

7. Convert the point plots of Problem 5 into line graphs.

8. Convert the point plots of Problem 6 into line graphs.

9. Construct a line graph of each of the following relations:
   a. $y = 5 - \dfrac{3}{x}$ (let x range from 1 to 6)
   b. $y = -x^2 + 6$ (let x range from −3 to 3)
   c. $y = 3x$ (let x range from −2 to 4)

10. Construct a line graph of each of the following relations:
    a. $y = 3 - \dfrac{2}{x}$ (let x range from 1 to 6)
    b. $y = -x^2 + 8$ (let x range from −3 to 3)
    c. $y = -2x$ (let x range from −2 to 4)

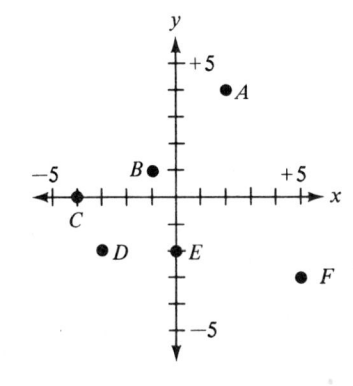

FIGURE 7.14

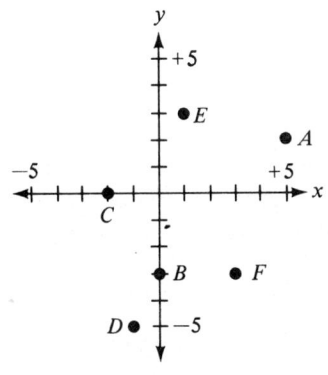

FIGURE 7.15

SECTION 7.5 CONSTRUCTING THE GRAPH OF A LINEAR EQUATION

We introduced the linear equation $ax + by = c$ in Section 7.2. Now we wish to construct a line graph that will represent the mathematical relation between x and y in this equation. To construct this graph we select representative points to be plotted. Since there are infinitely many values of x that we could select, this would seem to be an unending task. However, it is the case that the points corresponding to solutions always form a straight line. So it is only necessary to determine two x, y pairs, plot the corresponding points, and join them with a straight line. The entire mathematical relation is then displayed as a line graph.

EXAMPLE Graph the linear equation $x + y = 3$.

SOLUTION Solve the linear equation for y to find $y = 3 - x$. Then we select two values for x and compute the corresponding y values.

$$\text{When } x = 0, \quad y = 3$$

$$\text{When } x = 3, \quad y = 0$$

We have simply made a table with two entries:

x	0	3
y	3	0

Now we plot these two points and connect them with a straight line as shown in Figure 7.16. Although two points are sufficient to determine a straight line, finding a third point serves as a check on the accuracy of this line. To complete this example we let $x = 1$ and find $y = 2$. We verify that the point (1, 2) lies on the graph in Figure 7.16.

NOTE: The arrows added to the ends of the plotted line indicate that it continues indefinitely in both directions.

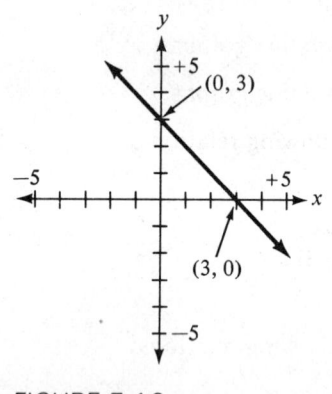

FIGURE 7.16

EXAMPLE Graph the linear equation $3x - 2y = 6$.

CONSTRUCTING THE GRAPH OF A LINEAR EQUATION 7.5

SOLUTION Solve for y.

$$3x - 2y = 6$$
$$-2y = 6 - 3x$$
$$\frac{-2y}{-2} = \frac{6}{-2} - \frac{3x}{-2}$$
$$y = -3 + \frac{3}{2}x$$

We construct a table of values.

x	0	2	4
y	-3	0	3

Then we plot these points and construct the line graph as shown in Figure 7.17. We recommend plotting all three points at the outset so that a check is available when the straight line is drawn.

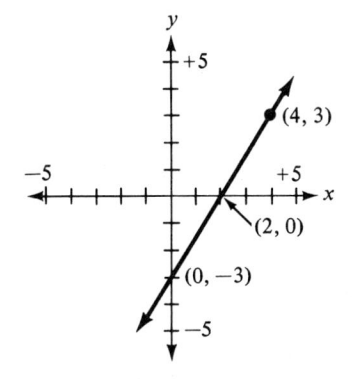

FIGURE 7.17

RULES FOR GRAPHING A LINEAR EQUATION

Step 1. Solve the equation for y.

Step 2. Construct a table of values (three values).

Step 3. Plot the points in the table.

Step 4. Connect the plotted points with a straight line.

There are two special cases in which this procedure has to be changed slightly.

CASE 1: THE COEFFICIENT OF y IN THE LINEAR EQUATION IS ZERO.

When y does not appear (coefficient zero), we are dealing with an equation of the form $ax = c$. We can view this as a linear equation in two variables in which y can take on any value whatsoever. On the other hand, x must satisfy the equation $ax = c$.

EXAMPLE Graph the linear equation $2x = 8$.

SOLUTION We solve this equation for x to find $x = 4$. Since y can take on any value at all, we can construct a table as follows:

258   **7 LINEAR FUNCTIONS AND GRAPHING**

x	4	4	4
y	-1	0	2

Note that x can only have the value 4. We construct the graph as shown in Figure 7.18. The result is a vertical line four units to the right of the y-axis.

CASE 2: THE COEFFICIENT OF x IN THE LINEAR EQUATION IS ZERO.

When x does not appear (coefficient zero), we have an equation of the form $by = c$. We can view this as a linear equation in two variables in which x can take on any value whatsoever. The variable y must satisfy the equation $by = c$.

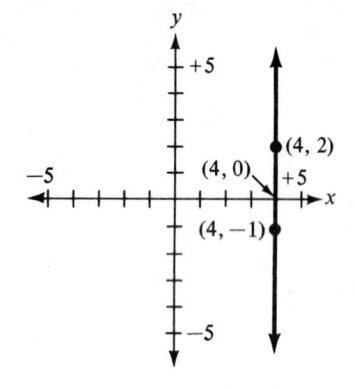

FIGURE 7.18

EXAMPLE   Graph the linear equation $2y = -3$.

SOLUTION   We solve this equation for y to find

$$y = -\frac{3}{2}$$

Since x can take on any value, we construct a table as follows:

x	-1	0	2
y	$-\dfrac{3}{2}$	$-\dfrac{3}{2}$	$-\dfrac{3}{2}$

Note that y can only have the value $-\frac{3}{2}$. The graph is shown in Figure 7.19. The result is a horizontal line $\frac{3}{2}$ units below the x-axis.

FIGURE 7.19

We conclude that if one of the coefficients (of x or y) is zero, the line is parallel to that axis. In particular, we should observe that $y = 0$ is the equation of the x-axis and $x = 0$ is the equation of the y-axis.

PERFORMANCE OBJECTIVE 10

You should now be able to graph a linear equation by finding at least two solution pairs for the equation.

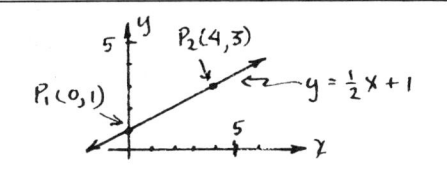

EXERCISE SET 7.5

Construct a graph of the following linear equations. Generate a table with three ordered pairs before graphing each line.

1. $-x + y = 3$
2. $2x - y = 0$
3. $3x - 5y = 15$
4. $5x + 12y = 60$
5. $3y = 4$
6. $-3y = 8$
7. $x = -4$
8. $x = 3$
9. $4x + 3y = 12$
10. $2x + 5y = 10$

SECTION 7.6 GRAPHING A LINE USING ITS CHARACTERISTICS (SLOPE, INTERCEPTS, POINTS)

In this section we examine the characteristics of a line drawn on a rectangular coordinate system. The purpose of this examination is to isolate certain properties of the line that will assist us in graphing it. These properties are the line's intercepts and slope. We begin with a definition of intercepts and a procedure for finding them.

DEFINITION

INTERCEPTS OF A LINE

The *x*-intercept is the *x*-value at which the line crosses the *x*-axis.

The *y*-intercept is the *y*-value at which the line crosses the *y*-axis.

RULES FOR FINDING INTERCEPTS OF A LINE

We find the *x*-intercept by letting $y = 0$ in the equation of the line and solving for *x*.

We find the *y*-intercept by letting $x = 0$ in the equation of the line and solving for *y*.

EXAMPLE Determine the *x* and *y*-intercepts of the line $3x - 2y = 6$ that we graphed earlier (Figure 7.17). The graph of this line with the intercepts marked is shown in Figure 7.20.

SOLUTION To find the *x*-intercept let $y = 0$ and solve for *x* in the equation $3x - 2y = 6$.

$$3x - 2(0) = 6$$
$$3x = 6$$
$$x = 2$$

The line should then intercept the *x*-axis at the value 2. This agrees with the graphical result.

To find the *y*-intercept let $x = 0$ and solve for *y*.

$$3(0) - 2y = 6$$
$$-2y = 6$$
$$y = -3$$

The line should then intercept the *y*-axis at the value -3. Again this agrees with the graphical result.

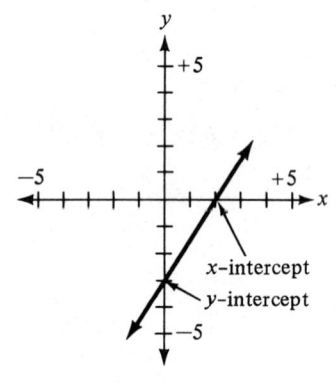

FIGURE 7.20

EXAMPLE Find the *x* and *y*-intercepts of the line whose equation is $4x + y = 8$.

SOLUTION The *x*-intercept is obtained from $4x + (0) = 8$ so that $x = 2$.

The *y*-intercept is obtained from $4(0) + y = 8$ so that $y = 8$.

The line can then be graphed as shown in Figure 7.21.

The intercepts are easily found when the equation is given. These points are then useful in constructing the graph.

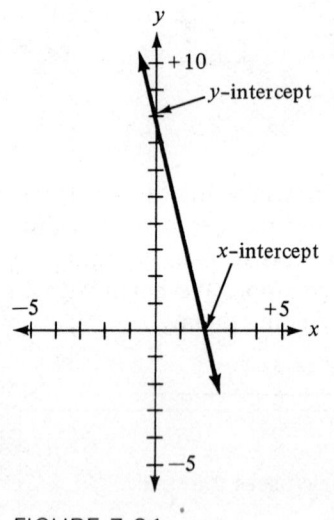

FIGURE 7.21

EXAMPLE Graph the line $-4x + 3y = 12$ using the intercepts.

SOLUTION

$-4x + 3(0) = 12$
$-4x = 12$
$x = -3$ (the x-intercept)

$-4(0) + 3y = 12$
$3y = 12$
$y = 4$ (the y-intercept)

The line is plotted in Figure 7.22

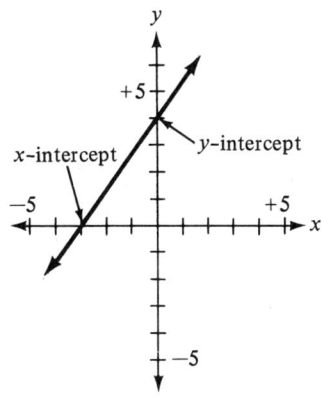

FIGURE 7.22

In this method we do not need to solve for y to produce a table. However, the use of a check point is still recommended.

Again, we consider the special cases in which the linear equation is missing either x or y.

EXAMPLE Find the intercepts of the line $3y = 8$.

SOLUTION Since we cannot set $y = 0$ in this equation, the line has no x-intercept. Its y-intercept is obtained by solving $3y = 8$.

$$y = \frac{8}{3} \quad \text{(y-intercept)}$$

The equation is plotted in Figure 7.23.

It is clear that the line cannot intercept the x-axis since it is parallel to that axis. Thus, the x-intercept is not defined.

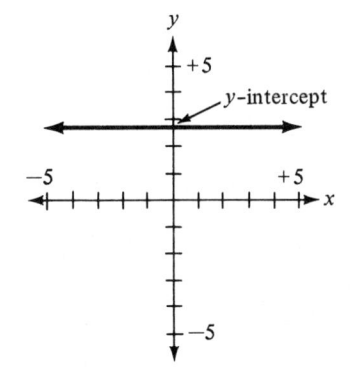

FIGURE 7.23

EXAMPLE Find the intercepts of the line $4x = 2$.

SOLUTION Since we cannot set $x = 0$ in this equation (or since the line is parallel to the y-axis), there is no y-intercept.

The x-intercept is found by solving $4x = 2$.

$x = \frac{1}{2}$ is the x-intercept.

This equation will graph as a line parallel to the y-axis and one-half unit to its right.

Another special case occurs when the equation has the form $ax + by = 0$. Both the x and y-intercepts are zero in this case. Another point is needed to generate the graph of the line.

EXAMPLE Graph the equation $3x - 2y = 0$.

SOLUTION We first locate the intercepts. Set $y = 0$ to obtain $3x - 2(0) = 0$ so that $x = 0$. Then set $x = 0$ to obtain $3(0) - 2y = 0$ so that $y = 0$.

The x and y-intercepts yield a single point, the origin. We need to generate a table with at least one more pair of values. Solve for y to obtain

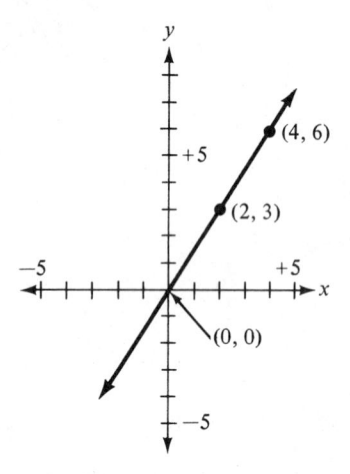

$$y = \frac{3}{2}x$$

x	0	2	4
y	0	3	6

FIGURE 7.24

The line is plotted in Figure 7.24.

PERFORMANCE OBJECTIVE 11

You should now be able to evaluate the x and y-intercepts of a line and use these to assist in graphing the line.

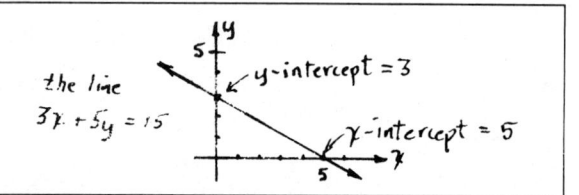

Next we consider a second important characteristic of a line in the rectangular coordinate system. With each line that is not vertical, we associate a number that represents the degree of inclination or steepness of that line. We call this number the **slope** of the line and we denote the slope by the letter m. Consider the line in Figure 7.25 on which the points (x_1, y_1) and (x_2, y_2) have been labeled.

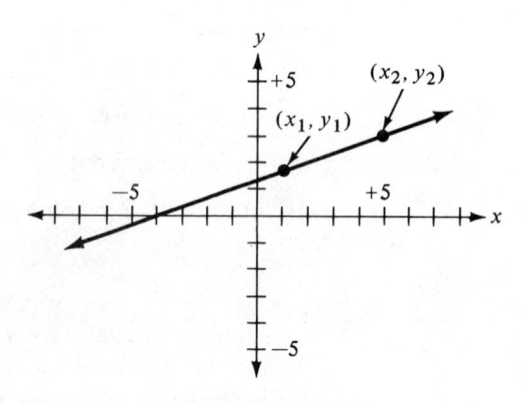

FIGURE 7.25

GRAPHING A LINE USING ITS CHARACTERISTICS 7.6

We can determine the slope by using the following formula:

FORMULA

SLOPE OF A LINE

$$m \text{ (slope)} = \frac{y_2 - y_1}{x_2 - x_1}$$

where (x_1, y_1) and (x_2, y_2) are two points on the line.

The slope of a line stays the same no matter what pair of points on the line is used to determine its value. The slope is often thought of as the rise of the line divided by the run of the line. That is,

$$m = \frac{\text{rise}}{\text{run}}$$

In Figure 7.26, the rise and run of the line are labeled.

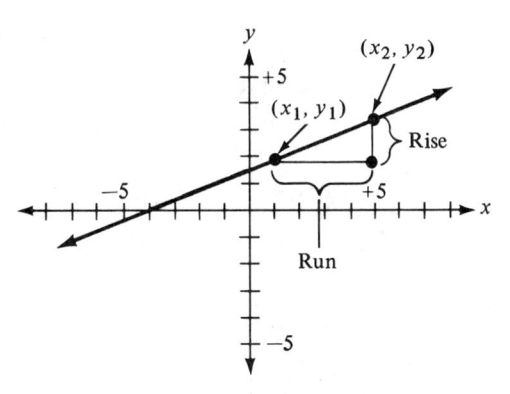

FIGURE 7.26

The equivalence of these two forms of the slope can be verified by considering Figures 7.27(a) and (b).

(a)

The coordinates of this point are (x_2, y_1)

(b)

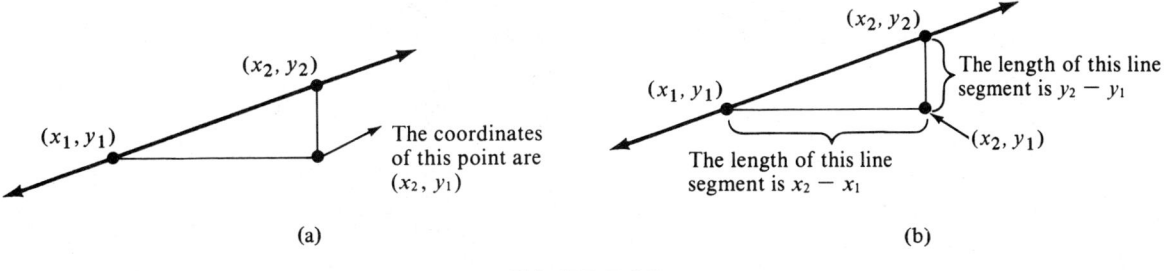

The length of this line segment is $y_2 - y_1$

The length of this line segment is $x_2 - x_1$

FIGURE 7.27

Thus rise $= y_2 - y_1$ and run $= x_2 - x_1$ so that

$$m = \frac{y_2 - y_1}{x_2 - x_1} = \frac{\text{rise}}{\text{run}}$$

EXAMPLE Find the slope of the line in Figure 7.28.

SOLUTION Two points on the line are $(0, -4)$ and $(2, 0)$. Using the equation for m we write

$$m = \frac{y_2 - y_1}{x_2 - x_1} = \frac{0 - (-4)}{2 - 0} = \frac{4}{2} = 2$$

Note that on the graph we can get from $(0, -4)$ to $(2, 0)$ by moving two units to the right and four units up. Using the rise/run definition, we would find

$$m = \frac{\text{rise}}{\text{run}} = \frac{+4}{+2} = 2$$

which agrees with the previous result.

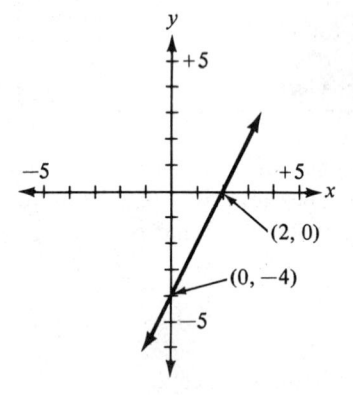

FIGURE 7.28

EXAMPLE The points $(1, -2)$ and $(4, 4)$ also lie on the straight line in Figure 7.28. Show the rise and run between these points and verify that the slope does not change.

SOLUTION The solution is shown in Figure 7.29.

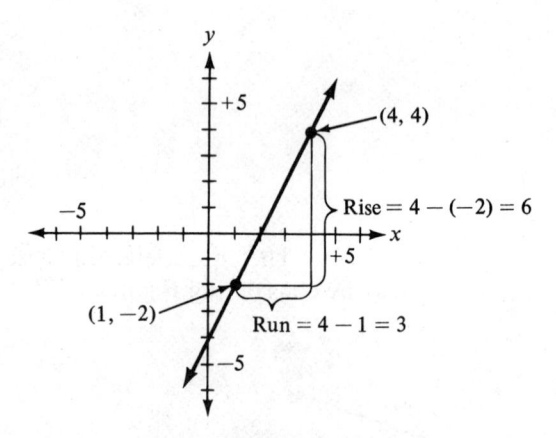

FIGURE 7.29

$$m = \frac{\text{rise}}{\text{run}} = \frac{6}{3} = 2$$

The slope is the same for both pairs of points.

GRAPHING A LINE USING ITS CHARACTERISTICS 7.6

EXAMPLE Find the slope of the line that passes through the points $(-5, 4)$ and $(2, 1)$.

SOLUTION Using the equation

$$m = \frac{y_2 - y_1}{x_2 - x_1}$$

and identifying $(2,1)$ as the rightmost point, we find

$$x_2 = 2, \quad x_1 = -5$$
$$y_2 = 1, \quad y_1 = 4$$

Then

$$y_2 - y_1 = 1 - (+4) = -3 \text{ units of rise}$$
$$x_2 - x_1 = 2 - (-5) = 7 \text{ units of run}$$

The rise, run, and slope are displayed in Figure 7.30.

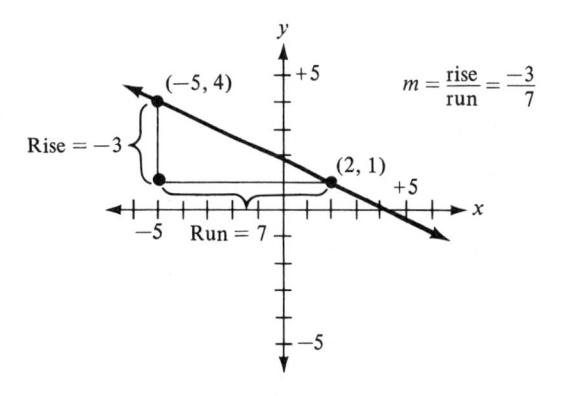

FIGURE 7.30

A rise of -3 indicates that in going from $(-5, 4)$ to $(2, 1)$ we moved down three units in the y direction. A run of $+7$ means we moved 7 units to the right.

The following examples deal with the special cases of horizontal and vertical lines.

EXAMPLE Find the slope of the horizontal line $y = 3$.

SOLUTION We select two points on the line such as $(0, 3)$ and $(2, 3)$. Then

$$m = \frac{y_2 - y_1}{x_2 - x_1} = \frac{3-3}{2-0} = \frac{0}{2} = 0$$

The slope of a horizontal line is zero.

EXAMPLE Find the slope of the vertical line $x = 1$.

SOLUTION We select two points on the line such as $(1, 2)$ and $(1, 5)$. If we try to use the slope formula, we get

$$m = \frac{5-2}{1-1} = \frac{3}{0}$$

which is not defined. We say that the line has no slope.

OBSERVATIONS ABOUT SLOPE

1. A line that rises as x increases has positive slope.
2. A line that drops or falls as x increases has negative slope.
3. The slope of a horizontal line is always zero.
4. A vertical line has no slope.

We use our knowledge of slope to assist in sketching the graph of a line. We need to know one point on the line as well as the slope to graph the line.

EXAMPLE Sketch the graph of the line through $(-2, 1)$ with slope $\frac{3}{5}$.

SOLUTION We plot the point $(-2, 1)$ in the coordinate system. Since the slope of the line is $\frac{3}{5}$, we know that y changes three units for every change of five units in x. Starting at $(-2, 1)$, we move five units to the right (that is, in the positive x direction) and three units up (that is, in the positive y direction). The point we arrive at must also lie on the line. So we connect this point with the point $(-2, 1)$ and we have graphed the desired line. See Figure 7.31.

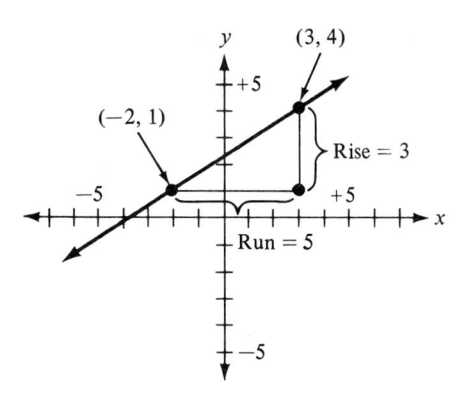

FIGURE 7.31

EXAMPLE Graph the line through (0, 2) with slope $-\frac{2}{3}$.

SOLUTION We plot (0, 2). We can write the slope

$$-\frac{2}{3} \quad \text{as} \quad \frac{-2}{3}$$

So y changes two units in the negative direction for every change of three units in the positive x direction. Starting at (0, 2), we move three units to the right and two units down to arrive at the point (3, 0). We connect (0, 2) and (3, 0) to complete the graph in Figure 7.32.

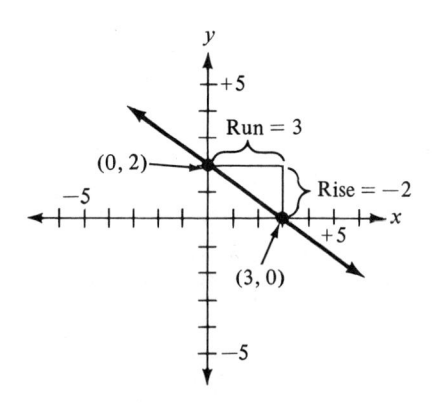

FIGURE 7.32

EXAMPLE Graph the line through (1, −2) with slope 3.

SOLUTION We can write the slope as

$$\frac{3}{1}$$

So y changes three units in the positive direction for every change of one unit in the positive x direction. Starting at $(1, -2)$, we move one unit to the right and three units up. We arrive at the point $(2, 1)$, which we connect with $(1, -2)$ to obtain the line shown in Figure 7.33.

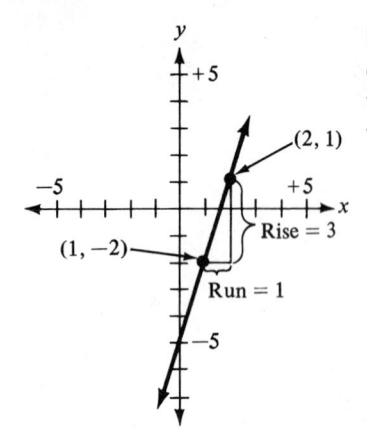

FIGURE 7.33

PERFORMANCE OBJECTIVE 12

You should now be able to sketch the graph of a line given a point on the line and its slope.

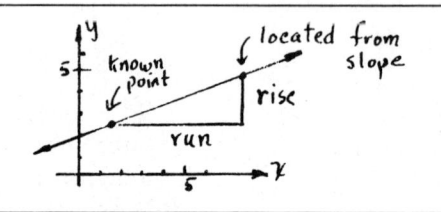

EXERCISE SET 7.6

1. Construct a graph of the following linear equations. Identify each x and y-intercept and use these to graph the equation when they produce two distinct points.

   a. $4x - 7y = 28$ b. $2x = 1$
   c. $-3x + y = 0$ d. $5y = -15$
   e. $4x + 9y = 12$

2. Construct a graph of the following linear equations. Identify each x and y-intercept and use these to graph the equation when they produce two distinct points.

   a. $3x - 6y = 18$ b. $5x = 3$
   c. $-4x + 3y = 0$ d. $2y = -10$
   e. $7x + 6y = 14$

3. Solve for the x and y-intercepts of the following linear equations. Do not plot the equations.
   a. $3x = 18$
   b. $-5x + y = -12$
   c. $4y = -9$
   d. $2x + 7y = 14$
   e. $5x + 7y = 0$

4. Solve for the x and y-intercepts of the following linear equations. Do not plot the equations.
   a. $5x = 12$
   b. $-4x + y = -9$
   c. $7y = -10$
   d. $5x + 4y = 20$
   e. $2x + 9y = 0$

5. Graph the lines determined by the following slopes and points:
   a. The line through $(-1, -2)$ with slope $m = 3$
   b. The line through $(0, 3)$ with slope $m = -\frac{2}{3}$
   c. The line through $(4, -5)$ with slope 0
   d. The line through $(0, 0)$ with slope $\frac{1}{2}$
   e. The line through $(0, 5)$ with slope -3
   f. The line through $(-2, -3)$ with slope $\frac{7}{4}$

6. Graph the lines determined by the following slopes and points:
   a. The line through $(-3, -1)$ with slope $m = 4$
   b. The line through $(0, 5)$ with slope $m = -\frac{3}{2}$
   c. The line through $(2, -3)$ with slope 0
   d. The line through $(0, 0)$ with slope $\frac{2}{3}$
   e. The line through $(0, 3)$ with slope -1
   f. The line through $(-4, -2)$ with slope $\frac{5}{3}$

SECTION 7.7 FINDING THE EQUATION OF A LINE FROM ITS CHARACTERISTICS

In the preceding sections, we either knew the equation of a line that we were graphing, or we knew enough information about the line to enable us to graph it without exhibiting its equation. In this section, our emphasis will be on finding the equation of a line given certain information about the line. We consider certains standard forms for the equation of a straight line.

DEFINITION

> **EQUATION OF A LINE (SLOPE-INTERCEPT FORM)**
>
> $y = mx + b$ is the slope-intercept form of the equation of a straight line.
>
> m is the slope; b is the y-intercept.

If we know the slope and y-intercept of a line, we write its equation by substituting the slope in place of m and the y-intercept in place of b in the equation $y = mx + b$.

EXAMPLE Find the equation of the line with slope -3 and y-intercept 1.

SOLUTION We identify m as equal to -3 and b as equal to 1. Then we substitute into $y = mx + b$.

$$y = (-3)x + (1) = -3x + 1$$

The desired equation is $y = -3x + 1$.

EXAMPLE Find the equation of the line with slope $\frac{1}{2}$ and y-intercept -5.

SOLUTION We identify $m = \frac{1}{2}$ and $b = -5$.

$$y = \left(\frac{1}{2}\right)x + (-5) = \frac{1}{2}x - 5$$

$y = \frac{1}{2}x - 5$ is the equation.

EXAMPLE Find the equation of the line with slope 0 and y-intercept 4.

SOLUTION We identify $m = 0$ and $b = 4$.

$$y = (0)x + (4) = 4$$

$y = 4$ is the equation.

After we have found the slope-intercept form for the equation of a line, we can rewrite it in the general form

$$ax + by = c$$

if we so desire. Also, if we are given an equation in the general form $ax + by = c$, we can rewrite it in the slope-intercept form by solving for y. We can then read the slope and y-intercept from the equation of the line.

EXAMPLE Write $6x + 2y = 3$ in slope-intercept form and find the slope and y-intercept of the line.

SOLUTION
$$6x + 2y = 3$$
$$2y = -6x + 3$$
$$y = \frac{-6}{2}x + \frac{3}{2}$$
$$y = -3x + \frac{3}{2}$$

The slope is -3 (the coefficient of x) and the y-intercept is $+\frac{3}{2}$.

PERFORMANCE OBJECTIVE 13

You should now be able to write the equation of a line given the slope and the y-intercept.

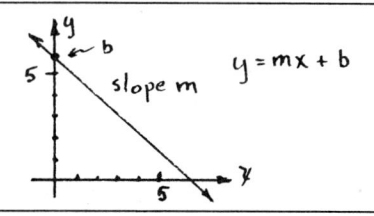

We consider another special form of the equation of the straight line.

DEFINITION

EQUATION OF A LINE (POINT-SLOPE FORM)

$y - y_1 = m(x - x_1)$ is the point-slope form of the equation of a line.

m is the slope; (x_1, y_1) is the point.

We use this form to write the equation of a line if we know the slope and a point on the line.

EXAMPLE Write an equation of the line through $(-1, 4)$ with slope -2.

SOLUTION We identify $x_1 = -1$, $y_1 = 4$ and $m = -2$. Using the point-slope form we get

$$y - 4 = -2(x - (-1))$$

We should simplify this as much as possible.

$$y - 4 = -2(x + 1)$$
$$y = -2x - 2 + 4$$
$$y = -2x + 2$$

EXAMPLE Write an equation of the line through $(0, 3)$ with slope $\frac{1}{2}$.

SOLUTION Using the point-slope form, we get

$$y - 3 = \frac{1}{2}(x - 0)$$
$$y - 3 = \frac{1}{2}x$$
$$y = \frac{1}{2}x + 3$$

NOTE: We could have used the slope-intercept form to start with since we knew the y-intercept was 3.

In the previous examples, we learned how to find an equation of a line given the slope of the line and a point on the line. If we are given two points on the line, we will first use the two points to find the slope of the line and then the point-slope form to find the equation of the line.

EXAMPLE Find the equation of the line through the points $(-5, -6)$ and $(3, 10)$.

SOLUTION The slope of the line is

$$m = \frac{10 - (-6)}{3 - (-5)} = \frac{10 + 6}{3 + 5} = \frac{16}{8} = 2$$

Using the slope and one of the two points, say (3, 10), we write

$$y - 10 = 2(x - 3) \quad \text{(the point-slope form)}$$
$$y - 10 = 2x - 6$$
$$y = 2x - 6 + 10$$
$$y = 2x + 4$$

If we had used the point $(-5, -6)$ instead of $(3, 10)$, we would have obtained the same equation. This is shown as follows:

$$y - (-6) = 2(x - (-5))$$
$$y + 6 = 2(x + 5)$$
$$y + 6 = 2x + 10$$
$$y = 2x + 10 - 6$$
$$y = 2x + 4$$

PERFORMANCE OBJECTIVE 14

You should now be able to write the equation of a line given a point and the slope or two points.

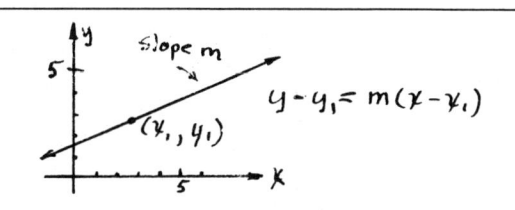

EXERCISE SET 7.7

1. Find equations of the following lines:
   a. The line through $(-2, 0)$ with slope $m = 3$
   b. The line through $(3, 2)$ with slope 0
   c. The line through $(5, -\frac{3}{2})$ with slope -2
   d. The line through $(-5, -2)$ with slope $-\frac{4}{7}$

2. Find equations of the following lines:
   a. The line through $(-4, 0)$ with slope $m = 5$
   b. The line through $(5, 1)$ with slope 0
   c. The line through $(2, -\frac{7}{3})$ with slope -3
   d. The line through $(-4, -3)$ with slope $-\frac{2}{3}$

3. Write each of the equations obtained in Problem 1 in slope-intercept form and determine the y-intercept.

4. Write each of the equations obtained in Problem 2 in slope-intercept form and determine the y-intercept.

5. Write the following equations in slope-intercept form and state the slope and y-intercept of the line given by the equation:
   a. $3x - 7y = 14$ b. $x + 2y = -4$ c. $2x + 7y = 0$
   d. $3y = 8$ e. $-4x + 9y = 3$

6. Write the following equations in slope-intercept form and state the slope and y-intercept of the line given by the equation:
   a. $2x - 5y = 15$ b. $x + 3y = -6$ c. $5x + 4y = 0$
   d. $2y = 7$ e. $-3x + 8y = 4$

7. Find equations of the following lines:
   a. The line through (0, 5) and (4, 3)
   b. The line through (−4, 2) and (5, −1)
   c. The line through (3, 2) and (0, 0)
   d. The line through (−3, 2) and (4, 2)
   e. The line through (3, −2) and (3, 2)

8. Find equations of the following lines:
   a. The line through (0, 2) and (1, 5)
   b. The line through (−6, 4) and (3, −3)
   c. The line through (5, 2) and (0, 0)
   d. The line through (−5, 4) and (6, 3)
   e. The line through (2, −4) and (2, 4)

SECTION 7.8   VARIATION

The notion of a linear function may be extended to functions that involve the independent variable to powers other than the first or to functions of more than one variable. As an example, we know that the area of a circle depends on the radius raised to the second power. In mathematical notation we may write

$$A = f(r) = kr^2$$

where A is area, r is radius, and k is called the **constant of proportionality** between A and r.

If we can establish the value of the constant of proportionality, we will have an expression that relates radius to area for any value of radius we may choose. In this case we know that the constant of proportionality equals pi (π) so that we may write

$$A = \pi r^2$$

Another way of describing this relation is to say that A varies directly as the square of r. This statement tells us what happens to A when r is changed. For example, if r is doubled, A will be quadrupled.

EXAMPLE The volume of a sphere varies as the cube of the radius. Determine the constant of proportionality between volume and radius.

SOLUTION From Chapter 3 we know that the volume of a sphere is given by

$$V = \frac{4}{3}\pi r^3$$

We compare this with the variation equation

$$V = f(r) = kr^3$$

Since $V = kr^3 = \frac{4}{3}\pi r^3$, k must be $\frac{4}{3}\pi$.

EXAMPLE The value of a quantity S depends upon t according to the relation $S = kt^2$. Find k, the constant of proportionality, given that $S = 4$ when $t = 0.5$.

SOLUTION Substitute the given values for the independent variable t and the dependent variable S into the given relation.

$$S = kt^2$$
$$4 = k(0.5)^2$$
$$4 = k(0.25)$$

Solve for k.

$$k = \frac{4}{0.25} = 16$$

The constant of proportionality is 16 and the relation becomes $S = 16t^2$.

The preceding example illustrates the value of knowing the variation of the dependent variable with respect to the independent variable. Knowing that variation and one set of corresponding values of the independent and dependent variables, the constant of proportionality may be determined. Once it is determined, prediction of dependent variable values can be made.

EXAMPLE The restoring force exerted by a spring is known to vary as the first power of its extension from its equilibrium position. That is,

$$F = kx$$

The restoring force exerted by the spring when extended 1.50×10^{-2} m is 400 N. What force will the spring exert when it is extended 2.5×10^{-2} m?

SOLUTION First determine the constant of proportionality from the given data. We substitute into $F = kx$ to find

$$400 = k(1.50 \times 10^{-2})$$
$$k = 400/(1.50 \times 10^{-2}) = 2.67 \times 10^4$$

Then write $F = (2.67 \times 10^4)x$. Substitute the given x to find the unknown F value.

$$F = (2.67 \times 10^4)(2.5 \times 10^{-2}) = 668 \text{ N}$$

We now extend this variation to examples involving more than one variable.

The rotational inertia I of a uniform disk about an axis through its center is known to be

$$I = \frac{1}{2}mr^2$$

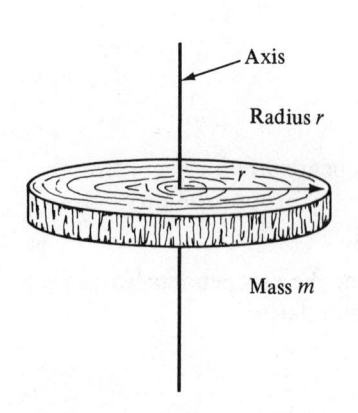

FIGURE 7.34

where m is the mass of the disk and r is its radius. See Figure 7.34.
The inertia I is dependent upon both mass and radius. We could write

$$I = f(m, r)$$

indicating that I is a function of (depends upon) both m and r. Knowing the equation above, we may be more specific and say that I varies as the first power of m and as the square of r.

EXAMPLE The rotational inertia I of a long thin rod about its end is known to vary as the first power of the mass and the square of the rod length. Calculate the constant of proportionality given that $I = 0.60$ kg-m^2 when $m = 0.80$ kg and $l = 1.5$ m.

SOLUTION Substitute the given data into the relation

$$I = kml^2$$
$$0.60 = k(0.80)(1.5)^2$$

Solve for k.

$$k = \frac{0.60}{(0.80)(1.5)^2} = 0.33$$

EXAMPLE The pressure P of a fixed quantity of an ideal gas is related to both the Kelvin temperature (T) and volume (V) according to the relation

$$P = kT/V$$

Find k given that $P = 1.5 \times 10^5$ N/m^2, $T = 300°$ Kelvin, and $V = 0.020$ m^3.

SOLUTION Substitute the given data into the relation $P = kT/V$.

$$1.5 \times 10^5 = k(300)/0.020$$
$$k = \frac{(1.5 \times 10^5)(0.020)}{300} = 10$$

PERFORMANCE OBJECTIVE 15

You should now be able to evaluate a constant of proportionality given a variation and specific values of the independent and dependent variables.

$S = kr^2$
Given $S = 113$ when $r = 3.00$,
then $k = \frac{S}{r^2} = \frac{113}{3^2} = 12.6$

We are now prepared to discuss specific kinds of variation.

DEFINITION

DIRECT VARIATION

Direct variation with respect to an independent variable implies that the dependent variable varies according to some positive power of the independent variable.

Consider the following relations:

$y = kx$ (here y varies directly as the first power of x. We often say that y is proportional to x)

$s = kt^2$ (here s varies directly as the second power of t)

DEFINITION

INVERSE VARIATION

Inverse variation with respect to an independent variable implies that the dependent variable varies according to the reciprocal of some positive power of the independent variable.

Consider the following relations:

$y = kx$ (here y varies inversely as the first power of x)

$z = k/x^2$ (here z varies inversely as the second power of x)

Direct and inverse variations may be combined in a single expression. Consider the following examples:

$z = k\dfrac{x}{y}$ (here z varies directly as the first power of x and inversely as the first power of y)

$I = kml^2$ (Here I varies directly as the first power of m and directly as the second power of l. This is sometimes called a **joint variation**)

EXAMPLE   Write an equation that describes the following variation: z varies directly as the square of x and inversely as the first power of y.

SOLUTION   The direct variation is expressed by placing an x^2 in the numerator; the inverse variation by placing a y in the denominator. We write

$$z = k\frac{x^2}{y}$$

EXAMPLE   Describe the variations expressed in the following equations:

1. $z = kx^3/y$
2. $z = k/\sqrt{y}$

SOLUTION

1. z varies directly as the third power of x and inversely as the first power of y.
2. z varies inversely as the square root of y.

EXAMPLE Calculate the constant of proportionality when the dependent variable z varies as the square of x and inversely as the cube of y. We are given that $z = 8$ when $x = 4$ and $y = 2$.

SOLUTION First, we must develop the relation from its description. We find

$$z = kx^2/y^3$$

Next we substitute into this expression

$$8 = k(4)^2/(2)^3$$

Solve for k.

$$8 = \frac{k(16)}{8}$$

$$k = \frac{(8)(8)}{16} = \frac{64}{16} = 4$$

EXAMPLE The gravitational force between two masses varies directly as the first power of each mass and inversely as the square of the distance between the masses. If the constant of proportionality is 6.67×10^{-11}, calculate the gravitational force between two masses (150 kg and 250 kg) separated by a distance of 5.0 m.

SOLUTION We write the given relation as

$$F = km_1m_2/r^2$$

Substitute the given values.

$$F = (6.67 \times 10^{-11})(150)(250)/(5)^2$$
$$= 1.0 \times 10^{-7} \text{ N}$$

It should be noted that not all functions can be expressed as variations. The variation terminology can be used to describe the function when the function consists of a single term. More complex expressions may not be expressible as variations. For example, in the general equation of a straight line, $y = mx + b$, y does not vary directly as the first power of x because of the presence of the constant term b. If b were zero, the expression would be $y = mx$ (single term) and y would vary directly as the first power of x. The constant of proportionality in this case would be m, the slope of the line.

PERFORMANCE OBJECTIVE 16	
You should now be able to describe a single term function as a variation and write the function from its variation.	*z varies directly as x^2 and inversely as the square root of y. We write $z = k \dfrac{x^2}{\sqrt{y}}$.*

EXERCISE SET 7.8

1. Determine the constant of proportionality in each variation.
   a. $y = kx^2$, $y = 2$ when $x = 4$
   b. $z = kx/y^2$, $z = 6$ when $x = 2$, $y = 0.5$
   c. $V = kT/P$, $V = 0.04$ when $T = 300°$, $P = 2 \times 10^5$

2. Determine the constant of proportionality in each variation.
   a. $z = ky^3$, $z = 2$ when $y = 0.5$
   b. $m = kI/l^2$, $m = 0.5$ when $I = 24$, $l = 2.0$
   c. $T = k\dfrac{\sqrt{m}}{\sqrt{c}}$, $T = 0.15$ when $m = 0.25$, $c = 439$

3. Predict the dependent variable value in the following expressions for the given independent variable value(s):
   a. $y = kx^3$. Determine y when $x = 2.5$ given that $y = 5$ when $x = 3$.
   b. $A = kr^2h$. Determine A when $r = 2.00$, $h = 1.50$ given that $A = 56.6$ when $r = 3.00$, $h = 2.00$.
   c. $T = k(\sqrt{L})(\sqrt{C})$. Determine T when $L = 3.00 \times 10^{-3}$, $C = 6.00 \times 10^{-6}$ given that $T = 1.26 \times 10^{-3}$ when $L = 20.0 \times 10^{-3}$, $C = 2.00 \times 10^{-6}$.

4. Predict the dependent variable value in the following expressions for the given independent variable value(s):
   a. $y = k\sqrt{x}$. Determine y when $x = 5$ given that $y = 7.3$ when $x = 2.1$.

b. $f = k\dfrac{\sqrt{c}}{\sqrt{m}}$. Determine f when $c = 330$, $m = 0.820$ given that $f = 4.03$ when $c = 385$, $m = 0.600$.

c. $F = km^2/r^2$. Determine F when $m = 1.5 \times 10^{20}$, $r = 5.0 \times 10^8$ given that $F = 2.96 \times 10^{15}$ when $m = 2.0 \times 10^{20}$, $r = 3.0 \times 10^7$.

5. Describe the variation of the dependent variable in the following expressions. Use the terminology introduced in this section. The constant of proportionality is k.

   a. $z = kx^2/\sqrt{y}$ b. $A = krh$ c. $y = k/x^2$

6. Describe the variation of the dependent variable in the following expressions. Use the terminology introduced in this section. The constant of proportionality is k.

   a. $V = kr^2h$ b. $z = k\sqrt{x}/y^2$

7. Write equations that describe the following variations. Choose appropriate symbols for the variables.

   a. The area varies directly as the product of base and height to the first power.

   b. The volume varies directly as the product of the base radius squared and height.

   c. z varies directly as x to the third power and inversely as the square root of y.

8. Write equations that describe the following variations. Choose appropriate symbols for the variables.

   a. z varies directly as the square root of x and inversely as y to the fourth power.

   b. The frequency varies directly as the square root of the spring constant and inversely as the square root of the mass.

   c. The acceleration varies directly as the mass and inversely as the square root of the distance.

SECTION 7.9   SUMMARY

In this chapter we have introduced linear equations in two variables and the concept of a function. We have also developed techniques for the graphic representation of mathematical relations on a rectangular coordinate system. Emphasis has been placed on understanding the graph of the general linear equation and its characteristics in the rectangular coordinate system.

The new concepts and definitions from this chapter are summarized here.

Independent variable: a variable whose value can be any number that makes sense in the expression.

Dependent variable: a variable whose value depends on the value of one or more independent variables.

Domain: the set of values the independent variable can have.

Range: the set of values the dependent variable can have.

Function: a set of ordered pairs (x, y) such that each domain value x is associated with exactly one range value y. Sometimes this term refers to the formula or method used to associate quantities in the domain with quantities in the range.

Linear equation in two variables: an equation that can be written in the form $ax + by = c$, where a, b, and c are real numbers and either $a \neq 0$ or $b \neq 0$.

Data table: an organized presentation of data or selected values in a mathematical relation. The data table displays values of the independent variable(s) and the corresponding dependent variable value.

Rectangular coordinate system: two dimensional coordinate system formed by the perpendicular intersection of two number lines at their respective zero points.

Coordinates: an ordered pair of numbers that locate a point in a rectangular coordinate system.

Point plot: a plot of ordered pairs of numbers that correspond to observed data or representative points in a mathematical relation.

Line graph: a line or curve representing a continuous relation between an independent and dependent variable.

Graph of a linear equation: a representation of the equation $ax + by = c$ in a rectangular coordinate system. This representation is a straight line.

Intercepts: the name given to those values at which a line graph intersects the x and y-axes.

Slope: the name given the inclination of the graph of a straight line. Slope can be determined as the ratio of rise to run between two points on the line.

Slope-intercept form: a form of the equation of a straight line in which the y-intercept and slope appear explicitly.

Point-slope form: a form of the equation of a straight line in which the slope and a selected point appear explicitly.

Constant of proportionality: a value in a variation that relates the numerical values of the independent variables to the dependent variable. It represents a constant ratio between dependent and independent variable values.

Direct variation: a variation in which the dependent variable varies as some positive power of the independent variable.

Inverse variation: a variation in which the dependent variable varies as the reciprocal of some positive power of the independent variable.

CHAPTER 7 PRACTICE TEST

1. (PO-1) For $x = -6$ find the corresponding range value in $y = 3x + 10$ and write your answer as an ordered pair.

2. (PO-2) Is the set of ordered pairs $(-3, 1)$, $(-2, 3)$, $(1, 7)$, $(0, 5)$, $(-3, 2)$ a function? Explain your answer.

3. (PO-2, 3) If $f(x) = x^2 - 4$, is $f(x)$ a function? Explain your answer.

4. (PO-4) Determine whether or not $3x + y^2 = 5$ is a linear equation in two variables. Explain your answer.

(PO-5) Determine three solutions to each of the following linear equations in two variables:

5. $\frac{2}{3}x - \frac{1}{4}y = 2$
6. $6x + 0y = 3$

(PO-6) Solve the following linear equations for y, if possible, to determine how the function processes the given values of the independent variable. Write the result of the processing as ordered pairs. If the equation does not represent a function, so state.

7. $x - 4y = 0$; $-4, 0, \frac{1}{2}$
8. $4x + 2y = 5$; $0, \frac{5}{4}, 1$

9. (PO-7) Construct a data table displaying the values of production units P versus hours of shift h. The relation between P and h is given by

$$P = -h^2 + 8h + 20$$

Let h take on integer values from 1 to 8.

10. (PO-8) Plot the following ordered pairs on a rectangular coordinate system:

$$(4, -2), \quad (-1, 5), \quad (2, 0), \quad (-5, -3), \quad (0, -3)$$

11. (PO-8) Write the ordered pairs corresponding to the points shown in Figure 7.35.

12. (PO-9) Construct a line graph to represent the following relations:

   a. $y = \dfrac{6}{x}$ (let x range from 1 to 15)

   b. $y = x^2 - 2x - 3$ (let x range from -2 to 4)

13. (PO-10) Construct a graph of the following linear equations:

   a. $2x - 7y = 14$ b. $2x + y = 7$

14. (PO-11) Construct a graph of the following linear equations by plotting the x and y-intercepts:

   a. $x - 3y = 6$ b. $-7x + 2y = 14$

15. (PO-12) Graph the lines determined by the following slopes and points:

   a. The line through $(-1, 2)$ with a slope of -2

   b. The line through $(1, -5)$ with a slope of $\frac{7}{3}$

16. (PO-13, 14) Write the equations of the following lines:

   a. The line with slope 3 and y-intercept -6

   b. The line with slope $-\frac{3}{2}$ and passing through the point $(-1, 5)$

   c. The line passing through the points $(-2, -2)$ and $(3, 1)$

17. (PO-15) Determine the constant of proportionality for $z = ky^3/x$ given that $z = 8$ when $x = 6$, $y = 2$.

18. (PO-15) Predict the dependent variable value in $y = kx^3$ for $x = 1.5$ given that $y = 4$ when $x = 2$.

19. (PO-16) Describe the variation of the dependent variable in $y = kx^3/y$ where k is the constant of proportionality.

20. (PO-16) Write an equation to describe the following variation. Choose appropriate symbols for the variables.

   The acceleration varies inversely as the cube of the distance and directly as the mass.

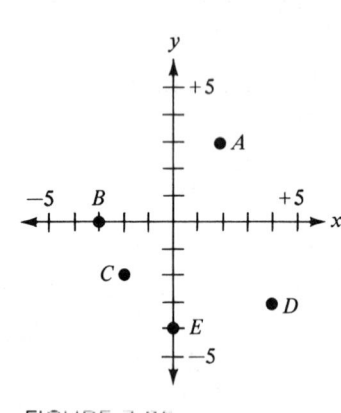

FIGURE 7.35

CHAPTER 8

BASIC TOPICS IN TRIGONOMETRY

8.1 Constructing a Triangle
8.2 Angles Defined by Ratios
8.3 Solving a Right Triangle Using the Trigonometric Ratios
8.4 Solving a Right Triangle with the Aid of a Calculator
8.5 Solving Applied Problems Involving Right Triangles
8.6 Solving Oblique Triangles
8.7 Trigonometry in Surveying
8.8 Summary

Trigonometry began as a study of the triangle. The triangle is the simplest closed figure that can be constructed using straight lines.

Trigonometry has expanded into areas where its applications may seem to have nothing to do with triangles. For example, we use expressions from trigonometry to describe the form of radio and television waves and the behavior of electrical circuits. We will treat these extensions in a later chapter of this text. Here we will deal with the solution of the triangle using trigonometric ratios.

SECTION 8.1 CONSTRUCTING A TRIANGLE

When we look at a triangle we see three sides and three angles. Suppose we wish to construct a triangle to a set of specifications. We use a protractor, a straightedge, and a compass in the constructions that follow. In describing the size of angles we will dispense with the words "the measure of angle . . . ," which we introduced in Chapter 3. When we say an angle = 56°, we will assume that this describes the measure of the angle.

EXAMPLE Construct a triangle in which two of the angles are 30° and 90° respectively.

SOLUTION We first construct the 90° angle using a protractor. See Figure 8.1.

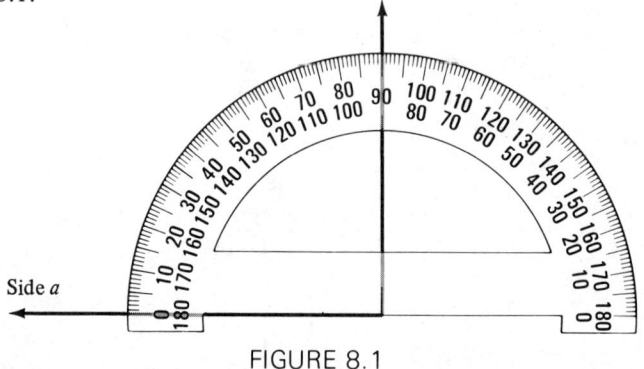

FIGURE 8.1

Next we move along side *a* to some point *P* and construct the 30° angle. See Figure 8.2.

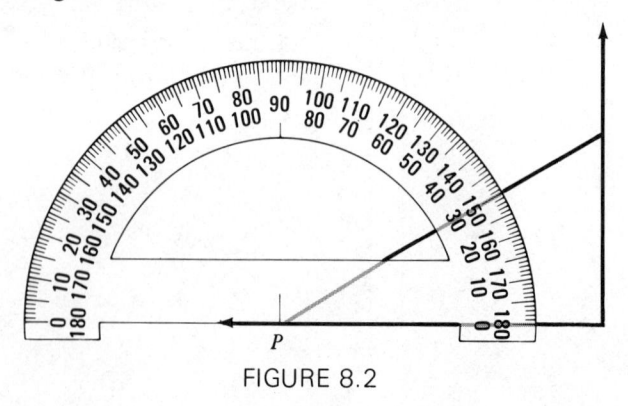

FIGURE 8.2

Our triangle is shown in Figure 8.3.

CONSTRUCTING A TRIANGLE 8.1   287

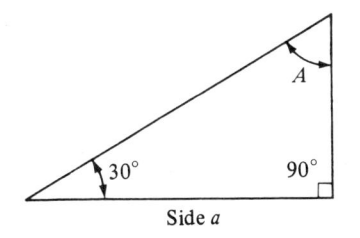

FIGURE 8.3

Notice that we had no choice as to the third angle A. If we measure, we should find angle A equal to 60° since the sum of the interior angles of a triangle equals 180°. However, we could have chosen any point P along side a at which to construct the 30° angle. That is, the triangles in Figure 8.4 also satisfy the original specifications.

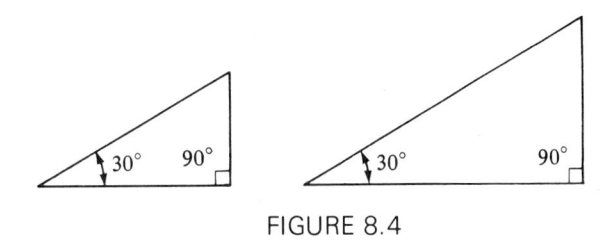

FIGURE 8.4

Thus the specification of two angles alone does not fix a unique triangle.

FIGURE 8.5

EXAMPLE   Construct a triangle in which two of the sides are 6 cm and 3.5 cm.

SOLUTION   We lay off one of the sides and draw an arc of length equal to the other side as shown in Figure 8.5. This construction leads to many possible triangles that meet the specifications. Two such triangles are shown in Figure 8.6.

You may begin to suspect that to determine a unique triangle, more than two characteristics need to be specified. In fact, three characteristics are required. But not just any three characteristics will produce a unique figure. For example, specifying three angles will not determine a unique triangle.

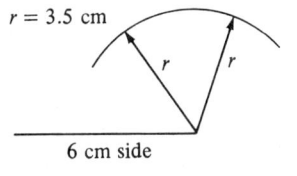

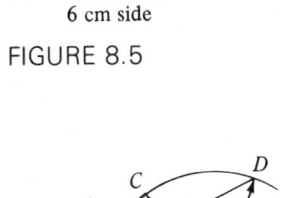

FIGURE 8.6

EXAMPLE   Construct a triangle with sides of 3.5 cm and 6 cm and included angle of 75°. (An included angle is formed by the two sides.)

SOLUTION   We begin by laying off the 6 cm side and the 75° angle as illustrated in Figure 8.7.

Next we measure 3.5 cm to form the line segment $\overline{BC}$ and connect the end of this length with A to form the triangle. See Figure 8.8.
In constructing this triangle we had no option other than to select an end of segment $\overline{AB}$ at which to construct the 75° angle. Constructing the 75° angle at A would have resulted in a triangle congruent with the one constructed in Figure 8.8. We thus have a unique triangle.

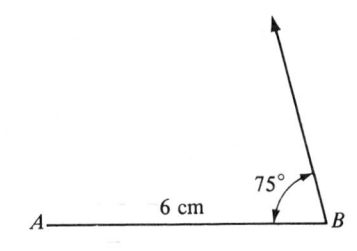

FIGURE 8.7

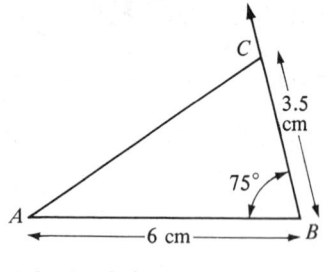

FIGURE 8.8

EXAMPLE Construct a triangle with sides of 6 cm, 3.5 cm, and 4.5 cm.

SOLUTION We begin by laying off the 6 cm side. Swing arcs of length 3.5 cm and 4.5 cm from its two ends. That is, use a compass centered at each end to produce intersecting arcs as shown in Figure 8.9. Draw the other two sides of the triangle to their intersection point and the specifications are met. See Figure 8.10.

This method will always result in a unique triangle providing that the length of any given side does not exceed the sum of the lengths of the other two. Can you see why this restriction is necessary?

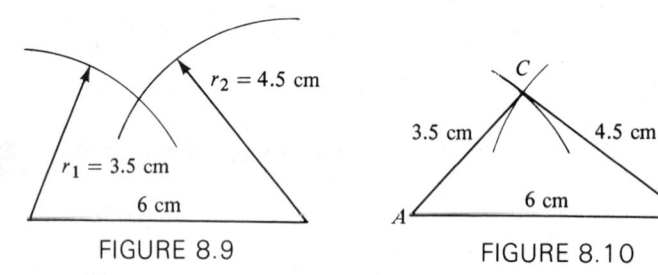

FIGURE 8.9

FIGURE 8.10

PERFORMANCE OBJECTIVE 1

You should now be able to construct a triangle given information about three of its parts.

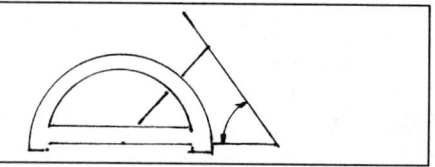

EXERCISE SET 8.1

1. Construct the triangles specified (given three side lengths).
   a. 4.8, 1.6, and 4.2 cm
   b. 5.0, 5.0, and 4.0 cm
   c. 1.0, 1.5, and 1.8 in.

2. Construct the triangles specified (given three side lengths).
   a. 5.6, 1.8, and 4.8 cm
   b. 4.0, 4.0, and 5.2 cm
   c. 1.5, 1.75, and 2.25 in.

3. Construct the triangles specified (given two side lengths and an included angle).
   a. 5.1 and 3.6 cm and included angle of 66°
   b. 3.8 and 5.2 cm and included angle of 105°
   c. 1.25 and 2.25 and included angle of 35°

4. Construct the triangles specified (given two side lengths and an

included angle).

  a. 6.1 and 4.2 cm and included angle of 72°
  b. 4.1 and 4.9 cm and included angle of 115°
  c. 1.75 and 2.50 and included angle of 29°

5. Construct the triangles specified (given two angles and an included side length).

  a. 35° and 45° and an included (common) side of 4.1 cm
  b. 70° and 20° and an included side of 1.75 in.
  c. 100° and 30° and an included side of 2.6 cm

6. Construct the triangles specified (given two angles and an included side length).

  a. 30° and 60° and an included side of 5.2 cm
  b. 44° and 85° and an included side of 2.25 in.
  c. 110° and 40° and an included side of 3.1 cm

SECTION 8.2 ANGLES DEFINED BY RATIOS

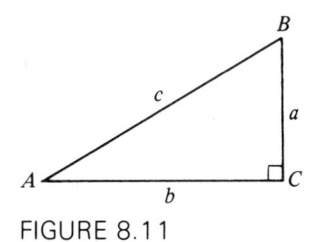

FIGURE 8.11

c = hypotenuse
a = side opposite angle A
b = side adjacent to angle A

FIGURE 8.12

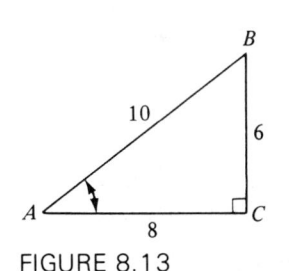

We discussed the right triangle briefly in Chapter 3. The right triangle is of particular interest because it occurs so frequently in practical problems and because of the simple relation between the lengths of its sides. That relation is expressed in the Pythagorean theorem.

The right triangle is also important in developing an alternate measure of an angle. Up to this point we have measured angles in degrees. We will see that an angle can be defined in terms of the ratio of the sides of a right triangle of which it is a part. Consider angle A in right triangle ABC pictured in Figure 8.11. In Figure 8.12 the relation of this angle to the sides of the triangle is shown.

EXAMPLE In triangle ABC (Figure 8.13) write the following:

a. The length of the side adjacent to angle A
b. The length of the side opposite angle A
c. The length of the hypotenuse

NOTE: In the problems that follow we omit units for measurement of the triangle's sides. The units may be considered arbitrary.

SOLUTION

a. Adjacent side = 8 units long
b. Opposite side = 6 units long
c. Hypotenuse = 10 units long

FIGURE 8.13

EXAMPLE In triangle ABC (Figure 8.14) write the following:

a. The length of the side adjacent to angle B
b. The length of the side opposite angle B
c. The length of the hypotenuse

SOLUTION

a. Note that side $\overline{BC}$ is adjacent (next to) angle B. Thus the adjacent side is 7.5 units long.

b. The opposite side is 13 units long.

c. The hypotenuse is 15 units long.

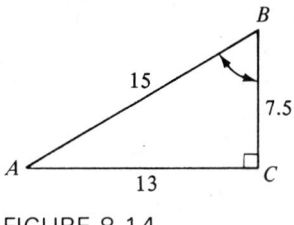

FIGURE 8.14

The terms "adjacent side" and "opposite side" will only be used in relation to the two acute angles in a right triangle.

PERFORMANCE OBJECTIVE 2

You should now be able to identify the adjacent and opposite sides for the acute angles in a right triangle.

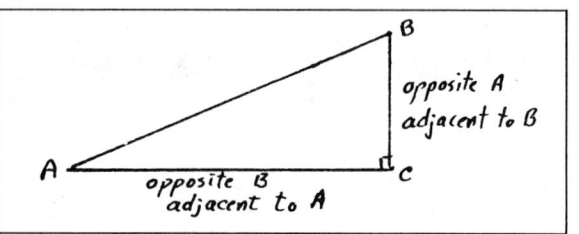

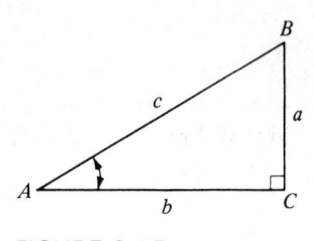

FIGURE 8.15

Now we are ready to define certain properties of an angle using the right triangle. Consider the triangle ABC in Figure 8.15. The size of angle A is related to the relative sizes of the sides a, b, and c of the triangle. We can form six ratios of the sides of this triangle that are related to the size of angle A. These ratios are frequently called the **trigonometric ratios.** We will consider only three of these ratios now.

We start with the ratio of the opposite side to the hypotenuse $\left(\dfrac{a}{c}\right)$. In Figure 8.16 we see two triangles, ABC and $AB'C'$, which are similar to one another. Therefore

$$\frac{a}{c} = \frac{a'}{c'}$$

since the ratio of corresponding sides in similar triangles must be equal.

We conclude that the ratio of the opposite side to the hypotenuse does not depend upon the triangle size but rather upon the size of angle A. We call this trigonometric ratio the **sine** of the angle A.

FIGURE 8.16

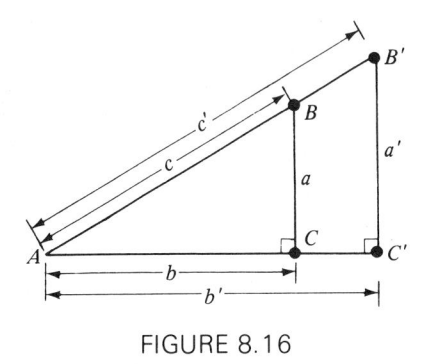

DEFINITION

SINE OF AN ANGLE IN A RIGHT TRIANGLE

The **sine** of an acute angle in a right triangle is given as the ratio of the opposite side to the hypotenuse.

EXAMPLE Calculate the sine of angle A in the right triangle in Figure 8.17.

SOLUTION The sine of angle $A = \dfrac{\text{opposite side}}{\text{hypotenuse}} = \dfrac{6.10}{9.30} = 0.656$.

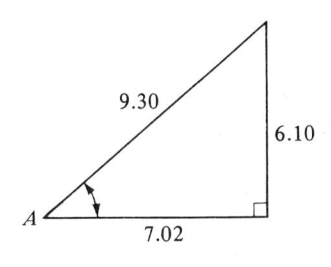

FIGURE 8.17

But what good is this value? We have said that the value of this trigonometric ratio must be another measure of the size of the angle. This implies that we should be able to convert the value of the sine into a value of the angle in other measures. This can be done with a set of trigonometric tables or with a calculator. We show a portion of a trigonometric table here.

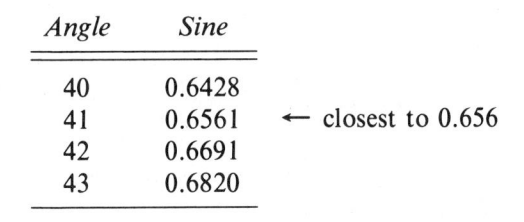

Angle	Sine
40	0.6428
41	0.6561 ← closest to 0.656
42	0.6691
43	0.6820

From the table we identify the sine value of 0.6561 as the number closest to our value of 0.656 and read the corresponding angle of 41°. We have determined that angle A is 41° based upon information about the sides of a right triangle containing that angle. It is clear that we now have a method of calculating angles in a right triangle given some information about that triangle.

Let us look again at the two triangles in Figure 8.16. We note the equalities

$$\frac{b}{c} = \frac{b'}{c'}$$

Each term represents the ratio of an adjacent side to a hypotenuse. This ratio must also depend on the size of angle A. This trigonometric ratio is called the **cosine** of the angle A.

DEFINITION

COSINE OF AN ANGLE IN A RIGHT TRIANGLE

The **cosine** of an acute angle in a right triangle is given as the ratio of the adjacent side to the hypotenuse.

EXAMPLE Calculate the cosine of angle A in Figure 8.17.

SOLUTION

The cosine of angle $A = \dfrac{\text{adjacent side}}{\text{hypotenuse}} = \dfrac{7.02}{9.30} = 0.755$

To relate the cosine value in the previous example to the size of angle A in degree measure, we need to expand the trigonometric table to include a second column. Consider the portion of the table that follows.

Angle	Sine	Cosine	
40	0.6428	0.7660	
41	0.6561	0.7547	← closest to 0.755
42	0.6691	0.7431	
43	0.6820	0.7314	

We identify 0.7547 as closest to the value of 0.755 and read the corresponding angle of 41°. We note that this ratio also allows us to determine the value of angle A. We will consider one more ratio in the triangles of Figure 8.16. This ratio is

$$\frac{a}{b} = \frac{a'}{b'}$$

which must hold since the triangles are similar.

This ratio, being independent of triangle size, must depend upon the size of angle A. This trigonometric ratio is called the **tangent** of angle A.

DEFINITION

TANGENT OF AN ANGLE IN A RIGHT TRIANGLE

The **tangent** of an acute angle in a right triangle is given as the ratio of the opposite side to the adjacent side.

EXAMPLE Calculate the tangent of angle A in Figure 8.17.

SOLUTION

The tangent of angle $A = \dfrac{\text{opposite side}}{\text{adjacent side}} = \dfrac{6.10}{7.02} = 0.869$

To relate this result to the size of angle A in degree measure, consider the expanded table that follows.

Angle	Sine	Cosine	Tangent
40	0.6428	0.7660	0.8391
41	0.6561	0.7547	0.8693 ← closest to 0.869
42	0.6691	0.7431	0.9004
43	0.6820	0.7314	0.9325

We again identify the angle as 41°.

PERFORMANCE OBJECTIVE 3

You should now be able to write the trigonometric ratios (sine, cosine, and tangent) for an acute angle in a right triangle.

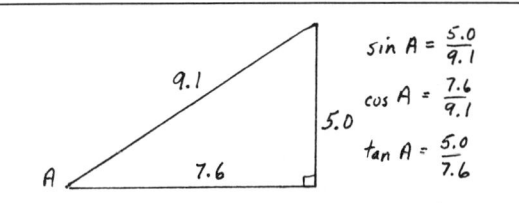

$\sin A = \dfrac{5.0}{9.1}$

$\cos A = \dfrac{7.6}{9.1}$

$\tan A = \dfrac{5.0}{7.6}$

We conclude that, with a knowledge of any two sides of a right triangle, we can determine the acute angles by computing a value for the sine, cosine, or tangent. We use this value to enter a table and find the measure of the angle in degrees. A complete table is included as an appendix to this chapter.

FIGURE 8.18

EXAMPLE Calculate angle A in the right triangle of Figure 8.18.

SOLUTION

Step 1. First we note the relation of the known sides to the unknown angle. Values of the opposite side and the hypotenuse are known.

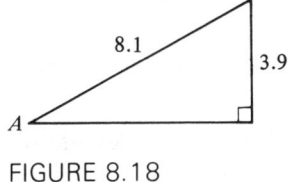

Step 2. Next we find the trigonometric ratio that involves those two sides. That ratio is the sine. We recall that the sine is equal to the opposite side divided by the hypotenuse.

Step 3. We calculate the sine.

$$\text{sine } A = \frac{3.9}{8.1} = 0.48$$

Step 4. Under "sine" in the trigonometric table we locate the value closest to 0.48. We find a value of 29° for the angle.

28	0.4695
29	0.4848
30	0.5000

EXAMPLE Calculate angle B in the right triangle of Figure 8.19.

SOLUTION

Step 1. We have an opposite side (4.8) and an adjacent side (7.1) given.

Step 2. The trigonometric ratio that involves these sides is the tangent.

Step 3. We calculate the tangent.

$$\text{tangent } B = \frac{4.8}{7.1} = 0.68$$

Step 4. We find B in the trigonometric tables.

Angle	Sine	Cosine	Tangent
33			0.6494
34			0.6745
35			0.7002

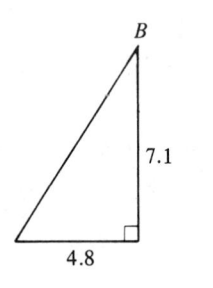

FIGURE 8.19

B equals 34° to the nearest degree.

Before we continue to solve problems to identify angle values using a table, we should understand how tables listing trigonometric ratio values can be arranged. In some tables certain properties of the ratios are used to avoid making a lengthy table. To see how this is done, consider the triangle in Figure 8.20.

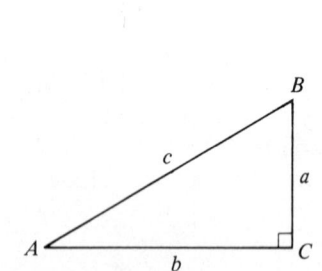

FIGURE 8.20

Note that sine $A = \frac{a}{c}$ and that cosine $B = \frac{a}{c}$. These two angles add to 90°. In mathematical terms, they are **complementary angles**.

Suppose we were constructing a table of sines and cosines. Consider the special case of $a = 3.0$, $b = 4.0$, $c = 5.0$.

$$\begin{array}{c c c} & \text{sine} & \text{cosine} \\ A & \dfrac{a}{c} = 0.6 & \dfrac{b}{c} = 0.8 \end{array}$$

These are also appropriate entries for the sine and cosine of B but in the reverse order. That is,

$$\begin{array}{c c c} & \text{sine} & \text{cosine} \\ B & \dfrac{b}{c} = 0.8 & \dfrac{a}{c} = 0.6 \end{array}$$

We can find the proper values for both angles A and B by simply interchanging the headings for the larger angle. We can make an entry as follows:

$$\begin{array}{c c c} & \text{sine} & \text{cosine} \\ A & 0.6 & 0.8 \qquad 90 - A = B \\ & \text{cosine} & \text{sine} \end{array}$$

Bottom labels refer to angle B while top labels refer to angle A. Our table only needs to run to 45° as a result. To handle the tangent function we should note that the tangents of complementary angles are reciprocals of one another. Thus we can abbreviate the tangent table by addition of a column entitled 1/tangent. The table at the chapter end contains the three functions for angles from 0 to 90°.

Henceforth, we will follow common practice and abbreviate these trigonometric ratios to their first three letters, that is, sine becomes sin, cosine becomes cos, and tangent becomes tan.

EXAMPLE Calculate angle A in the triangle of Figure 8.21.

SOLUTION $\sin A = \dfrac{5.9}{7.5} = 0.787$

From the table we do not find a value near 0.787 in the leftmost sin column. Therefore, we look in the sin column running upward. There we find the values

$$0.7880 \rightarrow 52°$$
$$0.7771 \rightarrow 51°$$

It appears that 0.7880 is closer to our value; so we would report 52° as the answer to the nearest degree.

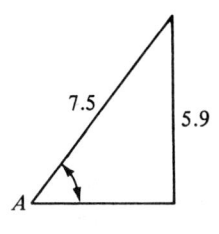

FIGURE 8.21

EXAMPLE Calculate angle B in the triangle of Figure 8.22.

SOLUTION $\tan B = \dfrac{5.41}{3.82} = 1.416$

Values of tangent greater than 1 are associated with the bottom heading. Starting upward on the tan column, we find

$$1.428 \to 55°$$
$$1.376 \to 54°$$

We report an angle of 55° to the nearest degree.

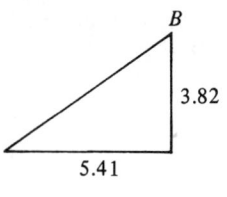

FIGURE 8.22

PERFORMANCE OBJECTIVE 4

You should now be able to evaluate the acute angles of a right triangle given information about any two sides.

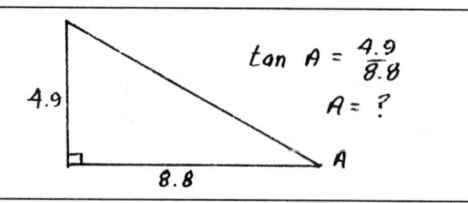

$\tan A = \dfrac{4.9}{8.8}$

$A = ?$

EXERCISE SET 8.2

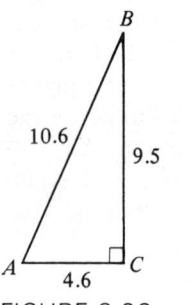

FIGURE 8.23

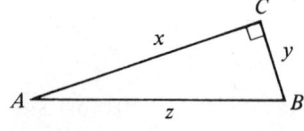

FIGURE 8.24

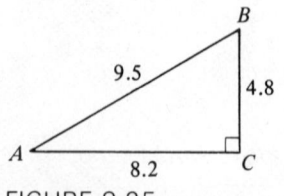

FIGURE 8.25

1. Identify sides of a right triangle in relation to the angles indicated.
   a. Report the length of the following (see Figure 8.23):

      The adjacent side to angle A
      The opposite side to angle A
      The adjacent side to angle B
      The hypotenuse

   b. Report the letter designation of the following (see Figure 8.24):

      The adjacent side to angle B
      The opposite side to angle B
      The opposite side to angle A
      The hypotenuse

2. Identify sides of a right triangle in relation to the angles indicated.
   a. Report the length of the following (see Figure 8.25):

      The adjacent side to angle A
      The opposite side to angle A
      The adjacent side to angle B
      The hypotenuse

ANGLES DEFINED BY RATIOS 8.2   **297**

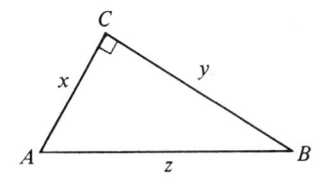

FIGURE 8.26

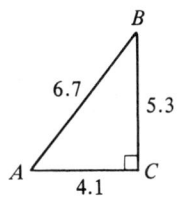

FIGURE 8.27

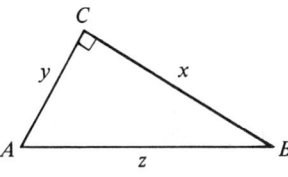

FIGURE 8.28

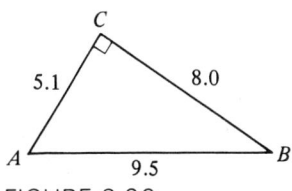

FIGURE 8.29

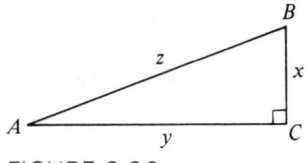

FIGURE 8.30

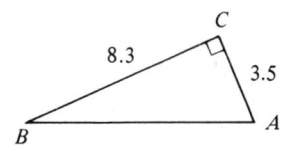

FIGURE 8.31

FIGURE 8.32

b. Report the letter designation of the following (see Figure 8.26):

The adjacent side to angle B
The opposite side to angle B
The opposite side to angle A
The hypotenuse

3. Write the trigonometric ratios indicated (leave in fraction form).

   a. The sine of angle A
      The cosine of angle A
      The tangent of angle B
      (See Figure 8.27.)

   b. The sine of angle B
      The cosine of angle B
      The tangent of angle A
      (See Figure 8.28.)

4. Write the trigonometric ratios indicated (leave in fraction form).

   a. The sine of angle A
      The tangent of angle B
      The cosine of angle B
      (See Figure 8.29.)

   b. The sine of angle B
      The cosine of angle A
      The tangent of angle A
      (See Figure 8.30.)

5. Calculate the following using a table (to the nearest degree):

   a. Angle A in Figure 8.27
   b. Angle B in Figure 8.27
   c. Angle A in Figure 8.31
   d. Angle B in Figure 8.32

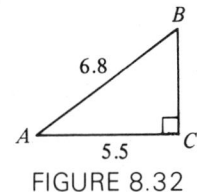

6. Calculate the following using a table (to the nearest degree):

   a. Angle A in Figure 8.29
   b. Angle B in Figure 8.30

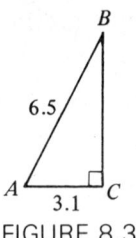

8 BASIC TOPICS IN TRIGONOMETRY

FIGURE 8.33

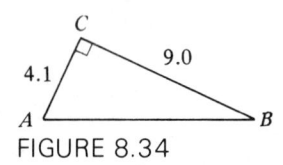

FIGURE 8.34

c. Angle A in Figure 8.33
d. Angle B in Figure 8.34

7. Calculate the unknown angles in the triangles shown.
   a. Angle A (Figure 8.35)
   b. Angle A (Figure 8.36)
   c. Angle A (Figure 8.37)

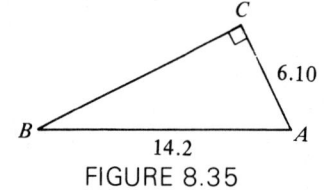

FIGURE 8.35

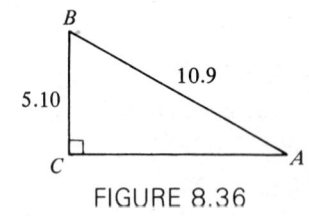

FIGURE 8.36

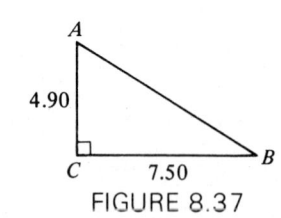

FIGURE 8.37

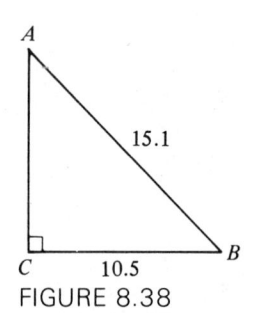

FIGURE 8.38

8. Calculate the unknown angles in the triangles shown.
   a. Angle A (Figure 8.38)
   b. Angle A (Figure 8.39)
   c. Angle A (Figure 8.40)

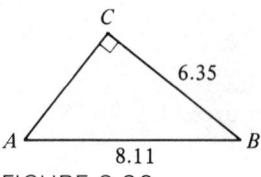

FIGURE 8.39

9. Using the standard right triangle given in Figure 8.41, evaluate the unknown angles in each case.
   a. Given $a = 4.9$, $b = 7.5$, determine angle B.
   b. Given $c = 15.5$, $a = 5.2$, determine angle A.
   c. Given $a = 9.1$, $c = 12.3$, determine angle B.

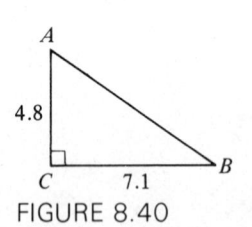

10. Using the standard right triangle given in Figure 8.42, evaluate the unknown angles in each case.
    a. Given $a = 6.1$, $b = 8.3$, determine angle B.
    b. Given $c = 9.5$, $a = 4.2$, determine angle A.
    c. Given $a = 15.1$, $c = 20.2$, determine angle B.

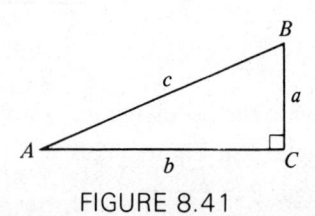

FIGURE 8.40

FIGURE 8.41

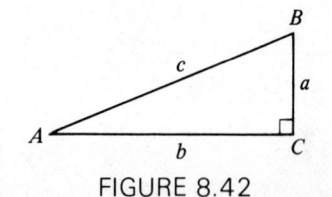

FIGURE 8.42

SECTION 8.3 SOLVING A RIGHT TRIANGLE USING THE TRIGONOMETRIC RATIOS

In the last section we considered a special class of problems in which we were given information about two sides of a right triangle and asked to evalute its acute angles. This type of problem is a part of a more general problem of solving a right triangle. By *solving a right triangle* we mean evaluating all unknown sides and angles.

We noted earlier that one would expect to have to know three things about a triangle in order to construct it. If the triangle is a right triangle, we already know one thing so that only two additional items of information should be necessary to construct or solve a right triangle. Those two things cannot be the other two angles. At least one side must be given.

EXAMPLE Solve the triangle in Figure 8.43 for angles A, B, and side c.

SOLUTION Note that we have enough information to find the tangent of either angle A or angle B.

$$\tan B = \frac{\text{opposite}}{\text{adjacent}} = \frac{6.10}{12.9} = 0.473$$

Using the table, we find B equal to 25° to the nearest degree. This requires that

$$A = 90° - 25° = 65°$$

The length of c can be obtained using the Pythagorean theorem or by using one of the trigonometric ratios.

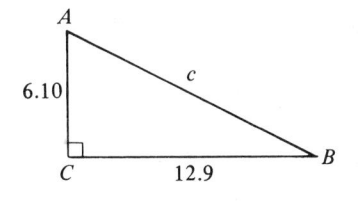

FIGURE 8.43

Pythagorean theorem	*Trigonometric ratio*
$c = \sqrt{6.1^2 + 12.9^2}$ $= \sqrt{203.6}$ $c = 14.3$	$\sin B = \dfrac{\text{opposite side}}{\text{hypotenuse}} = \dfrac{6.10}{c}$ Rearranging, we find $c = \dfrac{6.10}{\sin B} = \dfrac{6.10}{0.423} = 14.4$

These answers differ in the third digit. This discrepancy results from the use of a table that can only be read to the nearest degree.

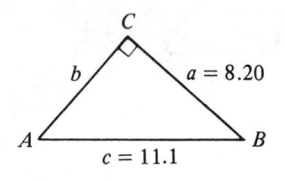

FIGURE 8.44

EXAMPLE Solve the triangle in Figure 8.44 for angles A, B, and for side b.

SOLUTION Note that $\sin A = \dfrac{a}{c} = \dfrac{8.20}{11.1} = 0.739$

Using the table, we find $A = 48°$ to the nearest degree. Then

$$B = 90° - A = 90° - 48° = 42°$$

The length of b is obtained by using the Pythagorean theorem or by using a trigonometric ratio.

Pythagorean theorem	Trigonometric ratio
$b^2 = c^2 - a^2$ $= 11.1^2 - 8.2^2$ $= 55.97$ $b = \sqrt{55.97}$ $= 7.48$	$\tan B = \dfrac{\text{opposite side}}{\text{adjacent side}} = \dfrac{b}{8.20}$ Solve for b to find $b = 8.20 \tan B = 8.20(0.900)$ $= 7.38$

Again note a discrepancy in these answers that results from rounding to the nearest degree.

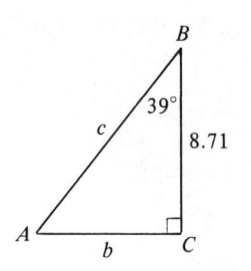

FIGURE 8.45

EXAMPLE Solve the triangle in Figure 8.45 for angle A and sides b and c.

SOLUTION Note that we cannot use the Pythagorean Theorem immediately since we only know the length of one side. However, angle A is known because the sum of the angles must equal 180°. Thus

$$A = 90° - 39° = 51° \quad \text{(the sum of the acute angles must equal 90°)}$$

We can now find both b and c using the trigonometric ratios.

$$\tan 39° = \frac{\text{opposite side}}{\text{adjacent side}} = \frac{b}{8.71}$$

We solve this equation for b.

$$b = 8.71 \tan 39° = 8.71(0.810) = 7.06$$

to three significant digits.

At this point we have a choice of using the Pythagorean Theorem or another trigonometric ratio to find c. The use of ratios is recommended because fewer operations are involved.

$$\cos 39° = \frac{\text{adjacent side}}{\text{hypotenuse}} = \frac{8.71}{c}$$

We solve this equation for c.

$$c = \frac{8.71}{\cos 39°} = \frac{8.71}{0.777} = 11.2$$

PERFORMANCE OBJECTIVE 5

You should now be able to use trigonometric tables to solve a right triangle given either two sides or one side and an angle.

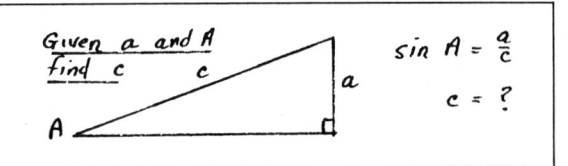

Given a and A
find c

$\sin A = \frac{a}{c}$

$c = ?$

EXERCISE SET 8.3

1. Solve for angles A, B, and side c in the following right triangles:

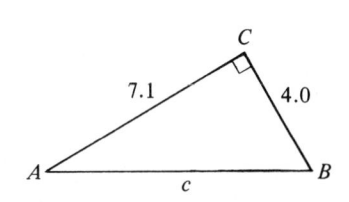

    a. (triangle with sides 7.1, 4.0, and c; right angle at C)

    b. (triangle with sides 20.9, 13.4, and c; right angle at C)

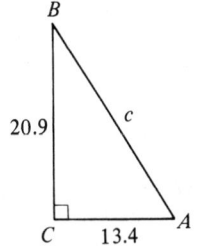

2. Solve for angles A, B, and side c in the following right triangles:

    a. (triangle with sides 4.1, 7.5, and c; right angle at C)

    b. (triangle with sides 5.5, 4.6, and c; right angle at C)

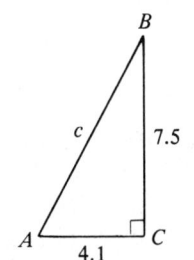

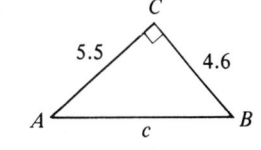

3. Solve for angles A, B, and side b in the following right triangles:

   a. Triangle with B at top, C at bottom-left (right angle), A at bottom-right. BC = 3.91, BA = 10.5, CA = b.

   b. Triangle with B at top, A at bottom-left, C at bottom-right (right angle). AB = 8.92, BC = 8.45, AC = b.

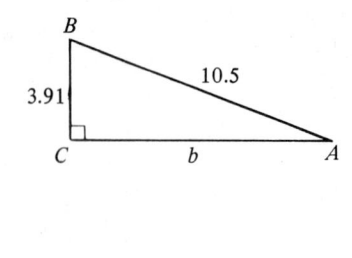

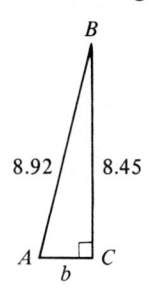

4. Solve for angles A, B, and side b in the following right triangles:

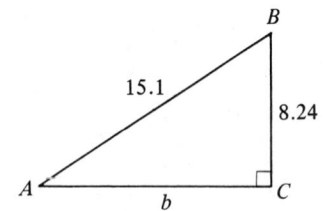

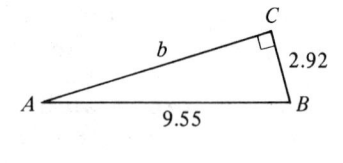

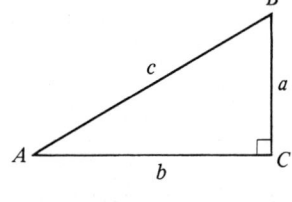

   a. Triangle with B at top, A at bottom-left, C at bottom-right (right angle). AB = 15.1, BC = 8.24, AC = b.

   b. Triangle with C at top-right, A at left, B at bottom-right (right angle). AC = b, CB = 2.92, AB = 9.55.

5. Consider the general right triangle shown in Figure 8.46. Solve for all unknowns using the given information.

   a. $a = 6.9$, angle $B = 56°$
   b. $b = 4.8$, angle $A = 38°$
   c. $c = 9.2$, angle $A = 78°$
   d. $a = 1.8$, $b = 3.6$
   e. $c = 9.5$, $a = 3.8$

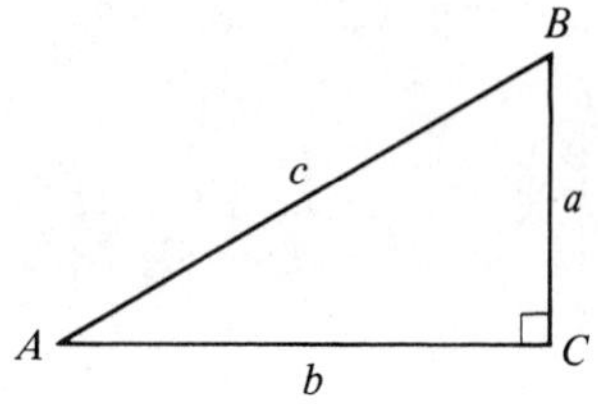

FIGURE 8.46

6. Consider the general right triangle shown in Figure 8.47. Solve for all unknowns using the given information.

   a. $a = 5.2$, angle $B = 61°$
   b. $a = 7.1$, angle $A = 40°$
   c. $c = 5.2$, angle $A = 48°$
   d. $a = 2.2$, $b = 3.1$
   e. $c = 15.1$, $a = 7.23$

FIGURE 8.47

SECTION 8.4 SOLVING A RIGHT TRIANGLE WITH THE AID OF A CALCULATOR

The use of a scientific calculator simplifies the problem of solving a right triangle. The calculator helps with the numerical calculation and serves as the trigonometric table.

There are two types of problems that we must be able to handle in order to use the calculator. These are

1. Determine the trigonometric ratio of an angle when the angle is known.
2. Determine the angle when its trigonometric ratio is known.

We illustrate the first type of problem with examples.

EXAMPLE   Find sin 69.0°.

SOLUTION   We enter the angle, then press the $\boxed{\text{SIN}}$ key and read 0.934 to three significant digits. The keystroke sequence is

$$69 \;\boxed{\text{SIN}}$$

EXAMPLE   Find cos 23.2°.

SOLUTION   Enter the angle, then press the $\boxed{\text{COS}}$ key and read 0.919 to three significant digits. The keystroke sequence is

$$23.2 \;\boxed{\text{COS}}$$

EXAMPLE   Find tan 53.0°.

SOLUTION   Enter the angle, then press the $\boxed{\text{TAN}}$ key and read 1.33 to three significant digits. The keystroke sequence is

$$53 \;\boxed{\text{TAN}}$$

CAUTION: If your calculator accepts angles in measures other than degrees, be sure it is in degree mode.

Before we go to the second type of problem, we recall that we defined the function in an earlier chapter as a sort of "processor." We feed a number into the processor and it turns out a second number that has some relation to the first. The input to the processor is the known independent quantity. The output is the unknown dependent quantity.

As an example, consider the "processor" or function we have called sine. If we enter 30° into the processor, it returns the number 0.5. If we enter 45°, 0.707 is returned. In working with functions on the calculator and on the computer, it is convenient to think of a diagram such as we introduced in Chapter 7. See Figure 8.48.

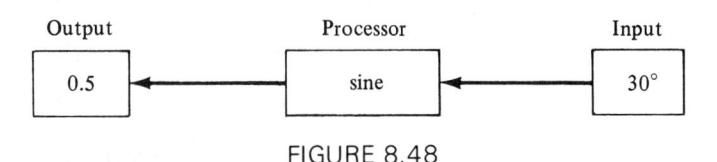

Output ← Processor ← Input
 0.5 ← sine ← 30°

FIGURE 8.48

We may write the result of the processor in Figure 8.48 in equation form as $0.5 = \sin 30°$.

This right-to-left process indicates the steps we follow to use a table or calculator (or computer) as we evaluate the function. For example, to use the table we follow these steps.

Step 1. Identify the angle (select the input to the processor).

Step 2. Identify the column for the sine function (select the correct processor).

Step 3. Read down the column until the number corresponding to 30° is found (identify the output).

To use the calculator we follow these steps.

Step 1. Enter the angle (select the input).

Step 2. Press the function key (select the correct processor).

Step 3. Read the register (identify the output).

This view of the functions may help us understand what we are doing when we use the function keys on the calculator. Having selected the input value (known quantity), we press the function key to activate the processor. The calculator performs the necessary operations to return the output value (the unknown or dependent quantity).

Now we consider the second type of problem in which the value of the trigonometric function is known.

EXAMPLE Find the angle whose sine is 0.751.

SOLUTION Enter the ratio, then press $\boxed{\text{INV}}$ and $\boxed{\text{SIN}}$ in that order.

$$0.751 \;\boxed{\text{INV}}\; \boxed{\text{SIN}}$$

and read 48.7° to three significant digits.

EXAMPLE Find the angle whose cosine is 0.800.

SOLUTION Enter the ratio, then press $\boxed{\text{INV}}$ and $\boxed{\text{COS}}$ in that order.

$$0.8 \;\boxed{\text{INV}}\; \boxed{\text{COS}}$$

and read 36.9°.

EXAMPLE Find the angle whose tangent is 1.79.

SOLUTION Enter the ratio, then press [INV] and [TAN] in that order.

$$1.79 \;\; \boxed{INV} \;\; \boxed{TAN}$$

and read 60.8°.

Finding an angle given one of its trigonometric functions is the reverse of what we were doing at the beginning of this section. We need a processor that will invert the operation. In this section, we illustrated the sine processor in Figure 8.48. The processor that reverses the action of the sine processor is shown in Figure 8.49.

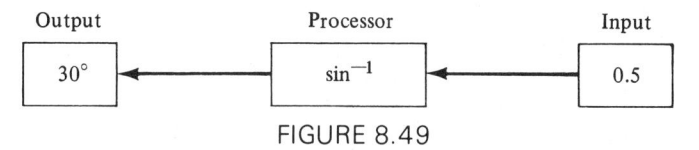

```
Output                 Processor              Input
 30°     ←───────      sin⁻¹      ←───────    0.5
```

FIGURE 8.49

The processor in Figure 8.49 is called the **inverse sine function**. It is usually indicated as $\sin^{-1}$ where the -1 represents inverse (not -1 power). In equation form, we can write the result of this processor as $30° = \sin^{-1} 0.5$.

Again, this right-to-left process indicates the steps we follow in using a calculator as we evaluate the inverse function. For example, in using the calculator, we follow these steps.

Step 1. Enter 0.5 (select the input).

Step 2. Press the inverse key (INV) and then the sine (SIN) key (select the correct processor).

Step 3. Read the register (identify the output).

We view the pressing of the inverse and sine keys as parts of the single step of selecting the correct processor. Were it not for the need to save space on a calculator keyboard, we would use a separate key. Then the inverse process would appear symmetric to the original function process.

It is unfortunate that we read from left to right but solve equations from right to left. This results in the reversal of the information in the equation from the sequence of steps in the solution. For example, we may write the equation for x as

$$x = \sin^{-1} 0.5 \quad \text{or} \quad x = \boxed{SIN} \; \boxed{INV} \; 0.5$$

To solve this equation we follow the sequence

$$0.5 \;\; \boxed{INV} \;\; \boxed{SIN}$$

8 BASIC TOPICS IN TRIGONOMETRY

We should be aware that restrictions exist on the input values to some of the inverse function processors. For example, an attempt to solve

$$x = \sin^{-1} 2$$

leads to an error message on the calculator. There is no angle whose sine equals 2. Such error messages serve as a warning that we have attempted to solve a problem that has no real solution.

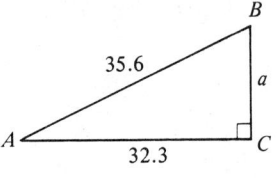

EXAMPLE Solve the right triangle shown in Figure 8.50.

SOLUTION We have two angles and one side to determine.

$$\cos A = \frac{32.3}{35.6} = 0.907$$
$$A = \cos^{-1} 0.907 = 24.9°$$
$$B = 90° - 24.9° = 65.1°$$

$$\sin A = \frac{a}{35.6}$$

FIGURE 8.50

so that

$$a = 35.6 \sin A = 35.6 \sin 24.9°$$
$$= 15.0 \quad \text{(to three significant digits)}$$

This whole process can be accomplished with the following sequence:

32.3 $\boxed{\div}$ 35.6 $\boxed{=}$ $\boxed{\text{INV}}$ $\boxed{\text{COS}}$ and read 24.864683 $\boxed{\text{STO}}$
$\boxed{\text{SIN}}$ $\boxed{\times}$ 35.6 $\boxed{=}$ and read 14.968968
90 $\boxed{-}$ $\boxed{\text{RCL}}$ $\boxed{=}$ and read 65.135317

Rounding these gives $A = 24.9°$, $B = 65.1°$, and $a = 15.0$.

EXAMPLE Solve the right triangle shown in Figure 8.51.

SOLUTION

$$\sin A = \frac{8.91}{c} \text{ so that } c = \frac{8.91}{\sin A} \quad \text{(rearrange the equation)}$$

$$\tan A = \frac{8.91}{b} \text{ so that } b = \frac{8.91}{\tan A} \quad \text{(again rearrange)}$$

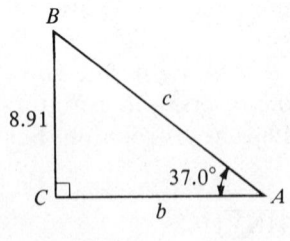

$$B = 90° - 37° = 53°$$

FIGURE 8.51

SOLVING A RIGHT TRIANGLE WITH A CALCULATOR 8.4

The preceding equations are solved as follows:

$$8.91 \; \boxed{\div} \; 37 \; \boxed{\text{SIN}} \; \boxed{=} \; 14.805214$$
$$8.91 \; \boxed{\div} \; 37 \; \boxed{\text{TAN}} \; \boxed{=} \; 11.823969$$

The answers to three significant digits are $c = 14.8$, $b = 11.8$, and $B = 53.0°$.

PERFORMANCE OBJECTIVE 6

You should now be able to use a calculator to solve a right triangle given either two sides or one side and an angle.

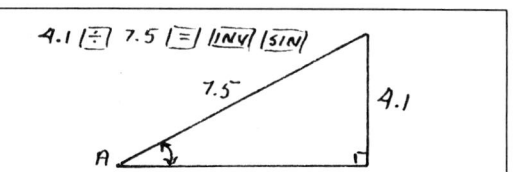

$4.1 \; \boxed{\div} \; 7.5 \; \boxed{=} \; \boxed{\text{INV}} \; \boxed{\text{SIN}}$

EXERCISE SET 8.4

1. Use the calculator to complete the following table:

Angle →	79°	8°	42°	59°	5°	30°
sin						
cos						
tan						

2. Use the calculator to complete the following table:

Angle →	46°	6°	82°	33°	61°	15°
sin						
cos						
tan						

3. Use the calculator to complete the following table:

Trigonometric ratio	Angle A
$\tan A = 1.8$	_____
$\cos A = 0.91$	_____
$\sin A = 0.83$	_____
$\cos A = 0.15$	_____
$\tan A = 1.0$	_____
$\sin A = 0.23$	_____
$\tan A = 0.45$	_____

4. Use the calculator to complete the following table:

Trigonometric ratio	Angle A
$\cos A = 0.81$	_____
$\tan A = 2.5$	_____
$\sin A = 0.56$	_____
$\sin A = 0.05$	_____
$\tan A = 0.30$	_____
$\cos A = 0.02$	_____
$\sin A = 0.99$	_____

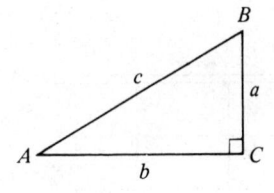

FIGURE 8.52

5. Consider the standard right triangle given in Figure 8.52. Use the calculator to solve this right triangle for all unknown parts using the given data.
   a. Angle $A = 59°$, side $a = 4.9$
   b. Angle $B = 32°$, side $a = 8.1$
   c. Side $c = 9.6$, side $b = 4.2$
   d. Side $a = 3.2$, side $b = 5.1$

6. Consider the standard right triangle given in Figure 8.53. Use the calculator to solve this right triangle for all unknown parts using the given data.
   a. Angle $A = 64°$, side $a = 5.2$
   b. Angle $B = 39°$, side $a = 9.5$
   c. Side $c = 10.5$, side $b = 5.1$
   d. Side $a = 4.6$, side $b = 6.2$

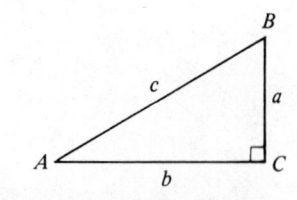

FIGURE 8.53

SECTION 8.5 SOLVING APPLIED PROBLEMS INVOLVING RIGHT TRIANGLES

The following examples illustrate the value of using trigonometric ratios in the solution of practical problems.

EXAMPLE Determine the distance across a lake (between markers C and B in Figure 8.54). We have available a compass, a tape measure, and a trigonometric function table.

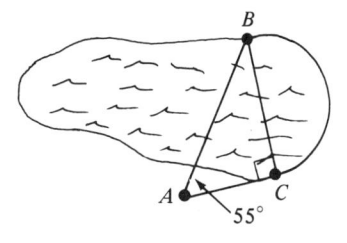

FIGURE 8.54

SOLUTION With the compass, we site along an angle 90° from $\overline{CB}$ and locate some object that is not too distant from C. For example, we locate a tree at point A. Next we measure the distance to the tree as 314 m. We move to the tree and measure an angle of 55° between lines $\overline{AB}$ and $\overline{AC}$. With this data we calculate the length of $\overline{BC}$ as follows: Knowing that $\tan 55° = \dfrac{BC}{AC}$ and that $AC = 314$ m, we write $\tan 55° = \dfrac{BC}{314}$. We look up $\tan 55°$ in the table and find $\tan 55° = 1.428$. Substituting, we find

$$1.428 = \frac{BC}{314}$$

To solve for BC, multiply both sides by 314.

$$314(1.428) = 314\left(\frac{BC}{314}\right)$$

Finally, $BC = 314(1.428) = 448.4 = 448$ m to three significant digits.

With the use of trigonometric ratios, we have found an unknown side of a right triangle from a knowledge of one angle and one side.

EXAMPLE Figure 8.55 gives an end view of the rafter construction in a roof. Rafters are butted to a center beam. Calculate the angle of the rafter cut where it joins the center beam.

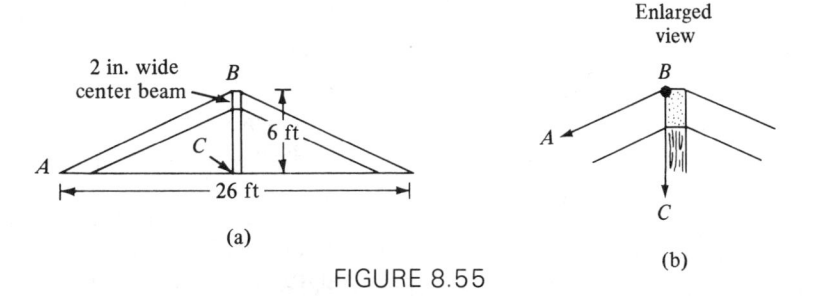

FIGURE 8.55

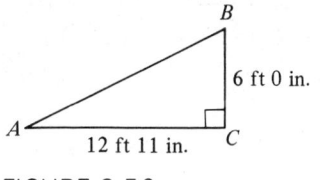

FIGURE 8.56

SOLUTION Consider right triangle ABC given in Figure 8.56.

$$\tan B = \frac{\text{opposite side}}{\text{adjacent side}} = \frac{12 \text{ ft } 11 \text{ in.}}{6 \text{ ft } 0 \text{ in.}} = 2.153$$

Using the calculator, we find $B = 65.1°$.

The steel square is a tool that depends upon right angle trigonometry. It is primarily used in framing roofs. With the use of the square, and in some cases roof framing tables, carpenters are able to avoid measurement of angles.

EXAMPLE The rafters on the lean-to in Figure 8.57 are to be cut so their ends are vertical. Use the steel square to mark the cut.

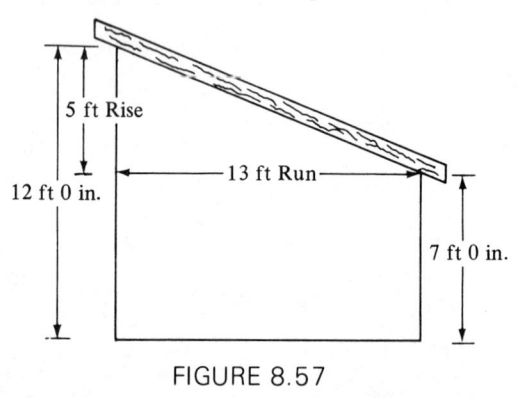

FIGURE 8.57

SOLUTION The roof rises 5 ft for a run of 13 ft. If we apply the steel square to the rafter and mark along the shorter arm (tongue) of the square, the angle is determined. (See Figure 8.58.)

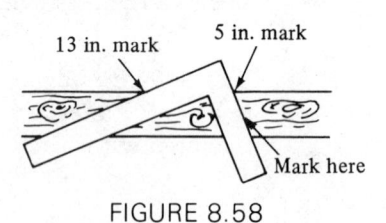

FIGURE 8.59

FIGURE 8.58

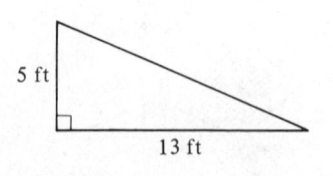

FIGURE 8.60

This method depends upon the similarity of the small triangle (Figure 8.59) with the larger triangle representing the actual building (Figure 8.60).

There are some special right triangles that appear frequently in practical problems. It may save time to memorize the ratio of sides of these triangles. Refer to Figure 8.61.

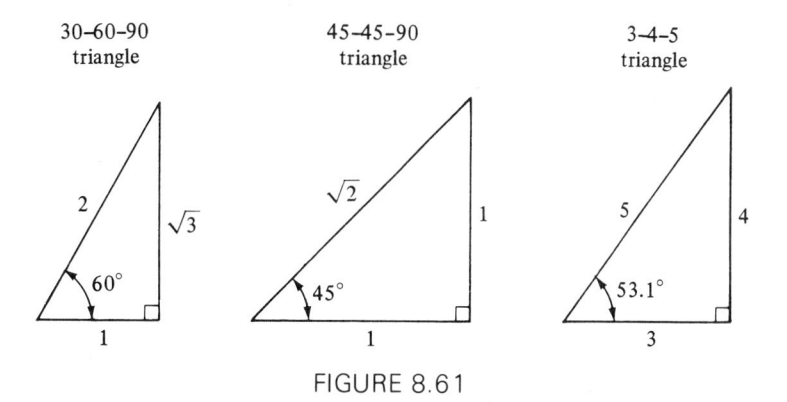

SOLVING APPLIED PROBLEMS INVOLVING RIGHT TRIANGLES 8.5

30-60-90 triangle

45-45-90 triangle

3-4-5 triangle

FIGURE 8.61

PERFORMANCE OBJECTIVE 7

You should now be able to use trigonometric ratios to solve applied problems involving right triangles.

$$h = b \tan A$$

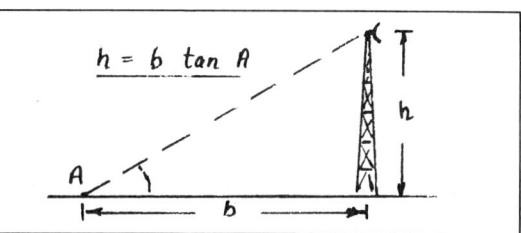

This section is concluded with a number of exercises in which you must interpret the problem. This requires that you identify the known parts of a triangle and the relation of the unknowns to those parts. It is important to be able to solve such practical exercises using trigonometric functions. This ability comes with practice.

EXERCISE SET 8.5

1. A 30 ft ladder must make an angle of at least 70° with the ground so the person climbing it will not slip. What is the maximum distance the base of the ladder should be placed from the building? See Figure 8.62.

2. A triangular lot is to have a retaining wall built on the short side. If the angle between this side and the longest side is 65°, and if the longest side is 76 m long, calculate the length of the retaining wall. See Figure 8.63.

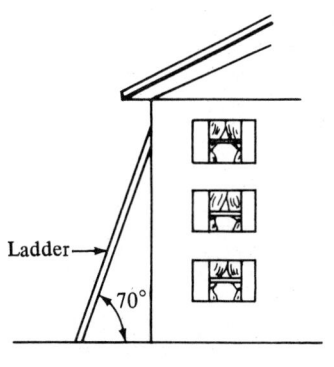

FIGURE 8.62

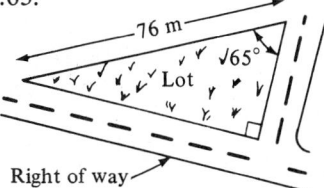

FIGURE 8.63

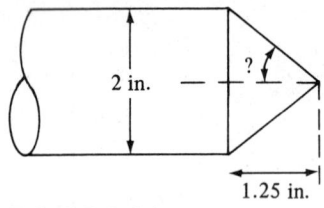

FIGURE 8.64

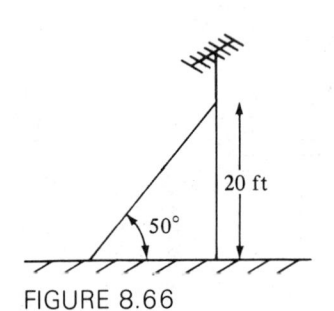

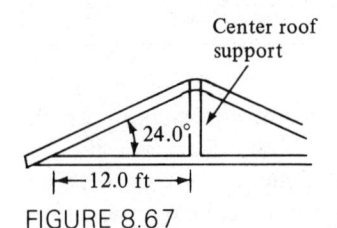

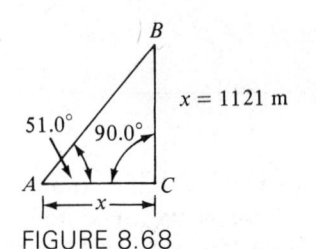

3. A 2 in. diameter cylindrical rod is to be tapered for 1.25 in. on the end. Calculate the angle of the taper (between the rod axis and the tapered face). See Figure 8.64.

4. A 7 mm diameter drill has a 1.1 mm taper on the end. Calculate the angle of the taper (between the drill axis and the tapered face). See Figure 8.65.

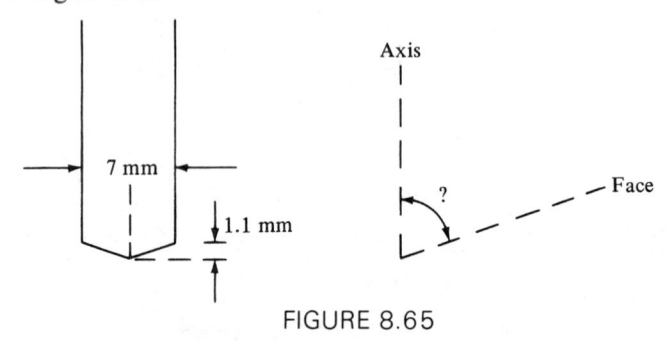

FIGURE 8.65

FIGURE 8.66

5. A guy wire is to be attached 20 ft up from the base of a pole. How much wire is needed if the wire is to make an angle of 50° with the ground? See Figure 8.66.

6. Calculate the length of the center roof support in Figure 8.67.

7. Calculate the straight line distance between markers C and B (Figure 8.68). The angles and sides have been measured by a surveyor. (Assume all points are at the same elevation).

8. Considering Figure 8.68, calculate the distance from A to B if $AC = 973.24$ m, angle $A = 49.51°$, and angle $C = 90.00°$.

FIGURE 8.67

In the following problems no sketch has been prepared. Start by drawing a sketch on which the known information is identified.

9. A guy wire on a television antenna is to be attached 30 ft from the ground. If the wire is to make an angle of 55° with the ground, how long must it be? Assume the ground is horizontal.

10. Calculate the length of a cross brace that makes an angle of 70° with the horizontal and rises (has a vertical extent of) 18 ft.

11. Find the length of a diagonal in a rectangular metal plate. The plate has one side equal to 5.53 cm. The diagonal makes an angle of 67° with this side.

12. A metal plate has two sides equal to 7.42 cm and one side of length 3.45 cm. Calculate the length of a line scribed between the middle of the short side and the opposite angle vertex.

13. Find the length of the other side of the rectangular plate in Problem 11.

FIGURE 8.68

14. Find all angles in the triangular plate in Problem 12.

15. A solar home in cross section is to be constructed in the shape of a right triangle with the hypotenuse on the earth's surface. The short side of the triangle is to be panelled in glass, have a length equal to 20 ft, and rise at an angle of 71° with the horizontal. Calculate the other two dimensions of the triangular cross section.

16. A shelter has front and back heights of 2.5 m and 1.9 m. The front and back walls are vertical and are 2.1 m apart. Calculate the angle the roof line makes with the horizontal.

SECTION 8.6 SOLVING OBLIQUE TRIANGLES

Information about unknowns in right triangles is fairly easily obtained using the trigonometric ratios. If the triangle we are solving is not a right triangle, it is called an **oblique triangle**. We then have a more difficult problem.

We will use the law of sines or law of cosines to solve oblique triangles. These laws are derived here. The derivations are excellent exercises in applying the techniques of solving right triangles. However, we stress an understanding of the laws and the conditions under which they may be useful in solving oblique triangles rather than the derivations themselves.

First we derive the **law of sines**. Consider the construction in Figure 8.69. Triangles ABE and AEC have been formed with triangle ABC by dropping a perpendicular from A to side $\overline{BC}$.

From the right triangle ABE we write

$$\sin B = \frac{AE \text{ (opposite side length)}}{AB \text{ (hypotenuse length)}}$$

From the right triangle AEC we write

$$\sin C = \frac{AE}{AC}$$

Both of these equations have AE in them. From the first,

$$AE = AB \sin B$$

From the second,

$$AE = AC \sin C$$

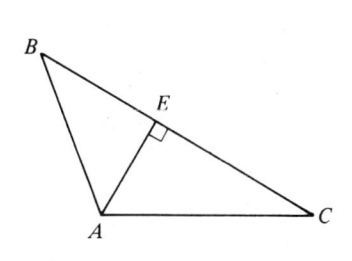

FIGURE 8.69

Then, $AB \sin B = AC \sin C$. Rearranging this equation, we find

$$\frac{\sin B}{AC} = \frac{\sin C}{AB}$$

Notice that $\overline{AC}$ is the side opposite angle B and $\overline{AB}$ is the side opposite angle C. It appears that the ratio of the sine of an angle to the side opposite is a constant in a triangle. We would need to complete the proof by showing that

$$\frac{\sin A}{BC} = \frac{\sin C}{AB}$$

To show this, use the construction in Figure 8.70. Consider triangles DBA and DBC in the proof and proceed as before. You will need to use the fact that the sine of an angle and its supplement have the same value. That is, the sine of angles DAB and CAB are equal. This will be established in Chapter 13. See if you can complete the proof.

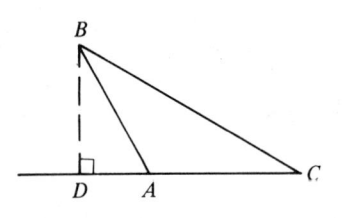

FIGURE 8.70

LAW OF SINES

In words: In any triangle the ratio of the sine of an angle to the length of the opposite side is a constant.

In symbols:

$$\frac{\sin A}{a} = \frac{\sin B}{b} = \frac{\sin C}{c}$$

See Figure 8.71.

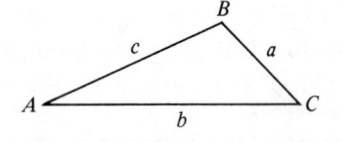

FIGURE 8.71

In using the law of sines to solve a triangle, we must have information about either

1. Two angles and one side, or
2. Two sides and one angle (with the angle opposite one of the sides)

EXAMPLE Solve the triangle in Figure 8.72.

SOLUTION There are three unknowns left in the triangle. These are labeled as angles A and B and side b. The law of sines allows us to write

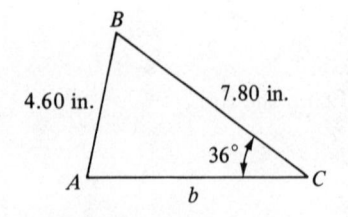

FIGURE 8.72

$$\frac{\sin A}{7.80} = \frac{\sin 36°}{4.60}$$

$$\sin A = \frac{7.80 \sin 36°}{4.60} = 0.997$$

$$A = 85.3° \quad \text{(using the calculator)}$$

$$B = 180° - 36° - 85.3° = 58.7°$$

Again use the sine law.

$$\frac{b}{\sin 58.7°} = \frac{4.60}{\sin 36°} \quad \text{(Note that we invert the sine law equation to keep the unknown in the numerator.)}$$

$$b = \frac{4.60 \sin 58.7°}{\sin 36°} = 6.69 \quad \text{(to three significant digits)}$$

Finally, the answers are $A = 85.3°$, $B = 58.7°$, and $b = 6.69$ in. The value of 85.3° was obtained with the calculator.

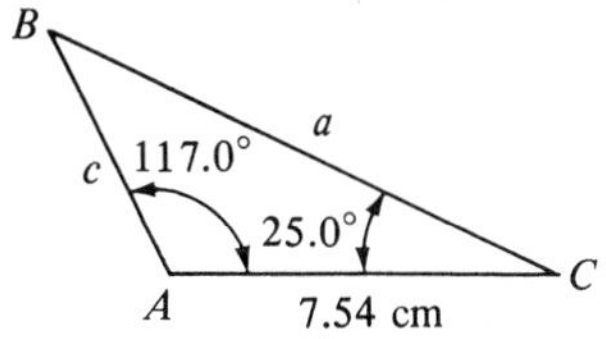

EXAMPLE Solve the triangle in Figure 8.73.

SOLUTION The three unknowns in the triangle are labeled as angle B and sides a and c. We note that

$$B = 180° - 25.0° - 117.0° = 38.0°$$

Using the sine law (inverted) we find

$$\frac{c}{\sin 25°} = \frac{7.54}{\sin 38°} \quad \text{so that} \quad c = \frac{7.54 \sin 25°}{\sin 38°} = 5.18$$

Using the sine law again yields

$$\frac{a}{\sin 117°} = \frac{7.54}{\sin 38°} \quad \text{so that} \quad a = \frac{7.54 \sin 117°}{\sin 38°} = 10.9$$

The unknowns are then $B = 38.0°$, $c = 5.18$ cm and $a = 10.9$ cm.

FIGURE 8.73

PERFORMANCE OBJECTIVE 8

You should now be able to use the law of sines to solve oblique triangles.

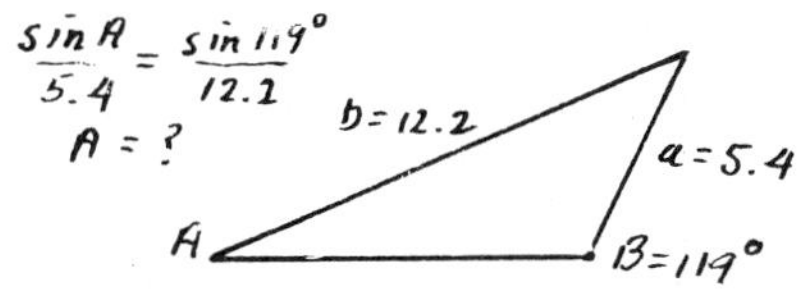

$$\frac{\sin A}{5.4} = \frac{\sin 119°}{12.1}$$

$A = ?$, $b = 12.2$, $a = 5.4$, $B = 119°$

8 BASIC TOPICS IN TRIGONOMETRY

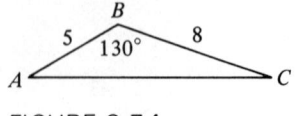

FIGURE 8.74

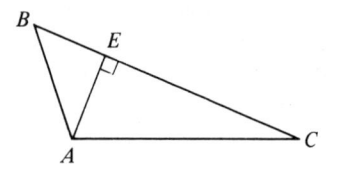

FIGURE 8.75

In discussing the law of sines we included a restriction that if one angle only were given, it must be opposite one of the known sides. This restriction rules out a situation such as that shown in Figure 8.74.

The triangle in Figure 8.74 cannot be solved using the law of sines. The **law of cosines** is needed to solve a problem such as this. To establish the law of cosines, consider the construction in Figure 8.75.

In Figure 8.75 triangles ABE and AEC have been formed with triangle ABC by dropping a perpendicular from angle A to side $\overline{BC}$. From the right triangle ABE we write

$$AB^2 = AE^2 + BE^2 \quad \text{(Pythagorean theorem)}$$

$AE = AC \sin C$ (definition of sine of C in triangle AEC)

$EC = AC \cos C$ (definition of cosine of C in triangle AEC)

From the properties of a straight line $BE = BC - EC$. Substituting for EC yields

$$BE = BC - AC \cos C$$

Finally, we take the expressions for AE and BE and substitute into $AB^2 = AE^2 + BE^2$ to obtain

$$\begin{aligned} AB^2 &= (AC \sin C)^2 + (BC - AC \cos C)^2 \\ &= AC^2\sin^2 C + BC^2 - 2(BC)(AC) \cos C + AC^2\cos^2 C \\ &\quad \text{(Note that } \cos^2 C = (\cos C)^2\text{)} \\ &= AC^2\sin^2 C + AC^2\cos^2 C + BC^2 - 2(BC)(AC) \cos C \end{aligned}$$

But $AC^2\sin^2 C + AC^2\cos^2 C$ is just the sum of the squares of the sides of right triangle AEC so that it must equal the hypotenuse squared. Then

$$AB^2 = AC^2 + BC^2 - 2(BC)(AC) \cos C$$

This is the law of cosines.

LAW OF COSINES

In words: The square of any side of a triangle equals the sum of the squares of the other two sides minus twice the product of these sides and the cosine of the angle included between them.

In symbols:

$$a^2 = b^2 + c^2 - 2bc \cos A$$
$$b^2 = a^2 + c^2 - 2ac \cos B$$
$$c^2 = a^2 + b^2 - 2ab \cos C$$

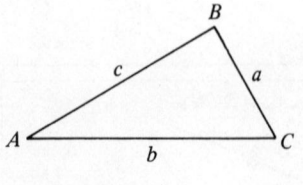

FIGURE 8.76

See Figure 8.76.

As noted earlier, the cosine law gives us a method of solving a triangle involving two known sides and the included angle. It also allows us to solve a triangle when three sides are known. Let us summarize the use of the laws of sines and cosines with respect to the solution of a triangle.

1. To solve a triangle we must know the value of three parts of the triangle. These parts cannot be the three angles because three angles do not determine a unique triangle.

2. The starting point for the solution of a triangle depends on the known information. Note the following cases:

Given information	Start with
Two sides and one angle (Not the included angle)	Law of sines
Two sides and the included angle	Law of cosines
Three sides	Law of cosines
Two angles and one side	Law of sines

The reason we say "start with" is that it may be necessary to use both relations to find all parts of the triangle.

EXAMPLE Find the unknown side and angles of the triangle in Figure 8.77.

SOLUTION We must start with the law of cosines. Call the side opposite the 120° angle side a.

$$a^2 = 5^2 + 7^2 - 2(5)(7) \cos 120°$$

(If you are not using a calculator, you will need to use the fact that supplementary angles have cosine values that differ only in sign. That is, $\cos 120° = -\cos 60°$.)

$$a^2 = 5^2 + 7^2 - 2(5)(7)(-0.5) = 25 + 49 + 35 = 109$$
$$a = \sqrt{109} = 10.4$$

Now use the law of sines to evaluate angle B. The law of cosines could be used; but, whenever possible, use the law of sines. It involves simpler operations.

$$\frac{\sin B}{7} = \frac{\sin 120°}{10.4} \quad (\sin 120° = \sin 60° \text{ in case you are working from a table})$$

$$\sin B = \frac{7 \sin 60°}{10.4} = \frac{7(0.866)}{10.4} = 0.583$$

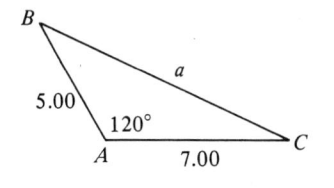

$$B = 35.7°$$

FIGURE 8.77

Finally, $C = 180° - 120° - 35.7° = 24.3°$

EXAMPLE Find the unknown side and angles of the triangle in Figure 8.78.

SOLUTION Start with the law of sines.

$$\frac{\sin A}{10.5} = \frac{\sin 72°}{14} =$$

$$\sin A = \frac{10.5 \sin 72°}{14} = 0.7133$$

so that $A = 46°$ to the nearest degree. Then

$$B = 180° - 72° - 46° = 62°$$

We can find the third side by using the law of sines again.

$$\frac{b}{\sin B} = \frac{14}{\sin 72°} \quad \text{so that} \quad b = \frac{14 \sin B}{\sin 72°} = \frac{14 \sin 62°}{\sin 72°} = 12.997$$

$b = 13$ to two significant figures. The triangle is completely solved.

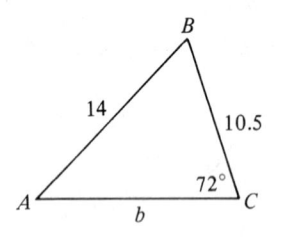

FIGURE 8.78

PERFORMANCE OBJECTIVE 9

You should now be able to use the law of cosines to solve oblique triangles.

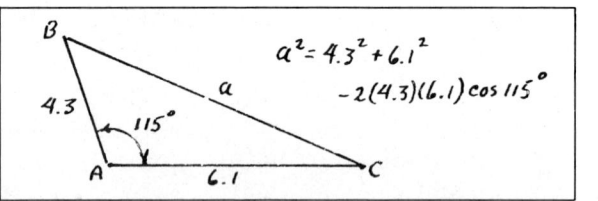

$a^2 = 4.3^2 + 6.1^2 - 2(4.3)(6.1) \cos 115°$

You should be cautioned that, without a picture of a triangle, it may not always be possible to uniquely determine the sides and angles. As an example, suppose you are given two sides and a non-included angle. Drawn in Figure 8.79 are two distinct triangles satisfying the given conditions.

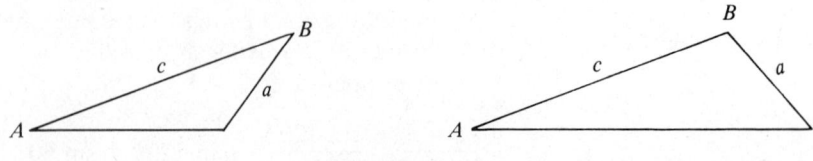

FIGURE 8.79

This ambiguity only occurs when $a < c$ and $a > c \sin A$. The reason for these conditions should be obvious if you swing an arc of radius a centered at B.

SOLVING OBLIQUE TRIANGLES 8.6   319

EXERCISE SET 8.6

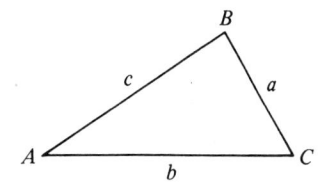

FIGURE 8.80

1. State the first law (sines or cosines) you would use in solving a triangle in which the following information is given. Consider that the standard triangle is labeled as shown in Figure 8.80.
   a. Given a, b, and c
   b. Given a, c, and C
   c. Given a, c, and B
   d. Given A, C, and b

2. Consider the standard triangle in Figure 8.80. State the first law (sines or cosines) you would use in solving a triangle given the following information:
   a. Given a, b, and C
   b. Given a, c, and A
   c. Given a, b, and c
   d. Given A, B, and c

3. Solve for the unknown sides in the following oblique triangles:

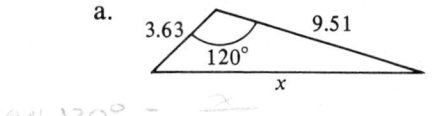

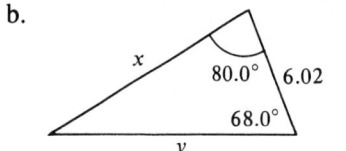

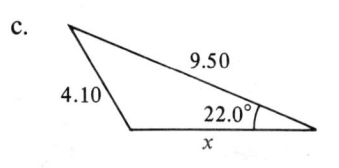

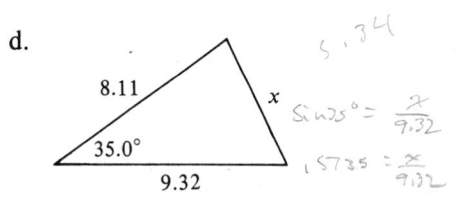

   a. sides 3.63 and 9.51 with included angle 120°, opposite side x

   b. angles 80.0° and 68.0°, side 6.02 opposite unknown, with x and y as other sides

   c. side 4.10, side 9.50, angle 22.0°, opposite side x

   d. side 8.11, angle 35.0°, side 9.32, unknown x

4. Solve for the unknown sides in the following oblique triangles:

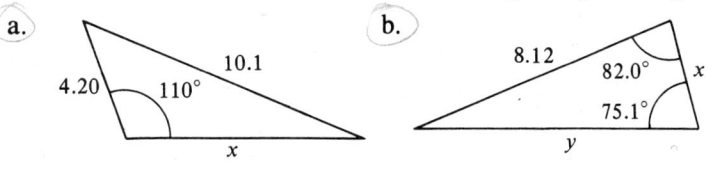

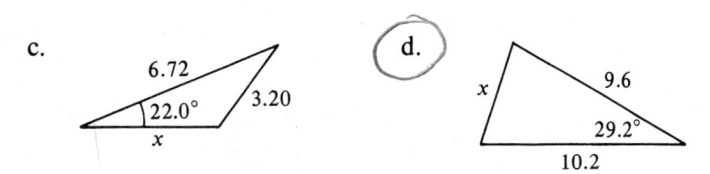

   a. side 4.20, angle 110°, side 10.1, unknown x

   b. side 8.12, angles 82.0° and 75.1°, unknowns x and y

   c. side 6.72, angle 22.0°, side 3.20, unknown x

   d. side 9.6, angle 29.2°, side 10.2, unknown x

5. Solve for the unknown angles in the following triangles:

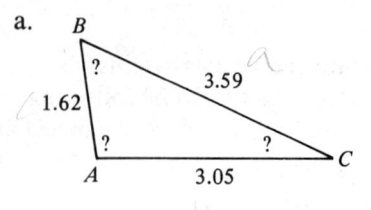

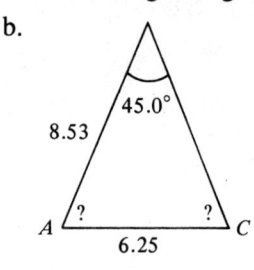

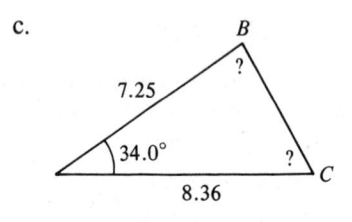

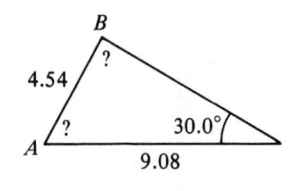

a. Triangle with B at top, sides: AB = 1.62, BC = 3.59, AC = 3.05. Angles at A, B, C all unknown.

b. Triangle with angle at top = 45.0°, side from top-left = 8.53, base AC = 6.25. Angles at A and C unknown.

c. Triangle with B at top, AB = 7.25, BC = 8.36 (actually base), angle at A = 34.0°, base = 8.36. Angles at B and C unknown.

d. Triangle with B at top, AB = 4.54, angle at C = 30.0°, base AC = 9.08. Angles at A and B unknown.

6. Solve for the unknown angles in the following triangles:

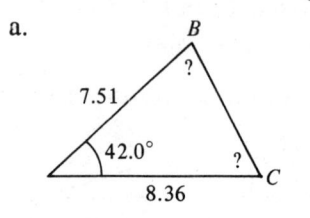

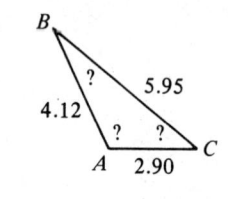

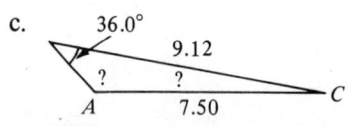

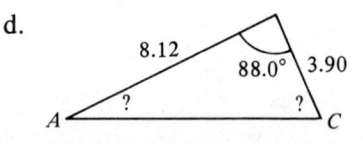

a. Triangle with B at top, AB = 7.51, angle at A = 42.0°, AC = 8.36. Angles at B and C unknown.

b. Triangle with AB = 4.12, BC = 5.95, AC = 2.90. All angles unknown.

c. Triangle with angle at A = 36.0°, AB = 9.12, AC = 7.50. Angles at B and C unknown.

d. Triangle with AB = 8.12, angle at B = 88.0°, BC = 3.90. Angles at A and C unknown.

7. Two observers measure the angle of elevation to an airborne object as 16° and 24° (see Figure 8.81). The observers are separated by 5 mi on an east-west line and both see the object due east of their positions. Determine the straight line distance of the object from each observer. Assume both observers are at the same level.

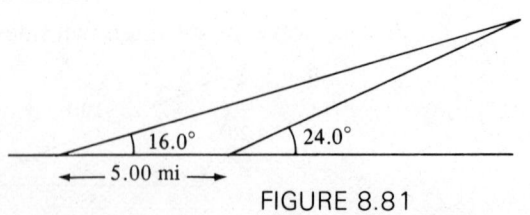

FIGURE 8.81

8. Two observers in fire towers spot a forest fire. The angle between the fire and the other tower location is measured by each observer. These angles are 78° and 46°. If the observers are 38.4 km apart, how far is the fire from each observer?

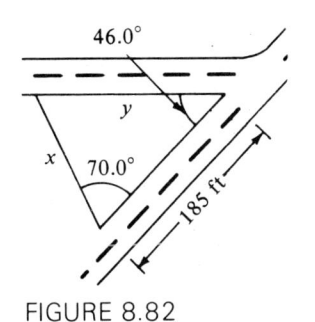

9. A corner lot is laid out in the form of a triangle as shown in Figure 8.82. Calculate the missing dimensions (x and y).

10. A triangular metal plate has its longest side of length 6.15 cm. The angles at each end of this side are 52° and 34°. Calculate the length of the other two sides of the plate.

FIGURE 8.82

SECTION 8.7   TRIGONOMETRY IN SURVEYING

Some think that surveying is the ultimate test of whether you can use trigonometry to solve practical problems. The surveyor deals with angles, distances, and coordinates. We will consider briefly two types of problems that the surveyor encounters, conducting a traverse and a triangulation. Although the techniques of conducting a survey require special training, the calculations involve rather basic trigonometry.

Shown in Figure 8.83 is a layout of a traverse. A **traverse** is a series of straight lines whose lengths and bearings are accurately measured. The traverse usually closes on itself and has as its purpose the fixing of certain stations that may be used for further survey work.

EXAMPLE Five additional stations are to be established in the 1000 m square tract shown in Figure 8.83. The coordinates of station A are given as North 0, East 0 (N0, E0). Find the coordinates of station B.

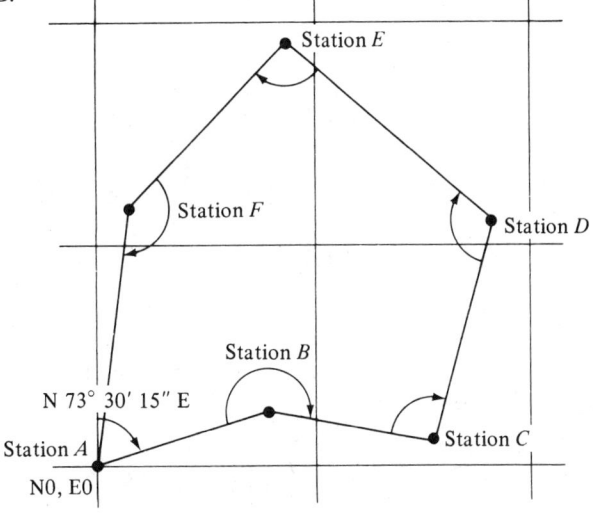

FIGURE 8.83

SOLUTION A bearing on the line from A to B is measured as N 73°30′15″ E. This designates an angle of 73°30′15″ east of north. The distance from A to B is measured as 437.376 m. Note that bearings are angles measured from the north-south direction.

Consider the right triangle that has as its sides the line AB, the northerly change in going from A to B, and the easterly change in going from A to B. See Figure 8.84.

We note that the northerly change (ΔN) is given by

$$\Delta N = 437.376 \cos 73°30′15″$$

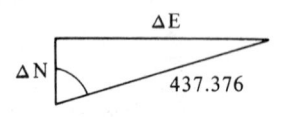

FIGURE 8.84

To evaluate this expression we must convert an angle in degrees, minutes, and seconds into decimal form so that we can use the calculator. Or we could use a trigonometric table that can be read to the second. The following procedure will allow a calculator solution:

15 [÷] 60 [+] 30 [=] [÷] 60 [+] 73 [=] yields 73.504167

Press [STO] to store this value. Complete the calculation as follows:

[COS] [×] 437.376 [=] and read 124.191

This is the N coordinate of station B. The E coordinate is given by

[RCL] [SIN] [×] 437.376 [=] and read 419.374

to six digits. The coordinates of station B are N 124.191 E 419.374 in meters.

EXAMPLE A forward angle of 211°35′20″ is measured to station C as shown in Figure 8.83. Compute the bearing to station C (from B) and the coordinates of station C if the measured distance from B to C is 397.492 m.

SOLUTION We can compute the new bearing by adding the measured forward angle to the original bearing and subtracting off a multiple of 90° to get an angle less than 90°.

$$73°30′15″ + 211°35′20″ = 285°5′35″$$

Subtract off 270° (the largest multiple of 90 in 285).

$$285°5′35″ - 270° = 15°5′35″$$

The bearing is then S 74°54′25″ E.

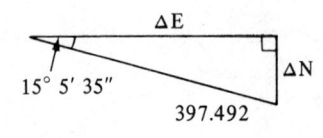

FIGURE 8.85

To find the coordinates of C, consider Figure 8.85.

TRIGONOMETRY IN SURVEYING 8.7

We follow the procedure outlined in the previous problem.

35 [÷] 60 [+] 5 [=] [÷] 60 [+] 15 [=] [STO][COS][x]
397.492 [=] and read 383.7802.

[RCL] [SIN] [x] 397.492 [=] and read 103.50194.

The coordinates are then

N = 124.191 − 103.502 = 20.689 N
E = 419.374 + 383.780 = 803.154 E   or   (N20.689, E803.154)

EXAMPLE Find the coordinates of station C by finding the straight line distance from A to C and the bearing of this straight line.

SOLUTION The straight line distance from A to C is one side of an oblique triangle. See Figure 8.86.

Two sides and an angle of this triangle are known. We can use the law of cosines to find the third side x.

$$x^2 = (437.376)^2 + (397.492)^2 - 2(437.376)(397.492) \cos 148°24'40''$$
$$= 191{,}297.77 + 157{,}999.89 - 347{,}706.92(-0.8518285)$$
$$= 645{,}484.32$$
$$x = 803.42039$$

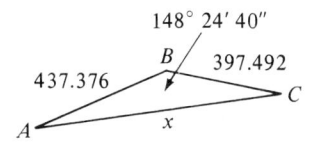

FIGURE 8.86

To find the bearing of this line, we must find angle A in Figure 8.87. Use the law of sines.

$$\frac{\sin A}{397.492} = \frac{\sin 148°24'40''}{803.42039}$$
$$\sin A = 0.2591601$$
$$A = 15.020234°$$

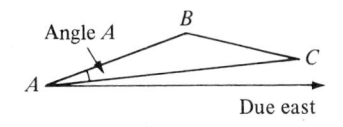

FIGURE 8.87

The line from A to B is $16.495833°$ north of east so that the bearing angle is

$$16.495833° - 15.020234° = 1.475599°$$

north of east. Then

$$N = 803.42039 \sin(1.475599°) = 20.689 \text{ N}$$
$$E = 803.42039 \cos(1.475599°) = 803.154 \text{ E}$$

These values agree with the values obtained in the previous example. This provides a check on the work.

324 8 BASIC TOPICS IN TRIGONOMETRY

A surveyor would continue to work forward in this manner and predict finally the coordinates of station A. Any discrepancy would have to be adjusted after the survey path is closed.

A second method used by surveyors to establish known stations is **triangulation.** Triangulation starts with a known base line and proceeds to a check base line while establishing intermediate stations. Triangulation does not require the measurement of distances between stations. Only angles are measured.

EXAMPLE Consider Figure 8.88. On base line $\overline{AB}$ the coordinates of station B are N 0, E 0. The angles from each end of the base line to station C are given. We also know that the bearing from B to A is N 9° 30′ 15″ E. Find the coordinates of station C.

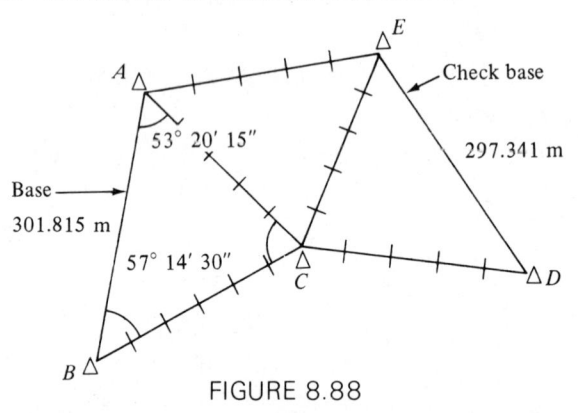

FIGURE 8.88

SOLUTION We need to find the length and bearing of either $\overline{AC}$ or $\overline{BC}$. The bearing of $\overline{BC}$ is seen to be

$$9°30'15'' + 57°14'30'' \quad \text{or} \quad N\ 66°44'45''\ E$$

The length of $\overline{BC}$ can be established using the law of sines. The angle at C must be

$$180° - 57°14'30'' - 53°20'15'' = 69°25'15''$$

$$\frac{BC}{\sin 53°20'15''} = \frac{301.815}{\sin 69°25'15''}$$

Converting these angles to decimal form, we find

$$\frac{BC}{\sin 53.3375°} = \frac{301.815}{\sin 69.420833°}$$

$$BC = \frac{(301.815)\sin 53.3375°}{\sin 69.420833°} = 258.60839\ \text{m}$$

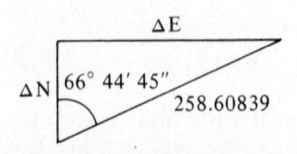

FIGURE 8.89

Next calculate the northerly and easterly changes (see Figure 8.89).

We find

$$\Delta N = 258.60839 \cos 66°44'45'' = 102.101 \text{ m}$$

$$\Delta E = 258.60839 \sin 66°44'45'' = 237.600 \text{ m}$$

The coordinates of station C are: N 102.101 E 237.600 m. As a check we could calculate the length of $\overline{AC}$ and its bearing.

We note that surveyors generally follow a fixed procedure to compute station coordinates. Also, computers are becoming more commonplace in the field. Although the computer may save tedious calculations, it does not eliminate a need for the surveyor to understand the basic trigonometry involved in the survey process.

PERFORMANCE OBJECTIVE 10

You should now be able to use bearings and distances to find the coordinates of a point.

EXERCISE SET 8.7

1. Use the surveying diagram in Figure 8.83 to compute the coordinates of station D given that an angle of $92°15'30''$ is measured at C. The distance CD is 520.346 m.

2. Use the surveying diagram of Figure 8.83 to compute the coordinates of station F (from A). Assume the bearing to F from A is N $9°35'10''$ and the distance AF is 564.302 m.

3. Use triangle BCD (Figure 8.83) and the coordinates of station B to predict the coordinates of station D. Your answer should agree with Problem 1.

4. Use triangle ABF (Figure 8.83) and the coordinates of station B to predict the coordinates of station F. Your answer should agree with Problem 2.

5. Calculate the coordinates of station C in Figure 8.88 using side $\overline{AC}$. Your answer should agree with that obtained in the example.

6. A marker is known to have a ΔN of 345.340 m and a ΔE of 218.211 m from station A. Calculate the bearing and distance to the marker from A.

SECTION 8.8 SUMMARY

The new concepts and terms introduced in this chapter are highlighted here.

Trigonometry: a study of the properties of the triangle and the functions associated with an angle.

Right triangle: a triangle in which one of the angles is equal to 90°.

Oblique triangle: a triangle other than a right triangle.

Sine: a measure of an angle that equals a ratio of the side opposite the angle to the hypotenuse in a right triangle.

Cosine: a measure of an angle that equals a ratio of the side adjacent to the angle to the hypotenuse in a right triangle.

Tangent: a measure of an angle that equals a ratio of the side opposite the angle to the side adjacent to the angle in a right triangle.

Trigonometric function: a general name for the sine, cosine, and tangent. The function processes an input angle into a trigonometric ratio. The trigonometric table or calculator serves as the processor.

Inverse trigonometric function: a general name for the inverse sine, inverse cosine, and inverse tangent. The inverse function processes a trigonometric ratio into an angle. The trigonometric table or calculator serves as the processor.

Law of sines: a statement that the ratio of the sine of an angle in a triangle to its opposite side has a constant value.

Law of cosines: a statement that the square of any side of a triangle is equal to the sum of the squares of the other two sides minus twice the product of the other two sides and the cosine of their included angle. The Pythagorean theorem is a special case of the law of cosines.

CHAPTER 8 APPENDIX—TRIGONOMETRIC TABLE

Angle	sin	cos	tan	1/tan	Angle
0	0.0000	1.0000	0.0000		90
1	0.0175	0.9998	0.0175	57.290	89
2	0.0349	0.9994	0.0349	28.636	88
3	0.0523	0.9986	0.0524	19.081	87
4	0.0698	0.9976	0.0699	14.301	86
5	0.0872	0.9962	0.0875	11.430	85
6	0.1045	0.9945	0.1051	9.514	84
7	0.1219	0.9925	0.1228	8.144	83
8	0.1392	0.9903	0.1405	7.115	82
9	0.1564	0.9877	0.1584	6.314	81
10	0.1736	0.9848	0.1763	5.671	80
11	0.1908	0.9816	0.1944	5.145	79
12	0.2079	0.9781	0.2126	4.705	78
13	0.2250	0.9744	0.2309	4.331	77
14	0.2419	0.9703	0.2493	4.011	76
15	0.2588	0.9659	0.2679	3.732	75
16	0.2756	0.9613	0.2867	3.487	74
17	0.2924	0.9563	0.3057	3.271	73
18	0.3090	0.9511	0.3249	3.078	72
19	0.3256	0.9455	0.3443	2.904	71
20	0.3420	0.9397	0.3640	2.747	70
21	0.3584	0.9336	0.3839	2.605	69
22	0.3746	0.9272	0.4040	2.475	68
23	0.3907	0.9205	0.4245	2.356	67
24	0.4067	0.9135	0.4452	2.246	66
25	0.4226	0.9063	0.4663	2.145	65
26	0.4384	0.8988	0.4877	2.050	64
27	0.4540	0.8910	0.5095	1.963	63
28	0.4695	0.8829	0.5317	1.881	62
29	0.4848	0.8746	0.5543	1.804	61
30	0.5000	0.8660	0.5774	1.732	60
31	0.5150	0.8572	0.6009	1.664	59
32	0.5299	0.8480	0.6249	1.600	58
33	0.5446	0.8387	0.6494	1.540	57
34	0.5592	0.8290	0.6745	1.483	56
35	0.5736	0.8192	0.7002	1.428	55
36	0.5878	0.8090	0.7265	1.376	54
37	0.6018	0.7986	0.7536	1.327	53
38	0.6157	0.7880	0.7813	1.280	52
39	0.6293	0.7771	0.8098	1.235	51
40	0.6428	0.7660	0.8391	1.192	50
41	0.6561	0.7547	0.8693	1.150	49
42	0.6691	0.7431	0.9004	1.111	48
43	0.6820	0.7314	0.9325	1.072	47
44	0.6947	0.7193	0.9657	1.036	46
45	0.7071	0.7071	1.0000	1.000	45
Angle	cos	sin	1/tan	tan	Angle

Top labels are for 0–45; bottom labels are for 45–90.

CHAPTER 8

PRACTICE TEST

1. (PO-1) Construct a triangle to the following specifications:
   a. Sides of length 4.3 cm and 6.1 cm and included angle of 78°
   b. Angles of 70° and 43° and a common side of length 5.3 cm

2. (PO-2) Identify the following for the triangles shown in Figure 8.90.

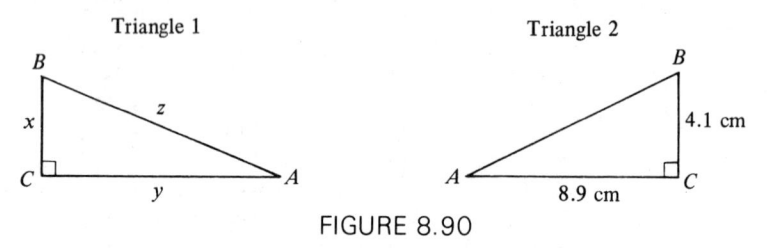

   FIGURE 8.90

   a. The side opposite angle A in triangle 1
   b. The side adjacent to angle B in triangle 1
   c. The hypotenuse of triangle 1
   d. The length of the side adjacent to angle A in triangle 2
   e. The length of the side opposite angle A in triangle 2

3. (PO-3) Using the triangles in Figure 8.90, write the following trigonometric ratios:
   a. The sine of angle A in triangle 1
   b. The cosine of angle B in triangle 1
   c. The tangent of angle A in triangle 2
   d. The tangent of angle B in triangle 2

4. (PO-4) Determine the unknown angles in the triangles in Figure 8.91.
   a. Angle A in triangle 1
   b. Angle B in triangle 2
   c. Angle A in triangle 2

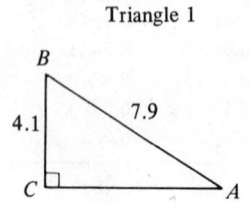

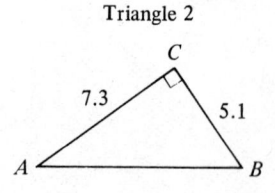

   FIGURE 8.91

CHAPTER 8 PRACTICE TEST 329

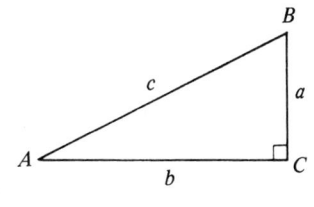

FIGURE 8.92

5. (PO-5, 6) Consider the general right triangle shown in Figure 8.92. Solve for all unknowns using the given information.
   a. $a = 4.35$, angle $B = 64.0°$
   b. $a = 5.1$, $c = 7.4$
   c. $c = 9.5$, angle $A = 34°$

6. (PO-7) What is the length of a diagonal on a rectangular steel plate if the diagonal makes an angle of 36° with a side of length 8.0 in.? What is the area of the plate?

7. (PO-7) A rod 2 cm in diameter is to be tapered to a point in a distance of 1.5 cm along the rod. What will be the angle of the taper with respect to the rod axis?

8. (PO-8, 9) Consider the general oblique triangle shown in Figure 8.93. Solve for all unknowns using the given information.
   a. $a = 4.9$, $b = 5.6$, $c = 8.1$
   b. $A = 39°$, $B = 62°$, $b = 5.9$
   c. $c = 10.1$, $a = 6.3$, $B = 72°$

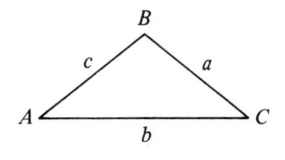

FIGURE 8.93

9. (PO-10) Calculate the coordinates of station B given that the bearing angle from A to B is N 42°25′34″ E and the distance from A to B is 1678.342 m. The coordinates of station A are (N 0, E 0).

CHAPTER 9

LINEAR SYSTEMS

9.1 Solving Systems of Linear Equations Graphically
9.2 Solving Systems of Linear Equations Algebraically
9.3 Solving Systems of Linear Equations Using Determinants
9.4 Applications of Linear Equations in Two Variables
9.5 Solving Larger Systems of Linear Equations
9.6 Summary

Frequently a problem is so complex that it is more easily solved by breaking it into parts. In order to solve the problem, we must treat the parts as a system. In this chapter we study systems in which the parts are linear equations, hence "linear systems." An example of a linear system is an electrical circuit analysis involving more than one loop.

Three methods for solving systems of equations will be discussed. We will restrict our examples to systems of two or three linear equations but the methods presented here can also be used for larger systems. You will note that as we deal with more equations, the solution becomes more complex and tedious. The computer is a valuable tool for solving larger systems of equations. It is useful since it has the capabililty to handle many variables quickly and accurately.

SECTION 9.1 SOLVING SYSTEMS OF LINEAR EQUATIONS GRAPHICALLY

Before proceeding with the method for solving systems of linear equations graphically, let's recall that a linear equation in two variables was defined in Chapter 7 to be any equation of the form

$$ax + by = c$$

where a, b, and c are real numbers. If we are working with any pair of linear equations in two variables, we define the pair to be a **linear system** of equations.

DEFINITION

LINEAR SYSTEM OF EQUATIONS (IN TWO VARIABLES)

A pair of linear equations in two variables is a **linear system.**

As an example, we would write the linear system of equations containing $a_1x + b_1y = c_1$ and $a_2x + b_2y = c_2$ as

$$a_1x + b_1y = c_1$$
$$a_2x + b_2y = c_2$$

NOTE: The subscripts are used on a, b, and c to indicate different constants.

We have already discussed solutions of linear equations. A solution to a linear system in an ordered pair of numbers that satisfies both equations in the system. That is, a solution of a linear system is a common solution of the two equations that make up the system.

A linear system may have no solutions, one solution, or infinitely many solutions. If a linear system has no solutions, we call the system **inconsistent;** if it has one solution, we call the system **independent;** and if it has infinitely many solutions, we call the system **dependent.**

EXAMPLE Determine if (2, 5) is a solution of the system

$$3x + y = 11$$
$$8x - 3y = 1$$

SOLUTION

$3(2) + 5 = 11$; so, (2, 5) is a solution of $3x + y = 11$.

$8(2) - 3(5) = 1$; so, (2, 5) is a solution of $8x - 3y = 1$.

EXAMPLE Determine if $(-1, 1)$ is a solution of the system

$$4x + 7y = 3$$
$$x - y = 2$$

SOLUTION Although $(-1, 1)$ is a solution of $4x + 7y = 3$, it is not a solution of $x - y = 2$.

So, $(-1, 1)$ is not a solution of the linear system.

PERFORMANCE OBJECTIVE 1	
You should now be able to recognize a linear system of equations and a solution for the system.	$(2, -3)$ is a solution of the system $x - y = 5$ $2x + y = 1$ since $2 - (-3) = 2 + 3 = 5$ and $2(2) + (-3) = 4 - 3 = 1$

The graph of a linear system of two equations in two variables will be either two straight lines or a single line. The graph will be a single line if the system is a dependent system. If the graph consists of two lines, the following possibilities exist:

1. The lines intersect in a point, giving an independent system.
2. The lines do not intersect, giving an inconsistent system.

If the graph is a pair of lines that intersect in a point, the ordered pair associated with the point of intersection is the solution of the independent system. The problem with trying to determine the solution of such a system graphically is that it can be very difficult to determine the exact coordinates of the solution from the graph.

EXAMPLE Solve the following system of equations graphically:

$$x + y = 4$$
$$2x - y = 5$$

SOLUTION We graph the two linear equations and obtain a pair of intersecting lines. See Figure 9.1.

334   **9 LINEAR SYSTEMS**

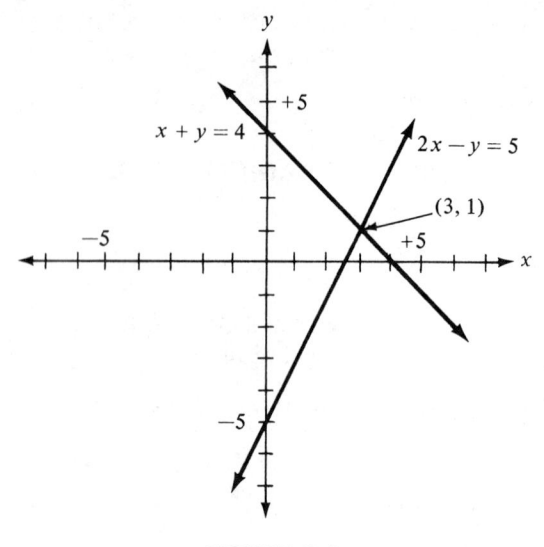

FIGURE 9.1

The point of intersection of the lines is given by (3, 1). So (3, 1) is the solution of the independent system.

If the graph of a linear system is a pair of parallel lines, the lines have no points in common and the system has no solutions. Two distinct lines are parallel if they have the same slope and different y-intercepts.

EXAMPLE   Solve the following system of equations:

$$y = \frac{1}{3}x + 1$$
$$x - 3y = 6$$

SOLUTION   The graph of the system is shown in Figure 9.2.

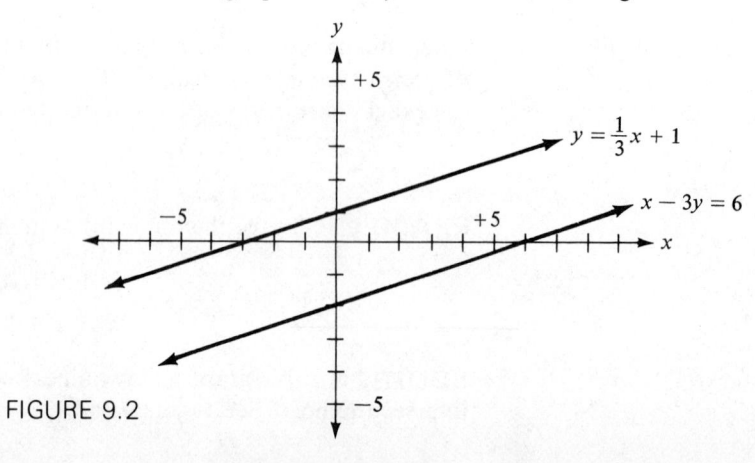

FIGURE 9.2

If we look at the graph of the two lines, they seem to be parallel. To prove that the lines are really parallel and thus that the system is inconsistent and has no solutions, we can verify that the lines have the same slope but different y-intercepts.

If the graph of a linear system is a single line, that is, if both equations have the same line as their graph, then the system has infinitely many solutions. Every point on the line represents a solution to the dependent system.

EXAMPLE Solve the system

$$y = \frac{1}{2}x + 3$$
$$-x + 2y = 6$$

SOLUTION Two solutions of the first equation are (0, 3) and (−6, 0). These are also solutions of the second equations. Thus, the graph of both equations is the same line and the system has infinitely many solutions. We can write the solution of the dependent system as

$$y = \frac{1}{2}x + 3$$

or as

$$-x + 2y = 6$$

The system is shown in Figure 9.3.

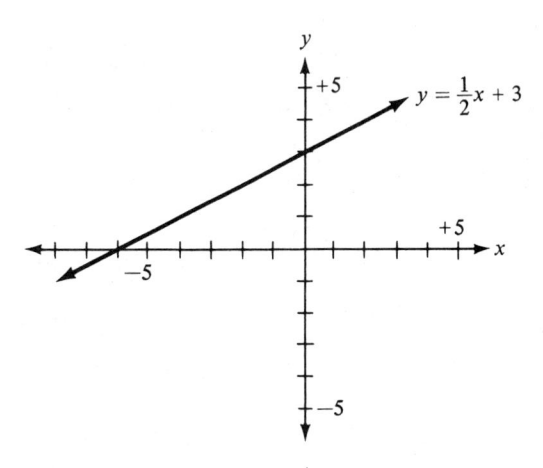

FIGURE 9.3

PERFORMANCE OBJECTIVE 2

You should now be able to solve a linear system by graphing.

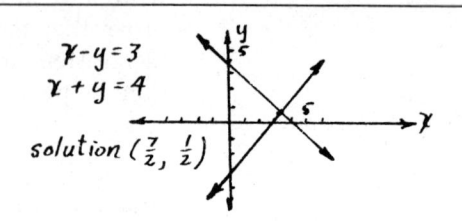

$x - y = 3$
$x + y = 4$

solution $(\frac{7}{2}, \frac{1}{2})$

EXERCISE SET 9.1

1. Determine whether or not the following are systems of linear equations in two variables:

   a. $y = x$
   $2x + 4y = 9$

   b. $x + y = 6$
   $x^2 - 3y = 0$

2. Determine whether or not the following are systems of linear equations in two variables:

   a. $4x - y = 5$
   $2y + 3z = 8$

   b. $x + y = 0$
   $y = -6x + 1$

3. In each of the following determine if the ordered pair is a solution of the given system:

   a. $3x + y = 3$
   $5x - 2y = 2$ $(4, -9)$

   b. $9x + 4y = 6$
   $12x - 4y = 1$ $(\frac{1}{3}, \frac{3}{4})$

4. In each of the following determine if the ordered pair is a solution of the given system:

   a. $3x + 5y = 5$
   $6x - 10y = 2$ $(\frac{2}{3}, \frac{3}{5})$

   b. $y = -x$
   $4x + y = 0$ $(0, 0)$

5. Solve the following system of equations graphically:

   a. $x - y = 5$
   $y = -2x + 4$

   b. $0x + 4y = 6$
   $3x + 0y = 9$

   c. $2x - 5y = 10$
   $8x = 20y + 40$

6. Solve the following systems of equations graphically:

   a. $4x + y = 8$
   $y = 3x + 1$

   b. $x - y = 0$
   $3x + 0y = -6$

   c. $-x + 3y = 3$
   $6y = 2x - 18$

SECTION 9.2 SOLVING SYSTEMS OF LINEAR EQUATIONS ALGEBRAICALLY

So far we have only solved linear systems of equations graphically. We need to consider algebraic methods for solving linear systems. These will be more precise methods than graphing. They include the substitution method and the addition method.

SUBSTITUTION METHOD FOR SOLVING LINEAR SYSTEMS

Step 1. Solve one equation for one variable in terms of the other.

Step 2. Substitute the resulting expression for this variable in the second equation.

Step 3. Solve the resulting linear equation in one variable.

Step 4. Use this numerical solution to find the other variable.

EXAMPLE Solve the following linear system by substitution:

$$4x + 3y = -6$$
$$-2x + y = 8$$

SOLUTION Solve for y in terms of x in the second equation.

$$-2x + y = 8 \quad \text{so } y = 2x + 8$$

Substitute $2x + 8$ for y in the first equation.

$$4x + 3(2x + 8) = -6$$

Solve this linear equation in one variable for x.

$$4x + 6x + 24 = -6$$
$$10x + 24 = -6$$
$$10x = -6 + (-24)$$
$$10x = -30$$
$$x = -3$$

Substitute -3 for x in $y = 2x + 8$.

$$y = 2(-3) + 8$$
$$y = -6 + 8$$
$$y = 2$$

So, the solution of the system is $(-3, 2)$ and the system is independent.

EXAMPLE Solve the following linear system by substitution:

$$2x - 6y = 5$$
$$4x + 7y = -9$$

SOLUTION Solve the first equation for x in terms of y.

$$2x - 6y = 5 \quad \text{so} \quad 2x = 6y + 5 \quad \text{and} \quad x = 3y + \frac{5}{2}$$

Substitute $3y + \frac{5}{2}$ in place of x in the second equation and solve for y.

$$4\left(3y + \frac{5}{2}\right) + 7y = -9$$
$$12y + 10 + 7y = -9$$
$$19y + 10 = -9$$
$$19y = -9 + (-10)$$
$$19y = -19$$
$$y = -1$$

Substitute -1 in place of y in $x = 3y + \frac{5}{2}$.

$$x = 3(-1) + \frac{5}{2}$$
$$x = -3 + \frac{5}{2}$$
$$x = -\frac{1}{2}$$

So, the solution of the system is $(-\frac{1}{2}, -1)$ and the system is independent.

EXAMPLE Solve the following system by substitution:

$$x - 2y = 6$$
$$2y = x + 8$$

SOLUTION Solving the first equation for x gives

$$x = 2y + 6$$

Substituting $2y + 6$ for x in the second equation and simplifying yields the obviously false statement that $0 = 14$.

$$2y = x + 8$$
$$2y = 2y + 6 + 8$$
$$2y = 2y + 14$$
$$2y + (-2y) = 2y + (-2y) + 14$$
$$0 = 14$$

So there can be no solution to the given system of equations, that is, the system is inconsistent.

EXAMPLE Solve the following system by substitution:

$$y = \frac{1}{2}x + 3$$
$$-x + 2y = 6$$

Note that we have already solved this system graphically and found it to have infinitely many solutions.

SOLUTION The first equation is already solved for y. Substituting $\frac{1}{2}x + 3$ for y in the second equation yields the statement $6 = 6$, which is always true.

$$-x + 2y = 6$$
$$-x + 2\left(\frac{1}{2}x + 3\right) = 6$$
$$-x + x + 6 = 6$$
$$6 = 6$$

This tells us that the two equations in the system have exactly the same solutions, that is, the system is dependent. Recall that we can write the solution of the system as

$$y = \frac{1}{2}x + 3 \quad \text{or as} \quad -x + 2y = 6$$

PERFORMANCE OBJECTIVE 3	
You should now be able to solve a linear system using the substitution method.	$x - y = 3$ $x + y = 4$ $\begin{cases} x = 3 + y \\ (3 + y) + y = 4 \\ y = \frac{1}{2} \\ x = 3 + \frac{1}{2} = \frac{7}{2} \end{cases}$

So far we have solved linear systems of equations graphically and algebraically using the substitution method. Now we will consider the algebraic addition method.

ADDITION METHOD FOR SOLVING LINEAR SYSTEMS

Step 1. Multiply one or both equations by whatever number it takes to eliminate a variable when the equations are added. Then add the equations.

Step 2. Solve the resulting linear equation in one variable.

Step 3. Use this numerical solution to find the other variable.

EXAMPLE Solve the following linear system by the addition method:

$$8x - 4y = 3$$
$$5x + 2y = 3$$

SOLUTION If we multiply the second equation by 2 and then add the two equations, we will eliminate the variable y. Multiplying the second equation by 2 gives the system

$$8x - 4y = 3$$
$$10x + 4y = 6$$

Adding these two equations yields

$$18x = 9$$

So, $x = \frac{9}{18} = \frac{1}{2}$. We solve for y by substituting $\frac{1}{2}$ in place of x in either of the original equations. Substituting $\frac{1}{2}$ for x in the first equation gives

$$8\left(\frac{1}{2}\right) - 4y = 3$$
$$4 - 4y = 3$$
$$-4y = -1$$
$$y = \frac{1}{4}$$

So, the solution of the system is $(\frac{1}{2}, \frac{1}{4})$.

EXAMPLE Solve the following linear system by the addition method:

$$3x - 10y = -8$$
$$2x - 3y = 2$$

SOLUTION We multiply the first equation by 2 and the second equation by -3. This gives the system

$$6x - 20y = -16$$
$$-6x + 9y = -6$$

Adding these two equations yields

$$-11y = -22$$
$$y = \frac{-22}{-11} = 2$$

We substitute 2 in place of y in the second equation to get

$$2x - 3(2) = 2$$
$$2x - 6 = 2$$
$$2x = 8$$
$$x = 4$$

So, the solution of the system is (4, 2).

If we are trying to use the addition method to solve a system that has no solution, at some point we will arrive at an obviously false statement, just as we did in using the substitution method.

Likewise, if we are using the addition method on a system with infinitely many solutions, we will arrive at a statement that is always true.

PERFORMANCE OBJECTIVE 4	
You should now be able to solve a linear system using the addition method.	$x + 2y = 4 \qquad x + 2y = 4$ $2x - y = 3 \qquad \underline{4x - 2y = 6}$ $\qquad\qquad\qquad\quad 5x = 10 \Rightarrow x = 2$ $\qquad\qquad 2(2) - y = 3 \Rightarrow y = 1$

EXERCISE SET 9.2

Solve the following systems algebraically by the substitution method:

1. a. $2x - 2y = 4$
   $3x + 2y = 6$

   b. $y = \frac{1}{2}x - 3$
   $2y = x + 12$

   c. $4y - 3x = 17$
   $x - 2y + 7 = 0$

   d. $3x + 2y = -6$
   $5x - 2y = -10$

e. $-4x + y = -9$
$2y = 8x - 18$

2. a. $2x + 7y = -8$
$2x = 3y + 12$

b. $x - y = 0$
$3x + 0y = -6$

c. $4x = y - 2$
$8x - 2y + 4 = 0$

d. $2x + y = 3$
$6x + 5y = 11$

e. $-x + 3y = 3$
$6y = 2x - 18$

Solve the following systems algebraically by the addition method:

3. a. $2x + y = 3$
$6x + 5y = 11$

b. $2x - 4y = 14$
$2y = x - 7$

c. $2x - 5y = 12$
$9x + 4y = 1$

d. $y + \dfrac{x}{2} = \dfrac{3}{2}$

$y + \dfrac{x}{3} = \dfrac{1}{3}$

e. $3x = 2y + 4$
$-6x + 4y = 7$

4. a. $5x + 7y = 6$
$-10x + 3y = -46$

b. $x - y = 5$
$y = -2x + 4$

c. $2x - 5y = 10$
$8x = 20y + 40$

d. $\dfrac{x}{5} + \dfrac{3y}{2} = -6$

$\dfrac{x}{5} + \dfrac{y}{4} = -\dfrac{7}{4}$

e. $y = \dfrac{1}{2}x - 3$
$2y = x + 12$

SECTION 9.3 SOLVING SYSTEMS OF LINEAR EQUATIONS USING DETERMINANTS

When solving a system of two linear equations in two variables algebraically using the addition method, we often needed to multiply one or both of the equations by some number so that when we added the equations the x-term or the y-term would be eliminated.

If we look at the system

$$a_1x + b_1y = c_1$$
$$a_2x + b_2y = c_2$$

and multiply the first equation by b_2 and the second equation by b_1, we have

$$a_1b_2x + b_1b_2y = b_2c_1$$
$$a_2b_1x + b_1b_2y = b_1c_2$$

If we then subtract and solve for x, we have

$$a_1b_2x + b_1b_2y = b_2c_1$$
$$-a_2b_1x - b_1b_2y = -b_1c_2$$
$$\overline{(a_1b_2 - a_2b_1)x = b_2c_1 - b_1c_2}$$

$$x = \frac{b_2c_1 - b_1c_2}{a_1b_2 - a_2b_1}$$

$$x = \frac{c_1b_2 - c_2b_1}{a_1b_2 - a_2b_1}$$

If we follow the same procedure but multiply the first equation by a_2 and the second equation by a_1 and solve for y, we have

$$a_1a_2x + b_1a_2y = c_1a_2$$
$$a_1a_2x + b_2a_1y = c_2a_1$$

$$a_1a_2x + b_1a_2y = c_1a_2$$
$$-a_1a_2x - b_2a_1y = -c_2a_1$$
$$\overline{(b_1a_2 - b_2a_1)y = c_1a_2 - c_2a_1}$$

$$y = \frac{c_1a_2 - c_2a_1}{b_1a_2 - b_2a_1}$$

Notice that the denominator in the preceding expression is very similar to the denominator in the expression that was found when we solved for x. In order to make these denominators the same, multiply the numerator and denominator of the expression for y by -1 and rearrange the letters and terms. This yields

$$y = \frac{(-1)(c_1a_2 - c_2a_1)}{(-1)(b_1a_2 - b_2a_1)} = \frac{a_1c_2 - a_2c_1}{a_1b_2 - a_2b_1}$$

There is another way to express the numerators and the denominators of the expressions for x and y. That is to write them as the evaluation of a **second order determinant.**

DEFINITION

SECOND ORDER DETERMINANT

The **second order determinant** is an array of four numbers that form two rows and two columns. The symbol for the second order determinant is

$$\begin{vmatrix} & \\ & \end{vmatrix}$$

In order to evaluate any second order determinant, we find the product of the elements of the first diagonal (upper left times lower right) and subtract the product of the elements in the second diagonal (lower left times upper right).

EVALUATION OF A SECOND ORDER DETERMINANT

$$\begin{vmatrix} a & b \\ c & d \end{vmatrix} = ad - bc$$

Consider the determinant $\begin{vmatrix} a_1 & b_1 \\ a_2 & b_2 \end{vmatrix}$. If we evaluate this determinant, we have

$$\begin{vmatrix} a_1 & b_1 \\ a_2 & b_2 \end{vmatrix} = a_1 b_2 - a_2 b_1$$

which is the denominator of the expressions for x and y. In the same way the numerator for x is

$$\begin{vmatrix} c_1 & b_1 \\ c_2 & b_2 \end{vmatrix} = c_1 b_2 - c_2 b_1$$

and the numerator for y is

$$\begin{vmatrix} a_1 & c_1 \\ a_2 & c_2 \end{vmatrix} = a_1 c_2 - a_2 c_1$$

So we can write x and y as follows:

$$x = \frac{\begin{vmatrix} c_1 & b_1 \\ c_2 & b_2 \end{vmatrix}}{\begin{vmatrix} a_1 & b_1 \\ a_2 & b_2 \end{vmatrix}} \quad \text{and} \quad y = \frac{\begin{vmatrix} a_1 & c_1 \\ a_2 & c_2 \end{vmatrix}}{\begin{vmatrix} a_1 & b_1 \\ a_2 & b_2 \end{vmatrix}}$$

You should notice the pattern that is used in forming the determinants for x and y. For the determinant in the denominator of both x and y, the coefficients of x (a_1 and a_2) and the coefficients of y (b_1 and b_2) are used. For the numerator of x, the coefficients of x (a_1 and a_2) are replaced by the constants (c_1 and c_2). For the numerator of y, the coefficients of y (b_1 and b_2) are replaced by the constants (c_1 and c_2).

The use of determinants to solve systems of linear equations is known as **Cramer's Rule.** Note that the equations in the system must be written in the form $ax + by = c$ before we can apply Cramer's Rule.

EXAMPLE   Use Cramer's Rule to solve

$$4x - 2y = 8$$
$$3x + y = -4$$

SOLUTION

$$x = \frac{\begin{vmatrix} c_1 & b_1 \\ c_2 & b_2 \end{vmatrix}}{\begin{vmatrix} a_1 & b_1 \\ a_2 & b_2 \end{vmatrix}} = \frac{\begin{vmatrix} 8 & -2 \\ -4 & 1 \end{vmatrix}}{\begin{vmatrix} 4 & -2 \\ 3 & 1 \end{vmatrix}}$$

$$= \frac{8(1) - (-4)(-2)}{4(1) - (3)(-2)} = \frac{8 - 8}{4 + 6} = \frac{0}{10} = 0$$

$$y = \frac{\begin{vmatrix} a_1 & c_1 \\ a_2 & c_2 \end{vmatrix}}{\begin{vmatrix} a_1 & b_1 \\ a_2 & b_2 \end{vmatrix}} = \frac{\begin{vmatrix} 4 & 8 \\ 3 & -4 \end{vmatrix}}{\begin{vmatrix} 4 & -2 \\ 3 & 1 \end{vmatrix}}$$

$$= \frac{4(-4) - (3)(8)}{4(1) - (3)(-2)} = \frac{-16 - 24}{4 + 6} = \frac{-40}{10} = -4$$

The solution is $(0, -4)$.

9 LINEAR SYSTEMS

EXAMPLE Use Cramer's Rule to solve

$$\frac{x}{7} - \frac{y}{6} = -\frac{1}{2}$$

$$\frac{x}{3} - \frac{y}{4} = -\frac{1}{2}$$

SOLUTION Before forming the determinants, we rewrite the system so that the fractions have been eliminated.

$$42\left(\frac{x}{7}\right) - 42\left(\frac{y}{6}\right) = 42\left(-\frac{1}{2}\right) \rightarrow 6x - 7y = -21$$

$$12\left(\frac{x}{3}\right) - 12\left(\frac{y}{4}\right) = 12\left(-\frac{1}{2}\right) \rightarrow 4x - 3y = -6$$

$$x = \frac{\begin{vmatrix} -21 & -7 \\ -6 & -3 \end{vmatrix}}{\begin{vmatrix} 6 & -7 \\ 4 & -3 \end{vmatrix}}$$

$$= \frac{(-21)(-3) - (-6)(-7)}{(6)(-3) - (4)(-7)} = \frac{63 - 42}{-18 + 28} = \frac{21}{10}$$

$$y = \frac{\begin{vmatrix} 6 & -21 \\ 4 & -6 \end{vmatrix}}{\begin{vmatrix} 6 & -7 \\ 4 & -3 \end{vmatrix}}$$

$$= \frac{6(-6) - (4)(-21)}{6(-3) - (4)(-7)} = \frac{-36 + 84}{-18 + 28} = \frac{48}{10} = \frac{24}{5}$$

The solution is $\left(\frac{21}{10}, \frac{24}{5}\right)$.

EXAMPLE Use Cramer's Rule to solve

$$15y = 9x + 1$$
$$3x = 5y + 1$$

SOLUTION We must first rewrite the equations in the form $ax + by = c$. This yields the system

$$-9x + 15y = 1$$
$$3x - 5y = 1$$

$$x = \frac{\begin{vmatrix} 1 & 15 \\ 1 & -5 \end{vmatrix}}{\begin{vmatrix} -9 & 15 \\ 3 & -5 \end{vmatrix}}$$

$$= \frac{1(-5) - (1)(15)}{(-9)(-5) - (3)(15)} = \frac{-5 - 15}{45 - 45} = \frac{-20}{0}$$

Since $\frac{-20}{0}$ is undefined, the system is inconsistent and therefore has no solution.

NOTE: If the evaluation of x or y gives

$$\frac{0}{0}$$

then the system is dependent and there are infinitely many solutions.

PERFORMANCE OBJECTIVE 5

You should now be able to solve a linear system using Cramer's Rule.

$2x - y = 3$
$x + 4y = 5$

$x = \frac{\begin{vmatrix} 3 & -1 \\ 5 & 4 \end{vmatrix}}{\begin{vmatrix} 2 & -1 \\ 1 & 4 \end{vmatrix}} = \frac{17}{9}$

$y = \frac{\begin{vmatrix} 2 & 3 \\ 1 & 5 \end{vmatrix}}{\begin{vmatrix} 2 & -1 \\ 1 & 4 \end{vmatrix}} = \frac{7}{9}$

EXERCISE SET 9.3

Use Cramer's Rule to solve each system of equations.

1. $3x + 2y = -6$
   $5x - 2y = -10$

2. $2x - 2y = 4$
   $3x + 2y = 6$

3. $5x + 7y = 6$
   $-10x + 3y = -46$

4. $2x - 5y = 12$
   $9x + 4y = 1$

5. $2x + 7y = -8$
   $-2x + 3y = -12$

6. $4y - 3x = 17$
   $x - 2y + 7 = 0$

7. $\dfrac{x}{2} + \dfrac{y}{3} = 8$
   $\dfrac{2x}{3} + \dfrac{3y}{2} = 17$

8. $y = \dfrac{1}{2}x - 3$
   $2y = x + 12$

9. $-4x + y = -9$
$-8x + 2y = -18$

10. $2x - 4y = 14$
$2y = x - 7$

11. $3x - 2y = 4$
$-6x + 4y = 7$

12. $2x + y = 3$
$6x + 5y = 11$

13. $-3x + y = -5$
$4x + y = 9$

14. $x - y = 5$
$y = -2x + 4$

15. $\dfrac{x}{5} + \dfrac{3y}{2} = -6$
$\dfrac{x}{5} + \dfrac{y}{4} = \dfrac{-7}{4}$

16. $0x + 4y = 6$
$3x + 0y = 9$

17. $-4x + y = 2$
$8x - 2y = -4$

18. $-x + 3y = 3$
$6y = 2x - 18$

19. $y + \dfrac{x}{2} = \dfrac{3}{2}$
$y + \dfrac{x}{3} = \dfrac{1}{3}$

20. $4x + y = 8$
$y = 3x + 1$

SECTION 9.4 APPLICATIONS OF LINEAR EQUATIONS IN TWO VARIABLES

Recall that the key step in solving practical application problems is to translate problems stated in words into mathematical equations. The general procedure for solving application problems using a system of two linear equations in two variables follows.

APPLICATION PROBLEMS INVOLVING A SYSTEM OF TWO LINEAR EQUATIONS IN TWO VARIABLES

Step 1. Read the problem as many times as it takes to understand the problem thoroughly.

Step 2. Identify the two quantities being sought and represent one of the quantities by the variable x and the other quantity by the variable y.

Step 3. Write a linear system of two equations that must be satisfied by the variables x and y.

Step 4. Solve the system of equations algebraically by the substitution method, addition method, or Cramer's Rule and clearly state the answer to the problem.

Step 5. Check the answer by making sure the values for x and y satisfy the conditions in the stated word problem.

For each type of application problem that follows we will work at least one example identical to an example in Section 6.3. In Section 6.3 we used a procedure similar to the one just outlined to solve application problems by means of a single linear equation in one variable. Here we solve application problems by means of two linear equations in two variables. The five steps in the procedure will be labeled in the first two examples.

EXAMPLE The difference of two numbers is 5. If three times the smaller number is seven more than the larger number, what are the two numbers?

SOLUTION

Step 1. Make sure that you understand the problem.

Step 2. Let x = the smaller number.
Let y = the larger number.

Step 3. We need to make use of the information in the problem to write a system of two linear equations in the variables x and y. Since the difference of the two numbers is 5, we have

$$y - x = 5$$

Note that we have subtracted the smaller number from the larger number. To show that three times the smaller number is seven more than the larger number, we write

$$3x = y + 7$$

So, our system of equations is

$$y - x = 5$$
$$3x = y + 7$$

Step 4. We solve the system by substitution as follows:

Since $y - x = 5$, we have $y = x + 5$.

Substituting $x + 5$ in place of y in the second equation gives

$$3x = x + 5 + 7$$
$$3x = x + 12$$
$$2x = 12$$
$$x = 6$$

Then, $y = x + 5 = 6 + 5 = 11$. So, the solution of the system is (6, 11) and the numbers being sought are 6 and 11.

Step 5. We check to make sure that our answers 6 and 11 satisfy the conditions of the problem. The difference between 6 and 11 is 5. Three times 6 is 18 and 18 is seven more than 11. Thus, our solution is correct.

EXAMPLE One number is four times another number. Find the two numbers given that the smaller number is 24 less than the larger number.

SOLUTION

Step 1. Reread the problem until you understand it.

Step 2. Let $x =$ the smaller number.
Let $y =$ the larger number.

Step 3. Since the larger number is four times the smaller number, we can write

$$y = 4x$$

Since the smaller number is 24 less than the larger number, we have

$$x = y - 24$$

So, our system of equations is

$$y = 4x$$
$$x = y - 24$$

Step 4. We can solve the system by substituting $4x$ in place of y in the second equation to get

$$x = 4x - 24$$
$$-3x = -24$$
$$x = 8$$

Then, $y = 4x = 4(8) = 32$. The numbers are 8 and 32.

Step 5. Since $4(8) = 32$ and $32 - 24 = 8$, our solution checks.

PERFORMANCE OBJECTIVE 6	
You should now be able to solve number problems using a system of two linear equations in two variables.	Let $x =$ smaller number Let $y =$ larger number $y = x + 7$ $2x = y + 1$

APPLICATIONS OF LINEAR EQUATIONS IN TWO VARIABLES 9.4

To solve word problems that involve geometric figures such as rectangles and triangles, we will use certain facts that we learned in Chapter 3 on Applied Geometry. Among those facts are the following:

1. The perimeter P of a rectangle is equal to twice its width w plus twice its length l. That is,

$$P = 2w + 2l$$

2. The sum of the measures of the angles in a triangle is 180°.

When we solve geometric word problems, it sometimes helps to draw a sketch of the geometric figure.

EXAMPLE The length of a rectangle is 6 ft less than three times its width. The perimeter of the rectangle is 140 ft. Find the dimensions of the rectangle.

SOLUTION

Let $x =$ the width of the rectangle in feet.
Let $y =$ the length of the rectangle in feet.

The following system satisfies the conditions of the problem:

$$y = 3x - 6$$
$$2x + 2y = 140$$

Substituting $3x - 6$ in place of y in the second equation gives

$$2x + 2(3x - 6) = 140$$

So we have

$$2x + 6x - 12 = 140$$
$$8x - 12 = 140$$
$$8x = 152$$
$$x = 19$$

Then, $y = 3x - 6 = 3(19) - 6 = 57 - 6 = 51$. The solution of the system is (19, 51). So, the rectangle is 19 ft wide and 51 ft long. We should check to see that these dimensions satisfy the conditions in the problem.

EXAMPLE A gardener wants to enclose a rectangular plot of ground with fencing and then subdivide the plot with a fence parallel to the shorter side. If the width of the plot is to be $\frac{3}{4}$ of its length and if the gardener has 102 ft of fencing to work with, what will be the dimensions of the enclosed plot?

SOLUTION

Let $x =$ the width of the plot.

Let $y =$ the length of the plot.

See Figure 9.4.

We can set up the following system of equations:

$$x = \frac{3}{4}y$$
$$3x + 2y = 102$$

Note that the $3x$ in the second equation takes into account the fence that divides the plot.

We solve the system by substitution as follows:

$$3\left(\frac{3}{4}y\right) + 2y = 102$$
$$\frac{9}{4}y + 2y = 102$$
$$9y + 8y = 408$$
$$17y = 408$$
$$y = 24$$

$$x = \frac{3}{4}y = \frac{3}{4}(24) = 18$$

The plot is 18 ft wide and 24 ft long.

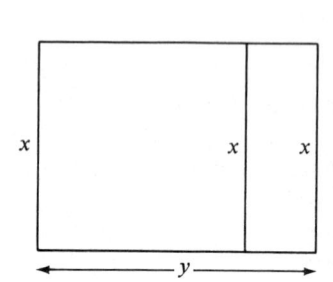

FIGURE 9.4

PERFORMANCE OBJECTIVE 7	
You should now be able to solve geometric problems using a system of two linear equations in two variables.	$y = 3x + 5$ $x + x + y = 180$

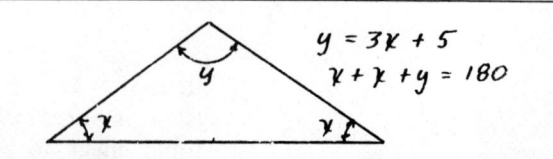

To solve word problems that involve combining components together to form a mixture, we will make use of the following facts:

1. The amount of a substance in the final mixture is equal to the sum of all the amounts of that substance in the components that make up the mixture.

2. The value of the final mixture is equal to the sum of the values of the components in the mixture.

EXAMPLE A chemist wants to mix an 18% iodine solution and a 6% iodine solution to make 60 oz of a 10% iodine solution. How much of each solution must he use to make the mixture?

SOLUTION

Let $x =$ the number of ounces in the 18% solution.

Let $y =$ the number of ounces of the 6% solution.

The following table summarizes the information given in the problem:

Component	Unit Value	Amount (oz)	Total Value (oz)
18%	0.18	x	$0.18x$
6%	0.06	y	$0.06y$
Mixture	0.10	60	$0.1(60) = 6$

The following system satisfies the conditions of the problem:

$$x + y = 60$$
$$0.18x + 0.06y = 6$$

We clear the second equation of decimals to get the system

$$x + y = 60$$
$$18x + 6y = 600$$

We apply the addition method as follows: Multiply the first equation by -6 to get the system

$$-6x - 6y = -360$$
$$18x + 6y = 600$$

Add these two equations to get

$$12x = 240$$
$$x = 20$$

Since $x + y = 60$, we have $20 + y = 60$. That is, $y = 40$.

The solution of the system is (20, 40). So the chemist should use 20 oz of the 18% solution and 40 oz of the 6% solution.

EXAMPLE A candy store mixes candy worth $1.50/lb with candy worth $2.25/lb. If the mixture is to weigh 50 lb and sell for $1.95/lb, how many pounds of each kind of candy should be used to make the mixture?

SOLUTION

Let $x =$ the number of pounds of the $1.50 candy.

Let $y =$ the number of pounds of the $2.25 candy.

The information given in the problem is summarized in the following table:

Component	Unit Value	Amount	Total Value
$1.50 candy	$1.50	x	$1.50x$
$2.25 candy	$2.25	y	$2.25y$
Mixture	$1.95	50	$1.95(50) = $97.50

A linear system that satisfies the conditions of the problem is

$$x + y = 50$$
$$1.50x + 2.25y = 97.50$$

Multiplying the second equation by 100 yields the system

$$x + y = 50$$
$$150x + 225y = 9750$$

We multiply the first equation by -150 and complete the solution by using the addition method.

$$-150x - 150y = -7500$$
$$\underline{150x + 225y = 9750}$$
$$75y = 2250$$
$$y = 30$$

Then $x + 30 = 50$, so $x = 20$.

The mixture should contain 20 lb of the $1.50 candy and 30 lb of the $2.25 candy.

PERFORMANCE OBJECTIVE 8

You should now be able to solve mixture problems using a system of two linear equations in two variables.

Component	Unit value	Amount (ml)	Total value
40% acid solution	0.40	x	0.40x
Water	0	y	0
Mixture	0.25	104	0.25(104)

We will consider motion problems in which the speed or rate of travel is constant. Such problems are called uniform motion problems.

The basic relationship used in solving motion problems is that the distance traveled equals the rate times the time. That is,

$$D = R \cdot T$$

The rate of travel and the time must involve the same time unit.

EXAMPLE Two cars that are 230 mi apart are moving toward each other. If their speeds differ by 5 mph and if they will meet in 2 hr, what is the speed of each car?

SOLUTION

Let $x =$ the speed in mph of the slower car.

Let $y =$ the speed in mph of the faster car.

The following table summarizes the information given in the problem:

Object	Rate	Time	Distance
Slower car	x	2	$2x$
Faster car	y	2	$2y$

Since their speeds differ by 5 mph, we have

$$y - x = 5$$

The distances traveled by the two cars in the 2 hr are $2x$ and $2y$ mi. So, we can write the total distance of 230 mi as

$$2x + 2y = 230$$

We then have the following system to solve:

$$y - x = 5$$
$$2x + 2y = 230$$

We multiply the first equation by 2 to get

$$2y - 2x = 10$$
$$2y + 2x = 230$$

Adding the equations gives

$$4y = 240$$

Then $y = 60$.

Since $y - x = 5$ we get

$$60 - x = 5$$
$$-x = -55$$
$$x = 55$$

The solution of the system is (55, 60) and the speeds of the cars are 55 mph and 60 mph.

EXAMPLE Nancy belongs to a carpool and picks up Betty and then Joe on her way to work. She travels at a speed of 30 mph from her house to Betty's house and 40 mph from Betty's house to Joe's house. She travels a total distance of 8 mi to pick up both Betty and Joe and it takes her a total of 15 min ($\frac{1}{4}$ hr). How long does it take for Nancy to get to Betty's house? How long does it take for her to get from Betty's house to Joe's house?

SOLUTION

Let $x =$ the time it takes to get to Betty's house (in hours).

Let $y =$ the time between Betty's house and Joe's house (in hours).

The information given in the problem is summarized in the following table where the object column represents the two segments of Nancy's trip:

Object	Rate	Time	Distance
Trip to Betty's	30	x	$30x$
Betty's to Joe's	40	y	$40y$

We can represent the conditions in the problem with the following linear system.

$$x + y = \frac{1}{4}$$
$$30x + 40y = 8$$

We multiply the first equation by -40 and solve the system using the addition method.

$$-40x - 40y = -10$$
$$30x + 40y = 8$$
$$\overline{-10x = -2}$$
$$x = \frac{-2}{-10} = \frac{1}{5}$$

Then $\frac{1}{5} + y = \frac{1}{4}$. So, $y = \frac{1}{4} - \frac{1}{5} = \frac{5}{20} - \frac{4}{20} = \frac{1}{20}$.

Nancy takes $\frac{1}{5}$ hr (12 min) to get to Betty's house and $\frac{1}{20}$ hr (3 min) to get from Betty's house to Joe's house.

PERFORMANCE OBJECTIVE 9

You should now be able to solve motion problems using a system of two linear equations in two variables.

Object	Rate (mph)	Time (hr.)	Distance
speeding car	x	$\frac{1}{5}$	$\frac{1}{5}x$
police car	y	$\frac{1}{5}$	$\frac{1}{5}x$

EXERCISE SET 9.4

Evens

Solve the following word problems using a system of two linear equations in two variables. These problems were solved in Chapter 6 using a single linear equation in one variable.

1. The sum of two numbers is 35. Twice the smaller number exceeds the larger number by 4. Find the two numbers.

2. Find two numbers such that their sum is 72 and their difference is 14.

3. The difference of two numbers is 35. Find the two numbers if the larger is two more than four times the smaller.

4. The larger of two numbers is three times the smaller number. Find the two numbers if five times their sum is 100.

5. The length of a rectangle is four times its width. The perimeter of the rectangle is 150 ft. Find the dimensions of the rectangle.

6. Two sides of a rectangle differ by 3.5 cm. What are the dimensions of the rectangle if its perimeter is 67 cm?

7. The third angle of an isosceles triangle is 16° less than the sum of the two equal angles. Find the three angles of the triangle.

8. One acute angle of a right triangle is 12° more than twice the other acute angle. Find the acute angles of the right triangle.

9. If an alloy containing 25% copper is mixed with a 40% copper alloy to get 300 pounds of 30% alloy, how much of each must be used?

10. If an 85% acid solution is to be mixed with a 40% acid solution to get 150 ml. of a 58% solution, how many milliliters of each solution should be used to make the mixture?

11. A store mixes nuts worth $2.75/lb with nuts worth $1.75/lb. If the mixture is to weigh 30 lb and sell for $2.35/lb, how many pounds of each kind of nut should be used to make the mixture?

12. A grocer plans to mix her most expensive ($3.50/lb) and least expensive ($2.25/lb) coffee beans to get a blend of intermediate cost. If the mixture is to weigh 50 lb and sell for $3.00/lb, how many pounds of each kind of bean should she use to make the mixture?

13. Two cars start out from the same point and travel in opposite directions. The speed of one car is 15 mph faster than the speed of the other car. After four hours, the two cars are 380 mi apart. Find the speed of each car.

14. Two cars that are 600 mi apart are moving toward each other. If their speeds differ by 6 mph and if the cars are 123 mi apart after 4.5 hr, what is the speed of each car?

15. Mary lives in the country and works in the city. To go to work she drives her car from her home to the outskirts of the city and then takes a bus the rest of the way. The distance between Mary's home and her job is 24 mi and it takes her 36 min ($\frac{3}{5}$ hr) to get to work. If Mary drives her car 45 mph and if the bus travels at 30 mph, how much time does she spend in her car and how much time does she spend on the bus?

16. Consider Problem 15 again. Due to icy road conditions on a winter day, the car and bus speeds were greatly reduced and Mary spent 20 min traveling on the bus and 30 min in her car. If the bus speed was half of the car speed, what were the two speeds?

SECTION 9.5 SOLVING LARGER SYSTEMS OF LINEAR EQUATIONS

The techniques we have developed can be applied to linear systems involving more than two variables. To solve the simplest electrical circuit problem involving more than one loop, we need to be able to deal with at least three linear equations in three variables. We outline a method for the solution of a system of three linear equations using determinants. Consider the system

$$a_1 x + b_1 y + c_1 z = d_1$$
$$a_2 x + b_2 y + c_2 z = d_2$$
$$a_3 x + b_3 y + c_3 z = d_3$$

To solve this system we need to be able to evaluate a **third order determinant**.

DEFINITION

> **THIRD ORDER DETERMINANT**
>
> The **third order determinant** is an array of nine numbers that form three rows and three columns.

In order to evaluate a third order determinant, we use a procedure called **expansion by minors.** This method reduces the third order determinant to a sum of second order determinants.

EXPANSION BY MINORS (THIRD ORDER DETERMINANT)

$$\begin{vmatrix} a_1 & b_1 & c_1 \\ a_2 & b_2 & c_2 \\ a_3 & b_3 & c_3 \end{vmatrix} = a_1 \begin{vmatrix} b_2 & c_2 \\ b_3 & c_3 \end{vmatrix} - b_1 \begin{vmatrix} a_2 & c_2 \\ a_3 & c_3 \end{vmatrix} + c_1 \begin{vmatrix} a_2 & b_2 \\ a_3 & b_3 \end{vmatrix}$$

This is an expansion about the first row.

The second order determinants are called **minors** of the third order determinant. These minors are formed by taking all elements of the third order determinant except those in the same row and

9 LINEAR SYSTEMS

column as the element being considered. Note the minor associated with a_1, which is formed by striking out the row and column of a_1.

$$\begin{vmatrix} \cancel{a_1} & \cancel{b_1} & \cancel{c_1} \\ a_2 & b_2 & c_2 \\ a_3 & b_3 & c_3 \end{vmatrix} \leftarrow \text{minor of } a_1 = \begin{vmatrix} b_2 & c_2 \\ b_3 & c_3 \end{vmatrix}$$

Likewise, the minor associated with b_1 is formed by striking out the row and column of b_1.

$$\begin{vmatrix} \cancel{a_1} & \cancel{b_1} & \cancel{c_1} \\ a_2 & b_2 & c_2 \\ a_3 & b_3 & c_3 \end{vmatrix} \leftarrow \text{minor of } b_1 = \begin{vmatrix} a_2 & c_2 \\ a_3 & c_3 \end{vmatrix}$$

EXAMPLE Find the minor associated with 7 in the following determinant.

$$\begin{vmatrix} 2 & 7 & 1 \\ 3 & -2 & 4 \\ 0 & 1 & 3 \end{vmatrix}$$

SOLUTION This minor is obtained by striking out the row and column containing 7.

$$\begin{vmatrix} \cancel{2} & \cancel{7} & \cancel{1} \\ 3 & -2 & 4 \\ 0 & 1 & 3 \end{vmatrix} \qquad \text{The minor is} \qquad \begin{vmatrix} 3 & 4 \\ 0 & 3 \end{vmatrix}$$

EXAMPLE Expand the following determinant about the first row.

$$\begin{vmatrix} 1 & 2 & 4 \\ -2 & 1 & 3 \\ 0 & 1 & 4 \end{vmatrix}$$

SOLUTION The expansion is

$$1 \begin{vmatrix} 1 & 3 \\ 1 & 4 \end{vmatrix} - 2 \begin{vmatrix} -2 & 3 \\ 0 & 4 \end{vmatrix} + 4 \begin{vmatrix} -2 & 1 \\ 0 & 1 \end{vmatrix}$$

SOLVING LARGER SYSTEMS OF LINEAR EQUATIONS 9.5

Note that the second term in the expansion has a negative sign.

EXAMPLE Expand the following determinant about the first row:

$$\begin{vmatrix} 2 & -1 & 3 \\ 1 & 0 & 2 \\ -2 & 1 & 3 \end{vmatrix}$$

SOLUTION

$$2\begin{vmatrix} 0 & 2 \\ 1 & 3 \end{vmatrix} - (-1)\begin{vmatrix} 1 & 2 \\ -2 & 3 \end{vmatrix} + 3\begin{vmatrix} 1 & 0 \\ -2 & 1 \end{vmatrix}$$

Having expanded the third order determinant by minors, we can now evaluate the determinant.

EXAMPLE Evaluate the following determinant, which was used in a previous example:

$$\begin{vmatrix} 1 & 2 & 4 \\ -2 & 1 & 3 \\ 0 & 1 & 4 \end{vmatrix}$$

SOLUTION

$$1\begin{vmatrix} 1 & 3 \\ 1 & 4 \end{vmatrix} - 2\begin{vmatrix} -2 & 3 \\ 0 & 4 \end{vmatrix} + 4\begin{vmatrix} -2 & 1 \\ 0 & 1 \end{vmatrix}$$

$$= 1(1 \cdot 4 - 1 \cdot 3) - 2(-2 \cdot 4 - 0 \cdot 3) + 4(-2 \cdot 1 - 0 \cdot 1)$$
$$= 1(4 - 3) - 2(-8 - 0) + 4(-2 - 0)$$
$$= 1(1) - 2(-8) + 4(-2) = 1 + 16 - 8$$
$$= 9$$

EXAMPLE Evaluate the following determinant, which was used in a previous example:

$$\begin{vmatrix} 2 & -1 & 3 \\ 1 & 0 & 2 \\ -2 & 1 & 3 \end{vmatrix}$$

SOLUTION

$$2\begin{vmatrix} 0 & 2 \\ 1 & 3 \end{vmatrix} - (-1)\begin{vmatrix} 1 & 2 \\ -2 & 3 \end{vmatrix} + 3\begin{vmatrix} 1 & 0 \\ -2 & 1 \end{vmatrix}$$

$$= 2(0 \cdot 3 - 1 \cdot 2) + 1(1 \cdot 3 - (-2)(2)) + 3(1 \cdot 1 - (-2)(0))$$
$$= 2(0 - 2) + 1(3 + 4) + 3(1 + 0)$$
$$= 2(-2) + 1(7) + 3(1) = -4 + 7 + 3$$
$$= 6$$

We could expand the determinant about any row or column but we will use the first row. The signs may be different depending on the row or column selected. Knowing how to expand a third order determinant, we can write Cramer's Rule for the solution of a system of three linear equations. Consider again the system

$$a_1 x + b_1 y + c_1 z = d_1$$
$$a_2 x + b_2 y + c_2 z = d_2$$
$$a_3 x + b_3 y + c_3 z = d_3$$

We write the solution for x, y, and z in the following manner:

$$x = \frac{\begin{vmatrix} d_1 & b_1 & c_1 \\ d_2 & b_2 & c_2 \\ d_3 & b_3 & c_3 \end{vmatrix}}{\begin{vmatrix} a_1 & b_1 & c_1 \\ a_2 & b_2 & c_2 \\ a_3 & b_3 & c_3 \end{vmatrix}}, \quad y = \frac{\begin{vmatrix} a_1 & d_1 & c_1 \\ a_2 & d_2 & c_2 \\ a_3 & d_3 & c_3 \end{vmatrix}}{\begin{vmatrix} a_1 & b_1 & c_1 \\ a_2 & b_2 & c_2 \\ a_3 & b_3 & c_3 \end{vmatrix}}, \quad z = \frac{\begin{vmatrix} a_1 & b_1 & d_1 \\ a_2 & b_2 & d_2 \\ a_3 & b_3 & d_3 \end{vmatrix}}{\begin{vmatrix} a_1 & b_1 & c_1 \\ a_2 & b_2 & c_2 \\ a_3 & b_3 & c_3 \end{vmatrix}}$$

The denominator determinant for x, y, and z is formed by using the coefficients on the left side of the equations in the system. The numerator determinants for x, y, and z are the same determinant except that the column of constants (the values on the right side of the equations) replaces the column of coefficients of x, y, and z, respectively.

Thus, the solution of a system of three linear equations using this method will require the evaluation of four third order determinants. We illustrate this in the following example.

SOLVING LARGER SYSTEMS OF LINEAR EQUATIONS 9.5

EXAMPLE Solve the following linear system:

$$x - 2y + z = 7$$
$$3x + y + z = 3$$
$$-x + 3y + 3z = -1$$

SOLUTION We first determine the value of the determinant in the denominator. This determinant is

$$\begin{vmatrix} 1 & -2 & 1 \\ 3 & 1 & 1 \\ -1 & 3 & 3 \end{vmatrix}$$

We expand by minors to find

$$1\begin{vmatrix} 1 & 1 \\ 3 & 3 \end{vmatrix} - (-2)\begin{vmatrix} 3 & 1 \\ -1 & 3 \end{vmatrix} + 1\begin{vmatrix} 3 & 1 \\ -1 & 3 \end{vmatrix}$$

$$= 1(3 - 3) + 2(9 - (-1)) + 1(9 - (-1))$$
$$= 1(0) + 2(10) + 1(10)$$
$$= 0 + 20 + 10$$
$$= 30$$

Now we evaluate x. Note that the determinant in the numerator of x is identical to the one just evaluated except that the first column is the column of values on the right side of the equations.

$$x = \frac{\begin{vmatrix} 7 & -2 & 1 \\ 3 & 1 & 1 \\ -1 & 3 & 3 \end{vmatrix}}{30}$$

$$= \frac{7\begin{vmatrix} 1 & 1 \\ 3 & 3 \end{vmatrix} - (-2)\begin{vmatrix} 3 & 1 \\ -1 & 3 \end{vmatrix} + 1\begin{vmatrix} 3 & 1 \\ -1 & 3 \end{vmatrix}}{30}$$

$$= \frac{7(3 - 3) + 2(9 + 1) + 1(9 + 1)}{30}$$

$$= \frac{7(0) + 2(10) + 1(10)}{30} = \frac{0 + 20 + 10}{30} = 1$$

Next we evaluate y.

$$y = \frac{\begin{vmatrix} 1 & 7 & 1 \\ 3 & 3 & 1 \\ -1 & -1 & 3 \end{vmatrix}}{30}$$

$$= \frac{1\begin{vmatrix} 3 & 1 \\ -1 & 3 \end{vmatrix} - 7\begin{vmatrix} 3 & 1 \\ -1 & 3 \end{vmatrix} + 1\begin{vmatrix} 3 & 3 \\ -1 & -1 \end{vmatrix}}{30}$$

$$= \frac{1(9+1) - 7(9+1) + 1(-3+3)}{30}$$

$$= \frac{1(10) - 7(10) + 1(0)}{30} = \frac{10 - 70 + 0}{30} = -2$$

Finally, we evaluate z.

$$z = \frac{\begin{vmatrix} 1 & -2 & 7 \\ 3 & 1 & 3 \\ -1 & 3 & -1 \end{vmatrix}}{30}$$

$$= \frac{1\begin{vmatrix} 1 & 3 \\ 3 & -1 \end{vmatrix} - (-2)\begin{vmatrix} 3 & 3 \\ -1 & -1 \end{vmatrix} + 7\begin{vmatrix} 3 & 1 \\ -1 & 3 \end{vmatrix}}{30}$$

$$= \frac{1(-1-9) + 2(-3+3) + 7(9+1)}{30}$$

$$= \frac{1(-10) + 2(0) + 7(10)}{30} = \frac{-10 + 0 + 70}{30} = 2$$

The solutions are $x = 1$, $y = -2$, $z = 2$.

We should substitute these values back into all three equations to see that the equations are satisfied.

Consider the first equation.

$$x - 2y + z = 7$$
$$1 - 2(-2) + 2 = 7$$
$$1 + 4 + 2 = 7$$
$$7 = 7$$

The first equation is satisfied. Consider the second equation.

$$3x + y + z = 3$$
$$3(1) + (-2) + 2 = 3$$
$$3 - 2 + 2 = 3$$
$$3 = 3$$

The second equation is satisfied. Consider the third equation.

$$-x + 3y + 3z = -1$$
$$-1 + 3(-2) + 3(2) = -1$$
$$-1 - 6 + 6 = -1$$
$$-1 = -1$$

The third equation is satisfied.

The length of the last solution emphasizes the value of having a computer to deal with larger systems of equations. A computer solution would probably involve techniques other than Cramer's Rule.

PERFORMANCE OBJECTIVE 10

You should now be able to solve a linear system of three equations in three variables using Cramer's Rule.

$$x + y + z = -1$$
$$x - y - z = 5$$
$$-x + y - z = -1$$

$$x = \frac{\begin{vmatrix} -1 & 1 & 1 \\ 5 & -1 & -1 \\ -1 & 1 & -1 \end{vmatrix}}{\begin{vmatrix} 1 & 1 & 1 \\ 1 & -1 & -1 \\ -1 & 1 & -1 \end{vmatrix}} = \frac{8}{4} = 2$$

EXERCISE SET 9.5

1. Display the indicated minors in the following determinant:

$$\begin{vmatrix} 0 & -2 & 3 \\ 1 & 4 & -3 \\ 2 & 1 & 0 \end{vmatrix}$$

   a. The minor of the 0 in the top left corner
   b. The minor of -2 c. The minor of $+3$

2. Display the indicated minors in the following determinant:

$$\begin{vmatrix} 4 & -1 & 0 \\ 2 & 3 & -4 \\ 1 & 3 & -3 \end{vmatrix}$$

   a. The minor of 4   b. The minor of -1 c. The minor of 0

3. Evaluate the determinant in Problem 1.

4. Evaluate the determinant in Problem 2.

5. Solve the following systems using Cramer's Rule:
   a. $2x - y + z = -2$
      $4x - y - 3z = 3$
      $0x + y + 2z = 0$
   b. $x + 2y + 2z = 2$
      $x + y + z = 0$
      $x \quad\quad -2z = -0.4$

   (When a term is missing the coefficient is taken as zero.)

6. Solve the following systems using Cramer's Rule:
   a. $x - 5y + 3z = -3$ b. $6x + 3y - z = -1$
      $2x + y - z = 8$ $2x \quad\quad + z = 6$
      $3x - 2y + 0z = 9$ $5x - 3y - 2z = 0$

7. The currents in a circuit satisfy the following equations. Find i_1, i_2, and i_3.

$$i_1 - i_2 + i_3 = 0$$
$$-5i_1 - 10i_2 \quad\quad = -11$$
$$\quad\quad - 10i_2 - 5i_3 = -14$$

8. The currents in a circuit satisfy the following equations. Find i_1, i_2, and i_3.

$$-i_1 - i_2 + i_3 = 0$$
$$3i_1 - 5i_2 + 2i_3 = -1$$
$$\quad\quad - 8i_2 - 3i_3 = -9$$

SECTION 9.6

SUMMARY

In this section we highlight the new concepts and definitions introduced in this chapter.

System of linear equations: a pair of linear equations in two variables or a larger group of linear equations, such as three linear equations in three variables.

Inconsistent system: a linear system that has no solutions

Independent system: a linear system that has one solution.

Dependent system: a linear system that has infinitely many solutions.

Second order determinant: an array of four numbers that form two rows and two columns. The symbol for a determinant is $\begin{vmatrix} & \\ & \end{vmatrix}$.

Cramer's Rule: the method of using determinants to solve a system of linear equations.

Third order determinant: an array of nine numbers that form three rows and three columns.

CHAPTER 9

PRACTICE TEST

(PO-1) Determine whether or not the following are systems of linear equations in two variables:

1. $5x - 2y = 3$
   $4x + y^2 = 1$

2. $y = -x$
   $4y = 3x$

(PO-1) In each of the following, determine if the ordered pair is a solution of the given system:

3. $x + 3y = -5$
   $4x - 2y = 6$ $(-2, -1)$

4. $2x + 3y = 5$
   $x - 2y + 8 = 0$ $(-2, 3)$

(PO-2) Solve the following systems of equations graphically:

5. $x - 2y = 0$
   $2x - y = 6$

6. $3x + 2y = 10$
   $x - y = 0$

(PO-3) Solve the following systems algebraically by the substitution method:

7. $0x - 2y = 5$
   $9x - 4y = 28$

8. $2x - 3y = 12$
   $y = \frac{2}{3}x - 4$

9. $2x + y = 5$
   $4y - 3x = -2$

(PO-4) Solve the following systems algebraically by the addition method:

10. $3x + 2y = 7$
    $4x + 3y = 8$

11. $\frac{2}{3}x - \frac{1}{4}y = 1$
    $-5x + 3y = 6$

12. $5x - 4y = -9$
    $2x + 3y = \frac{25}{2}$

(PO-5, 10) Solve the following systems using Cramer's Rule:

13. $2x - y = 1$
    $3x + y = 6$

14. $3x - 4y = 6$
    $2x + 5y = 4$

15. $7y + z = -40$
    $2x + 2y - 2z = -34$
    $4x + 8y + 8z = 12$

Solve the following word problems using a system of two linear equations in two variables.

16. (PO-6) The difference of two numbers is seven. Find the two numbers if three times the larger exceeds four times the smaller by three.

17. (PO-7) Two sides of a triangle are the same length. The third side is 6 in. less than the sum of the equal sides. If the perimeter of the triangle is 46 in., find the three sides.

18. (PO-7) The length of a rectangular garden is 12 ft less than four times the width. If the perimeter of the garden is 256 ft, find the dimensions of the garden.

19. (PO-8) A chemist is going to dilute a 40% sulfuric acid solution with water to get a 10% solution. He needs 20 ℓ of the 10% solution. How much of the 40% solution and how much water should he use?

20. (PO-9) Mr. Jones drove for 2 hr to get to a business appointment in another city. On his return trip, he followed the same route but, due to traffic, he traveled 20 mph slower and it took him 1 hr longer. Find his speed going and his speed returning.

CHAPTER 10

RADICALS AND COMPLEX NUMBERS

10.1 The Simplification of Radicals
10.2 The Addition and Subtraction of Radicals
10.3 The Multiplication and Division of Radicals
10.4 The Definition of a Complex Number
10.5 The Addition and Subtraction of Complex Numbers
10.6 The Multiplication of Complex Numbers
10.7 The Division of Complex Numbers
10.8 Graphing a Complex Number
10.9 The j-Operator
10.10 Summary

The notion of a power of a number or expression is rather familiar to us. We understand x^3 to mean $x \cdot x \cdot x$. The concept of a root is not quite as easy to verbalize. However, powers and roots have a relation to each other such as was described in Chapter 8 for trigonometric functions and their inverses. We will develop this relation as we go along.

This chapter deals with roots. Roots of positive numbers lead to the definition of a radical. Roots of negative numbers lead, in certain cases, to the definition of the complex number. Thus, radicals and complex numbers fit together because of their connection with roots.

SECTION 10.1 THE SIMPLIFICATION OF RADICALS

If we see 4^3 we understand the number 3 as a direction to multiply 4 by itself three times, that is, $4^3 = 4 \cdot 4 \cdot 4$. We call the 3 an **exponent**. But what do we mean by $\sqrt[3]{4}$? Here again we have the symbol $\sqrt{}$, which is a direction to do something with the number 4. We call the $\sqrt[3]{4}$ a **radical**.

The symbol $\sqrt{}$ that is used to designate a radical is called the **radical sign.** It is a direction to find the root of a quantity. We can find such roots as square roots, cube roots, and fourth roots.

When a radical sign is used, the number whose root is required is written inside the radical sign and is called the **radicand**. The order of the root is called the **index** and, except for square roots, is shown by the small number in the opening of the radical sign. Some examples are

$\sqrt{8}$ is the square root of 8; the radicand is 8 and the index is 2
Note that we do not write the index 2 when we find the square root; we simply write the radical sign.

$\sqrt[3]{10}$ is the cube root of 10; the radicand is 10 and the index is 3

$\sqrt[4]{16}$ is the fourth root of 16; the radicand is 16 and the index is 4

DEFINITION

> **A RADICAL**
>
> A **radical** is a direction to find the root of a number or expression. The parts of the radical are
>
> sign ($\sqrt{}$)
> index (n) $\sqrt[n]{x}$
> radicand (x)

We define the nth root of a number as follows:

$$\sqrt[n]{a} = m \quad \text{if } \underbrace{m \cdot m \cdot m \ldots \cdot m}_{n \text{ times}} = a$$

This means that in order to find the square root of a number, we try to find two equal factors of the number whose product equals the number. If we can find these two equal factors, the square root becomes one of the factors. For example,

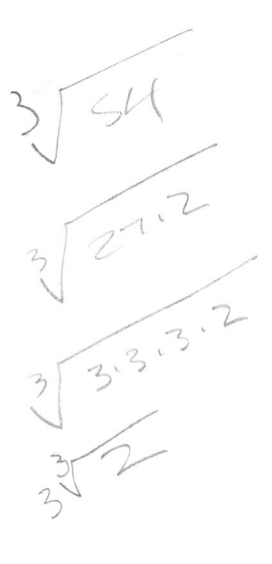

$$\sqrt{4} = 2 \quad \text{since } 2 \cdot 2 = 4$$
$$\sqrt{16} = 4 \quad \text{since } 4 \cdot 4 = 16$$

Note that we could have written $(-2)(-2) = 4$. Therefore, -2 is a valid square root of 4. However, in finding even roots, we will normally write only the positive root.

To find the cube root of a number, we try to find three equal factors of the number whose product equals the number. If we can find these three equal factors, the cube root becomes one of the factors. For example,

$$\sqrt[3]{8} = 2 \quad \text{since } 2 \cdot 2 \cdot 2 = 8$$
$$\sqrt[3]{216} = 6 \quad \text{since } 6 \cdot 6 \cdot 6 = 216$$

To find larger roots, we proceed in the same manner. For example,

$$\sqrt[4]{625} = 5 \quad \text{since } 5 \cdot 5 \cdot 5 \cdot 5 = 625$$
$$\sqrt[6]{64} = 2 \quad \text{since } 2 \cdot 2 \cdot 2 \cdot 2 \cdot 2 \cdot 2 = 64$$

When we are not able to determine the root of a number by inspection, we need a more systematic approach. Prime factoring the radicand often helps to make the root more obvious. For example,

$$\sqrt[3]{216} = \sqrt[3]{2 \cdot 2 \cdot 2 \cdot 3 \cdot 3 \cdot 3} = 2 \cdot 3 = 6$$

In working with radicals, we will not always be able to eliminate the radical sign completely as was done in the previous examples. However, we will want to simplify the radicals as much as possible.

We have previously reduced fractions and fractional algebraic expressions to lowest terms. Simplification of a radical has the same general purpose as reduction to lowest terms. The process of simplification is outlined here.

RULES FOR SIMPLIFYING A RADICAL

Step 1. Prime factor the radicand.

Step 2. Remove from the radicand repeated factors until the number of such factors remaining is less than the index.

Step 3. If the radicand is a numerical fraction or algebraic fraction, remove the denominator through multiplication by the appropriate factors.

The following examples illustrate the first two steps.

EXAMPLE Simplify $\sqrt{18}$.

SOLUTION

Step 1. $\sqrt{18} = \sqrt{2 \cdot 3 \cdot 3} = \sqrt{2 \cdot 3^2}$

Step 2. Since $\sqrt{3^2} = 3$ we write $3\sqrt{2}$

Step 3. Not Applicable

EXAMPLE Simplify $\sqrt[3]{48}$.

SOLUTION $\sqrt[3]{48} = \sqrt[3]{2 \cdot 2 \cdot 2 \cdot 2 \cdot 3}$. Since $\sqrt[3]{2 \cdot 2 \cdot 2} = 2$ and $\sqrt[3]{2 \cdot 3}$ cannot be simplified further, we write $\sqrt[3]{48} = 2\sqrt[3]{2 \cdot 3} = 2\sqrt[3]{6}$.

EXAMPLE Simplify $\sqrt[3]{x^2 y^6}$.

SOLUTION $\sqrt[3]{x^2 y^6} = \sqrt[3]{x^2(y^2)^3}$. Since $\sqrt[3]{(y^2)^3} = y^2$ and $\sqrt[3]{x^2}$ cannot be simplified further, we write

$$\sqrt[3]{x^2 y^6} = y^2 \sqrt[3]{x^2}$$

The following examples illustrate all three steps. To accomplish Step 3, we must multiply both the numerator and the denominator of the radicand by whatever factors are necessary to permit the removal of the denominator from the radical.

EXAMPLE Simplify $\sqrt{\frac{8}{3}}$.

SOLUTION

Step 1. $\sqrt{\frac{2 \cdot 2 \cdot 2}{3}}$

Step 2. $2\sqrt{\frac{2}{3}}$

Step 3. We need a perfect square in the denominator in order to remove the denominator from the radical. Multiplying both the numerator and denominator of the radicand by 3 will yield the desired perfect square.

$$2\sqrt{\frac{2 \cdot 3}{3 \cdot 3}} = \frac{2}{3}\sqrt{6}$$

EXAMPLE Simplify $\sqrt{\frac{y^3}{x}}$.

SOLUTION $\sqrt{\frac{y \cdot y \cdot y}{x}} = y\sqrt{\frac{y}{x}}$

THE SIMPLIFICATION OF RADICALS 10.1

To obtain a perfect square in the denominator, we will multiply both the numerator and the denominator of the radicand by x. We then complete the simplification.

$$y\sqrt{\frac{y \cdot x}{x \cdot x}} = \frac{y}{x}\sqrt{xy}$$

NOTE: These simplifications involving even roots are valid for positive values of the variable.

If a radical has a coefficient, the coefficient is carried throughout the simplification and is then multiplied by the factor removed from the radical. For example, to simplify $6\sqrt{75}$, we write

$$6\sqrt{75} = 6\sqrt{5 \cdot 5 \cdot 3} = 6 \cdot 5\sqrt{3} = 30\sqrt{3}$$

PERFORMANCE OBJECTIVE 1	$\sqrt[3]{x^8 y^5} = \sqrt[3]{(x^2)^3 \cdot x^2 \cdot y^3 \cdot y^2}$
You should now be able to simplify a radical.	$= x^2 y \sqrt[3]{x^2 y^2}$

Up to this time, we have used the radical sign as a direction to find a root. However, we noted that the process of taking a root is the inverse of the operation of raising to a power. If we start with 6 and raise it to the third power and then take the cube root, we should obtain 6 (what we started with). That is,

$$\sqrt[3]{6^3} = 6$$

From the laws of exponents, $(6^3)^x = 6^{3x}$. If $6^{3x} = 6^1$, then $3x$ must equal 1, so that $x = \frac{1}{3}$. Now compare the two expressions.

$$(6^3)^{1/3} = 6 \quad \text{and} \quad \sqrt[3]{6^3} = 6$$

It appears that taking the cube root is equivalent to raising a quantity to a fractional power if the laws of exponents are to be valid for fractions. Since we will want to use the laws of exponents for fractional exponents, we have an equivalent way of expressing the radical sign.

$\sqrt{}$ is equivalent to raising to the $\frac{1}{2}$ power

$\sqrt[3]{}$ is equivalent to raising to the $\frac{1}{3}$ power

$\sqrt[q]{}$ is equivalent to raising to the $\frac{1}{q}$ power

EQUIVALENCE OF RADICALS AND FRACTIONAL EXPONENT FORMS

For any real number a, $\sqrt[q]{a} = a^{1/q}$ where q is a positive integer.

EXAMPLE Convert the following radicals to equivalent expressions involving exponents:

$$\sqrt[3]{4}, \quad \sqrt[4]{xy}, \quad \sqrt[5]{x}, \quad \sqrt{y}$$

SOLUTION $4^{1/3}, \quad (xy)^{1/4}, \quad x^{1/5}, \quad y^{1/2}$

EXAMPLE Convert the following expressions involving fractional exponents to equivalent expressions involving radicals:

$$9^{1/3}, \quad x^{1/2}, \quad y^{1/4}, \quad (xy)^{1/2}$$

SOLUTION $\sqrt[3]{9}, \quad \sqrt{x}, \quad \sqrt[4]{y}, \quad \sqrt{xy}$

The quantity a in the statement

$$\sqrt[q]{a} = a^{1/q}$$

may itself be raised to a power. That is,

$$\sqrt[q]{a^p} = (a^p)^{1/q} = a^{p/q}$$

EXAMPLE Change $x^{2/5}$ to a radical.
SOLUTION $x^{2/5} = \sqrt[5]{x^2}$

EXAMPLE Change $\sqrt[4]{y^3}$ to fractional exponent form.
SOLUTION $\sqrt[4]{y^3} = y^{3/4}$

EXAMPLE Change $x^{5/3}$ to a radical and simplify.
SOLUTION $x^{5/3} = \sqrt[3]{x^5} = \sqrt[3]{(x^3)x^2} = x\sqrt[3]{x^2}$

EXAMPLE Change $(x^3y)^{1/4}$ to a radical.
SOLUTION $\sqrt[4]{x^3y}$

EXAMPLE Change $\sqrt[7]{y^3}$ to fractional exponent form.
SOLUTION $y^{3/7}$

THE ADDITION AND SUBTRACTION OF RADICALS 10.2

PERFORMANCE OBJECTIVE 2	
You should now be able to change fractional exponent forms to radicals and vice versa.	$\sqrt{x^5 y^3} = x^{\frac{5}{2}} y^{\frac{3}{2}}$

EXERCISE SET 10.1

Simplify the radicals in Exercises 1–25.

1. $\sqrt{64}$ 2. $\sqrt{81}$ 3. $\sqrt{289}$ 4. $\sqrt{729}$
5. $\sqrt[3]{343}$ 6. $\sqrt[3]{729}$ 7. $\sqrt{169}$ 8. $\sqrt{225}$
9. $\sqrt{98}$ 10. $\sqrt{108}$ 11. $\sqrt{192}$ 12. $\sqrt{80}$
13. $\sqrt{32}$ 14. $\sqrt{175}$ 15. $\sqrt{432}$ 16. $\sqrt{588}$
17. $\sqrt{162}$ 18. $\sqrt{147}$ 19. $\sqrt{507}$ 20. $\sqrt{500}$
21. $\sqrt[3]{54}$ 22. $\sqrt[3]{192}$ 23. $\sqrt{xy^3}$ 24. $\sqrt{x^2 y^2}$
25. $\sqrt{\dfrac{18}{5}}$

In Exercises 26–30, change each fractional exponent form to a radical.

26. $y^{5/9}$ 27. $y^{5/6}$ 28. $y^{3/2}$ 29. $y^{7/4}$
30. $x^{2/5}$

In Exercises 31–35, change each radical into an equivalent form involving fractional exponents.

31. $\sqrt{x^5}$ 32. $\sqrt[3]{y^5}$ 33. $\sqrt[5]{y^6}$ 34. $\sqrt{y^7}$
35. $\sqrt[3]{x^5 y^2}$

SECTION 10.2 THE ADDITION AND SUBTRACTION OF RADICALS

In this section and subsequent sections we develop procedures for combining radicals. The rules developed are equally valid for expressions involving fractional exponents. Where the word "index" is used in a rule, substitute "denominator of the exponent." Where the word "radicand" is used, substitute "quantity being raised to the fractional power." Some examples will involve fractional exponents.

RULES FOR ADDING RADICALS

Step 1. Combine coefficients of the addends that have the same indices and radicands.

Step 2. Write the answer as a product of the combined coefficient and the radical.

Step 3. Simply indicate the addition for those radicals that cannot be combined.

EXAMPLE $2\sqrt{3} + 4\sqrt{3}$

SOLUTION $2\sqrt{3} + 4\sqrt{3} = (2 + 4)\sqrt{3} = 6\sqrt{3}$

EXAMPLE $5\sqrt{x} + 8\sqrt{x}$

SOLUTION $5\sqrt{x} + 8\sqrt{x} = (5 + 8)\sqrt{x} = 13\sqrt{x}$

EXAMPLE $3\sqrt{2} + 4\sqrt{3}$

SOLUTION We cannot combine these terms because the radicals do not have the same radicand. We simply write the sum $3\sqrt{2} + 4\sqrt{3}$ without combining.

EXAMPLE $4\sqrt{x} + 5\sqrt[3]{x}$

SOLUTION We cannot combine these terms because the radicals do not have the same index. We simply write the sum $4\sqrt{x} + 5\sqrt[3]{x}$ without combining.

EXAMPLE $4x^{1/3} + 3x^{1/3}$

SOLUTION $4x^{1/3} + 3x^{1/3} = (4 + 3)x^{1/3} = 7x^{1/3}$

When we add terms involving radicals, it may be necessary to simplify the radicals before we combine the terms. For example,

$$4\sqrt{75} + 5\sqrt{12} = 4\sqrt{5 \cdot 5 \cdot 3} + 5\sqrt{2 \cdot 2 \cdot 3} = 4 \cdot 5\sqrt{3} + 5 \cdot 2\sqrt{3}$$
$$= 20\sqrt{3} + 10\sqrt{3} = (20 + 10)\sqrt{3} = 30\sqrt{3}$$

PERFORMANCE OBJECTIVE 3	$5\sqrt{20} + 2\sqrt{5} = 10\sqrt{5} + 2\sqrt{5}$
You should now be able to add radicals.	$= 12\sqrt{5}$

RULES FOR SUBTRACTING RADICALS

Step 1. Combine coefficients of the terms that have the same indices and radicands.

Step 2. Write the answer as a product of the combined coefficient and the radical.

Step 3. Simply indicate the subtraction for those radicals which cannot be combined.

EXAMPLE $5\sqrt{3} - 2\sqrt{3}$
SOLUTION $5\sqrt{3} - 2\sqrt{3} = (5-2)\sqrt{3} = 3\sqrt{3}$

EXAMPLE $-6\sqrt{xy} - 8\sqrt{xy}$
SOLUTION $-6\sqrt{xy} - 8\sqrt{xy} = (-6-8)\sqrt{xy} = -14\sqrt{xy}$

EXAMPLE $2\sqrt{x} - \sqrt{xy}$

SOLUTION We cannot combine these terms because the radicals do not have the same radicand. We simply write the difference $2\sqrt{x} - \sqrt{xy}$ without combining.

EXAMPLE $\sqrt{54} - 2\sqrt{24}$

SOLUTION We must first simplify the radicals before we can subtract the terms.

$$\sqrt{54} - 2\sqrt{24} = \sqrt{3\cdot 3\cdot 3\cdot 2} - 2\sqrt{2\cdot 2\cdot 2\cdot 3} = 3\sqrt{3\cdot 2} - 2\cdot 2\sqrt{2\cdot 3}$$
$$= 3\sqrt{6} - 4\sqrt{6} = (3-4)\sqrt{6} = -\sqrt{6}$$

PERFORMANCE OBJECTIVE 4	$3\sqrt{12} - 4\sqrt{3} = 6\sqrt{3} - 4\sqrt{3}$
You should now be able to subtract radicals.	$= 2\sqrt{3}$

EXERCISE SET 10.2

odds

Perform the indicated operations.

1. $8\sqrt{3} + 11\sqrt{3}$
2. $4\sqrt{5} + 7\sqrt{5}$
3. $2\sqrt{7} - 4\sqrt{7}$
4. $3\sqrt{11} - 6\sqrt{11}$
5. $2\sqrt{x} + 9\sqrt{xy}$
6. $4\sqrt{y} + 8\sqrt{yz}$
7. $5\sqrt{5} + 26\sqrt{5}$
8. $9\sqrt{3} + 15\sqrt{3}$

9. $4\sqrt{y} - 8\sqrt{y}$
10. $10\sqrt{x} - 9\sqrt{x}$
11. $15\sqrt{17} + 6\sqrt{17}$
12. $13\sqrt{13} + 8\sqrt{13}$
13. $-8\sqrt{5} + 3\sqrt{5} - 9\sqrt{5}$
14. $-9\sqrt{7} + 4\sqrt{7} - 12\sqrt{7}$
15. $4\sqrt{x} - 8\sqrt{x} + 9\sqrt{x} - 5\sqrt{y}$
16. $3\sqrt{x} - 9\sqrt{x} + 12\sqrt{x} - \sqrt{x}$
17. $4\sqrt{45} - 6\sqrt{20}$
18. $9\sqrt{24} - 6\sqrt{54}$
19. $\sqrt{x^2 y} - 4x\sqrt{y} + \sqrt{x^2 y}$
20. $\sqrt{xy^2} - 6y\sqrt{x} + 3\sqrt{xy^2}$

SECTION 10.3 THE MULTIPLICATION AND DIVISION OF RADICALS

Recall that radicals could be combined during addition and subtraction only if both the radicand and the index of the radicals agreed. In multiplication one of these restrictions is removed.

RULES FOR MULTIPLYING RADICALS

Step 1. Multiply the coefficients of the radicals.

Step 2. Multiply the radicands and combine under the same radical sign if possible.

Step 3. Simplify the resulting radical.

Note that the combination in Step 2 can only be performed when the index for the combined radicals is the same. If the index is not the same, the multiplication is simply indicated.

EXAMPLE $3\sqrt{5} \cdot 2\sqrt{10}$

SOLUTION

Step 1. $3 \cdot 2 = 6$

Step 2. $\sqrt{5} \cdot \sqrt{10} = \sqrt{50}$

Step 3. $6\sqrt{50} = 6\sqrt{5 \cdot 5 \cdot 2} = 6 \cdot 5\sqrt{2} = 30\sqrt{2}$

EXAMPLE $5\sqrt{x} \cdot 4\sqrt{x^3}$

SOLUTION

Step 1. $5 \cdot 4 = 20$

Step 2. $\sqrt{x} \cdot \sqrt{x^3} = \sqrt{x^4}$

Step 3. $20\sqrt{x^4} = 20x^2$

EXAMPLE $5\sqrt{3} \cdot 6\sqrt{27}$

SOLUTION

Step 1. $5 \cdot 6 = 30$

Step 2. $\sqrt{3}\sqrt{27} = \sqrt{81}$

Step 3. $30\sqrt{81} = 30 \cdot 9 = 270$

EXAMPLE $3\sqrt{2} \cdot 5\sqrt[3]{81}$

SOLUTION

Step 1. $3 \cdot 5 = 15$

Step 2. Since the indices are not the same, we cannot multiply the radicands. We can only indicate the multiplication by writing

$$\sqrt{2} \cdot \sqrt[3]{81}$$

Step 3. $15\sqrt{2} \cdot \sqrt[3]{81} = 15\sqrt{2} \cdot \sqrt[3]{3 \cdot 3 \cdot 3 \cdot 3} = 15\sqrt{2} \cdot 3\sqrt[3]{3}$
$= 45\sqrt{2} \cdot \sqrt[3]{3}$

EXAMPLE $5x^{1/2} \cdot 4x^{3/2}$

SOLUTION

Step 1. $5 \cdot 4 = 20$

Step 2. Since the denominators of the fractional exponents are the same, we may write

$$x^{1/2} \cdot x^{3/2} = (x \cdot x^3)^{1/2} = (x^4)^{1/2} = x^2$$

Step 3. $20x^2$

Note that the last example problem is identical to the second example in this section. We could have worked it more simply by using the laws of exponents to multiply the two powers of x together. The use of fractional exponents in place of radicals will always allow us to combine radicands, even if the indices do not agree.

EXAMPLE $3x^{1/4} \cdot 7x^{1/3}$

SOLUTION $3 \cdot 7 = 21$, $x^{1/4} \cdot x^{1/3} = x^{(1/4 + 1/3)} = x^{7/12}$. The answer is $21x^{7/12}$. This could be converted to radical form.

10 RADICALS AND COMPLEX NUMBERS

Fractional exponents have some advantages over radicals in the multiplication process.

PERFORMANCE OBJECTIVE 5

You should now be able to multiply radicals.

$$-5x^{\frac{2}{5}} \cdot 4x^{\frac{1}{2}} = -20x^{(\frac{2}{5}+\frac{1}{2})}$$
$$= -20x^{\frac{9}{10}}$$

When we divide by radicals, we use a procedure known as **rationalizing the denominator.** To rationalize the denominator, we multiply both the numerator and the denominator by a factor chosen so as to remove the radical sign(s) in the denominator.

RULES FOR DIVIDING RADICALS

Step 1. Rationalize the denominator.

Step 2. Simplify the resulting expression.

We will work two types of problems involving radical expressions in the divisor. The first type is one in which the divisor is a single term containing a radical. To rationalize this denominator, we ignore the coefficient and multiply the numerator and denominator by the radical in the denominator. We then simplify if possible. The problems that we will work will involve square roots only. The procedure is more complex for other roots.

EXAMPLE $\sqrt{2} \div \sqrt{3}$

SOLUTION $\dfrac{\sqrt{2}}{\sqrt{3}} = \dfrac{\sqrt{2}}{\sqrt{3}} \cdot \dfrac{\sqrt{3}}{\sqrt{3}} = \dfrac{\sqrt{6}}{\sqrt{9}} = \dfrac{\sqrt{6}}{3}$

EXAMPLE $5\sqrt{6} \div \sqrt{2}$

SOLUTION

$\dfrac{5\sqrt{6}}{\sqrt{2}} = \dfrac{5\sqrt{6}}{\sqrt{2}} \cdot \dfrac{\sqrt{2}}{\sqrt{2}} = \dfrac{5\sqrt{12}}{\sqrt{4}} = \dfrac{5\sqrt{2 \cdot 2 \cdot 3}}{2} = \dfrac{5 \cdot 2\sqrt{3}}{2} = 5\sqrt{3}$

EXAMPLE $6\sqrt{xy} \div 5\sqrt{x}$

SOLUTION $\dfrac{6\sqrt{xy}}{5\sqrt{x}} = \dfrac{6\sqrt{xy}}{5\sqrt{x}} \cdot \dfrac{\sqrt{x}}{\sqrt{x}} = \dfrac{6\sqrt{x^2y}}{5\sqrt{x^2}} = \dfrac{6x\sqrt{y}}{5x} = \dfrac{6\sqrt{y}}{5}$

THE MULTIPLICATION AND DIVISION OF RADICALS   10.3

EXAMPLE $(3\sqrt{6} - 2\sqrt{5}) \div 5\sqrt{10}$

SOLUTION

$$\frac{3\sqrt{6} - 2\sqrt{5}}{5\sqrt{10}} = \frac{3\sqrt{6} - 2\sqrt{5}}{5\sqrt{10}} \cdot \frac{\sqrt{10}}{\sqrt{10}} = \frac{3\sqrt{60} - 2\sqrt{50}}{5\sqrt{100}}$$

$$= \frac{3\sqrt{2 \cdot 2 \cdot 3 \cdot 5} - 2\sqrt{5 \cdot 5 \cdot 2}}{5 \cdot 10} = \frac{3 \cdot 2\sqrt{15} - 2 \cdot 5\sqrt{2}}{50}$$

$$= \frac{3\sqrt{15} - 5\sqrt{2}}{25}$$

The second type of problem involving a radical in the divisor is one in which the divisor has two terms. To rationalize this denominator, we multiply the numerator and denominator by the denominator modified so that the sign of the second term is changed. We then complete the problem by simplifying.

EXAMPLE $(\sqrt{xy} - \sqrt{y}) \div (\sqrt{x} - \sqrt{y})$

SOLUTION

$$\frac{\sqrt{xy} - \sqrt{y}}{\sqrt{x} - \sqrt{y}} = \frac{\sqrt{xy} - \sqrt{y}}{\sqrt{x} - \sqrt{y}} \cdot \frac{\sqrt{x} + \sqrt{y}}{\sqrt{x} + \sqrt{y}}$$

$$= \frac{\sqrt{x^2 y} - \sqrt{xy} + \sqrt{xy^2} - \sqrt{y^2}}{\sqrt{x^2} - \sqrt{xy} + \sqrt{xy} - \sqrt{y^2}}$$

$$= \frac{x\sqrt{y} - \sqrt{xy} + y\sqrt{x} - y}{x - y}$$

EXAMPLE $(\sqrt{10} - \sqrt{5}) \div (\sqrt{11} - \sqrt{3})$

SOLUTION

$$\frac{\sqrt{10} - \sqrt{5}}{\sqrt{11} - \sqrt{3}} = \frac{\sqrt{10} - \sqrt{5}}{\sqrt{11} - \sqrt{3}} \cdot \frac{\sqrt{11} + \sqrt{3}}{\sqrt{11} + \sqrt{3}}$$

$$= \frac{\sqrt{110} + \sqrt{30} - \sqrt{55} - \sqrt{15}}{\sqrt{121} + \sqrt{33} - \sqrt{33} - \sqrt{9}}$$

$$= \frac{\sqrt{110} + \sqrt{30} - \sqrt{55} - \sqrt{15}}{11 - 3}$$

$$= \frac{\sqrt{110} + \sqrt{30} - \sqrt{55} - \sqrt{15}}{8}$$

PERFORMANCE OBJECTIVE 6	$(\sqrt{12} - \sqrt{6}) \div (\sqrt{3} + \sqrt{5})$
You should now be able to divide radicals.	$= \dfrac{\sqrt{12} - \sqrt{6}}{\sqrt{3} + \sqrt{5}} \cdot \dfrac{\sqrt{3} - \sqrt{5}}{\sqrt{3} - \sqrt{5}}$
	$= \dfrac{6 - 2\sqrt{15} - 3\sqrt{2} + \sqrt{30}}{-2}$

When we deal with radicals involving pure numbers it is seldom worthwhile to go through the processes of combining and simplifying as described in this chapter. This is particularly true if a final numerical answer is sought. With a calculator it is usually easier to perform the indicated operations in whatever form we find the problem. For example, in a previous problem, we performed the division

$$(3\sqrt{6} - 2\sqrt{5}) \div 5\sqrt{10}$$

to obtain $\dfrac{3\sqrt{15} - 5\sqrt{2}}{25}$. The simplified form requires three fewer steps on the calculator than the original form. But these three steps take much less time than the simplification procedure. Hand simplification is primarily valuable as an exercise that develops skills in manipulating radicals involving variables.

Using the calculator, we work one of each type problem selected from previous examples.

EXAMPLE (Addition of radicals) Add $3\sqrt{2} + 4\sqrt{3}$.

SOLUTION

3 $\boxed{\times}$ 2 $\boxed{\text{INV}}$ $\boxed{x^2}$ (or $\sqrt{x}$) $\boxed{+}$ 4 $\boxed{\times}$ 3 $\boxed{\text{INV}}$ $\boxed{x^2}$ $\boxed{=}$

We read 11.17 as an answer.

EXAMPLE (Subtraction of radicals) Subtract $\sqrt{54} - 2\sqrt{24}$.

SOLUTION 54 $\boxed{\text{INV}}$ $\boxed{x^2}$ $\boxed{-}$ 2 $\boxed{\times}$ 24 $\boxed{\text{INV}}$ $\boxed{x^2}$ $\boxed{=}$

We read -2.449

The simplified form of this expression was $-\sqrt{6}$.

We enter this as

6 $\boxed{\text{INV}}$ $\boxed{x^2}$ $\boxed{+/-}$

and read -2.449.

EXAMPLE (Multiplication of radicals) Multiply $5\sqrt{3} \cdot 6\sqrt{27}$.

THE MULTIPLICATION AND DIVISION OF RADICALS 10.3

SOLUTION

5 $\boxed{\times}$ 3 $\boxed{\text{INV}}$ $\boxed{x^2}$ $\boxed{\times}$ 6 $\boxed{\times}$ 27 $\boxed{\text{INV}}$ $\boxed{x^2}$ $\boxed{=}$

We read 270.

EXAMPLE (Division of radicals) Divide $\dfrac{\sqrt{10} - \sqrt{5}}{\sqrt{11} - \sqrt{3}}$.

SOLUTION We first work out the denominator.

11 $\boxed{\text{INV}}$ $\boxed{x^2}$ $\boxed{-}$ 3 $\boxed{\text{INV}}$ $\boxed{x^2}$ $\boxed{=}$ $\boxed{\text{STO}}$

Next the numerator is evaluated.

10 $\boxed{\text{INV}}$ $\boxed{x^2}$ $\boxed{-}$ 5 $\boxed{\text{INV}}$ $\boxed{x^2}$ $\boxed{=}$

and then the quotient.

$\boxed{\div}$ $\boxed{\text{RCL}}$ $\boxed{=}$

We read 0.5845.
We may check this with the simplified form:

$$\frac{\sqrt{110} + \sqrt{30} - \sqrt{55} - \sqrt{15}}{8}$$

110 $\boxed{\text{INV}}$ $\boxed{x^2}$ $\boxed{+}$ 30 $\boxed{\text{INV}}$ $\boxed{x^2}$ $\boxed{-}$ 55 $\boxed{\text{INV}}$ $\boxed{x^2}$ $\boxed{-}$ 15 $\boxed{\text{INV}}$ $\boxed{x^2}$ $\boxed{=}$ $\boxed{\div}$ 8 $\boxed{=}$

This yields 0.5845 as before.

All of the examples so far have involved square roots. How do we find other roots? If your calculator has a $\boxed{y^x}$ key, then you can find other roots in the following manner:

EXAMPLE Find $\sqrt[3]{27}$.

SOLUTION 27 $\boxed{\text{INV}}$ $\boxed{y^x}$ 3 $\boxed{=}$

We read 3, which we recognize as the cube root of 27.

EXAMPLE Multiply $3\sqrt{2} \cdot 5\sqrt[3]{81}$. (Note that we were unable to multiply radicands in this problem earlier in the section.)

SOLUTION

3 $\boxed{\times}$ 2 $\boxed{\text{INV}}$ $\boxed{x^2}$ $\boxed{\times}$ 5 $\boxed{\times}$ 81 $\boxed{\text{INV}}$ $\boxed{y^x}$ 3 $\boxed{=}$

We read 91.78.

PERFORMANCE OBJECTIVE 7

You should now be able to perform operations on radicals involving pure numbers with the use of a calculator.

$3\sqrt{7} \times 6\sqrt{11}$

3 [x] 7 [INV] [x^y] [x] 6 [x] 11 [INV] [x^y]
[=] 157.949

EXERCISE SET 10.3

Multiply the radicals in Exercises 1–10.

1. $5\sqrt{3} \cdot 8\sqrt{28}$
2. $4\sqrt{2} \cdot \sqrt{27}$
3. $6\sqrt{2} \cdot -5\sqrt{18}$
4. $-3\sqrt{3} \cdot 8\sqrt{12}$
5. $5\sqrt[3]{10} \cdot \sqrt{6}$
6. $7\sqrt[3]{15} \cdot \sqrt{5}$
7. $3\sqrt{x} \cdot 6\sqrt{y} \cdot 2\sqrt{x^2 y}$
8. $4\sqrt{y} \cdot 5\sqrt{x^2} \cdot 3\sqrt{xy}$
9. $5\sqrt{2} \cdot 3\sqrt[3]{16}$
10. $3\sqrt{3} \cdot 6\sqrt[3]{32}$

Divide the radicals in Exercises 11–26.

11. $(6\sqrt{6} + 3\sqrt{3}) \div 9\sqrt{15}$
12. $(6\sqrt{5} + 2\sqrt{2}) \div 2\sqrt{10}$
13. $(2\sqrt{6} - 3\sqrt{3}) \div 2\sqrt{30}$
14. $(4\sqrt{6} - 2\sqrt{3}) \div 5\sqrt{21}$
15. $3 \div (\sqrt{x} - 2)$
16. $7 \div (\sqrt{x} - 5)$
17. $(4 - \sqrt{2}) \div \sqrt{6}$
18. $(9 + \sqrt{3}) \div \sqrt{15}$
19. $8\sqrt{27} \div 3\sqrt{8}$
20. $7\sqrt{12} \div \sqrt{7}$
21. $(3\sqrt{x} - 2\sqrt{xy}) \div (\sqrt{y} - \sqrt{x})$
22. $(6\sqrt{x} - 5\sqrt{xy}) \div (\sqrt{x} - \sqrt{y})$
23. $(\sqrt{7} + 2) \div (\sqrt{7} - 3)$
24. $(\sqrt{5} - 4) \div (\sqrt{5} + 7)$
25. $(1 - \sqrt{xy}) \div (\sqrt{x} + \sqrt{y})$
26. $(1 + \sqrt{xy}) \div (\sqrt{y} - \sqrt{x})$

27. Solve Problem 17 in Exercise Set 10.2 and Problems 9 and 23 in this section using a calculator.

28. Solve Problem 18 in Exercise Set 10.2 and Problems 10 and 24 in this section using a calculator.

In Exercises 29–34, perform the indicated operations and write the answers using fractional exponents and in radical form.

29. $5x^{1/3} \cdot 6x^{4/3}$
30. $3x^{2/3} \cdot 8x^{5/3}$
31. $3x^{1/4} \cdot x^{2/3}$
32. $6x^{1/5} \cdot x^{8/5}$
33. $(2x^{1/2} - 3y^{1/2}) \div (y^{1/2} - x^{1/2})$
34. $(4x^{1/2} - 5y^{1/2}) \div (y^{1/2} - x^{1/2})$

SECTION 10.4 THE DEFINITION OF A COMPLEX NUMBER

In the next chapter, we will be solving quadratic equations in one variable. The solution to a quadratic equation may involve finding the square root of a number. If that number is greater than or equal to zero, then its square root can be found in the real numbers. For example, $\sqrt{4} = 2$ and $\sqrt{12} = 2\sqrt{3}$. However, if the number is less than zero, its square root will not be a real number. So the solution to a quadratic equation (and to equations of higher degree) often cannot be found in the real numbers. This situation makes it necessary to extend our number system and define a new kind of number that includes the desired solution.

The new kind of number involves $\sqrt{-1}$. The number $\sqrt{-1}$, usually designated by j, satisfies the property that

$$\sqrt{-1} \cdot \sqrt{-1} = -1, \quad \text{that is, } j^2 = -1$$

Using the property that $j^2 = -1$, we can simplify certain expressions involving j. For example,

$$(-4j)(5j) = -20j^2 = -20(-1) = +20$$

and

$$(3j)^3 = 3^3 j^3 = 27j^3 = 27j^2 j = 27(-1)j = -27j$$

Using the number j, we can represent the square root of negative numbers as follows:

$$\sqrt{-n} = \sqrt{-1} \cdot \sqrt{n} = j\sqrt{n}$$

Numbers involving j are said to be in **complex form.**

EXAMPLE Express $\sqrt{-4}$ in complex form.
SOLUTION $\sqrt{-4} = \sqrt{-1} \cdot \sqrt{4} = j\sqrt{4} = 2j$

EXAMPLE Express $\sqrt{-12}$ in complex form.
SOLUTION $\sqrt{-12} = \sqrt{-1} \cdot \sqrt{12} = j\sqrt{12} = 2\sqrt{3}j$

We call any number of the form bj an **imaginary number,** provided that b is a real number. Examples of imaginary numbers are $2j$ and $2\sqrt{3}j$.

DEFINITION

> **IMAGINARY NUMBER**
>
> An **imaginary number** is a number of the form bj with b real and $j = \sqrt{-1}$.

We define a complex number to be any number of the form

$$z = a + bj$$

where a and b are real numbers. The real number a is called the **real part** of z and bj is called the **imaginary part** of z.

DEFINITION

> **COMPLEX NUMBER**
>
> A **complex number** is a number of the form $z = a + bj$ where a and b are real numbers.

Note that every real number can be written in the form $a + bj$ with $b = 0$. For example,

$$8 = 8 + 0j \quad \text{and} \quad -12 = -12 + 0j$$

Thus, real numbers are also complex numbers of a special kind. Also note that every imaginary number can be written in the form $a + bj$ with $a = 0$. Imaginary numbers are also complex numbers of a special kind.

We define $a - bj$ to be the **conjugate** of $a + bj$. Likewise, $a + bj$ is said to be the conjugate of $a - bj$. So to form the conjugate of a complex number, we change only the sign of the imaginary part of the number. For example, $6 - 3j$ is the conjugate of $6 + 3j$ and $-2 + 4j$ is the conjugate of $-2 - 4j$.

DEFINITION

> **CONJUGATE OF A COMPLEX NUMBER**
>
> The **conjugate** of a complex number is a related complex number in which the sign of the imaginary part is changed. The conjugate of $a + bj$ is $a - bj$.

EXAMPLE Find the conjugate of $8 - 5j$.

SOLUTION $8 + 5j$

EXAMPLE Find the conjugate of $-6 + 8j$.

SOLUTION $-6 - 8j$

THE ADDITION AND SUBTRACTION OF COMPLEX NUMBERS 10.5

PERFORMANCE OBJECTIVE 8	
You should now be able to identify a complex number and its conjugate.	$-2 + 4j$ is the conjugate of $-2 - 4j$.

EXERCISE SET 10.4

1. Evaluate each of the following:
   a. j^6
   b. $(-2j)^3$
   c. $(3j)(-2j)$
   d. j^4
   e. $(5j)^2$
   f. $(8j)^3$

2. Evaluate each of the following:
   a. j^5
   b. $(-4j)^3$
   c. $(-6j)(4j)$
   d. j^8
   e. $(7j)^2$
   f. $(2j)^3$

3. Simplify and express the following in terms of j: *adds*
   a. $\sqrt{-49}$
   b. $\sqrt{-9}$
   c. $\sqrt{-225}$
   d. $3 - \sqrt{-25}$
   e. $2 + \sqrt{-4}$
   f. $\dfrac{6 + \sqrt{-4}}{2}$

4. Simplify and express the following in terms of j:
   a. $\sqrt{-64}$
   b. $\sqrt{-81}$
   c. $\sqrt{-256}$
   d. $6 - \sqrt{-36}$
   e. $5 + \sqrt{-121}$
   f. $\dfrac{9 + \sqrt{-144}}{3}$

5. Find the conjugate of each of the following:
   a. $-3 - 4j$
   b. $6 + 2j$
   c. $-2 + 5j$
   d. $10 + 7j$
   e. $3 - 4j$
   f. $-2 - 5j$

6. Find the conjugate of each of the following:
   a. $-5 - 6j$
   b. $9 + j$
   c. $-4 + 7j$
   d. $12 + 6j$
   e. $9 - 5j$
   f. $-6 - 9j$

SECTION 10.5 THE ADDITION AND SUBTRACTION OF COMPLEX NUMBERS

The rules for adding and subtracting complex numbers are given in this section. We recommend the vertical format when adding and subtracting. Real and imaginary parts are easier to combine when they are lined up vertically.

RULES FOR ADDING AND SUBTRACTING COMPLEX NUMBERS

Step 1. Add (or subtract) the real parts to obtain the real part of the sum (or difference).

Step 2. Add (or subtract) the imaginary parts to obtain the imaginary part of the sum (or difference).

EXAMPLE $(6 + 2j) + (5 + 8j)$

SOLUTION
$$\begin{array}{r} 6 + 2j \\ +(5 + 8j) \\ \hline 11 + 10j \end{array}$$

EXAMPLE $(8 - 3j) + (4 + 6j)$

SOLUTION
$$\begin{array}{r} 8 - 3j \\ +(4 + 6j) \\ \hline 12 + 3j \end{array}$$

EXAMPLE $(4 + 5j) - (2 + 6j)$

SOLUTION
$$\begin{array}{r} 4 + 5j \\ -(2 + 6j) \end{array} \rightarrow \begin{array}{r} 4 + 5j \\ +(-2 - 6j) \\ \hline 2 - j \end{array}$$

Note that subtraction of complex numbers is performed in the same way as subtraction of polynomials.

EXAMPLE $(6 - 5j) - (8 - 8j)$

SOLUTION
$$\begin{array}{r} 6 - 5j \\ -(8 - 8j) \end{array} \rightarrow \begin{array}{r} 6 - 5j \\ +(-8 + 8j) \\ \hline -2 + 3j \end{array}$$

PERFORMANCE OBJECTIVE 9	
You should now be able to add or subtract complex numbers.	$(4 - 2j) + (-6 + 4j) = -2 + 2j$ $(3 + 5j) - (2 + 8j) = 1 - 3j$

EXERCISE SET 10.5

odd

Perform the indicated operations.

1. $(8 + 9j) + (5 + 4j)$
2. $(5 + 16j) + (4 + 3j)$
3. $(9 - 2j) + (-2 + j)$
4. $(8 - 5j) + (-4 + j)$
5. $(1 + 3j) + (5 + 2j)$
6. $(3 + 6j) + (7 + 2j)$
7. $(-5 + 11j) - (2 - 4j)$
8. $(-7 + 12j) - (4 - 6j)$
9. $(16 + 8j) - (2 - 8j)$
10. $(14 + 9j) - (-6 - j)$

SECTION 10.6 THE MULTIPLICATION OF COMPLEX NUMBERS

The rules for multiplying complex numbers parallel those for multiplication of binomials.

RULES FOR MULTIPLYING COMPLEX NUMBERS

Step 1. Arrange the complex numbers in vertical multiplication format.

Step 2. Find the product as if you were multiplying two binomials.

Step 3. Replace j^2 with -1 and simplify.

EXAMPLE $(2 + 3j)(5 + 4j)$

SOLUTION

$$\begin{array}{r} 2 + 3j \\ 5 + 4j \\ \hline 8j + 12j^2 \\ 10 + 15j \phantom{{}+12j^2} \\ \hline 10 + 23j + 12j^2 \end{array}$$

Replacing j^2 with -1 and simplifying yields

$$10 + 23j + 12(-1) = 10 + 23j + (-12)$$
$$= -2 + 23j$$

10 RADICALS AND COMPLEX NUMBERS

EXAMPLE $(-6 + 4j)(3 - 8j)$

SOLUTION

$$\begin{array}{r} -6 + 4j \\ 3 - 8j \\ \hline 48j - 32j^2 \\ -18 + 12j \\ \hline \end{array}$$

$$-18 + 60j - 32j^2 = -18 + 60j - 32(-1)$$
$$= -18 + 60j + 32$$
$$= 14 + 60j$$

EXAMPLE $(8 - 3j)(-4 - 5j)$

SOLUTION

$$\begin{array}{r} 8 - 3j \\ -4 - 5j \\ \hline -40j + 15j^2 \\ -32 + 12j \\ \hline \end{array}$$

$$-32 - 28j + 15j^2 = -32 - 28j + 15(-1)$$
$$= -32 - 28j - 15$$
$$= -47 - 28j$$

EXAMPLE $(5 - 6j)(5 + 6j)$

SOLUTION

$$\begin{array}{r} 5 - 6j \\ 5 + 6j \\ \hline 30j - 36j^2 \\ 25 - 30j \\ \hline 25 - 36j^2 \end{array}$$

$$25 - 36j^2 = 25 - 36(-1)$$
$$= 25 + 36$$
$$= 61$$

NOTE: The product of a complex number and its conjugate is a complex number with $b = 0$. That is, the product is a real number.

PERFORMANCE OBJECTIVE 10	
You should now be able to multiply complex numbers.	$\begin{array}{r}5 - 4j \\ -2 - 2j \\ \hline -10j + 8j^2 \\ -10 + 8j \\ \hline -10 - 2j + 8j^2 = -18 - 2j\end{array}$

EXERCISE SET 10.6

odd

Multiply the given complex numbers.

1. $(5 + 3j)(6 - 2j)$
2. $(3 + 5j)(4 - 3j)$
3. $(-2 - 3j)(5 - 3j)$
4. $(-6 - 2j)(3 - 8j)$
5. $(-2 - 4j)(-2 + 4j)$
6. $(-3 - 7j)(-3 + 2j)$
7. $(2 + 6j)(-8 - 4j)$
8. $(4 + 8j)(-6 - 3j)$
9. $(0 + 5j)(-2 - 10j)$
10. $(0 + 5j)(-4 - 12j)$

SECTION 10.7 THE DIVISION OF COMPLEX NUMBERS

The procedure for division of complex numbers points out the need to understand the conjugate of a complex number. The fact that the product of a number and its conjugate is real allows us to simplify the quotient. The procedure for division is outlined here.

RULES FOR DIVIDING COMPLEX NUMBERS

Step 1. Write the indicated quotient as a fraction.

Step 2. Multiply both the numerator and denominator of this fraction by the conjugate of the denominator.

Step 3. Replace j^2 by -1 and simplify.

Step 4. Write the result from Step 3 in the form $a + bj$.

EXAMPLE $(7 + 3j) \div (6 + 2j)$

SOLUTION

Step 1. $\dfrac{7 + 3j}{6 + 2j}$

Step 2. $\dfrac{7 + 3j}{6 + 2j} \cdot \dfrac{6 - 2j}{6 - 2j} = \dfrac{42 - 14j + 18j - 6j^2}{36 - 12j + 12j - 4j^2}$

$= \dfrac{42 + 4j - 6j^2}{36 - 4j^2}$

Step 3. $\dfrac{42 + 4j - 6j^2}{36 - 4j^2} = \dfrac{42 + 4j - 6(-1)}{36 - 4(-1)} = \dfrac{42 + 4j + 6}{36 + 4}$

$= \dfrac{48 + 4j}{40}$

Step 4. $\dfrac{48 + 4j}{40} = \dfrac{48}{40} + \dfrac{4j}{40} = \dfrac{6}{5} + \dfrac{j}{10}$

EXAMPLE $(2 + 2j) \div (2 - 2j)$

SOLUTION

Step 1. $\dfrac{2 + 2j}{2 - 2j}$

Step 2. $\dfrac{2 + 2j}{2 - 2j} \cdot \dfrac{2 + 2j}{2 + 2j} = \dfrac{4 + 8j + 4j^2}{4 - 4j^2}$

Step 3. $\dfrac{4 + 8j + 4j^2}{4 - 4j^2} = \dfrac{4 + 8j + 4(-1)}{4 - 4(-1)} = \dfrac{8j}{8}$

Step 4. $\dfrac{8j}{8} = j$ (or $0 + j$)

EXAMPLE $(8 - 6j) \div (-3 - 4j)$

SOLUTION

$$\dfrac{8 - 6j}{-3 - 4j} = \dfrac{8 - 6j}{-3 - 4j} \cdot \dfrac{-3 + 4j}{-3 + 4j} = \dfrac{-24 + 50j - 24j^2}{9 - 16j^2} = \dfrac{50j}{25} = 2j$$

PERFORMANCE OBJECTIVE 11	$\dfrac{2 + 5j}{-4 - 3j} = \dfrac{2 + 5j}{-4 - 3j} \cdot \dfrac{-4 + 3j}{-4 + 3j}$
You should now be able to divide complex numbers.	$= \dfrac{-8 - 14j + 15j^2}{16 - 9j^2} = \dfrac{-23 - 14j}{25}$

EXERCISE SET 10.7

Divide the given complex numbers.

1. $(2 + 4j) \div (-3 - 2j)$
2. $(3 + 6j) \div (-2 - 3j)$
3. $(8 - 3j) \div (0 - j)$
4. $(5 - 4j) \div (0 + j)$
5. $(4 + 2j) \div (-3 - 3j)$
6. $(8 + 3j) \div (-4 - 4j)$
7. $(6 + 5j) \div (6 - 5j)$
8. $(7 + 2j) \div (7 - 2j)$
9. $(9 - 4j) \div (9 - 2j)$
10. $(5 - 3j) \div (5 - j)$
11. $(16 - 32j) \div (2 - 3j)$
12. $(18 - 36j) \div (3 - 4j)$

SECTION 10.8 GRAPHING A COMPLEX NUMBER

Complex numbers may be represented either as points or vectors in a rectangular coordinate system. In many practical applications the vector representation is more useful.

In order to graph the complex number $z = a + bj$ as a point we represent it as the ordered pair (a, b). We then locate this ordered pair (a, b) as a point in the two dimensional complex plane called the z plane. To represent the complex number as a vector we simply connect the origin of the complex plane with the point (a, b). The tail of the vector is at the origin and the tip at (a, b).

We let the horizontal axis of the coordinate system represent the real part of the complex number. We label this axis the x-axis just as we did in our previous work with graphing. We let the vertical axis of the coordinate system represent the imaginary part of the complex number. We label this axis as the y_j-axis.

DEFINITION

COMPLEX z PLANE

The **complex z plane** is a plane defined by an axis of reals (horizontal) and an axis of imaginaries (vertical).

EXAMPLE Graph $z = 6 - 4j$.

SOLUTION To graph this complex number,

Step 1. Draw the coordinate system.

Step 2. Rewrite $z = 6 - 4j$ as $(6, -4)$.

Step 3. Locate 6 on the real axis.

Step 4. Locate $-4j$ on the imaginary axis.

Step 5. Label the point or sketch the vector. See Figure 10.1.

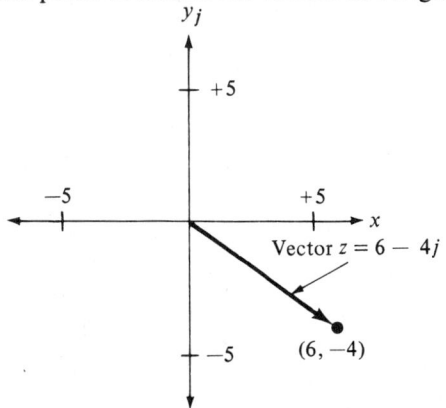

FIGURE 10.1

EXAMPLE Graph $z = -3 + 2j$.

SOLUTION The solution is shown in Figure 10.2

EXAMPLE Graph $z = -5 - j$.

SOLUTION The solution is shown in Figure 10.3

DEFINITION

MAGNITUDE OF A COMPLEX NUMBER

The **magnitude** of the complex number $z = a + bj$ is given by $|z| = \sqrt{a^2 + b^2}$.

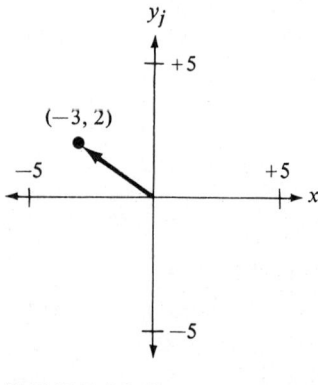

FIGURE 10.2

The magnitude is then the distance of the point (a, b) from the origin (the length of the vector z). We now calculate the magnitude of the complex numbers that were graphed previously.

EXAMPLE Find $|z|$ when $z = 6 - 4j$.

SOLUTION

$$|z| = \sqrt{(6)^2 + (-4)^2} = \sqrt{36 + 16} = \sqrt{52} = \sqrt{2 \cdot 2 \cdot 13} = 2\sqrt{13}$$

EXAMPLE Find $|z|$ when $z = -3 + 2j$.

SOLUTION

$$|z| = \sqrt{(-3)^2 + (2)^2} = \sqrt{9 + 4} = \sqrt{13}$$

EXAMPLE Find $|z|$ when $z = -5 - j$.

SOLUTION

$$|z| = \sqrt{(-5)^2 + (-1)^2} = \sqrt{25 + 1} = \sqrt{26}$$

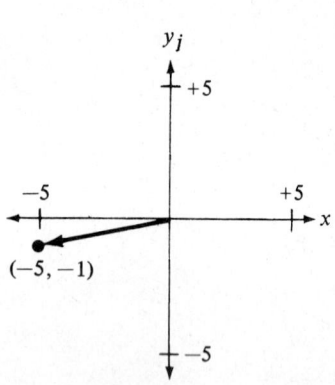

FIGURE 10.3

The complex number in the form $z = a + bj$ is said to be in rectangular form because it is easily represented in a rectangular system of coordinates.

PERFORMANCE OBJECTIVE 12

You should now be able to graph a complex number and find its magnitude.

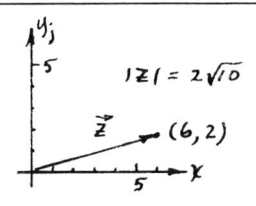

EXERCISE SET 10.8

Graph each of the following complex numbers and find its magnitude:

1. $z = 6 - 5j$
2. $z = 5 - 3j$
3. $z = 10 + 5j$
4. $z = 8 + 4j$
5. $z = 4 - 2j$
6. $z = -2 - 5j$
7. $z = 3 - 3j$
8. $z = 4 - 4j$
9. $z = -5 + 4j$
10. $z = -6 + 5j$

SECTION 10.9 THE j-OPERATOR

Although you may see $\sqrt{-1}$ written as i in mathematics texts, j is preferable in most technical work. For example, the symbol i in electrical work customarily represents current. Also, in technical work, we may hear the expression "j-operator." The symbol $j(\sqrt{-1})$ acts as an operator in the complex plane defined in Section 10.8. The j-operator rotates a vector to which it is applied by 90° in a counter-clockwise direction. For example, consider the sketches in Figure 10.4 in which the vectors 4 and $4j$ are plotted. Note that $4j$ is rotated counterclockwise by 90° from 4.

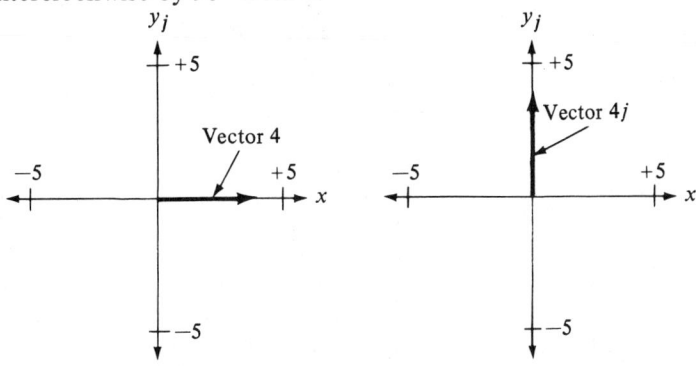

FIGURE 10.4

We consider a more general vector **A** in the complex plane. Let **A** = 3 + 4j and **B** = j**A**. Then

$$\mathbf{B} = j(3 + 4j) = 3j + 4(j^2) = 3j + 4(-1) = -4 + 3j$$

Vectors **A** and **B** are plotted in Figure 10.5.

Note that the application of j has rotated vector **A** 90° counterclockwise. To emphasize this property of j we frequently write the j before the quantity. For vector **A** above, we would write **A** = 3 + j4 while **B** would be written **B** = −4 + j3.

This procedure is more than just a mathematical exercise. It turns out that when we solve ac circuit problems the imaginary parts of such solutions correspond to circuit properties that are shifted in phase by 90° from the real parts. Thus, the j-operator serves the practical purpose of representing phase shifts.

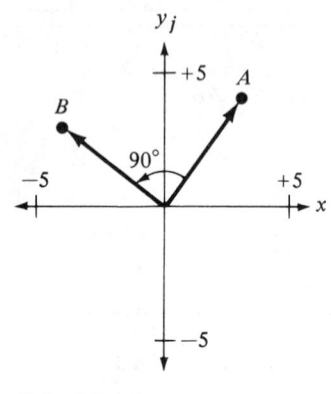

FIGURE 10.5

DEFINITION

THE j-OPERATOR

The **j-operator** applied to a vector in the complex plane rotates that vector by 90° counterclockwise.

Successive operations of j correspond to successive rotations. Consider a vector along the positive axis of reals, say **A**. Apply the j-operator twice.

$$j \cdot j\mathbf{A} = j^2\mathbf{A} = (-1)\mathbf{A} = -\mathbf{A}$$

But −**A** is along the negative axis of reals. See Figure 10.6 where this is shown graphically.

Additional applications of the j-operator give the following results:

Three times: $j^3\mathbf{A} = j^2(j\mathbf{A}) = (-1) \cdot (j\mathbf{A}) = -j\mathbf{A}$

Four times: $j^4\mathbf{A} = j^2 \cdot j^2 \cdot \mathbf{A} = (-1) \cdot (-1) \cdot \mathbf{A} = \mathbf{A}$

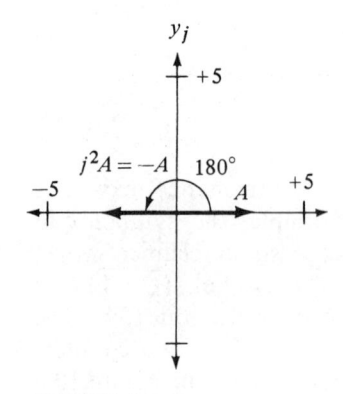

FIGURE 10.6

See Figure 10.7 where these rotations are displayed.

Thus, successive operations of j correspond to the successive powers of $\sqrt{-1}$.

The notion of j as an operator allows us to think of the vector representation of a complex number in a different and useful way. Consider the complex number z = a + bj, which we have represented as a vector. Using the j-operator, we can think of z as the sum of a real vector (a) along the x-axis and a real vector (b) that has been rotated 90° by the operator j. See Figure 10.8.

THE j-OPERATOR 10.9

FIGURE 10.7

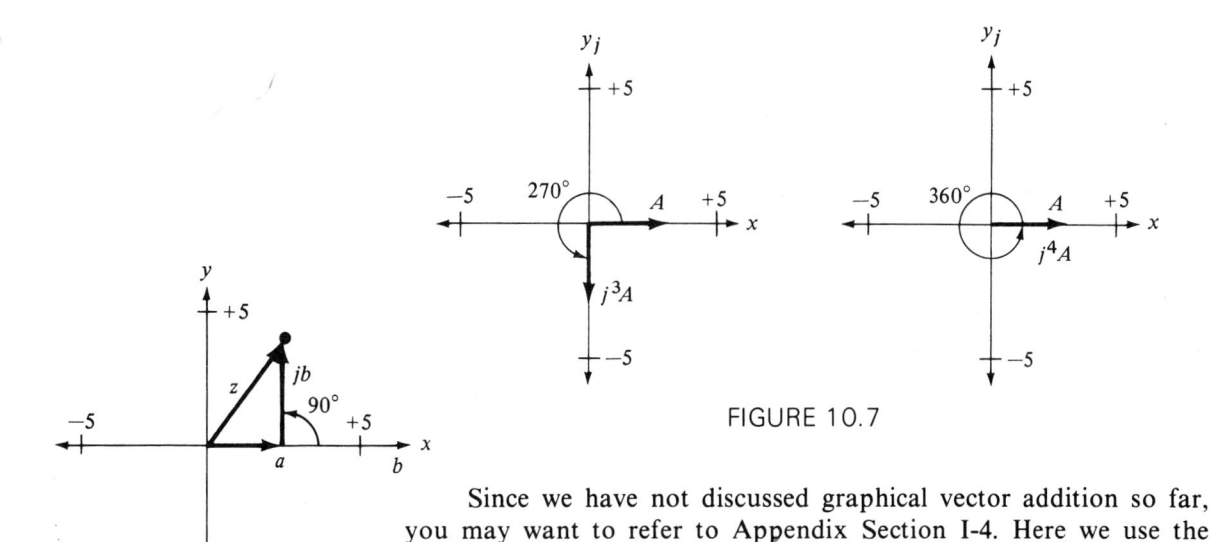

FIGURE 10.8

Since we have not discussed graphical vector addition so far, you may want to refer to Appendix Section I-4. Here we use the procedure to add vectors. We simply place them tail to tip with the sum represented by a vector drawn from the tail of the first to the tip of the last.

EXAMPLE Represent the vector $z = 3 - j6$ using j-operator notation.

SOLUTION See Figure 10.9.

EXAMPLE Represent the following vectors using j-operator notation:

a. $z = -3 + j5$
b. $z = -4 - j2$
c. $z = -j2$

SOLUTION See Figure 10.10.

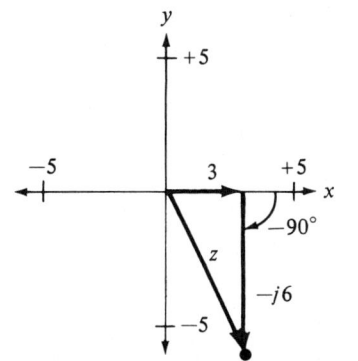

FIGURE 10.9

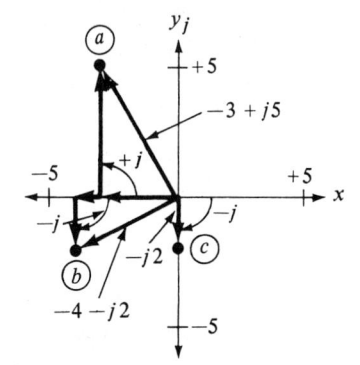

FIGURE 10.10

We summarize the properties of the successive operation of the *j*-operator on a vector.

Operator	Rotation	Mathematical expression
j	90° ccw	$\sqrt{-1}$
j^2	180° ccw	-1
j^3	270° ccw	$-\sqrt{-1}$
$-j$	90° cw	$-\sqrt{-1}$

PERFORMANCE OBJECTIVE 13

You should now be able to represent a complex vector using the *j*-operator notation.

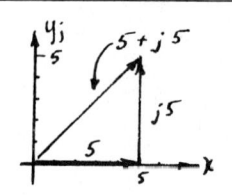

Numbers involving *j*-operators may be added or subtracted in the same manner as the complex numbers. That is, those parts containing no *j*-operators may be combined since they are along the *x*-axis while those parts containing a *j*-operator may be combined because they are along the *y*-axis.

EXAMPLE Form the sum of z_1 and z_2 where $z_1 = 4 - j5$ and $z_2 = -1 + j8$.

SOLUTION

$$z_1 = 4 - j5$$
$$z_2 = -1 + j8$$
$$z_1 + z_2 = 3 + j3$$

The answer is $3 + j3$.

We can investigate whether this result agrees with what we expect to find using graphical methods. See Figure 10.11.

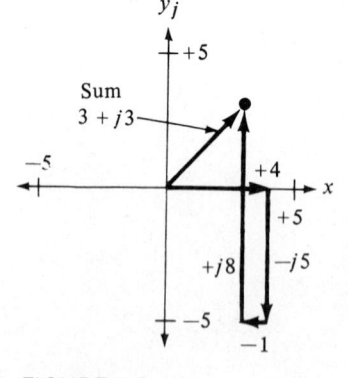

FIGURE 10.11

It appears that the rule for combining complex numbers mathematically agrees with the concept of vector addition.

PERFORMANCE OBJECTIVE 14	$z_1 = -2 + j7$
You should now be able to add or subtract numbers involving j-operators.	$z_2 = -3 - j2$
	$z_1 + z_2 = -5 + j5$

EXERCISE SET 10.9

Represent each of the following vectors graphically using j-operator notation (Exercises 1–10)

1. $z = 5 + j4$
2. $z = 6 + j3$
3. $z = -3 - j3$
4. $z = -5 - j5$
5. $z = -j3$
6. $z = -j2$
7. $z = 2 - j5$
8. $z = 4 - j6$
9. $z = -5 + j4$
10. $z = -3 + j2$

Perform the indicated operations graphically (Exercises 11–20)

11. $(3 + j2) + (2 - j4)$
12. $(4 + j3) + (3 - j2)$
13. $(-5 - j6) + (-3 + j3)$
14. $(-8 - j2) + (5 + j4)$
15. $(4 + j5) - (4 - j5)$
16. $(3 + j5) - (6 - j6)$
17. $(-2 - j) - (-5 + j3)$
18. $(-3 - j) - (-4 + j5)$
19. $(4 - j3) + (2 + j4) - (3 - j6)$
20. $(2 - j5) + (4 + j6) - (5 - j7)$

SECTION 10.10

SUMMARY

In this section we highlight the new concepts and definitions introduced in the chapter.

Radical sign: the symbol $\sqrt{}$

Radicand: the number inside the radical sign whose root is to be found.

Index: the order of the root.

Simplest form of a radical: a form of the radical in which both repeated radicand factors in multiples of the index and the denominator, if any, are removed from under the radical sign.

Fractional exponent form: a method of expressing a radical in which the index is indicated as the denominator of a fraction. The fraction becomes an exponent of the radicand according to the following rule:

$$\sqrt[q]{a^p} = a^{p/q}$$

Imaginary number: a number of the form bj with b real and $j = \sqrt{-1}$

Complex number: a number of the form $z = a + bj$ where a and b are real numbers.

Conjugate: a related complex number in which the sign of the imaginary part has been changed.

Complex z plane: a plane defined by an axis of reals (horizontal) and an axis of imaginaries (vertical).

Magnitude of a complex number: a real number given by $|z| = \sqrt{a^2 + b^2}$ where $z = a + bj$. The magnitude represents the length of the complex vector z.

j-operator: the square root of -1 considered as an operator in the complex plane. Multiplication of a vector by this quantity (j) corresponds to rotating the vector 90° counterclockwise in the complex plane.

CHAPTER 10

PRACTICE TEST

Perform the indicated operations and simplify (Exercises 1–19).

1. (PO-1, 3) $-4\sqrt{6} + 2\sqrt{54}$
2. (PO-1, 3) $3\sqrt{7} + 4\sqrt{28}$
3. (PO-2) Change $x^{5/4}$ to a radical
4. (PO-1, 4) $8\sqrt{12} - 5\sqrt{36}$
5. (PO-1) $\sqrt{125} - \sqrt{320}$
6. (PO-1, 5) $-6\sqrt{5} \cdot 3\sqrt{72}$
7. (PO-5) $4\sqrt{11} \cdot -5\sqrt{2}$
8. (PO-1, 6) $(\sqrt{3} + 8) \div (2\sqrt{5} + \sqrt{11})$
9. (PO-1, 6) $(4 - \sqrt{13}) \div (-\sqrt{3} + 7)$

CHAPTER 10 PRACTICE TEST

10. (PO-8) $(-4j)^5$
11. (PO-8, 9) $(3 - \sqrt{-36}) + (-2 + \sqrt{-99})$
12. (PO-9) $(-6 + 4j) + (3 - 12j)$
13. (PO-9) $(-3 - 2j) + (-4 - 7j)$
14. (PO-9) $(16 + 9j) - (-4 + 6j)$
15. (PO-9) $(-12 + 4j) - (3 - 2j)$
16. (PO-10) $(-3 - 4j) \cdot (-3 + 4j)$
17. (PO-10) $(2 - 6j) \cdot (5 + 9j)$
18. (PO-11) $(10 - 3j) \div (-2 + 4j)$
19. (PO-11) $(6 + 2j) \div (-4 - 5j)$
20. (PO-12) Graph and find the magnitude of $z = -5 + 6j$.
21. (PO-13) Represent $4 + j3$ using j-operator notation.
22. (PO-13) Represent $-2 - j5$ using j-operator notation.

Add or subtract the following complex numbers, considered as vectors, graphically:

23. (PO-14) $(-3 - j5) + (2 + j4)$
24. (PO-14) $(-4 + j6) - (-3 - j)$
25. (PO-14) $(2 + j4) - (3 - j6) + (-4 - j2)$

CHAPTER 11

QUADRATIC EQUATIONS

11.1 The Solution of a Quadratic Equation by Factoring
11.2 The Solution of a Quadratic Equation Using the Quadratic Formula
11.3 The Solution of a Quadratic Equation Using the Calculator
11.4 Summary

In this chapter we introduce quadratic equations in one variable. The emphasis is on three methods of solving these quadratic equations.

If the coefficients of the terms are small integers, it may be worthwhile to attempt a solution by factoring (first method). If the coefficients are large integers, it will be necessary to use the quadratic formula (second method) to arrive at a solution in most cases. The main advantage of the quadratic formula is that it will work in every case, even when the solutions involve radicals and complex numbers.

The calculator (third method) will be used to solve quadratic equations in conjunction with the quadratic formula.

SECTION 11.1 THE SOLUTION OF A QUADRATIC EQUATION BY FACTORING

Quadratic equations occur frequently in mathematics, engineering, business, and the sciences. Consequently, it is important to be able to solve such equations. To solve a quadratic equation, it will often be necessary to write the equation in the general form $ax^2 + bx + c = 0$, and to identify the values of a, b, and c.

DEFINITION

QUADRATIC EQUATION IN ONE VARIABLE

A **quadratic equation in one variable** is one that can be written in the general form

$$ax^2 + bx + c = 0$$

where a, b and c are any real numbers, x is the variable, and $a \neq 0$.

Note that the equation must contain a term in which the variable is raised to the second power in order to be quadratic. No higher degree terms appear in the equation.

EXAMPLE Identify a, b, and c in the equation $3x^2 + 3 = 2x$.

SOLUTION Write the equation in general form by moving all terms to the left side and arranging them in descending powers of x.

$$3x^2 + (-2x) + 3 = 2x + (-2x)$$
$$3x^2 - 2x + 3 = 0$$

Reading from the equation, we find $a = 3$, $b = -2$, and $c = 3$.

To solve a quadratic equation means to find the values of the variable, which, when substituted into the equation in place of x, result in true statements.

EXAMPLE Verify that 3 and -2 are both solutions or roots of the equation

$$x^2 - x - 6 = 0$$

SOLUTION Substituting 3 into the equation yields

$$(3)^2 - 3 - 6 = 0$$
$$9 - 3 - 6 = 0$$
$$0 = 0$$

This is a true statement so 3 is a solution.

Substituting −2 in the equation yields

$$(-2)^2 - (-2) - 6 = 0$$
$$4 + 2 - 6 = 0$$
$$0 = 0$$

This is a true statement so −2 is a solution.

We were able to identify two roots of the quadratic equation $x^2 - x - 6 = 0$. They will be the only solutions to this equation. The number of solutions of a polynomial equation will be equal to the degree of the polynomial. For example, when we studied linear or first degree equations in one variable, the equation had just one solution. For quadratic equations in one variable, we will always find two solutions. These solutions may be repeated (the same solution twice) or complex, but the two solutions are always present.

PERFORMANCE OBJECTIVE 1	$\left. \begin{array}{c} x^2 - x + 3 = 0 \\ -3x^2 = 0 \\ 5x^2 + 2y - 4 = 0 \end{array} \right\}$ quadratic equations in one variable
You should now be able to identify a quadratic equation in one variable.	

To solve a quadratic equation by factoring, use the following procedure.

RULES FOR SOLVING A QUADRATIC EQUATION BY FACTORING

Step 1. Write the equation in general form.

Step 2. Find the factors of the polynomial.

Step 3. Set each factor equal to zero and solve the resulting equations.

EXAMPLE Solve $x^2 - x - 6 = 0$ by factoring.

SOLUTION

Step 1. The equation is already in general form.

Step 2. We factor the trinomial $x^2 - x - 6$.

$$x^2 - x - 6 = (x - 3)(x + 2)$$

Step 3. We set each factor equal to zero and solve the resulting equations.

$$x - 3 = 0, \quad x + 2 = 0$$

So $x = 3$ and $x = -2$.

Therefore, $x = 3$ and $x = -2$ are the solutions of the quadratic equation $x^2 - x - 6 = 0$.

We include additional examples to illustrate the factoring technique for solving quadratic equations.

EXAMPLE Solve $x^2 - 6x = 0$ (constant term is missing).

SOLUTION

$x^2 - 6x = 0$
$x(x - 6) = 0$

Set $x = 0$ and $x - 6 = 0$, which yields $x = 0$ and $x = 6$.

Therefore, $x = 0$ and $x = 6$ are the solutions. When the constant term is missing, $x = 0$ will always be a solution.

EXAMPLE Solve $x^2 - 12x + 32 = 0$.

SOLUTION

$x^2 - 12x + 32 = 0$
$(x - 8)(x - 4) = 0$

Set $x - 8 = 0$ and $x - 4 = 0$, which yields $x = 8$ and $x = 4$.

Therefore, $x = 8$ and $x = 4$ are the solutions.

EXAMPLE Solve $2x^2 + 5x + 2 = 0$.

THE SOLUTION OF A QUADRATIC EQUATION 11.1

SOLUTION
$$2x^2 + 5x + 2 = 0$$
$$2x^2 + 4x + x + 2 = 0$$
$$2x(x + 2) + 1(x + 2) = 0$$
$$(2x + 1)(x + 2) = 0$$

Set $2x + 1 = 0$ and $x + 2 = 0$, which yields $2x = -1$ and $x = -2$. Solving $2x = -1$ for x gives

$$x = -\frac{1}{2}$$

Therefore, $x = -\frac{1}{2}$ and $x = -2$ are the solutions.

EXAMPLE Solve $x^2 - 36 = 0$ (linear term missing).

SOLUTION
$$x^2 - 36 = 0$$
$$(x - 6)(x + 6) = 0$$

Set $x - 6 = 0$ and $x + 6 = 0$, which yields $x = 6$ and $x = -6$.

Therefore, $x = 6$ and $x = -6$ are the solutions. When the linear term is missing, the solutions will always differ only in the sign.

EXAMPLE Solve $4(x^2 - 2x) = 21$.

SOLUTION We first eliminate the parentheses from the equation by using the Distributive Law. We then apply the procedure for solving quadratic equations by factoring.

$$4(x^2 - 2x) = 21$$
$$4x^2 - 8x = 21$$
$$4x^2 - 8x - 21 = 0$$
$$4x^2 + 6x - 14x - 21 = 0$$
$$2x(2x + 3) - 7(2x + 3) = 0$$
$$(2x - 7)(2x + 3) = 0$$

Set $2x - 7 = 0$ and $2x + 3 = 0$, which yields $2x = 7$ and $2x = -3$. Solving for x gives

$$x = \frac{7}{2}, \quad x = -\frac{3}{2}$$

Therefore, $x = \frac{7}{2}$ and $x = -\frac{3}{2}$ are the solutions.

EXAMPLE Solve $\frac{1}{2}x^2 = \frac{2}{5}x^2 + 10$ (fractional coefficients).

SOLUTION To solve an equation such as this, clear the equation of all fractions. Since the LCD is 10, we multiply all terms by 10.

$$10\left(\frac{1}{2}x^2\right) = 10\left(\frac{2}{5}x^2\right) + 10(10)$$
$$5x^2 = 4x^2 + 100$$
$$x^2 - 100 = 0$$
$$(x - 10)(x + 10) = 0$$

Set $x - 10 = 0$ and $x + 10 = 0$, which yields $x = 10$ and $x = -10$.

Therefore, $x = 10$ and $x = -10$ are the solutions. Note again that the linear term was missing.

EXAMPLE Solve $-12 - 17x - 5x^2 = 0$.

SOLUTION Although this equation is in the general form, it is best to arrange the terms in descending powers of x and then multiply by -1 so that the coefficient of x^2 is positive.

$$-5x^2 - 17x - 12 = 0$$
(terms rearranged)
$$5x^2 + 17x + 12 = 0$$
(multiplication by -1)
$$5x^2 + 5x + 12x + 12 = 0$$
$$5x(x + 1) + 12(x + 1) = 0$$
$$(5x + 12)(x + 1) = 0$$

Set $5x + 12 = 0$ and $x + 1 = 0$, which yields $5x = -12$ and $x = -1$. From $5x = -12$,

$$x = -\frac{12}{5}$$

Therefore, $x = -\frac{12}{5}$ and $x = -1$ are the solutions.

EXAMPLE Solve $\dfrac{7}{x - 3} - \dfrac{1}{2} = \dfrac{3}{x - 4}$ (equation is not a polynomial equation).

SOLUTION Although this equation does not look like a quadratic equation, it can be made to look quadratic by clearing the equation of fractions. We multiply all terms by the LCD, which is $2(x - 3)(x - 4)$.

$$\frac{2(x-3)(x-4)(7)}{x-3} - \frac{2(x-3)(x-4)}{2} = \frac{2(x-3)(x-4)(3)}{x-4}$$

$$2(x-4)(7) - (x-3)(x-4) = 2(x-3)(3)$$
$$14(x-4) - (x-3)(x-4) = 6(x-3)$$
$$14x - 56 - (x^2 - 7x + 12) = 6x - 18$$
$$14x - 56 - x^2 + 7x - 12 = 6x - 18$$
$$-x^2 + 21x - 68 = 6x - 18$$
$$-x^2 + 15x - 50 = 0$$
$$x^2 - 15x + 50 = 0$$

This equation is now in general form. Factor to obtain

$$(x - 5)(x - 10) = 0$$

Set $x - 5 = 0$ and $x - 10 = 0$, which yields $x = 5$ and $x = 10$.

Therefore, $x = 5$ and $x = 10$ are the solutions.

NOTE: When we convert rational algebraic expressions to quadratics, we examine the solutions to insure that the rational algebraic expression is defined for all solutions found. For the preceding example, the rational algebraic expression is not defined for $x = 3$ and for $x = 4$. However, since the solution set does not contain 3 or 4, there is no problem.

PERFORMANCE OBJECTIVE 2	
You should now be able to solve a quadratic equation by factoring.	$2x^2 + 5x = 12$ $2x^2 + 5x - 12 = 0$ $(2x - 3)(x + 4) = 0$ $x = \frac{3}{2}, x = -4$

EXERCISE SET 11.1

Solve each of the following:

1. $x^2 + 6x = 55$
2. $x^2 + 3x = 28$
3. $x^2 = 121$
4. $x^2 = 100$

5. $3x^2 - 27x = 0$
6. $5x^2 - 55x = 0$
7. $x^2 + 4x = 12$
8. $x^2 + 10x = 24$
9. $x^2 - 2x = 63$
10. $x^2 - 3x = 40$
11. $2(x^2 + 5) = 9x$
12. $3(x^2 + 4) = 13x$
13. $\dfrac{5}{x+4} = 4 + \dfrac{3}{x-2}$
14. $\dfrac{6}{x+3} = 2 + \dfrac{1}{x-4}$
15. $\dfrac{9}{2}x^2 + 8 = 12x$
16. $\dfrac{8}{3}x^2 + 6 = 17x$
17. $2(4x^2 - 5x) + 3 = 0$
18. $4(3x^2 - 8x) + 3 = 0$
19. $2(4x^2 + 11x) = -15$
20. $3(4x^2 + 12x) = -15$

SECTION 11.2 THE SOLUTION OF A QUADRATIC EQUATION USING THE QUADRATIC FORMULA

In mathematics there are many formulas that prove to be time-saving devices in that they provide a routine for solving frequently encountered problems. The **quadratic formula** is such a device.

The quadratic equations we solved in the last section of this chapter were selected so as to be factorable. Not all quadratics are so readily factored. The quadratic formula arises from the need to solve quadratic equations that cannot be factored or for which the factors are not readily apparent. The formula deals with the coefficients of the variables and the constant term in the quadratic equation. It contains a radical that may have a negative radicand. But, since we have studied complex numbers, this will not be a problem. With the formula, we can truly solve any quadratic equation.

The quadratic formula can be derived using a technique known as completing the square. The derivation of the formula is as follows:

Step 1. Write the general form of the quadratic equation.

$$ax^2 + bx + c = 0$$

Step 2. Move the constant to the right side of the equation.

$$ax^2 + bx = -c$$

Step 3. Divide each term in the equation by a.

$$x^2 + \dfrac{b}{a}x = -\dfrac{c}{a}$$

Step 4. Add $\frac{b^2}{4a^2}$ to both sides of the equation. This step is called **completing the square** since it yields a perfect square trinomial on the left side of the equation.

$$x^2 + \frac{b}{a}x + \frac{b^2}{4a^2} = -\frac{c}{a} + \frac{b^2}{4a^2}$$

Step 5. Factor the trinomial on the left side of the equation and combine the fractions on the right side.

$$\left(x + \frac{b}{2a}\right)\left(x + \frac{b}{2a}\right) = \frac{b^2 - 4ac}{4a^2}$$

Step 6. Write the left side as the square of a binomial.

$$\left(x + \frac{b}{2a}\right)^2 = \frac{b^2 - 4ac}{4a^2}$$

Step 7. Take the square root of both sides.

$$x + \frac{b}{2a} = \pm\frac{\sqrt{b^2 - 4ac}}{2a}$$

Step 8. Solve for x.

$$x = -\frac{b}{2a} \pm \frac{\sqrt{b^2 - 4ac}}{2a}$$

Step 9. Combine the terms on the right to obtain the desired formula.

FORMULA

QUADRATIC FORMULA

$x = \dfrac{-b \pm \sqrt{b^2 - 4ac}}{2a}$ where a, b, and c are taken from the equation $ax^2 + bx + c = 0$.

RULES FOR USING THE QUADRATIC FORMULA

Step 1. Write the equation in general form.

Step 2. Identify a, b, and c.

Step 3. Substitute the values of a, b, and c into the formula.

Step 4. Evaluate and simplify the result.

Note that the quadratic formula gives both solutions to a quadratic equation. This happens because of the $\pm$ sign in the numerator of the formula.

11 QUADRATIC EQUATIONS

To illustrate the use of the quadratic formula, we will rework some of the previous examples using the formula rather than factoring. We will also work some new examples.

EXAMPLE Solve $x^2 - x - 6 = 0$.

SOLUTION

Step 1. $x^2 - x - 6 = 0$

Step 2. $a = 1, b = -1, c = -6$

Step 3. $x = \dfrac{-(-1) \pm \sqrt{(-1)^2 - 4(1)(-6)}}{2(1)}$

is obtained by substitution of the values of a, b, and c into

$$x = \frac{-b \pm \sqrt{b^2 - 4ac}}{2a}$$

Step 4. $x = \dfrac{1 \pm \sqrt{1 + 24}}{2}$

$x = \dfrac{1 \pm \sqrt{25}}{2}$

$x = \dfrac{1 \pm 5}{2}$

Then, $x = \frac{6}{2} = 3$ and $x = -\frac{4}{2} = -2$.

Thus, $x = 3$ and $x = -2$ are the solutions. Note that these are the same roots we found previously.

EXAMPLE Solve $\dfrac{7}{x - 3} - \dfrac{1}{2} = \dfrac{3}{x - 4}$.

SOLUTION Clearing this equation of fractions yields the following quadratic equation in general form. (See this example in Section 11.1.)

$$x^2 - 15x + 50 = 0$$

$a = 1, b = -15, c = 50$

Use $x = \dfrac{-b \pm \sqrt{b^2 - 4ac}}{2a}$.

$x = \dfrac{-(-15) \pm \sqrt{(-15)^2 - 4(1)(50)}}{2(1)}$

$x = \dfrac{15 \pm \sqrt{225 - 200}}{2}$

$x = \dfrac{15 \pm \sqrt{25}}{2}$

Therefore, $x = \dfrac{15 + 5}{2} = \dfrac{20}{2} = 10$

and $x = \dfrac{15 - 5}{2} = \dfrac{10}{2} = 5.$

Thus, $x = 5$ and $x = 10$ are the solutions.

EXAMPLE Solve $4(x^2 - 3x) = -9$.

SOLUTION We first eliminate the parentheses from the equation and write the equation in general form. We then use the quadratic formula.

$$4(x^2 - 3x) = -9$$
$$4x^2 - 12x = -9$$
$$4x^2 - 12x + 9 = 0$$

$a = 4, b = -12, c = 9$

Use $x = \dfrac{-b \pm \sqrt{b^2 - 4ac}}{2a}$.

$$x = \dfrac{-(-12) \pm \sqrt{(-12)^2 - 4(4)(9)}}{2(4)}$$

$$x = \dfrac{12 \pm \sqrt{144 - 144}}{8}$$

$$x = \dfrac{12 \pm \sqrt{0}}{8}$$

$$x = \dfrac{12 \pm 0}{8}$$

Therefore, $x = \dfrac{12 + 0}{8} = \dfrac{12}{8} = \dfrac{3}{2}$

and $x = \dfrac{12 - 0}{8} = \dfrac{12}{8} = \dfrac{3}{2}.$

Thus, $x = \dfrac{3}{2}$ is the only solution (a repeated root).

EXAMPLE Solve $\dfrac{x}{2} = \dfrac{10}{9x}$.

SOLUTION We clear this equation of fractions by multiplying by the LCD 18x. This yields a quadratic equation, which we will solve using the quadratic formula.

$$\frac{x}{2} = \frac{10}{9x}$$

$$18x\left(\frac{x}{2}\right) = 18x\left(\frac{10}{9x}\right)$$

$$9x^2 = 20$$

$$9x^2 - 20 = 0$$

$a = 9, b = 0, c = -20$

Use $x = \dfrac{-b \pm \sqrt{b^2 - 4ac}}{2a}$.

$$x = \frac{-0 \pm \sqrt{0^2 - 4(9)(-20)}}{2(9)}$$

$$x = \frac{0 \pm \sqrt{720}}{18}$$

$$x = \frac{\pm 12\sqrt{5}}{18}$$

$$x = \frac{+12\sqrt{5}}{18} = \frac{2\sqrt{5}}{3}$$

$$x = -\frac{12\sqrt{5}}{18} = -\frac{2\sqrt{5}}{3}$$

Therefore, $x = \dfrac{2\sqrt{5}}{3}$ and $x = -\dfrac{2\sqrt{5}}{3}$ are the solutions.

EXAMPLE Solve $x^2 = 12x - 37$.

SOLUTION

$x^2 - 12x + 37 = 0$

$a = 1, b = -12, c = 37$

Use $x = \dfrac{-b \pm \sqrt{b^2 - 4ac}}{2a}$.

$$x = \frac{-(-12) \pm \sqrt{(-12)^2 - 4(1)(37)}}{2(1)}$$

$$x = \frac{12 \pm \sqrt{144 - 148}}{2}$$

$$x = \frac{12 \pm \sqrt{-4}}{2}$$

$$x = \frac{12 \pm 2j}{2}$$

$$x = \frac{2(6 \pm j)}{2} = 6 \pm j$$

Therefore, $x = 6 + j$ and $x = 6 - j$ are the solutions.

PERFORMANCE OBJECTIVE 3

You should now be able to solve a quadratic equation using the quadratic formula.

$2x^2 + 12x - 5 = 0$
$a = 2 \quad b = 12 \quad c = -5$
$x = \dfrac{-12 \pm \sqrt{144 + 40}}{4} = \dfrac{-6 \pm \sqrt{46}}{2}$

EXERCISE SET 11.2

Solve the following quadratics using the quadratic formula:

1. $\dfrac{x^2 - 2}{3} = \dfrac{5}{6}$
2. $\dfrac{x^2 - 5}{2} = \dfrac{7}{4}$
3. $2(4x^2 + 11x) = -15$
4. $3(x^2 + 4) = 13x$
5. $9x^2 + 30 - 30x = 0$
6. $16x^2 + 36x = 0$
7. $\dfrac{5}{x + 4} = 4 + \dfrac{3}{x - 2}$
8. $\dfrac{6}{x + 3} = 2 + \dfrac{1}{x - 4}$
9. $16x^2 - 40x + 25 = 0$
10. $25x^2 - 70x + 49 = 0$

SECTION 11.3 THE SOLUTION OF A QUADRATIC EQUATION USING THE CALCULATOR

To solve the quadratic equation with the calculator we use the quadratic formula. The procedure is illustrated in the following examples.

11 QUADRATIC EQUATIONS

EXAMPLE Solve $3x^2 - 6x + 1 = 0$.

SOLUTION First identify a, b, and c. $a = 3$, $b = -6$, $c = 1$. Then set up the quadratic formula.

$$x = \frac{-(-6) \pm \sqrt{(-6)^2 - 4(3)(1)}}{2(3)} = \frac{6 \pm \sqrt{(-6)^2 - 4(3)(1)}}{2(3)}$$

Next evaluate and store the radical.

6 $\boxed{+/-}$ $\boxed{x^2}$ $\boxed{-}$ 4 $\boxed{\times}$ 3 $\boxed{\times}$ 1 $\boxed{=}$ $\boxed{\text{INV}}$ $\boxed{x^2}$ $\boxed{\text{STO}}$

To obtain one root use the following procedure:

6 $\boxed{+}$ $\boxed{\text{RCL}}$ $\boxed{=}$ $\boxed{\div}$ 2 $\boxed{\div}$ 3 $\boxed{=}$ yields 1.816

The second root is obtained as follows:

6 $\boxed{-}$ $\boxed{\text{RCL}}$ $\boxed{=}$ $\boxed{\div}$ 2 $\boxed{\div}$ 3 $\boxed{=}$ yields 0.1835

EXAMPLE Solve $5x^2 + 17x + 12 = 0$.

SOLUTION

$$x = \frac{-17 \pm \sqrt{(17)^2 - 4(5)(12)}}{2(5)}$$

17 $\boxed{x^2}$ $\boxed{-}$ 4 $\boxed{\times}$ 5 $\boxed{\times}$ 12 $\boxed{=}$ $\boxed{\text{INV}}$ $\boxed{x^2}$ $\boxed{\text{STO}}$

17 $\boxed{+/-}$ $\boxed{+}$ $\boxed{\text{RCL}}$ $\boxed{=}$ $\boxed{\div}$ 2 $\boxed{\div}$ 5 $\boxed{=}$ yields -1

17 $\boxed{+/-}$ $\boxed{-}$ $\boxed{\text{RCL}}$ $\boxed{=}$ $\boxed{\div}$ 2 $\boxed{\div}$ 5 $\boxed{=}$ yields -2.4

The examples used so far have involved real roots. If the radicand in the quadratic formula is negative, the procedure outlined will generate an error message when we attempt to take the square root. We illustrate a procedure for handling quadratics with complex roots.

if the roots are complex, they will be in the form

$$x = -\frac{b}{2a} + \frac{\sqrt{4ac - b^2}}{2a}j$$

and

$$x = -\frac{b}{2a} - \frac{\sqrt{4ac - b^2}}{2a}j$$

We notice that the real parts of the roots are the same and the imaginary parts differ only in sign. Thus, the roots are conjugates of one another.

SOLUTION OF QUADR. EQUATION USING THE CALCULATOR 11.3

EXAMPLE Solve $3x^2 + 2x + 1 = 0$.

SOLUTION

$$x = \frac{-2 \pm \sqrt{(2)^2 - 4(3)(1)}}{2(3)}$$

First evaluate the radicand.

2 $\boxed{x^2}$ $\boxed{-}$ 4 $\boxed{\times}$ 3 $\boxed{\times}$ 1 $\boxed{=}$ yields -8

The appearance of a negative number tells us that the roots are complex. At this stage we change the sign of the radicand, take its square root and divide by $2a$ to get the imaginary part of the roots.

$\boxed{+/-}$ $\boxed{INV}$ $\boxed{x^2}$ $\boxed{\div}$ 2 $\boxed{\div}$ 3 $\boxed{=}$ yields 0.4714

We now find the real part of each root by evaluating $-\dfrac{b}{2a}$.

2 $\boxed{+/-}$ $\boxed{\div}$ 2 $\boxed{\div}$ 3 $\boxed{=}$ yields -0.3333

as the real part of each root. Finally, we construct the two roots as follows:

$$-0.3333 + 0.4714j$$
$$-0.3333 - 0.4714j$$

EXAMPLE Solve $2x^2 + 3x + 5 = 0$.

SOLUTION

$$x = \frac{-3 \pm \sqrt{(3)^2 - 4(2)(5)}}{2(2)}$$

3 $\boxed{x^2}$ $\boxed{-}$ 4 $\boxed{\times}$ 2 $\boxed{\times}$ 5 $\boxed{=}$ yields -31

At this stage we know we have complex roots. We continue as follows:

$\boxed{+/-}$ $\boxed{INV}$ $\boxed{x^2}$ $\boxed{\div}$ 2 $\boxed{\div}$ 2 $\boxed{=}$ yields 1.3919 (the imaginary part)

3 $\boxed{+/-}$ $\boxed{\div}$ 2 $\boxed{\div}$ 2 $\boxed{=}$ yields -0.75 (the real part)

The complete roots are then

$$-0.75 + 1.3919j$$
$$-0.75 - 1.3919j$$

PERFORMANCE OBJECTIVE 4

You should now be able to use the calculator to solve a quadratic equation.

$$x = \frac{4 \pm \sqrt{(-4)^2 - 4(1)(4)}}{2}$$

4 [+/-] [x²] [-] 4 [x] 1 [x] 4 [=] [INV] [x²] [STO]
4 [+] [RCL] [=] [÷] 2 [=] yields 2
4 [-] [RCL] [=] [÷] 2 [=] yields 2

EXERCISE SET 11.3

Solve the following quadratic equations using the calculator (Exercises 1–10):

1. $4x^2 + 3x - 10 = 0$
2. $2x^2 + 5x - 12 = 0$
3. $x^2 - 4x - 21 = 0$
4. $x^2 - 12x + 32 = 0$
5. $x^2 + 2x + 4 = 0$
6. $x^2 + 3x + 5 = 0$
7. $10x^2 + 13x - 3 = 0$
8. $10x^2 + 9x - 7 = 0$
9. Problem 9 of Exercise set 11.2.
10. Problem 10 of Exercise set 11.2.

SECTION 11.4 SUMMARY

In this section we highlight the new concepts and definitions introduced in the chapter.

Quadratic equation in one variable: an equation of the form

$$ax^2 + bx + c = 0$$

where a, b, and c are any real numbers, x is the variable, and $a \neq 0$.

Quadratic formula: a formula for finding the roots of any quadratic. It has the form

$$x = \frac{-b \pm \sqrt{b^2 - 4ac}}{2a}$$

where a, b, and c are from the equation $ax^2 + bx + c = 0$.

CHAPTER 11 PRACTICE TEST

Solve the equations in Exercises 1–7.

1. (PO-2) $x^2 + 11x + 30 = 0$
2. (PO-2) $2x^2 + 2x - 12 = 0$
3. (PO-2) $6x^2 + 19x + 15 = 0$
4. (PO-2) $x^2 - 40 = -6x$
5. (PO-2) $-5x^2 + 31x - 30 = 0$
6. (PO-2) $4x^2 - 25 = 0$
7. (PO-2) $-8x^2 + 34x = 35$
8. (PO-3) In the equation $5x^2 - 6x + 4 = 0$, identify a, b, and c.
9. (PO-3) In the equation $-3x^2 + 5x = -12$, identify a, b, and c.
10. (PO-3) Solve $2x^2 + 4x - 1 = 0$ for x.
11. (PO-3) Solve $x^2 + 5x - 2 = 0$ for x.
12. (PO-3) Solve $2x^2 + 10x + 3 = 0$ for x.
13. (PO-3) Solve $15x^2 + 2x = 8$ for x.
14. (PO-3) Solve $\frac{1}{8}x^2 - \frac{1}{2}x + \frac{3}{2} = 0$ for x.
15. (PO-2, 3) Solve $-x^2 + 6x - 5 = 0$ for x.
16. (PO-1) Rewrite $\dfrac{x+3}{2x+8} - \dfrac{x-5}{3x-12} = 1$ in the form of a quadratic equation.
17. (PO-2, 3) Solve $\dfrac{3x}{x-1} + \dfrac{2}{x+3} = \dfrac{3x+1}{x^2+2x-3}$ for x.
18. (PO-2, 3) Solve $\dfrac{x-20}{x-7} + \dfrac{4x+4}{x^2-49} = 0$ for x.
19. (PO-2, 3) Solve $\dfrac{x-\frac{5}{4}}{3x} + \dfrac{x+3}{4x} = \dfrac{x-5}{2x}$ for x.
20. (PO-3) Solve $\dfrac{4x+2}{x^2-2x-3} + \dfrac{2x+1}{2x-6} + \dfrac{3x-1}{3x+3} = 0$ for x.
21. (PO-4) Obtain numerical values for the roots of the equations in Problems 11 and 13 by using the calculator.

CHAPTER 12

GRAPHING QUADRATIC EQUATIONS AND SYSTEMS

12.1 The Graph of a Quadratic Equation
12.2 Quadratic Systems
12.3 Applications of Quadratic Systems
12.4 Summary

In Chapter 9 we noted that linear systems often arise as a result of breaking a problem into parts so as to make it more manageable. We may also view the system as a set of interdependent variables. The whole system must be solved to determine the value of any single variable.

As an example, the electrical currents in different parts of a multiloop circuit can be described by equations in which the variables interact. The value of current in one part of the circuit depends upon the characteristics of the whole circuit and hence it is related to currents in other parts of the circuit.

Mechanical systems coupled by springs also give rise to systems of interrelated variables. If the system equations are at most second degree, the system is described as quadratic.

SECTION 12.1 THE GRAPH OF A QUADRATIC EQUATION

In Chapter 11 we studied quadratic equations in one variable. Here we begin by defining the **quadratic equation in two variables** after which we graph this equation in a rectangular coordinate system.

DEFINITION

> **QUADRATIC EQUATION IN TWO VARIABLES**
>
> A **quadratic equation in two variables** has the form
>
> $$y = ax^2 + bx + c$$
>
> where a, b, and c are any real numbers and $a \neq 0$.

Note that only the variable x appears with an exponent of two. This is the only type of quadratic equation in two variables we will consider in this text.

The graph of any quadratic equation in the form $y = ax^2 + bx + c$ is called a **parabola.** The general shape of the parabola and its possible orientations are illustrated in Figure 12.1. Some terminology associated with a parabola is also illustrated in this figure.

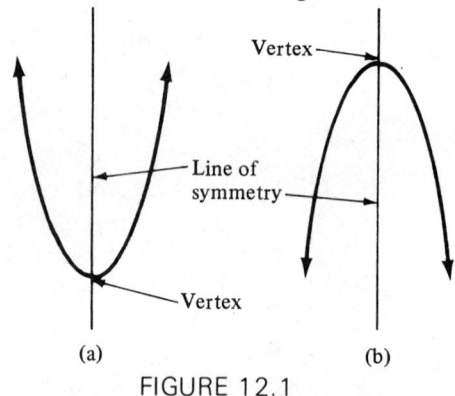

FIGURE 12.1

The extreme point (either maximum or minimum) on the graph of a parabola is called the **vertex** of the parabola. The vertical line that passes through the vertex and separates the parabola into two symmetrical parts is called the **line of symmetry** of the parabola.

When the vertex is the minimum point on the parabola, such as in Figure 12.1(a), we say that the parabola **opens upward.** When the vertex is the maximum point, as in Figure 12.1(b), we say that the parabola **opens downward.** We can determine whether a parabola will open upward or downward by inspecting its equation. If the coefficient of x^2, a, is positive, the graph opens upward. If the coefficient of x^2 is negative, the graph opens downward.

CHARACTERISTICS OF THE GRAPH OF $y = ax^2 + bx + c$

1. The graph is called a parabola.
2. The high or low point on the graph is called the vertex.
3. The graph is symmetric about a vertical line through the vertex.
4. The graph opens upward if $a > 0$.
5. The graph opens downward if $a < 0$.

PERFORMANCE OBJECTIVE 1

You should now be able to describe the characteristics of the graph of a quadratic equation.

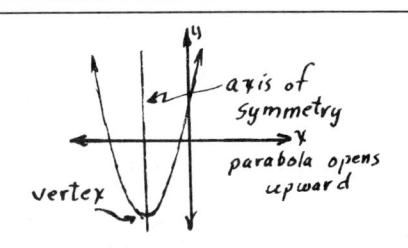

axis of symmetry
parabola opens upward
vertex

When we graph a quadratic equation, we first find the points that will be most useful in sketching the parabola. These points are the x-intercepts (if there are any), the y-intercept, and the vertex.

We find the x- and y-intercepts of a parabola in the same way as we find the x- and y-intercepts of a line when we graph linear equations. To find the y-intercept, we let $x = 0$ in the equation and solve for y. The parabola will always have one y-intercept. To find the x-intercepts, we let $y = 0$ in the equation and solve for x. When solving for x, we will be solving a quadratic equation and will obtain either one repeated real solution, two distinct real solutions, or two complex solutions. If the equation has complex solutions, the parabola will have no x-intercepts. If the solutions are real, the parabola will have either one or two x-intercepts.

The following examples illustrate the procedure for finding the x- and y-intercepts of the graph of a quadratic equation.

EXAMPLE Find the x- and y-intercepts of the equation $y = x^2$.

SOLUTION Letting $x = 0$ yields $y = (0)^2 = 0$. So the y-intercept is 0.

Letting $y = 0$ yields the equation $x^2 = 0$, which implies that $x = 0$ (a repeated real root). So the x-intercept is also 0.

EXAMPLE Find the x- and y-intercepts of the equation $y = x^2 + 2x$.

SOLUTION Letting $x = 0$ yields $y = (0)^2 + 2(0) = 0$. So the y-intercept is 0.

Letting $y = 0$ yields the quadratic equation $x^2 + 2x = 0$. We solve this equation by factoring as follows:

$$x^2 + 2x = 0$$
$$x(x + 2) = 0$$

Set $x = 0$ and $x + 2 = 0$, which yields $x = 0$ and $x = -2$. So the x-intercepts are 0 and -2.

EXAMPLE Find the x- and y-intercepts of the equation

$$y = -x^2 + 5x - 4$$

SOLUTION To find the y-intercept, we let $x = 0$.

$$y = -(0)^2 + 5(0) - 4$$
$$y = -4$$

is the y-intercept.

To find the x-intercepts, we let $y = 0$ and solve.

$$-x^2 + 5x - 4 = 0$$
$$x^2 - 5x + 4 = 0$$
$$(x - 1)(x - 4) = 0$$

Set $x - 1 = 0$ and $x - 4 = 0$, which yields $x = 1$ and $x = 4$. So the x-intercepts are 1 and 4.

EXAMPLE Find the x- and y-intercepts of the equation

$$y = 3x^2 - 10x - 8$$

SOLUTION To find the y-intercept, we let $x = 0$.

$$y = 3(0)^2 - 10(0) - 8$$
$$y = -8$$

is the y-intercept.

To find the x-intercepts, we let $y = 0$ and solve.

$$3x^2 - 10x - 8 = 0$$
$$(3x + 2)(x - 4) = 0$$

THE GRAPH OF A QUADRATIC EQUATION 12.1

Set $3x + 2 = 0$ and $x - 4 = 0$, which yields $3x = -2$ and $x = 4$. That is, $x = -\frac{2}{3}$ and $x = 4$. So the x-intercepts are $-\frac{2}{3}$ and 4.

EXAMPLE Find the x- and y-intercepts of the equation

$$y = x^2 + 4x + 8$$

SOLUTION Letting $x = 0$ yields $y = (0)^2 + 4(0) + 8 = 8$. So the y-intercept is 8.

When we let $y = 0$, we obtain the quadratic equation

$$x^2 + 4x + 8 = 0$$

which we cannot solve by factoring. We solve this equation by using the quadratic formula.

$$x^2 + 4x + 8 = 0$$

$$a = 1, b = 4, c = 8$$

Use $x = \dfrac{-b \pm \sqrt{b^2 - 4ac}}{2a}$.

$$x = \frac{-4 \pm \sqrt{4^2 - 4(1)(8)}}{2(1)}$$

$$= \frac{-4 \pm \sqrt{-16}}{2}$$

$$= \frac{-4 \pm 4j}{2}$$

$$= -2 \pm 2j$$

Since the equation has complex solutions, the parabola has no x-intercepts.

EXAMPLE Mr. Johnson has a new car. The EPA claims that he can estimate his miles per gallon by

$$M(x) = -\frac{1}{245}x^2 + \frac{2}{7}x + 14$$

where $M(x)$ is the miles per gallon and x is the car speed in miles per hour. Find the x- and y-intercepts for this quadratic equation.

SOLUTION To find the y-intercept, we let $x = 0$.

$$M(0) = -\frac{1}{245}(0)^2 + \frac{2}{7}(0) + 14$$

$M(0) = +14$ is the y-intercept.

To find the x-intercepts, we let $y = M(x) = 0$ and solve.

$$-\frac{1}{245}x^2 + \frac{2}{7}x + 14 = 0$$

Clear this equation of fractions to obtain

$$-x^2 + 70x + 3430 = 0 \quad \text{or} \quad x^2 - 70x - 3430 = 0$$

Since $x^2 - 70x - 3430$ cannot be factored, we use the quadratic formula to solve for x. When the coefficients are large numbers, it is generally easier to use the calculator.

First, identify a, b, and c.

$$a = 1, \; b = -70, \; c = -3430$$

Then set up the quadratic formula.

$$x = \frac{-(-70) \pm \sqrt{(-70)^2 - 4(1)(-3430)}}{2(1)}$$

Next evaluate and store the radical.

70 $\boxed{+/-}$ $\boxed{x^2}$ $\boxed{-}$ 4 $\boxed{\times}$ 3430 $\boxed{+/-}$ $\boxed{=}$ $\boxed{\text{INV}}$ $\boxed{x^2}$ $\boxed{\text{STO}}$

This procedure yields 136.45512 as the stored value.

To obtain one x-intercept, use the following procedure:

70 $\boxed{+}$ $\boxed{\text{RCL}}$ $\boxed{=}$ $\boxed{\div}$ 2 $\boxed{=}$ yields 103.2

The second x-intercept is obtained as follows:

70 $\boxed{-}$ $\boxed{\text{RCL}}$ $\boxed{=}$ $\boxed{\div}$ 2 $\boxed{=}$ yields -33.2.

So the x-intercepts are 103.2 and -33.2.

EXAMPLE According to a national survey, the demand for a certain item can be approximated by the equation $y = -5x^2 + 150$, where x is the price per item. Find the x- and y-intercepts for this quadratic equation.

SOLUTION To find the y-intercept, we let $x = 0$.

$$y = -5(0)^2 + 150$$

$y = 150$ is the y-intercept.

To find the x-intercepts, we let $y = 0$ and solve.

$$0 = -5x^2 + 150$$
$$5x^2 = 150$$
$$x^2 = 30$$
$$x = \pm\sqrt{30} = \pm 5.5$$

So the x-intercepts are $+5.5$ and -5.5.

PERFORMANCE OBJECTIVE 2	$y = x^2 - 3x - 4$
You should now be able to find the x- and the y-intercepts of a parabola.	Let $x = 0$ to find y-intercept $\Rightarrow y = -4$ Let $y = 0$ to find x-intercepts $0 = x^2 - 3x - 4 \Rightarrow x = 4, x = -1$

The technique for finding the vertex of a parabola involves rewriting the equation of the parabola in the form

$$y - k = a(x - h)^2$$

Then the vertex of the parabola is the ordered pair (h, k). To rewrite the equation in the desired form, we use the process known as completing the square.

Assuming that the original equation of the parabola is written in the form

$$y = ax^2 + bx + c$$

we perform the following steps to rewrite the equation and find the vertex of the parabola.

Step 1. Subtract c (the constant term) from both sides of the equation.

$$y - c = ax^2 + bx$$

Step 2. Factor a (the coefficient of x^2) from each term on the right side.

$$y - c = a\left(x^2 + \frac{b}{a}x\right)$$

Step 3. Form $\frac{1}{2}$ the coefficient of x, square it, and add this term inside the parentheses on the right side of the equation. To maintain the equality, add the product of this term and a to the left side.

$$y - c + \frac{b^2}{4a} = a\left(x^2 + \frac{b}{a}x + \frac{b^2}{4a^2}\right)$$

Note that the process of completing the square occurs in this step where the correct term is added to the right side to make it a perfect square.

Step 4. Since the quantity in parentheses is a perfect square, we write it as a squared factor.

$$y - c + \frac{b^2}{4a} = a\left(x + \frac{b}{2a}\right)^2$$

If we write this last equation as

$$y - \left(c - \frac{b^2}{4a}\right) = a\left[x - \left(-\frac{b}{2a}\right)\right]^2$$

we see that it is in the form

$$y - k = a(x - h)^2$$

with

$$h = -\frac{b}{2a} \quad \text{and} \quad k = c - \frac{b^2}{4a}$$

However, we can also determine h and k as in Step 5.

Step 5. Set each side of the equation equal to zero and solve for x and y to find the values of h and k, respectively.

$$a\left(x + \frac{b}{2a}\right)^2 = 0 \qquad y - c + \frac{b^2}{4a} = 0$$

$$x + \frac{b}{2a} = 0 \qquad\qquad y = c - \frac{b^2}{4a}$$

$$x = -\frac{b}{2a}$$

So $h = -\dfrac{b}{2a}$ and $k = c - \dfrac{b^2}{4a}$.

Step 6. Write the vertex (h, k).

The vertex is $\left(-\dfrac{b}{2a},\, c - \dfrac{b^2}{4a}\right)$.

The following examples illustrate the procedure for finding the vertex of a parabola. The quadratic equations in the examples are the same ones for which we have previously found the x- and y-intercepts.

THE GRAPH OF A QUADRATIC EQUATION 12.1

EXAMPLE Find the vertex of $y = x^2$.

SOLUTION This equation already has a perfect square on the right side. So the first four steps in the procedure for finding the vertex are not necessary.

Step 5. Set $x^2 = 0$ and $y = 0$, which yields $x = 0$ and $y = 0$. So h and k are both 0.

Step 6. The vertex is (0, 0).

EXAMPLE Find the vertex of $y = x^2 + 2x$.

SOLUTION

Step 1. The constant term is 0.
$$y - 0 = x^2 + 2x$$

Step 2. The coefficient of x^2 is 1; no factoring is required.
$$y - 0 = x^2 + 2x$$

Step 3. The coefficient of the x term is 2.
$$\frac{1}{2}(2) = 1, \quad 1^2 = 1, \quad 1(1) = 1$$
$$y - 0 + 1 = x^2 + 2x + 1$$

Step 4. Since the right side is now a perfect square, the equation can be written as
$$y + 1 = (x + 1)^2$$

Step 5. Set $(x + 1)^2 = 0$ and $y + 1 = 0$, which yields
$$x = -1 \quad \text{and} \quad y = -1$$
So $h = -1$ and $k = -1$.

Step 6. The vertex is $(-1, -1)$.

EXAMPLE Find the vertex of $y = -x^2 + 5x - 4$.

SOLUTION

Step 1. $y + 4 = -x^2 + 5x$

Step 2. $y + 4 = -1(x^2 - 5x)$

Step 3. $\frac{1}{2}(-5) = -\frac{5}{2}$, $(-\frac{5}{2})^2 = \frac{25}{4}$, $-1(\frac{25}{4}) = -\frac{25}{4}$

$$y + 4 - \frac{25}{4} = -1(x^2 - 5x + \frac{25}{4})$$

Step 4. $y - \frac{9}{4} = -1(x - \frac{5}{2})^2$

Step 5. Set $-1(x - \frac{5}{2})^2 = 0$ and $y - \frac{9}{4} = 0$, which yields

$$x = \frac{5}{2} \quad \text{and} \quad y = \frac{9}{4}$$

So $h = \frac{5}{2}$ and $k = \frac{9}{4}$.

Step 6. The vertex is $(\frac{5}{2}, \frac{9}{4})$. Note that we can omit Step 5 and write the vertex by inspecting the result in Step 4.

EXAMPLE Find the vertex of $y = 3x^2 - 10x - 8$.

SOLUTION

Step 1. $y + 8 = 3x^2 - 10x$

Step 2. $y + 8 = 3(x^2 - \frac{10}{3}x)$

Step 3. $\frac{1}{2}(-\frac{10}{3}) = -\frac{5}{3}$, $(-\frac{5}{3})^2 = \frac{25}{9}$, $3(\frac{25}{9}) = \frac{25}{3}$

$$y + 8 + \frac{25}{3} = 3(x^2 - \frac{10}{3}x + \frac{25}{9})$$

Step 4. $y + \frac{49}{3} = 3(x - \frac{5}{3})^2$

Step 5. Set $3(x - \frac{5}{3})^2 = 0$ and $y + \frac{49}{3} = 0$, which yields

$$x = \frac{5}{3} \quad \text{and} \quad y = -\frac{49}{3}$$

So $h = \frac{5}{3}$ and $k = -\frac{49}{3}$.

Step 6. The vertex is $(\frac{5}{3}, -\frac{49}{3})$.

EXAMPLE Find the vertex of $y = x^2 + 4x + 8$.

SOLUTION

Step 1. $y - 8 = x^2 + 4x$

Step 2. $y - 8 = x^2 + 4x$ (no factoring is required)

Step 3. $\frac{1}{2}(4) = 2$, $2^2 = 4$, $1(4) = 4$

$$y - 8 + 4 = x^2 + 4x + 4$$

Step 4. $y - 4 = (x + 2)^2$

Step 5. By inspection

Step 6. The vertex is $(-2, 4)$.

PERFORMANCE OBJECTIVE 3	
You should now be able to find the vertex of a parabola by completing the square.	$y = x^2 + 2x - 3$ $y + 3 = x^2 + 2x$ $y + 3 + 1 = x^2 + 2x + 1$ $y + 4 = (x+1)^2 \Rightarrow$ vertex is $(-1, -4)$

CRITICAL POINTS IN THE GRAPH $y = ax^2 + bx + c$

1. The y-intercept is $(0, c)$.
2. The x-intercepts are solutions of the equation $ax^2 + bx + c = 0$ (if they exist).
3. The vertex is $\left(-\dfrac{b}{2a}, c - \dfrac{b^2}{4a}\right)$.
4. The line of symmetry has the equation

$$x = -\frac{b}{2a}$$

A common application of the quadratic equation is to view the vertex as the point at which the dependent variable is maximized or minimized. For instance, if production is related to hours worked through a quadratic equation, we could determine the number of hours to be worked in order to maximize production. In the same way, we can determine the maximum or minimum of any dependent variable that is related to the independent variable through a quadratic equation.

The following examples illustrate the use of the vertex in this way.

EXAMPLE Find the vertex for the EPA equation that estimates Mr. Johnson's miles per gallon, that is,

$$M(x) = -\frac{1}{245}x^2 + \frac{2}{7}x + 14$$

SOLUTION

Step 1. Let $y = M(x)$. Then

$$y = -\frac{1}{245}x^2 + \frac{2}{7}x + 14$$

$$y - 14 = -\frac{1}{245}x^2 + \frac{2}{7}x$$

Step 2. $y - 14 = -\frac{1}{245}(x^2 - 70x)$

Step 3. $\frac{1}{2}(-70) = -35$, $(-35)^2 = 1225$, $-\frac{1}{245}(1225) = -5$

$$y - 14 - 5 = -\frac{1}{245}(x^2 - 70x + 1225)$$

Step 4. $y - 19 = -\frac{1}{245}(x - 35)^2$

Step 5. By inspection

Step 6. The vertex is (35, 19).

Mr. Johnson will attain his maximum miles per gallon (19) when he drives at 35 mph. We know this is a maximum because $a < 0$ in the quadratic equation (the parabola opens downward).

EXAMPLE Find the maximum or minimum for the demand equation

$$y = -5x^2 + 150$$

SOLUTION

Step 1. $y - 150 = -5x^2$

Step 2. $y - 150 = -5(x^2)$

The equation has a perfect square on the right side. So Steps 3 and 4 are not necessary.

Step 5. By inspection

Step 6. The vertex is (0, 150).

The maximum demand for the product is 150. Again, this is a maximum rather than a minimum since $a < 0$.

PERFORMANCE OBJECTIVE 4	
You should now be able to find the maximum or minimum value of the dependent variable in a quadratic equation.	$p = -x^2 + 10x - 20$ (p = price, x = demand) Since the vertex is at (5,5) the maximum price is 5. It is a maximum because $a < 0$.

THE GRAPH OF A QUADRATIC EQUATION 12.1

As we have already indicated, we begin the graphing of a quadratic equation by finding certain vital points (intercepts and vertex). Finding the x- and y-intercepts and the vertex of a parabola can yield from one to four distinct points. When the intercepts and vertex yield three or four points, we can get a fairly good sketch of the parabola without finding additional points. However, the more points we have the more accurate the graph is likely to be. When the intercepts and vertex yield only one or two distinct points, it is necessary for us to find some additional points to graph.

We can use the symmetry of a parabola in finding additional points on the graph. This will be illustrated in some of the following examples. The parabolas that we will graph are the same ones for which we have already found the intercepts and vertex.

EXAMPLE Graph $y = x^2$.

SOLUTION The x- and y-intercepts are both 0 and the vertex is (0, 0). In this case, the intercepts and vertex yield only the point (0, 0) on the graph.

We need to find some additional points that satisfy the equation. This is accomplished by substituting numbers for x and evaluating the quadratic in x to determine corresponding values for y. The ordered pairs (x, y) determined in this manner become points on the graph. A smooth curve is then drawn through the plotted points to get the desired graph.

Sometimes a table is used to help in identifying points to be graphed. Such a table is shown below and the resulting graph of $y = x^2$ is shown in Figure 12.2.

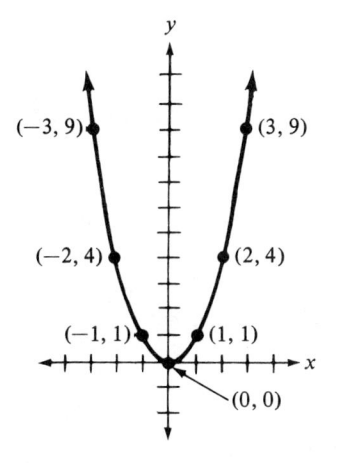

FIGURE 12.2

x	-3	-2	-1	0	1	2	3
y	9	4	1	0	1	4	9

Note that for each x-value to the left of the vertex there is another x-value the same distance to the right of the vertex with the same y-value. This symmetry is very useful in sketching the graph.

EXAMPLE Graph $y = x^2 + 2x$.

SOLUTION The x-intercepts are -2 and 0, the y-intercept is 0, and the vertex is $(-1, -1)$. So the intercepts and vertex yield the three points $(-2, 0)$, $(0, 0)$, and $(-1, -1)$. Some additional points on the graph are $(-3, 3)$ and $(1, 3)$.

The graph of $y = x^2 + 2x$ and its line of symmetry are shown in Figure 12.3.

12 GRAPHING QUADRATIC EQUATIONS AND SYSTEMS

EXAMPLE Graph $y = -x^2 + 5x - 4$.

SOLUTION The x-intercepts are 1 and 4, the y-intercept is -4, and the vertex is $(\frac{5}{2}, \frac{9}{4})$. So the intercepts and vertex yield the four points $(1, 0)$, $(4, 0)$, $(0, -4)$, and $(\frac{5}{2}, \frac{9}{4})$. The graph of $y = -x^2 + 5x - 4$ with its line of symmetry is shown in Figure 12.4.

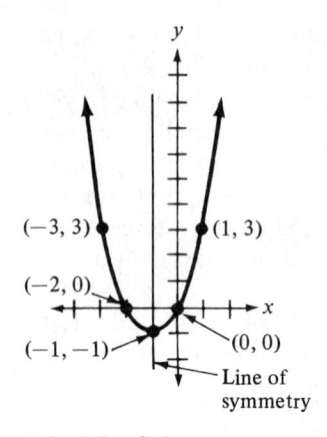

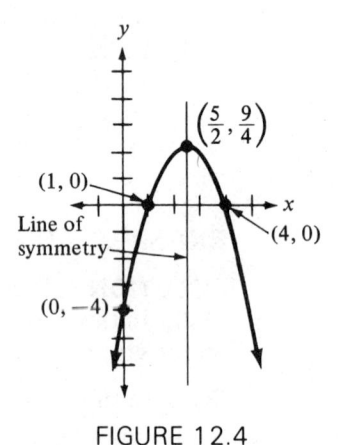

FIGURE 12.3

FIGURE 12.4

EXAMPLE Graph $y = 3x^2 - 10x - 8$.

SOLUTION The x-intercepts are $-\frac{2}{3}$ and 4, the y-intercept is -8, and the vertex is $(\frac{5}{3}, -\frac{49}{3})$. So the intercepts and vertex yield the four points $(-\frac{2}{3}, 0)$, $(4, 0)$, $(0, -8)$ and $(\frac{5}{3}, -\frac{49}{3})$.

The graph of $y = 3x^2 - 10x - 8$ with its line of symmetry is shown in Figure 12.5.

EXAMPLE Graph $y = x^2 + 4x + 8$.

SOLUTION The y-intercept is 8 and the vertex is $(-2, 4)$. There are no x-intercepts. So we know only two points on the graph: $(0, 8)$ and $(-2, 4)$. We must find some additional points that satisfy the equation. A table of points is generated.

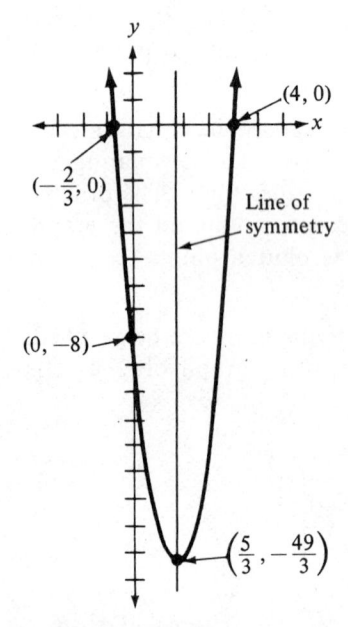

FIGURE 12.5

x	-5	-4	-3	-1	1
y	13	8	5	5	13

The graph is shown in Figure 12.6.

THE GRAPH OF A QUADRATIC EQUATION 12.1 435

EXAMPLE Graph the EPA equation $M(x) = -\frac{1}{245}x^2 + \frac{2}{7}x + 14$.

SOLUTION The x-intercepts are -33.2 and 103.2, the y-intercept is $+14$ and the vertex is $(35, 19)$. So the intercepts and vertex yield the four points $(-33.2, 0)$, $(103.2, 0)$, $(0, 14)$, and $(35, 19)$.

Since the negative x-intercept is meaningless (zero miles per gallon), a few more points around the vertex would help when we draw the graph. A table of points is generated.

x	15	20	25	30	40	45
y	17.4	18.1	18.6	18.9	18.9	18.6

The graph is shown in Figure 12.7.

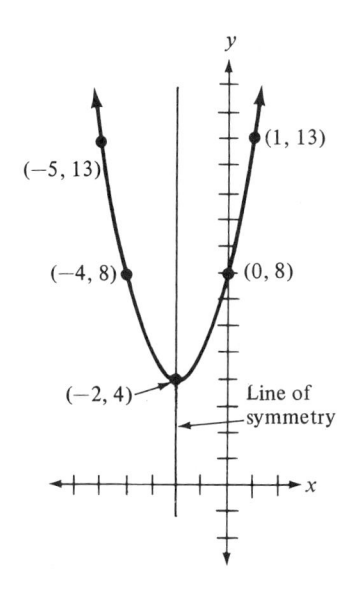

FIGURE 12.6

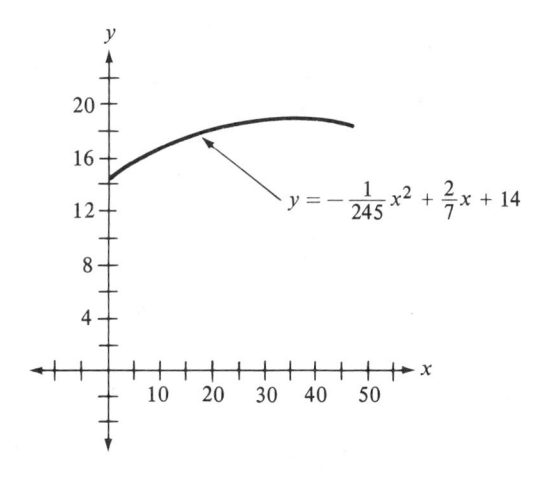

FIGURE 12.7

EXAMPLE Graph the demand equation $y = -5x^2 + 150$.

SOLUTION The x-intercepts are $+5.5$ and -5.5, the y-intercept is 150, and the vertex is $(0, 150)$. So the intercepts and the vertex yield the three points $(5.5, 0)$, $(-5.5, 0)$, and $(0, 150)$.

Since one of the x-intercepts is a negative number, it is meaningless as a figure to represent price. Therefore, a few more points would help when we draw the graph. A table of points is generated.

x	1	2	3	4	5
y	145	130	105	70	25

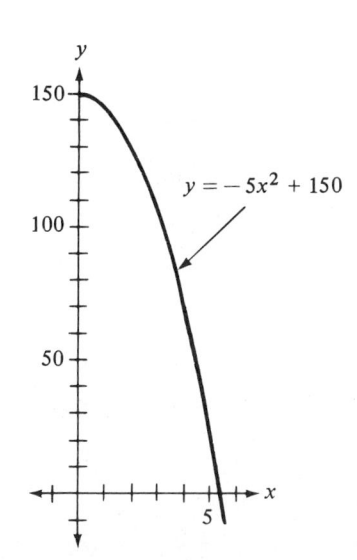

FIGURE 12.8

The graph is shown in Figure 12.8.

PERFORMANCE OBJECTIVE 5

You should now be able to graph a quadratic equation using the intercepts, the vertex, and symmetry.

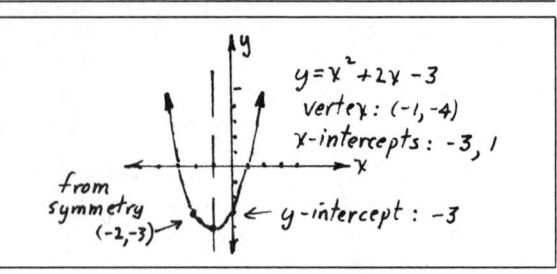

$y = x^2 + 2x - 3$
vertex: $(-1, -4)$
x-intercepts: $-3, 1$
y-intercept: -3
from symmetry $(-2, -3)$

We now take a look at the quadratic equation to determine whether it represents a mathematical function. Consider the general equation

$$y = ax^2 + bx + c$$

To generate ordered pairs using this equation, we substitute a value of x in the right side of the equation and calculate a value for y. As an example, let

$$y = 5x^2 - 3x + 1$$

We substitute $x = 2$ to obtain

$$y = 5(2)^2 - 3(2) + 1$$
$$= 20 - 6 + 1$$
$$= 15$$

The ordered pair is (2, 15), a point on the graph of this equation. Each value of x will produce one and only one value of y. This satisfies the requirement that a function must return a unique result as its output.

QUADRATIC FUNCTION

A quadratic of the form $y = ax^2 + bx + c$ defines a mathematical function.

Once more we emphasize the role of the function as a processor. Consider again the equation

$$y = 5x^2 - 3x + 1$$

If we enter 2 into the processor, the number 15 comes out. The algebraic form of the processor tells us what happens to the 2 as it goes through the processor.

Step 1. Multiply 2 by itself.

Step 2. Multiply 5 by the result of Step 1.

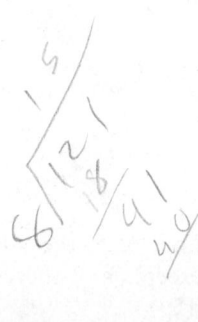

Step 3. Multiply 3 by 2.

Step 4. Subtract the result of Step 3 from the result of Step 2.

Step 5. Add 1 to the result of Step 4.

Each number entered for x gets processed the same way. We feel it is important to think of the function as a mechanism that converts input values into unique output values. As noted earlier, this concept is particularly valuable as we deal with functions on the calculator.

EXERCISE SET 12.1

Sketch the graph of the quadratics in Exercises 1–10. Identify intercepts and vertices.

1. $y = x^2 - x - 6$
2. $y = x^2 - 2x - 15$
3. $y = x^2 + 4x - 12$
4. $y = x^2 + 2x - 3$
5. $y = 2x^2 + 11x + 9$
6. $y = 3x^2 + 8x + 4$
7. $y = -x^2 + x + 12$
8. $y = -x^2 + 3x + 10$
9. $y = 5x^2 + 12x + 4$
10. $y = 5x^2 + 24x + 16$

11. A variable voltage is given by the formula $e = t^2 - 8t + 12$ where t is time in seconds. Find the minimum value for e in the interval $0 < t < 8$ sec.

12. The motion of an object is defined by $y = 20 + 1.5x - 0.16x^2$ where x and y are horizontal and vertical coordinates. Calculate the maximum vertical coordinate of the object.

SECTION 12.2 QUADRATIC SYSTEMS

For a system of equations to be called a **quadratic system,** at least one of the equations must be quadratic. The simplest quadratic system is one that contains a linear equation and a quadratic equation. A quadratic system of this kind can be solved by the method of substitution. In this method the linear equation is solved for one variable and the resulting expression is substituted into the quadratic equation. We will work four examples to illustrate the solution of quadratic systems containing a linear equation.

EXAMPLE Solve the system of equations

$$y = x + 6$$
$$y = x^2 + 4x - 4$$

SOLUTION Since the linear equation is solved for y, we can substitute into the quadratic equation for y and then solve the resulting equation.

Since $y = x + 6$, substitution gives

$$x + 6 = x^2 + 4x - 4$$
$$0 = x^2 + 4x - x - 4 - 6$$
$$0 = x^2 + 3x - 10$$

Solve the quadratic equation by factoring.

$$x^2 + 3x - 10 = 0$$
$$(x + 5)(x - 2) = 0$$

Set $x + 5 = 0$ and $x - 2 = 0$, which yields $x = -5$ and $x = 2$.

To complete the solution, these two values of x are substituted into either of the equations in the system to obtain y values of the solution.

Substitute into $y = x + 6$. For $x = -5$,

$$y = -5 + 6 = 1$$

For $x = 2$,

$$y = 2 + 6 = 8$$

Therefore, the solution set of this system is $(-5, 1), (2, 8)$.

If we graph the equations of this system, we can see the relationship between the graphs and the solution set. Note that the solution set corresponds to the intersection points of the graphs. See Figure 12.9.

EXAMPLE Solve the system of equations

$$x = 1$$
$$y = x^2 - 2x - 8$$

SOLUTION Since the linear equation is solved for x, we can substitute into the quadratic equation.

$$y = (1)^2 - 2(1) - 8$$
$$y = 1 - 2 - 8$$
$$y = -9$$

QUADRATIC SYSTEMS 12.2

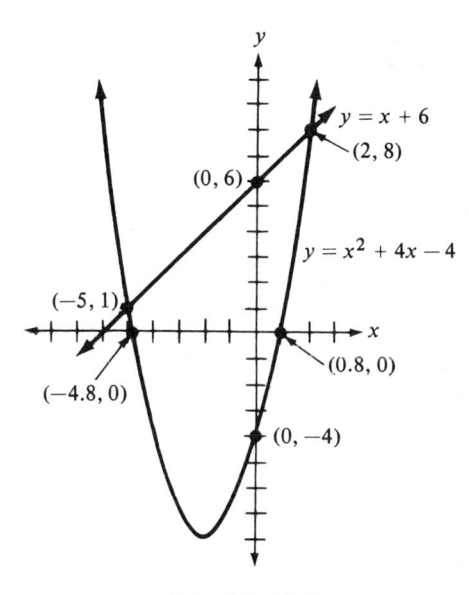

FIGURE 12.9

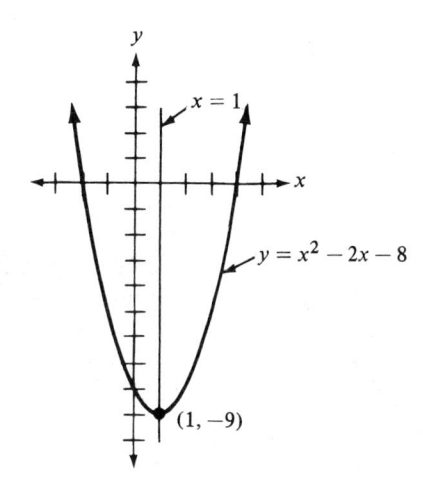

FIGURE 12.10

Therefore, since x can only equal 1 and the corresponding $y = -9$, we have only one solution to this system, that is, $(1, -9)$.

The graphs of the equations of this system are shown in Figure 12.10.

EXAMPLE Solve the system of equations

$$y - x = 0$$
$$y = x^2 + 4$$

SOLUTION Solve the linear equation for y.

$$y - x = 0$$
$$y = x$$

Substitute into the quadratic equation for y.

$$y = x^2 + 4$$
$$x = x^2 + 4$$
$$0 = x^2 - x + 4$$

Solve this quadratic equation for x.

$$x^2 - x + 4 = 0$$

$a = 1, b = -1, c = 4$

$$x = \frac{-(-1) \pm \sqrt{(-1)^2 - 4(1)(4)}}{2(1)}$$

$$x = \frac{1 \pm \sqrt{1 - 16}}{2}$$

$$x = \frac{1 \pm \sqrt{-15}}{2} = \frac{1 \pm j\sqrt{15}}{2}$$

Therefore, the solutions of this equation are not in the set of real numbers. We can say that there is no solution to the system. The interpretation of this result graphically is that the parts of the system do not intersect. Note the graph of the system in Figure 12.11.

EXAMPLE Solve the system of equations

$$y = 35x$$
$$y = -5x^2 + 150$$

Consider only solutions with positive x values.

SOLUTION Substitute the linear expression for y into the quadratic equation and solve.

$$35x = -5x^2 + 150$$
$$5x^2 + 35x - 150 = 0$$
$$x^2 + 7x - 30 = 0$$
$$(x - 3)(x + 10) = 0$$

Therefore, $x = -10$ and $x = 3$ are solutions.

$x = -10$ is ruled out by the conditions imposed in the problem statement. Consider the positive root. For $x = 3$, $y = 105$.

The only solution to the quadratic system with positive x is (3, 105). This solution is graphed in Figure 12.12.

From the four examples, you will note that a quadratic system with one linear equation and one quadratic equation can have no solutions, one solution or two solutions.

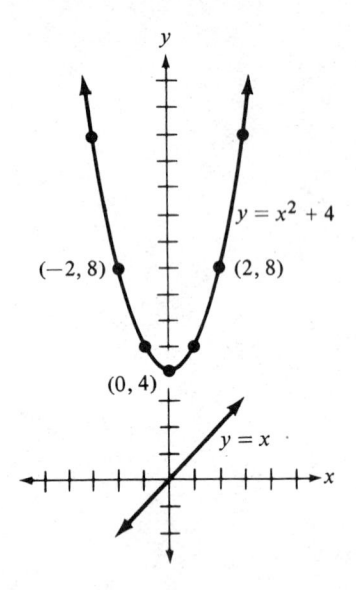

FIGURE 12.11

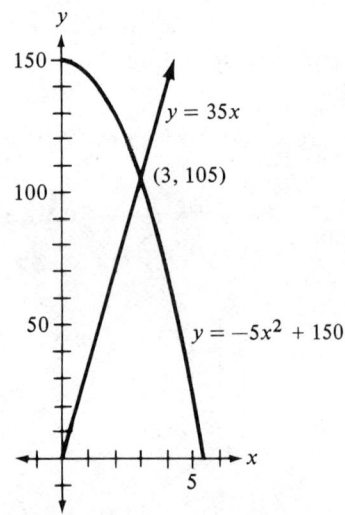

FIGURE 12.12

PERFORMANCE OBJECTIVE 6

You should now be able to solve a quadratic system in which one equation is linear.

System: $x + y = 2$, $y = x^2 + 3x + 5$
$y = -x + 2$
$-x + 2 = x^2 + 3x + 5 \Rightarrow x = -3, -1$
Solutions: $(-3, 5)$, $(-1, 3)$

QUADRATIC SYSTEMS 12.2

We will now consider quadratic systems in which both equations are quadratic. Three examples that illustrate the possible solutions will be worked.

EXAMPLE Solve the system of equations

$$y = x^2 - 7x + 10$$
$$y = -3x^2 + 12$$

SOLUTION Since both equations are solved for y, we can substitute an expression for y in one equation in place of y in the other equation. Therefore,

$$-3x^2 + 12 = x^2 - 7x + 10$$
$$-4x^2 + 7x + 2 = 0$$
$$4x^2 - 7x - 2 = 0$$

We can now solve this equation for x.

$$4x^2 - 7x - 2 = 0$$
$$4x^2 - 8x + x - 2 = 0$$
$$4x(x - 2) + (x - 2) = 0$$
$$(4x + 1)(x - 2) = 0$$

Set $4x + 1 = 0$ and $x - 2 = 0$, which yields $4x = -1$ and $x = 2$. Then

$$x = 2 \quad \text{and} \quad x = -\frac{1}{4}$$

These values of x are substituted into one of the original equations to obtain the y-values of our solution set. For $x = -\frac{1}{4}$,

$$y = -3\left(-\frac{1}{4}\right)^2 + 12 = -3\left(\frac{1}{16}\right) + 12 = -\frac{3}{16} + 12$$
$$= \frac{189}{16}$$

For $x = 2$,

$$y = -3(2)^2 + 12 = -3(4) + 12 = -12 + 12 = 0$$

Therefore, $(-\frac{1}{4}, \frac{189}{16})$ and $(2, 0)$ are the solutions of this system. The system is shown graphically in Figure 12.13.

12 GRAPHING QUADRATIC EQUATIONS AND SYSTEMS

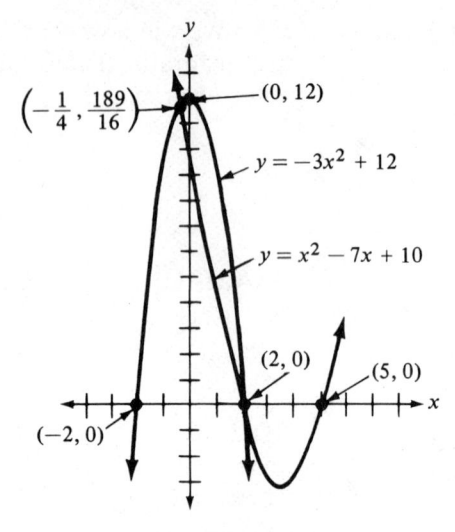

FIGURE 12.13

EXAMPLE Solve the system of equations

$$y = x^2 - x - 6$$
$$y = x^2 - 2$$

SOLUTION We will again substitute from one equation into the other. Therefore,

$$x^2 - 2 = x^2 - x - 6$$
$$x + 4 = 0$$
$$x = -4$$

Substitute to obtain y.

$$y = (-4)^2 - 2 = 16 - 2 = 14$$

Therefore, $(-4, 14)$ is the solution set of this system of equations. The system is graphed in Figure 12.14.

EXAMPLE Solve the system of equations

$$y = x^2 - 9x + 14$$
$$y = -x^2 - 5x - 4$$

SOLUTION We will again substitute from one equation into the other. Therefore,

$$-x^2 - 5x - 4 = x^2 - 9x + 14$$
$$-2x^2 + 4x - 18 = 0$$
$$x^2 - 2x + 9 = 0$$
$$x = \frac{-(-2) \pm \sqrt{(-2)^2 - 4(1)(9)}}{2(1)}$$
$$x = \frac{2 \pm \sqrt{4 - 36}}{2} = \frac{2 \pm \sqrt{-32}}{2}$$
$$= \frac{2 \pm 4\sqrt{2}j}{2} = 1 \pm 2\sqrt{2}j$$

Therefore, this system has no solutions. The graph of the system is shown in Figure 12.15.

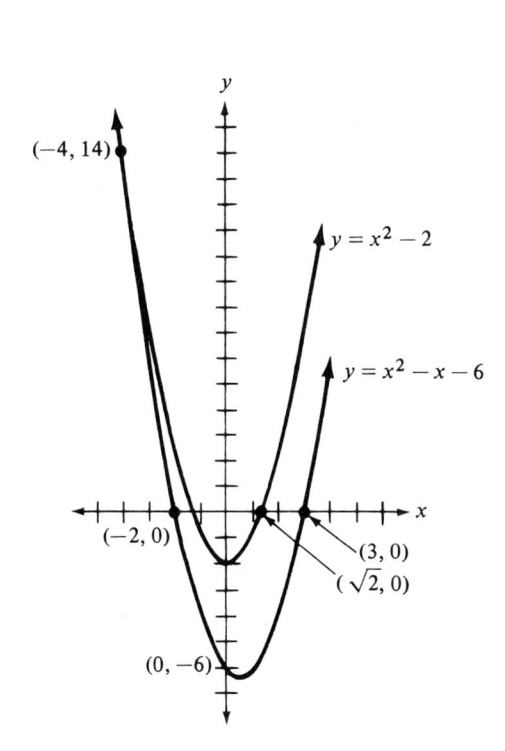

FIGURE 12.14

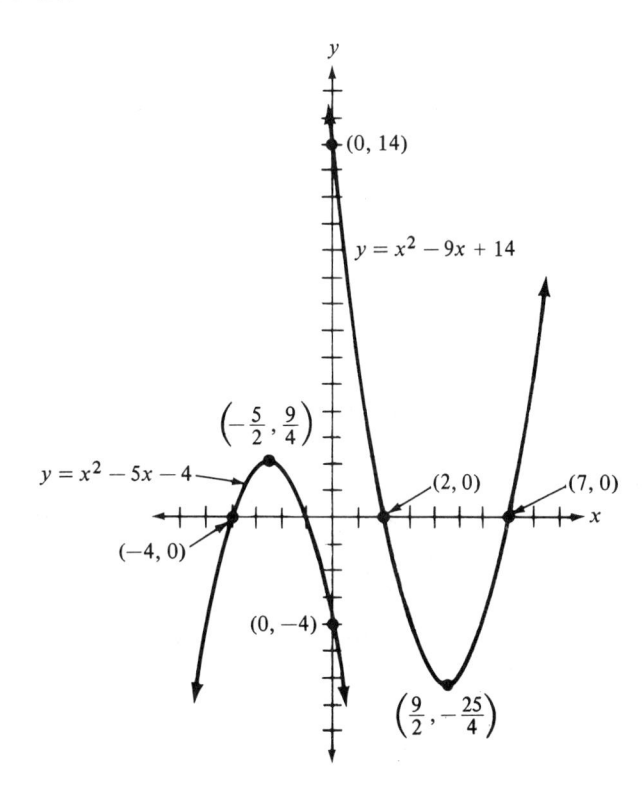

FIGURE 12.15

From the three examples, you should observe that a system in which both equations are quadratic can have no solutions, one solution or two solutions.

12 GRAPHING QUADRATIC EQUATIONS AND SYSTEMS

Handle this objective!

PERFORMANCE OBJECTIVE 7

You should now be able to solve a quadratic system in which both equations are quadratic.

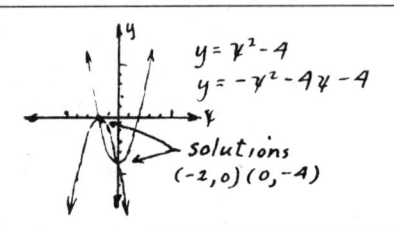

$y = x^2 - 4$
$y = -x^2 - 4x - 4$
solutions
$(-2, 0) (0, -4)$

EXERCISE SET 12.2

odds

Solve the following quadratic systems:

1. $x^2 - 2y = 2$
   $x - 3y = -1$

2. $x^2 - 2y = 2$
   $x - 5y = -3$

3. $x^2 - y = 1$
   $x - 2y = -1$

4. $x^2 + y = 1$
   $x - 3y = -1$

Make y = to

5. $4x^2 + y - 3 = 0$
   $8x + y - 7 = 0$

6. $3x^2 + y - 8 = 0$
   $18x + y - 35 = 0$

7. $x^2 - y - 4 = 0$
   $x - y - 3 = 0$

8. $x^2 - y - 8 = 0$
   $2x^2 - y - 6 = 0$

9. $x^2 + y = -4$
   $4x^2 + y = 17$

10. $x^2 + y = -18$
    $6x^2 + y = 37$

11. $y = x^2 - 5x + 3$
    $y = -x^2 + 5x - 5$

12. $y = x^2 - 8x + 9$
    $y = -x^2 + 7x - 9$

13. $y = x^2 + 8x + 15$
    $y - x^2 + 8x = 15$

14. $y = x^2 + 5x + 10$
    $y - x^2 + 5x = 10$

15. $y = x^2 + 7x + 10$
    $y = -x^2 - 9$

16. $y = x^2 + 5x + 4$
    $y = -x^2 - 3$

SECTION 12.3 APPLICATIONS OF QUADRATIC SYSTEMS

This section consists entirely of examples of applications of quadratic equations and systems. We will look at geometric problems, motion problems, and problems dealing with supply and demand.

In solving geometric problems, the independent variable most frequently represents the dimension of some physical object. The dependent variable may have the same dimension or a combination of the dimensions of the independent variable. For example, if we are evaluating the area of a circle, we might write

$$y = \pi x^2$$

In this case, x is the independent variable and represents the circle radius. The dependent variable y is the circle area and will have the dimension of length squared.

EXAMPLE The shape of the suspension cable on a bridge of uniform construction is given by

$$y = \frac{wx^2}{2T}$$

where w is the weight per unit length of the bridge. T is the horizontal component of tension in the cable. The weight of the cable is neglected in this expression. See Figure 12.16. Lightweight wires are to be attached between the road surface and the top of the opposite support as shown. Supporting collars are to be mounted on the cables at A and B. Determine the horizontal distance from the center of the bridge to the point of attachment of the collar. Assume $w = 4000$ lb/ft, $T = 2.8 \times 10^5$ lb.

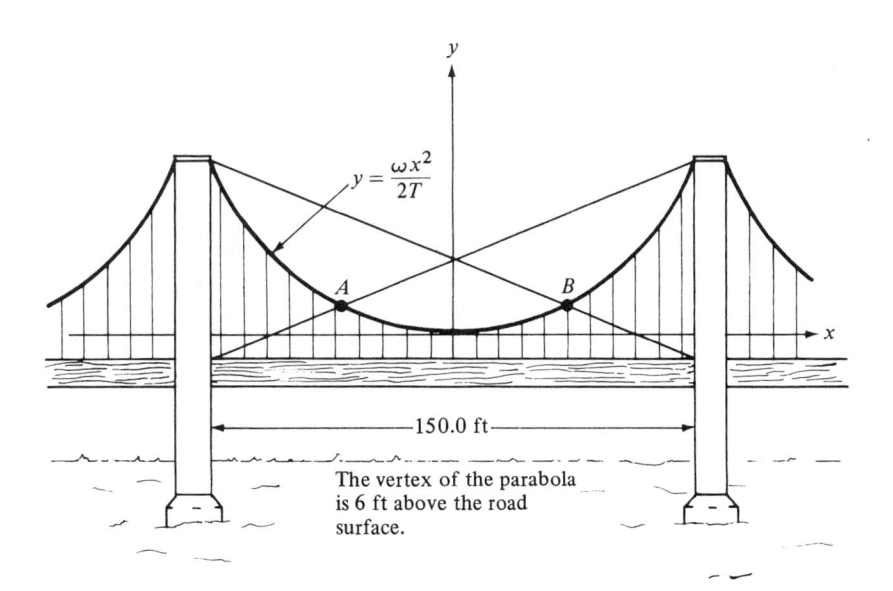

FIGURE 12.16

SOLUTION We first construct a graph of the equation. The coefficient of x^2 is obtained by substitution of the values of w and T.

$$\frac{w}{2T} = \frac{4000}{2(2.8 \times 10^5)} = 0.007143$$

The following table gives ordered pairs that satisfy $y = 0.007143x^2$:

x	10	20	30	40	50	60	70	80
y	0.714	2.86	6.43	11.43	17.86	25.71	35.00	45.72

The y coordinate of the top of the right pier is determined by substituting 75 into

$$y = 0.007143x^2$$

We find $y = 0.007143(75)^2 = 40.18$. The straight line between the top of the right pier and the opposite road surface passes through the points $(-75, -6)$ and $(75, 40.18)$.

The two curves are plotted in Figure 12.17.

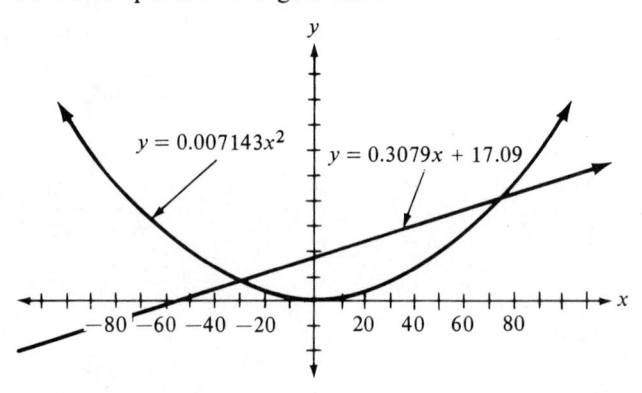

FIGURE 12.17

We now solve the system algebraically. The equation of the straight line may be determined from the two given points and from the point-slope form of the equation of the straight line.

$$\text{Slope} = \frac{\text{rise}}{\text{run}} = \frac{40.18 + 6}{150} = 0.3079$$

The equation is then

$$y - (-6) = 0.3079[x - (-75)]$$
$$y + 6 = 0.3079x + 23.09$$

APPLICATIONS OF QUADRATIC SYSTEMS 12.3

We now solve the system.

$$y = 0.007143x^2$$
$$y = 0.3079x + 17.09$$

We find

$$0.007143x^2 = 0.3079x + 17.09$$

or

$$0.007143x^2 - 0.3079x - 17.09 = 0$$

$$x = \frac{0.3079 \pm \sqrt{(0.3079)^2 - 4(0.007143)(-17.09)}}{2(0.007143)}$$

$$x = \frac{0.3079 \pm \sqrt{0.5831}}{2(0.007143)} = \frac{0.3079 \pm 0.7636}{2(0.007143)}$$

Choose the negative sign; then $x = -31.9 = -32$ ft to two significant digits.

The collars should be located 32 ft on either side of the center of the bridge.

EXAMPLE A rectangular plate is to be designed so that its dimensions are equal to the radius and diameter of a circular hole in another square plate. See Figure 12.18.

The larger square plate has a side length of 4.00 in. It is cut from the same material as the rectangular plate. If the square plate (with the hole) and the rectangular plate are to weigh the same, what is the radius of the circular hole?

Square plate   Rectangular plate

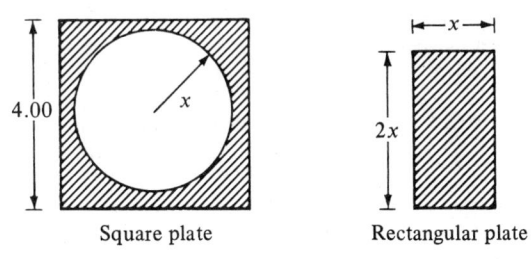

FIGURE 12.18

SOLUTION If the plates are to weigh the same, then their surface areas must be equal. The area of the square plate is

$$A_s = (4)^2 - \pi x^2$$

12 GRAPHING QUADRATIC EQUATIONS AND SYSTEMS

The area of the rectangular plate is $A_r = x(2x) = 2x^2$. We have a quadratic system in which both equations are quadratic. The quadratics are graphed in Figure 12.19. The following data table was used to produce the graph:

x	-3	-2	-1	0	1	2	3
A_s	-12.27	3.43	12.86	16	12.86	3.43	-12.27
A_r	18	8	2	0	2	8	18

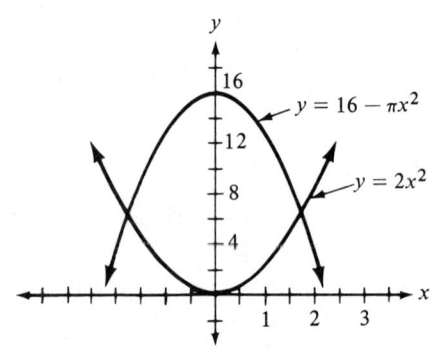

FIGURE 12.19

From the graph it appears there is a solution for positive x. Next solve the system algebraically. Let $A_s = A_r$ so that

$$16 - \pi x^2 = 2x^2$$
$$2x^2 + \pi x^2 = 16$$
$$(2 + \pi)x^2 = 16$$
$$x^2 = \frac{16}{2 + \pi} = 3.11$$
$$x = 1.76 \text{ in.}$$

We check this solution by substituting x into each of the area equations:

$$A_s = 16 - \pi(1.76)^2 = 6.27 \text{ in.}^2$$
$$A_r = 2(1.76)^2 = 6.20 \text{ in.}^2$$

Had we not rounded earlier these two answers would be identical. The solution is (1.76, 6.27).

EXAMPLE Templates of equal area are to be produced. One is to be a parallelogram with a height of 3.00 in. The other is to be a circle.

APPLICATIONS OF QUADRATIC SYSTEMS 12.3

If the base of the parallelogram and the diameter of the circle are to be equal, what are the dimensions of each template?

SOLUTION First draw a sketch showing the templates with the known information (Figure 12.20).

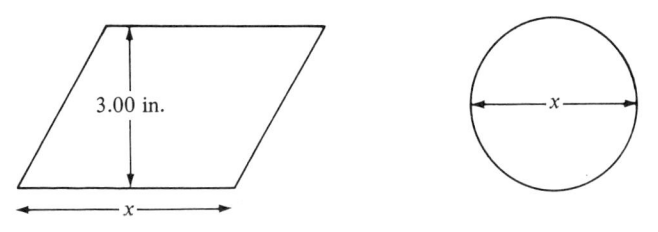

FIGURE 12.20

Write an expression for each area.

$$\text{Area (parallelogram)} = 3x; \quad \text{Area (circle)} = \frac{\pi x^2}{4}$$

Plot these equations on the same set of coordinate axes.

x	0	1	2	3	4	5
A_p	0	3	6	9	12	15
A_c	0	$\frac{\pi}{4}$	π	$\frac{9\pi}{4}$	4π	$\frac{25\pi}{4}$

See Figure 12.21.

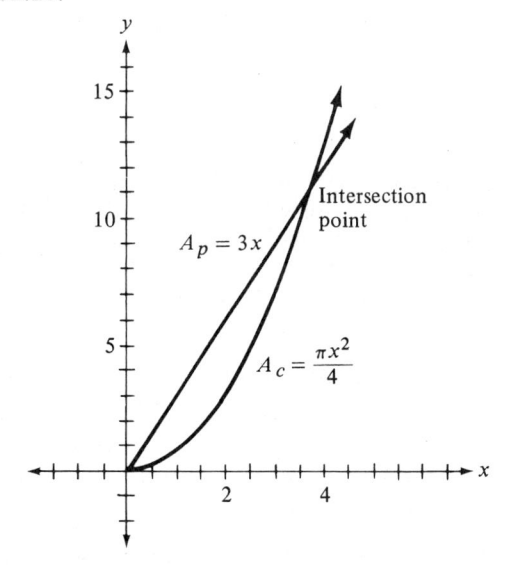

FIGURE 12.21

We note that we have an apparent solution that results from the intersection of the straight line and the quadratic. We solve the system algebraically. Let $A_p = A_c$. Then

$$\frac{\pi x^2}{4} = 3x$$

$$\frac{\pi x^2}{4} - 3x = 0$$

$$x\left(\frac{\pi x}{4} - 3\right) = 0$$

Solutions are then $x = 0$ and $x = \dfrac{12}{\pi} = 3.82$ in.

The base of the parallelogram and the circle diameter are then 3.82 in. in length. Verify that their areas are equal and determine that area.

PERFORMANCE OBJECTIVE 8

You should now be able to solve a geometric problem that involves a quadratic equation.

Motion problems differ from other quadratic systems in that the independent variable will generally be time. The position of an object that experiences a constant force (and hence a constant acceleration) is described by a coordinate that is quadratic in time. The position of an object that experiences zero force is described by a coordinate that is linear in time. Thus, vertical motion near the surface of the earth (assuming no friction) would be described by a quadratic since the gravitational force is constant. Horizontal motion near the earth's surface (again assuming no friction) would be described by a linear equation in time. In the following examples we consider the motion of objects under the influence of gravity and under the influence of an electrical field.

EXAMPLE A baseball is thrown vertically upward from the ground with an initial velocity of 32 ft/sec. Its displacement s from its starting point after t sec is given by

$$s = 32t - 16t^2$$

Find the maximum height to which the ball rises and the length of time it takes to reach the ground.

APPLICATIONS OF QUADRATIC SYSTEMS 12.3

SOLUTION We first graph the equation.

t	0	0.5	1.0	1.5	2.0
s	0	12	16	12	0

See Figure 12.22.

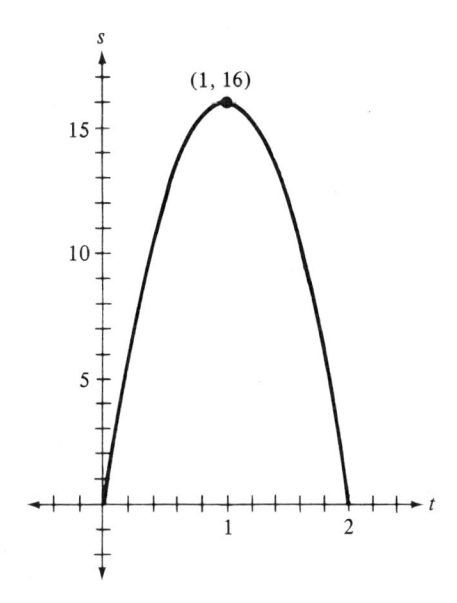

FIGURE 12.22

It appears from the graph that the maximum height is 16 ft. That is, the parabola vertex occurs at (1, 16). We may verify this by evaluating the ordered pair $\left(\dfrac{-b}{2a}, c - \dfrac{b^2}{4a}\right)$, which we know represents the vertex of the parabola $y = ax^2 + bx + c$.

The quadratic is $s = -16t^2 + 32t + 0$ so that $a = -16$, $b = 32$, and $c = 0$. Then

$$\frac{-b}{2a} = \frac{-32}{-32} = +1, \quad c - \frac{b^2}{4a} = 0 - \frac{(32)^2}{4(-16)} = +16$$

We also note that the graph indicates that at $t = 2$, $s = 0$. This means that the ball is back at ground level after 2 sec. We could verify this algebraically by finding the intercepts on the horizontal axis (the t-axis). We do this by solving

$$-16t^2 + 32t = 0$$

Factor to obtain

$$-16t(t-2) = 0$$

from which we set $-16t = 0$ and $t - 2 = 0$. The solutions are $t = 0$ and $t = 2$. The ball is at ground level at $t = 0$ (at the start) and at $t = 2$. Thus, it takes 2 sec to reach the ground as we predicted from the graph.

EXAMPLE The position of an object moving vertically in the gravitational field of the earth is given by

$$y = y_0 + v_0 t - 16t^2 \quad \text{in feet}$$

In this equation the independent variable t represents time in seconds. The starting position is y_0 and the starting velocity is v_0.

Requirement 1: Construct a plot of the position of the object in time from $t = 0$ to $t = 20$ sec. Assume $v_0 = 300$ ft/sec and $y_0 = 100$ ft.

SOLUTION

Requirement 1: First develop a data table using the given equation. This table follows.

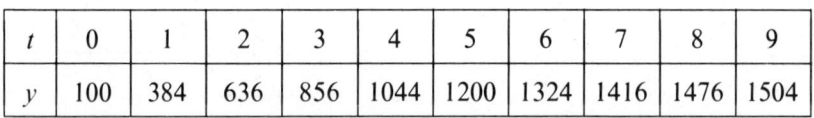

t	0	1	2	3	4	5	6	7	8	9
y	100	384	636	856	1044	1200	1324	1416	1476	1504

t	10	11	12	13	14	15	16	17	18	19
y	1500	1464	1396	1296	1164	1000	804	576	316	24

One final data pair is $(20, -300)$. The graph is shown in Figure 12.23.

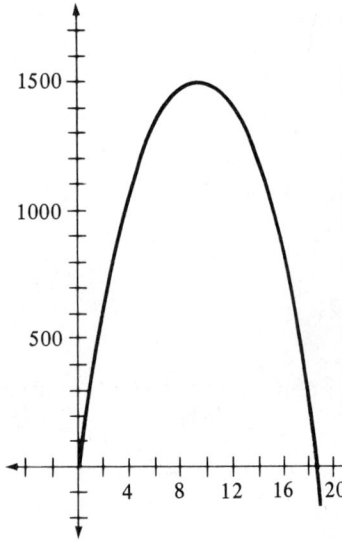

FIGURE 12.23

EXAMPLE

Requirement 2: Determine the maximum height to which the object will rise. At what time will the object reach this height?

SOLUTION

Requirement 2: The easiest way to do this is to evaluate the coordinates of the vertex as we did in Section 12.1. The y-coordinate will be the maximum height. The t-coordinate will be the time at which this occurs.

APPLICATIONS OF QUADRATIC SYSTEMS 12.3   453

The equation of the vertex is given by $\left(\dfrac{-b}{2a},\ c - \dfrac{b^2}{4a}\right)$.

In this problem we compare $y = -16t^2 + 300t + 100$ with the standard form of the parabola $y = ax^2 + bx + c$. We see that $a = -16$, $b = 300$, and $c = 100$. Then the vertex is at

$$\left(\dfrac{-300}{2(-16)},\ 100 - \dfrac{(300)^2}{4(-16)}\right) = \left(\dfrac{300}{32},\ 1506.25\right)$$
$$= (9.375,\ 1506.25)$$

EXAMPLE

Requirement 3: The position of a second object falling at constant speed through the atmosphere is given by

$$y = 2000 - 100t$$

Determine if the first object and this second object will ever be at the same height at the same time. If this occurs, at what height(s) will it occur?

SOLUTION

Requirement 3: We need to solve the equations

$$y = -16t^2 + 300t + 100 \quad \text{and}$$
$$y = 2000 - 100t$$

as a system. We see that we have a combination linear and quadratic system. We first make a graphic plot of $y = 2000 - 100t$ to see if it appears to intersect the parabola plot. A table of data values for $y = 2000 - 100t$ is shown.

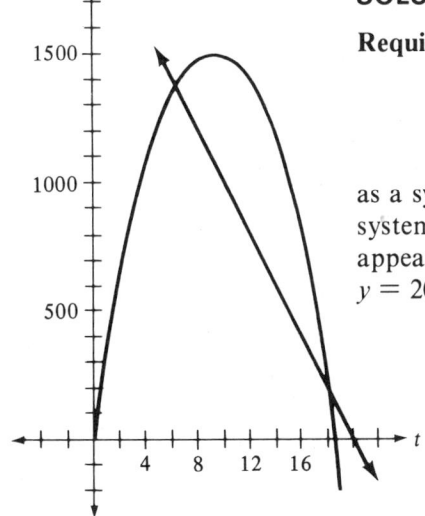

t	10	20
y	1000	0

FIGURE 12.24

A graph is shown in Figure 12.24.

There appear to be two solutions to the system of equations. To find the coordinates of these points equate the y-values.

$$-16t^2 + 300t + 100 = 2000 - 100t$$
$$-16t^2 + 400t - 1900 = 0$$

Use the quadratic formula to solve this equation.

$$t = \frac{-b \pm \sqrt{b^2 - 4ac}}{2a} = \frac{-400 \pm \sqrt{(400)^2 - 4(-16)(-1900)}}{2(-16)}$$

$$= \frac{-400 \pm 195.959}{-32} = 6.376 \text{ or } 18.62 \text{ sec}$$

Thus, the objects are at the same height at times of $t = 6.376$ sec and $t = 18.62$ sec. We evaluate the heights using the linear equation because it is simpler.

$$y = 2000 - 100t = 2000 - 100(6.376) = 2000 - 637.6$$
$$= 1362.4 \text{ ft} \quad \text{or} \quad 1360 \text{ ft}$$

$$y = 2000 - 100t = 2000 - 100(18.62) = 2000 - 1862$$
$$= 138 \text{ ft}$$

EXAMPLE The lateral position of a charged particle as it moves through a uniform electric field is given by the equation

$$y = 0.2 + 1.2 \times 10^7 \, t - 3.5 \times 10^{13} t^2 \text{ cm}$$

If the particle is in the field for 3×10^{-7} sec, calculate its lateral position upon exit. Also determine the maximum distance it gets from the lower (+) plate. We represent the equation of the lower plate by $y = 0$. See Figure 12.25.

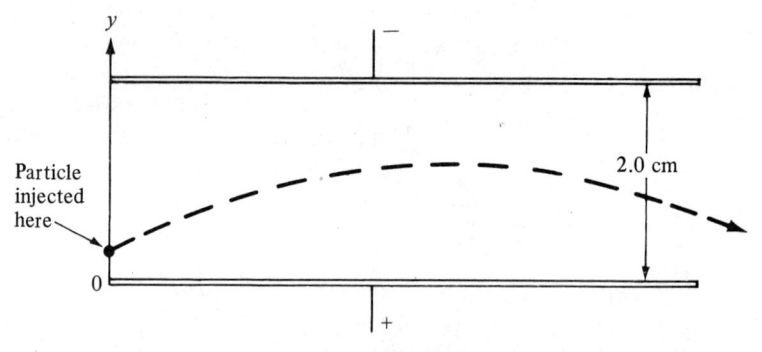

FIGURE 12.25

SOLUTION We first evaluate y for certain values of t and then plot the resulting data pairs.

$t \times 10^{-7}$	0.5	1.0	1.5	2.0	2.5	3.0
y	0.7125	1.05	1.2125	1.2	1.0125	0.65

See Figure 12.26 for the graph.

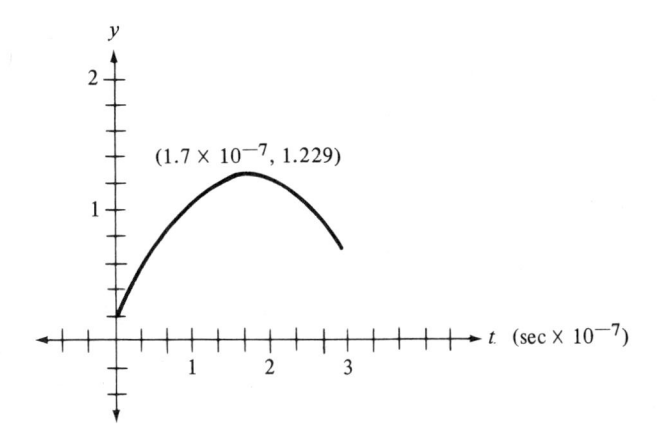

FIGURE 12.26

The curve appears to peak during the time the particle is between the plates. We solve for the vertex of the parabola. Note that

$$a = -3.5 \times 10^{13}$$
$$b = 1.2 \times 10^{7}$$
$$c = 0.2$$

Then the vertex is

$$\left(\frac{-b}{2a}, c - \frac{b^2}{4a}\right) = \left(\frac{-1.2 \times 10^7}{2(-3.5 \times 10^{13})}, 0.2 - \frac{(1.2 \times 10^7)^2}{4(-3.5 \times 10^{13})}\right)$$
$$= (1.7 \times 10^{-7}, 1.229)$$

The particle thus reaches a distance of 1.229 cm from the lower plate. The distance from the lower plate at $t = 3 \times 10^{-7}$ sec is obtained from the previous data table. Note the ordered pair $(3 \times 10^{-7}, 0.65)$. The particle is 0.65 cm from the lower plate as it exits.

PERFORMANCE OBJECTIVE 9

You should now be able to solve motion problems involving quadratic equations and systems.

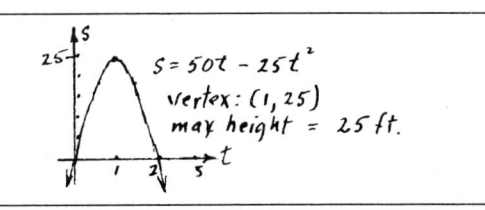

$s = 50t - 25t^2$
vertex: $(1, 25)$
max height $= 25$ ft.

12 GRAPHING QUADRATIC EQUATIONS AND SYSTEMS

In manufacturing processes, a very useful application of quadratic equations occurs in the discussion of supply and demand. When working with supply and demand, we are trying to find point(s) where supply (the quantity produced as a function of price) and demand (the quantity that the market will absorb as a function of price) are equal. We try to find this point to keep from having a supply of the product that either fails to meet or exceeds the demand. The point that we find as the solution to the system consisting of the supply and demand equations is called the **point of equilibrium.**

One feature of marketing applications of quadratic equations is that we are concerned only with the first quadrant. This is the case since the other three quadrants would involve either negative prices or quantities.

We will now consider an example in which we solve a quadratic system for the point of equilibrium.

EXAMPLE Suppose the supply of a certain item is given by the function

$$y = x^2 + 4x - 8$$

while the demand for the item is given by the function

$$y = x^2 - 6x + 12 \qquad \text{for } x < 3$$

Find the point of equilibrium.

SOLUTION Set the expressions for y equal to each other and solve for x.

$$x^2 - 6x + 12 = x^2 + 4x - 8$$
$$-10x = -20$$
$$x = 2$$

For $x = 2$, $y = (2)^2 - 6(2) + 12 = 4 - 12 + 12 = 4$. Therefore, the point of equilibrium is (2, 4). The system is graphed in Figure 12.27.

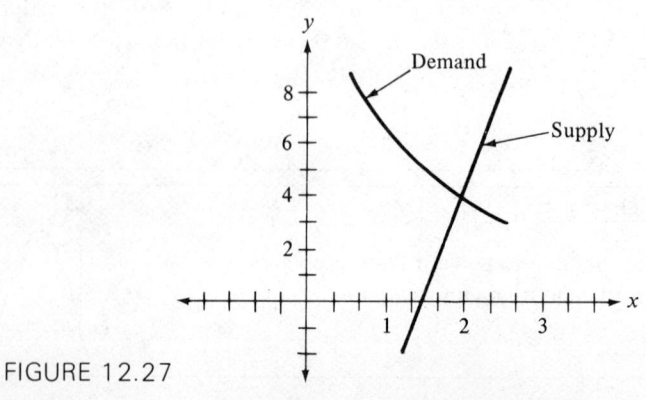

FIGURE 12.27

APPLICATIONS OF QUADRATIC SYSTEMS 12.3   457

PERFORMANCE OBJECTIVE 10

You should now be able to solve supply and demand problems involving quadratic equations.

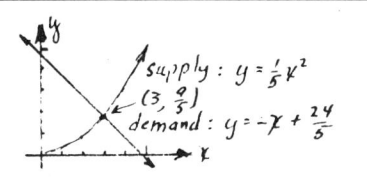

supply: $y = \frac{1}{5}x^2$
$(3, \frac{9}{5})$
demand: $y = -x + \frac{24}{5}$

EXERCISE SET 12.3

1. A trapezoid-shaped plate is to have the same area as a circular plate. The trapezoid has a height of 5.00 in., a long base = 8.00 in., and a short base that is equal to the diameter of the circle. Calculate the unknown dimension. See Figure 12.28.

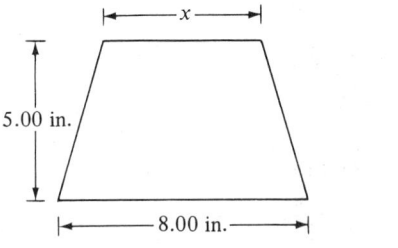

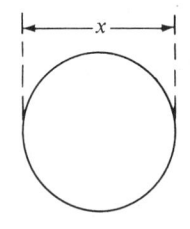

   FIGURE 12.28

2. Calculate the unknown dimension so that the two figures in Figure 12.29 have equal areas.

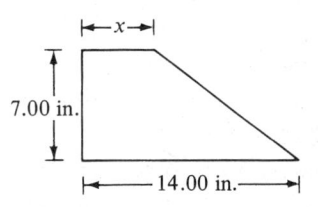

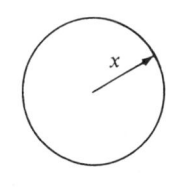

   FIGURE 12.29

3. The following equations represent the motion of an object thrown vertically under gravity and a second that is falling at a constant speed. Determine if there are any times at which the objects will be at the same height. Solve graphically and algebraically.

   a. $y = 200 + 400t - 16t^2$ and $y = 1500 - 80t$
   b. $y = 200 + 400t - 16t^2$ and $y = 2500 - 50t$

4. Find the time at which the following objects have their maximum y coordinate.

   a. $y = 3.5 + 55t - 4.9t^2$ m   b. $y = 40 - 55t - 16t^2$ ft

5. Suppose the supply and demand for a certain item are related to price by

$$\text{Supply: } y = \frac{2}{5}x \quad \text{Demand: } y = -\frac{2}{5}x^2 + 80$$

Find the point of equilibrium. Use both graphical and algebraic methods.

6. The supply and demand functions are given by

$$\text{Supply: } S(x) = 3x - 6, \ x \geq 2$$

$$\text{Demand: } D(x) = -x^2 + 6x + 10, \ x \geq 3$$

Find the point of equilibrium.

SECTION 12.4 SUMMARY

Key terms and concepts introduced in this chapter are reviewed.

Quadratic equation in two variables: an expression of the general form $y = ax^2 + bx + c$, where a cannot be zero.

Parabola: the curve that results when a quadratic equation in two variables is plotted. Characteristics of a parabola are

Line of symmetry: a vertical line passing through the vertex of a parabola. The parabola is symmetric about this line.

Vertex: the extreme value of a parabola in the y direction. The vertex is either the minimum value of y or the maximum value of y on the graph.

x-intercept(s): point(s) at which the parabola intercepts the x-axis.

y-intercept: point at which the parabola intercepts the y-axis.

Completing the square: a process of producing a perfect square trinomial involving the x variable in a quadratic equation.

Quadratic function: a name given the quadratic equation to indicate that the equation acts as a function.

Quadratic system: a system of two equations at least one of which is quadratic.

Maximization and minimization problems: common applications of the quadratic equation in which the vertex is viewed as the maximum or the minimum value of the dependent variable.

Point of equilibrium: the solution to a supply-demand system representing the point where supply and demand, as functions of price, are equal.

CHAPTER 12 PRACTICE TEST

1. (PO-1) Identify whether the graph could be the graph of a quadratic equation.

   a.

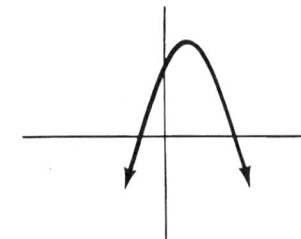

   b.

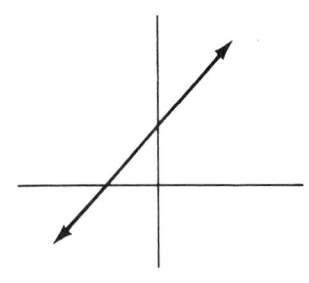

   c.

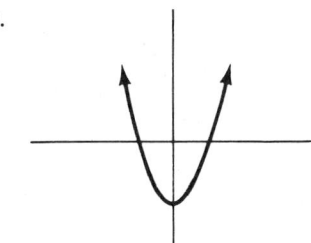

   d.

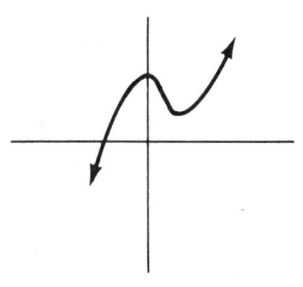

2. (PO-2, 3) For each of the following equations, complete the table provided.

Equation	x-intercept(s)	y-intercept	Vertex
$y = x^2 - 5x - 6$			
$y = 2x^2 - 4x + 4$			
$y = 2x^2 - 5x - 12$			

3. (PO-4) Graph $y - x^2 = -9$.

4. (PO-4) Graph $y = -x^2 + x + 20$.

5. (PO-5) Graph and solve the system

$$y + x^2 = 4$$

$$8x + 4y = 4$$

6. (PO-6) Graph and solve the system

$$y - x^2 + 4x + 12 = 0$$

$$y = x^2 + 4x - 12$$

7. (PO-6) Graph and solve the system

$$y = x^2 + x - 12$$

$$y = -x^2 + x + 12$$

8. (PO-7) If a variable voltage is given by

$$e = t^2 - 5t + 6$$

find the minimum value for e.

9. (PO-8) A rectangular plate with a circular hole in it is to have the same weight (and hence the same area) as a triangular plate of the same material. The dimensions of the triangular plate and the rectangular plate are shown in Figure 12.30.

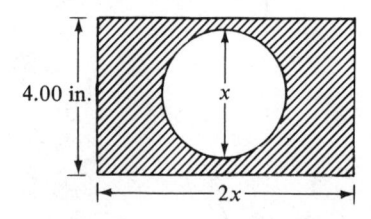

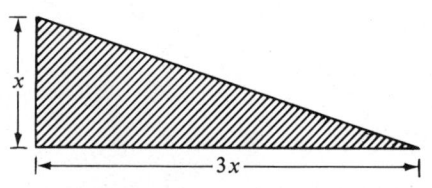

FIGURE 12.30

Find the unknown dimension x.

10. (PO-9) An object is thrown vertically upward from the top of a building 96 feet high with an initial velocity of 64 feet per second. The equation of its vertical position in time is given by

$$s = 96 + 64t - 16t^2$$

a. Plot the position of the object versus time.

b. Find its maximum height graphically and algebraically.

11. (PO-10) Suppose the supply and demand functions for a certain item are

Supply: $S(x) = 2x + 1$

Demand: $D(x) = -\dfrac{3}{5}x^2 + 80$

Find the point of equilibrium. Use both the graphic and algebraic methods.

CHAPTER 13

ADDITIONAL TOPICS IN TRIGONOMETRY

13.1 Angles: Measurement and Relation to Other Quantities
13.2 Trigonometric Functions (First Quadrant)
13.3 Trigonometric Functions (All Quadrants)
13.4 Vector Resolution
13.5 Vector Composition
13.6 Vector Addition
13.7 Statics Problems
13.8 Summary

We noted in Chapter 8 that trigonometry began as a study of triangles. However, it has developed applications far beyond the context of the triangle. In order to appreciate these applications, we need to broaden our definitions of trigonometric functions.

In this chapter we introduce additional measures of an angle. Three additional trigonometric functions are defined and the methods of finding values of the trigonometric functions over the full 360° circle are developed. We will find that the trigonometric functions are useful in resolution and addition of vectors. This leads us into the area of engineering statics.

SECTION 13.1 ANGLES: MEASUREMENT AND RELATION TO OTHER QUANTITIES

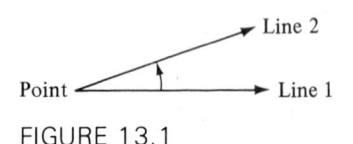

FIGURE 13.1

There are two contexts in which an angle is frequently used. The first is a geometric context in which an angle is formed by two straight lines emanating from a point in space. An example of the geometric angle is shown in Figure 13.1.

DEFINITION

THE PARTS OF AN ANGLE

1. The point is called the vertex of the angle.
2. Line 1 is called the initial side.
3. Line 2 is called the terminal side.

The second context occurs in a study of rotational motion where angle refers to amount of rotation (angular displacement). In Figure 13.2 the wheel starts in position 1 and rotates counterclockwise to position 2. Although we do not see an angle in a geometric sense (the initial and terminal sides are not visible at the same time), we can visualize an angle formed by a straight line drawn through the axis of rotation as the wheel rotates from position 1 to position 2.

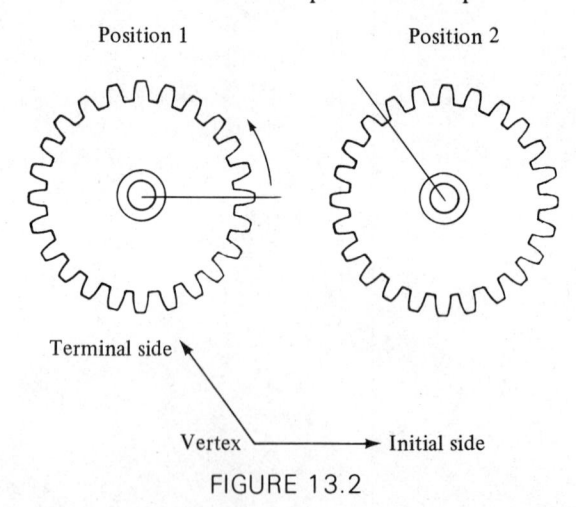

FIGURE 13.2

To complete the definition of an angle, we define a sense (positive or negative) for the angle. A small arc determines the direction of the rotation and hence the sign of the angle.

ANGLES: MEASUREMENT, RELATION TO OTHER QUANTITIES 13.1 **463**

By convention, angles formed by rotating the initial side counterclockwise into the terminal side are considered positive. A clockwise rotation is considered negative. A positive and a negative angle are shown in Figure 13.3.

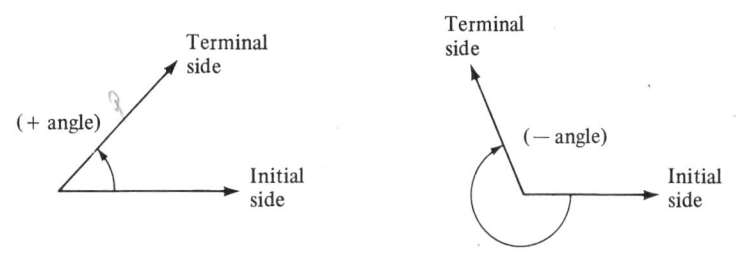

FIGURE 13.3

When using angles for drafting or machine shop purposes, we will not normally include a sign of the angle. In these cases, the drawing will treat both sides of the angle as equivalent. An example is shown in Figure 13.4.

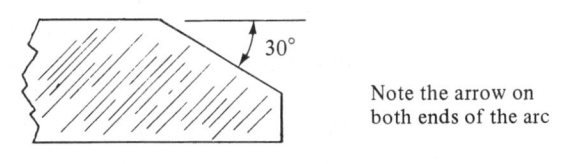

Note the arrow on both ends of the arc

FIGURE 13.4

PERFORMANCE OBJECTIVE 1	
You should now be able to draw and label the parts of a signed angle.	

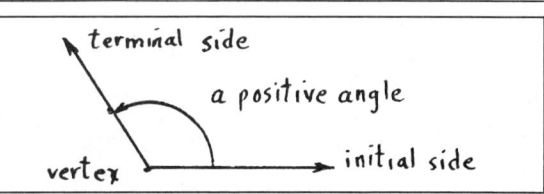

If we are asked to measure the length of a line, the answer provided depends on the unit of measure we choose. The same situation holds in measuring angles; that is, we have to cope with more than one system of units.

The three systems of angle measurement we will use in this chapter are **degrees, radians,** and **rotations.** Greater emphasis will be placed on the system of radian measurement since degrees and rotations (revolutions) are quite commonly used. Conversion between systems will also be emphasized.

464 13 ADDITIONAL TOPICS IN TRIGONOMETRY

DEFINITION

> **SYSTEMS OF ANGULAR MEASUREMENT**
>
> 1. Degree measure: This system divides the circle into 360 parts, each equal to one degree, the unit angle.
> 2. Rotation measure: This system uses a full circle as the unit angle.
> 3. Radian measure: This system defines the unit angle as that angle which subtends an arc equal in length to the arc radius. The arc is centered at the vertex of the angle.

To illustrate radian measure, consider Figure 13.5. We will use the Greek letter theta (θ) to designate the variable angle.

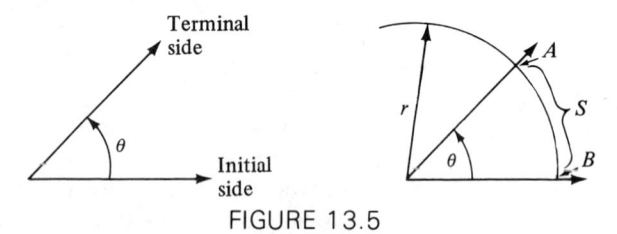

FIGURE 13.5

In the right-hand sketch in Figure 13.5 an arc of radius r is drawn. The vertex of the angle is the center of this arc. The arc intersects the sides of the angle at A and B. The arc length cut off by the angle is the distance AB measured along the arc (not the straight line distance). Call this arc length s.

DEFINITION

> **ANGLE θ IN RADIAN MEASURE**
>
> $\theta = \dfrac{s}{r}$, the ratio of arc length subtended to radius of the arc.

From this definition, we verify that the unit angle in radian measure, $\theta = 1$, is obtained when $s = r$. We are now able to find relations among the three systems of measurement. If 1° equals $\frac{1}{360}$ of a full circle, then one rotation = 360°. From plane geometry, we know that the circumference of a circle is the product of 2π and r (its radius). Thus, the radian measure of a full circle is

$$\theta = \frac{\text{circumference}}{\text{radius}} = \frac{2\pi r}{r} = 2\pi$$

Thus

$$2\pi \text{ radians} = 360° = 1 \text{ rotation}$$

ANGLES: MEASUREMENT, RELATION TO OTHER QUANTITIES 13.1

From these relations, we can work out conversion tables. See Tables 13.1, 13.2, and 13.3. These tables include the calculator steps for conversion.

TABLE 13.1

To convert degrees to

1. Radians: multiply by $\pi/180$

   angle $\boxed{\times}$ $\boxed{\pi}$ $\boxed{\div}$ 180 $\boxed{=}$

2. Rotations: divide by 360

   angle $\boxed{\div}$ 360 $\boxed{=}$

TABLE 13.2

To convert radians to

1. Degrees: multiply by $180/\pi$

   angle $\boxed{\times}$ 180 $\boxed{\div}$ $\boxed{\pi}$ $\boxed{=}$

2. Rotations: divide by 2π

   angle $\boxed{\div}$ 2 $\boxed{\div}$ $\boxed{\pi}$ $\boxed{=}$

TABLE 13.3

To convert rotations to

1. Degrees: multiply by 360

   angle $\boxed{\times}$ 360 $\boxed{=}$

2. Radians: Multiply by 2π

   angle $\boxed{\times}$ 2 $\boxed{\times}$ $\boxed{\pi}$ $\boxed{=}$

CAUTION: Some calculator instruction books describe a method of conversion that uses the trigonometric functions. These methods only work for a limited range of values of the angle. We recommend the preceding method of direction conversion on the calculator.

Since $\dfrac{180}{\pi}$ is approximately 57.3, there are approximately 57.3° in a radian. Also, since 2π is approximately 6.28, there are approximately 6.28 radians in a rotation. The following examples illustrate the use of the preceding tables.

EXAMPLE Convert 80° to radians.

SOLUTION

1. By hand calculation:

$$80 \times \frac{\pi}{180} = \frac{4\pi}{9} = 1.396 \text{ radians}$$

2. By calculator:

80 ☒×☒ ☒π☒ ☒÷☒ 180 ☒=☒ yields 1.396 to 4 significant digits

EXAMPLE Convert 1.5 radians to rotations.

SOLUTION

1. By hand calculation:

$$1.5 \div 2\pi = 0.75/\pi = 0.239$$

2. By calculator:

1.5 ☒÷☒ ☒π☒ ☒÷☒ 2 ☒=☒ yields 0.239 to three significant digits

EXAMPLE Convert 0.7 rotations to radians.

SOLUTION

1. By hand calculation:

$$0.7 \times 2\pi = 1.4\pi = 4.398 \text{ radians}$$

2. By calculator:

0.7 ☒×☒ 2 ☒×☒ ☒π☒ ☒=☒ yields 4.398 to four significant digits

---

PERFORMANCE OBJECTIVE 2

You should now be able to convert angle measurements from one system of units to another (degrees, radians, and rotations).

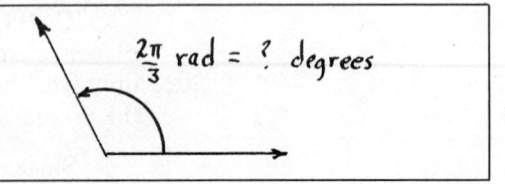

$\frac{2\pi}{3}$ rad = ? degrees

The definition of an angle in radian measure may be rearranged to produce a useful formula.

ANGLES: MEASUREMENT, RELATION TO OTHER QUANTITIES 13.1 **467**

FORMULA

> **RELATION AMONG ARC LENGTH, RADIUS, AND ANGLE**
>
> s(arc length) $= r$(radius) $\times \theta$(in radians)

Consider the following examples of the use of this formula.

EXAMPLE A baseball field is laid out as shown in Figure 13.6. How many feet of fence will be required to construct an outfield wall?

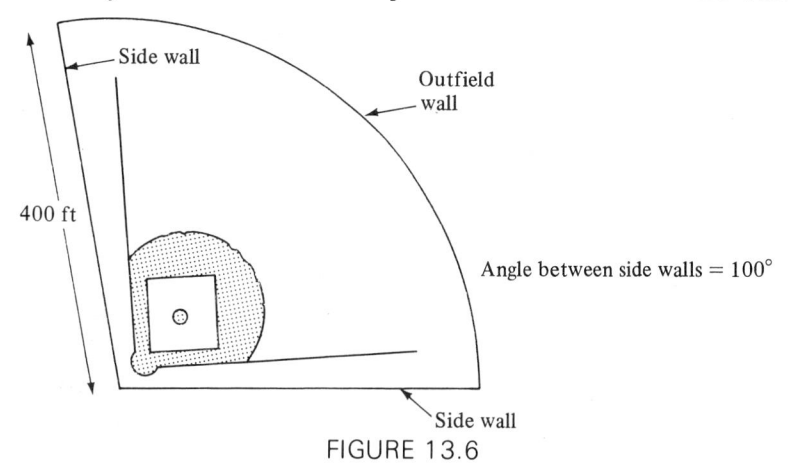

FIGURE 13.6

SOLUTION To solve this problem use the relation $s = r\theta$.
r is 400 ft and θ is 100°. But this equation was developed using θ in radian measure.
100° must be converted to radians by multiplying by $\dfrac{\pi}{180}$.

$$\theta = (100)\dfrac{\pi}{180} = \dfrac{5\pi}{9} \quad \text{and}$$

$$s = (400)\dfrac{(5)(3.14)}{9} = \dfrac{(2000)(3.14)}{9} = 698 \text{ ft}$$

EXAMPLE Calculate the perimeter of the metal plate in Figure 13.7. Sides AB and CD are on radii of a circle centered at O. AD and BC are arcs of circles centered at O.

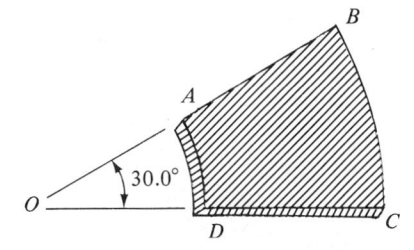

OA (distance to inner radius) = 2.50 in.

OB (distance to outer radius) = 5.80 in.

FIGURE 13.7

SOLUTION The perimeter is obtained by summing the lengths of each of the plate edges. Note that $OA = 2.50$ in. and $OB = 5.80$ in.

1. The smaller arc $= r\theta = (2.5)\left(\dfrac{\pi}{6}\right) = 1.309$ in.

2. The larger arc $= r\theta = (5.80)\left(\dfrac{\pi}{6}\right) = 3.037$ in.

3. Each radial side $= 5.80 - 2.50 = 3.30$ in.

4. The perimeter $= 2(3.30) + 1.309 + 3.037 = 10.946$ in., which we report as 10.9 in. to three significant digits.

The area of a sector of a circle is also determined from r and θ. Consider the complete circle in Figure 13.8.

We know from plane geometry that the area of this circle is πr^2. But what angle in radians is subtended by the arc of this circle? There are 2π radians in a rotation so that the angle is 2π radians. We can write

$$A \text{ (for area)} = \pi r^2 = 2\pi \times \dfrac{r^2}{2} = \dfrac{1}{2}r^2(2\pi)$$

This result, established for this special case, holds for all angles θ.

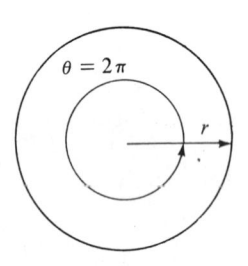

FIGURE 13.8

FORMULA

RELATION AMONG AREA, RADIUS, AND ANGLE

Area $= \dfrac{1}{2}$(radius)$^2 \times \theta$(in radians)

EXAMPLE Consider again the problem of the baseball field in Figure 13.6. How many square feet of sod should be purchased if the entire field is to be placed in grass?

SOLUTION We need to evaluate the area of the "pie-shaped" baseball field. Again, $r = 400$ ft and $\theta = \dfrac{5\pi}{9}$. Then

$$A = \dfrac{1}{2}(400)^2 \dfrac{5\pi}{9} = 139{,}626 \text{ ft}^2$$

Assuming that r and θ are known to three significant digits, we would write $A = 140{,}000$ ft^2

EXAMPLE Calculate the weight of a plate that is made of metal sheeting. The sheeting weighs 2.1 lb/ft^2. The dimensions of the plate are shown in Figure 13.9. Edges shown are portions of radii and arcs of circles centered at *O*.

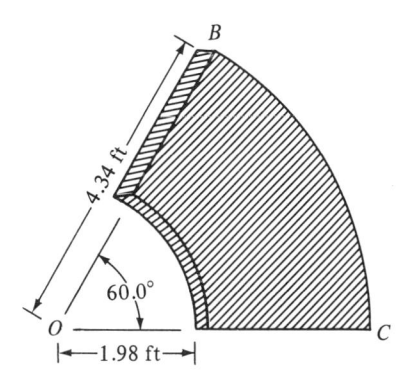

FIGURE 13.9

SOLUTION We calculate the area of the plate as if it extends to the center of the circle. See Figure 13.10. Then calculate the missing area.

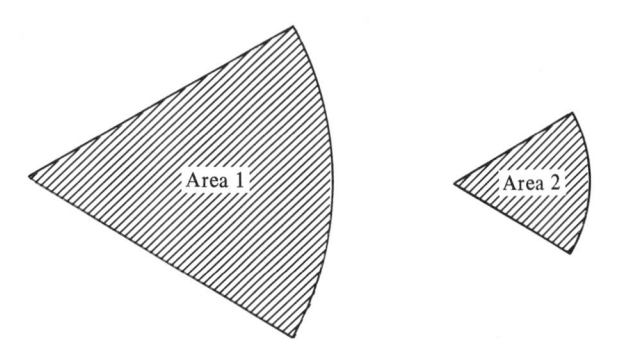

FIGURE 13.10

$$\text{Area 1} = \frac{1}{2}r^2\theta = \frac{1}{2}(4.34)^2\left(\frac{\pi}{3}\right) = 9.862 \text{ ft}^2$$

$$\text{Area 2} = \frac{1}{2}r^2\theta = \frac{1}{2}(1.98)^2\left(\frac{\pi}{3}\right) = 2.053 \text{ ft}^2$$

The net area of the plate is then

$$9.862 - 2.053 = 7.809 \text{ ft}^2$$

The weight of the plate will be the product of weight per unit area and area. Then

$$W = (2.1)(7.809) = 16.4 \text{ lb to three significant digits}$$

13 ADDITIONAL TOPICS IN TRIGONOMETRY

> **PERFORMANCE OBJECTIVE 3**
>
> You should now be able to use the relation among arc length, angle, and radius to solve numerical problems.
>
> *calculate area and perimeter*
> [sketch: shaded region with 30° angle at O, bounded by an arc of circle centered at O]

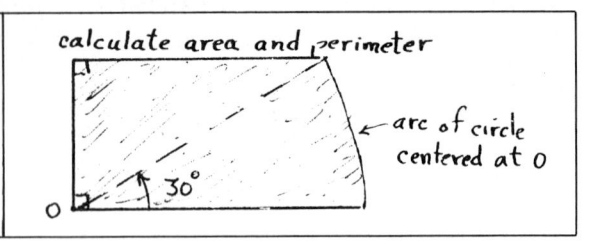

Consider a circular wheel in rotational motion about an axis through its center and perpendicular to the plane of the wheel. See Figure 13.11.

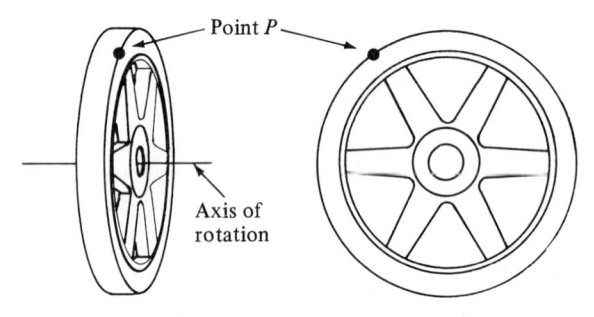

FIGURE 13.11

If we locate a point P on the circumference of the wheel of radius r, P will travel a distance of $r\theta$ as the wheel turns through an angle θ (measured in radians). The speed at which P travels this distance is obtained by dividing distance traveled by the time required to travel that distance. Thus, the speed, for which we will use the symbol v, is given by

$$v = \frac{\text{distance}}{\text{time}} = \frac{r\theta}{t}$$

The symbol t has been used to represent elapsed time. Consider the expression $\frac{r\theta}{t}$. We can separate it into two factors, r and $\frac{\theta}{t}$. What interpretation can be given to $\frac{\theta}{t}$? The angle turned through in t seconds is θ. Just as linear distance traveled divided by time is called linear speed, so angular distance traveled divided by time is called **angular speed**. Thus,

$$v = \text{radius} \cdot \text{angular speed}$$

with angular speed measured in radians per second.

Use the Greek letter omega (ω) to represent angular speed. Then,

$v = r\omega$. We have established a relation between linear speed and angular speed of a point in circular motion. In dealing with electrical problems we frequently call ω **angular frequency** rather than angular speed. In such problems it represents a rate of oscillation of a signal.

FORMULA

RELATION AMONG LINEAR SPEED, RADIUS, AND ANGULAR SPEED

$$v\text{(linear speed)} = r\text{(radius)} \times \omega\text{(angular speed)}$$

Note the similarity between the equations

$s = r\theta$ (linear distance on a circle related to angular distance)

and

$v = r\omega$ (linear speed on a circle related to angular speed).

It is now clear why the radian system of measurement is so convenient. These equations have their simplest form when the angle involved is measured in radians.

EXAMPLE   Calculate the linear speed of points A, B, and C on the rotating wheel in Figure 13.12. Radial distances of A, B, and C from the axis of rotation are $r_A = 1.20$ ft, $r_B = 2.70$ ft, and $r_C = 3.70$ ft. The wheel rotates two complete revolutions each second.

SOLUTION   We use the relation $v = r\omega$; however, ω must be known in radians per second. The angular speed is given as two revolutions per second. Each revolution equals 2π radians so that $\omega = 2 \cdot 2\pi = 4\pi$ radians per second. Then

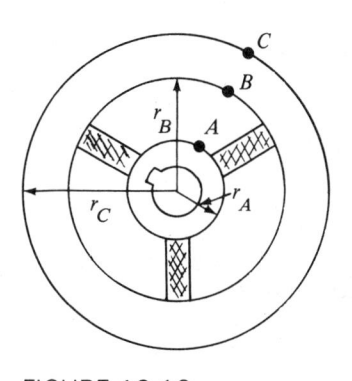

$$v_A \text{ (speed of point } A\text{)} = 1.20 \times 4\pi = 4.8\pi = 15.1 \text{ ft/sec}$$
$$v_B = 2.70 \times 4\pi = 10.8\pi = 33.9 \text{ ft/sec}$$
$$v_C = 3.70 \times 4\pi = 14.8\pi = 46.5 \text{ ft/sec}$$

FIGURE 13.12

PERFORMANCE OBJECTIVE 4

You should now be able to use the relation among angular speed, radius, and linear speed to solve numerical problems.

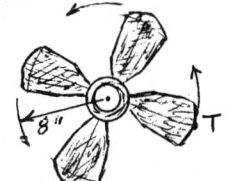

8 in radius fan blade
$W = 8\pi$ rad/s
$V_T = ?$

EXERCISE SET 13.1

1. Label the parts of the angles.
   a. b. c.

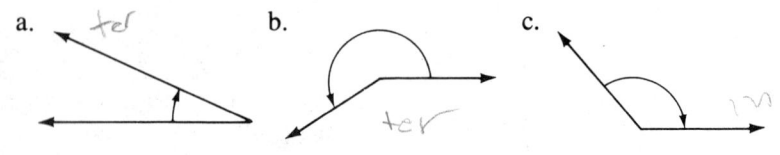

2. Label the parts of the angles.
   a. b. c.

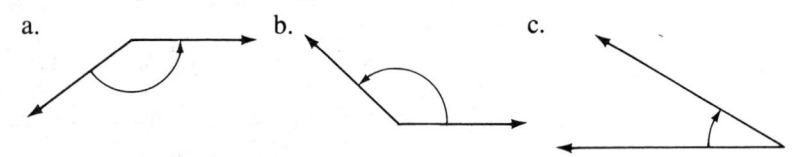

3. Draw an angle that would represent the rotation shown in Figure 13.13. Label the three parts of the angle and its sign.

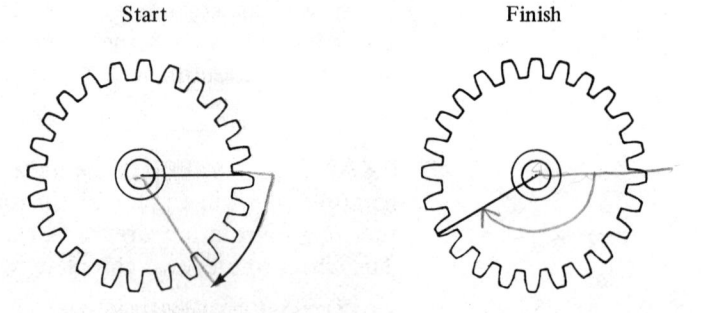

   Start   Finish

   FIGURE 13.13

4. Draw an angle that would represent the rotation shown in Figure 13.14. Label the three parts of the angle and its sign.

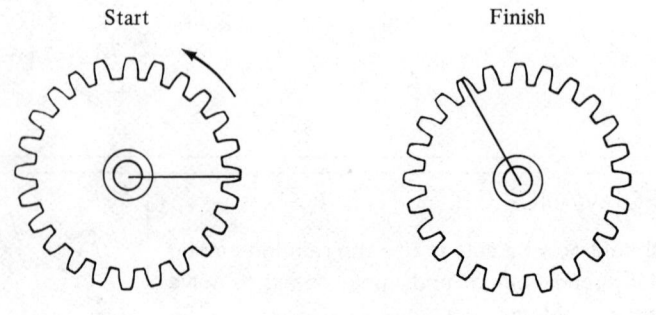

   Start   Finish

   FIGURE 13.14

5. Determine the signs of the following angles:

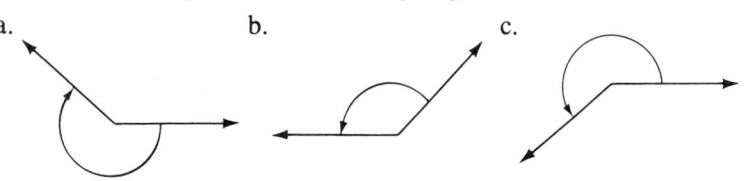

   a.　　　　　　　　　b.　　　　　　　　　c.

6. Determine the signs of the following angles:

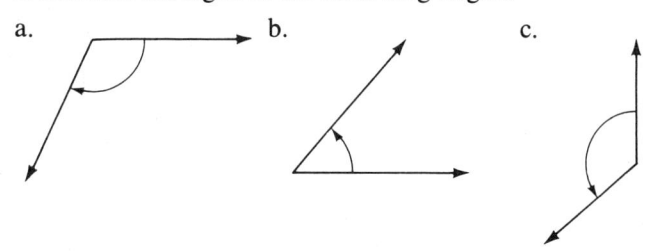

   a.　　　　　　　　　b.　　　　　　　　　c.

7. The minute hand on a clock moves from three to seven. Determine the value (to include mathematical sign) of the angle through which it turns. Express the answer in all three systems of units.

8. Calculate the angle through which the minute hand of a clock moves during the time interval 2:17 P.M. to 3:05 P.M. Express the answer in all three systems of units.

9. Complete the following table by giving the measure of the angle in each of the other two systems:

Degree	Radian	Rotation
135°		
	$\frac{5\pi}{6}$	
		$\frac{5}{8}$
	$\frac{11\pi}{6}$	
	$\frac{3\pi}{2}$	
210°		
		$\frac{7}{18}$

10. Complete the following table by giving the measure of the angle in each of the other two systems:

Degree	Radian	Rotation
75°		
	$\frac{5\pi}{8}$	
		$\frac{3}{4}$
	$\frac{7\pi}{6}$	
	π	
330°		
	$\frac{3\pi}{20}$	

11. Complete the following table, assuming that s is an arc of a circle of radius r and θ is an angle subtending that arc:

s	r	θ
	2 m	2 rad
2 m		2π rad
	2 m	180°
	3 ft	90°
	4 m	$\frac{\pi}{4}$ rad
π ft		30°

12. Complete the following table, assuming that s is an arc of a circle of radius r and θ is an angle subtending that arc:

s	r	θ
	3 m	3 rad
1.5 m		3 rad
	1.5 m	270°
	4 ft	180°
	3 m	$\frac{\pi}{4}$ rad
2 ft		120°

13. Complete the following table, assuming that A is the area of a sector of a circle of radius r and θ is the angle between the sides of that sector:

A	r	θ
	2 m	60°
2.7 m²		1.2 rad
13.5 m²	3 m	
	4 m	45°

14. Complete the following table, assuming that A is the area of a sector of a circle of radius r and θ is the angle between the sides of that sector:

A	r	θ
	3 m	60°
7.4 m²		1.1 rad
12.6 m²	2.9 m	
	3.5 m	60°

15. Find the weight of a thin metal plate in the shape shown in Figure 13.15. The material of which the plate is made weighs 2.25 lb/ft².

16. Find the weight of a thin metal plate in the shape shown in Figure 13.16. The material of which the plate is made weighs 75.2 N/m².

17. What is the length of the circular edge of the plate in Figure 13.15?

18. What is the perimeter of the plate in Figure 13.16?

19. Complete the entries in the following table. Assume that v is the speed of an object in circular motion at a distance r from the axis of rotation. The angular speed is ω.

v	r	ω
	2 m	3 rad/sec
4 m/sec		0.8 rad/sec
	1.5 ft	5 rev/sec
π m/sec		60°/sec
8 ft/sec	1.5 ft	

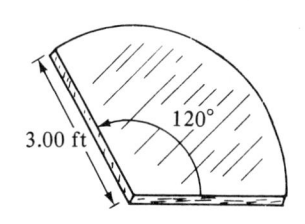

FIGURE 13.15

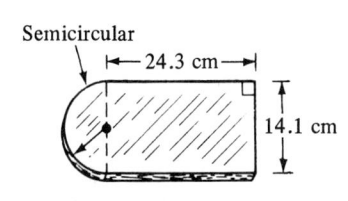

FIGURE 13.16

20. Complete the entries in the following table. Assume that v is the speed of an object in circular motion at a distance r from the axis of rotation. The angular speed is ω.

v	r	ω
	3 m	3 rad/sec
5 m/sec		0.7 rad/sec
	3.1 m	4 rev/sec
$\pi/2$ m/sec		45°/sec
7 ft/sec	1.4 ft	

21. An automobile fan of radius 8 in. is rotating with the engine at 12 revolutions per second. What is the linear speed of the tip of the fan blades?

22. Calculate the maximum speed of a conveyor belt when its driving wheel has a radius of 7 cm and a maximum angular speed of 10 revolutions per second.

SECTION 13.2 TRIGONOMETRIC FUNCTIONS (FIRST QUADRANT)

We now introduce the angle in relation to the x-y coordinate system. Consider the coordinate system in Figure 13.17. An angle has been placed in this system with the initial side along the positive x-axis. We again identify the angle with the Greek letter theta (θ). This angle is said to be in **standard position.**

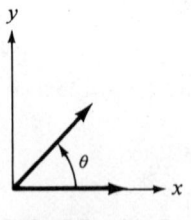

FIGURE 13.17

DEFINITION

STANDARD POSITION OF AN ANGLE

An angle is in **standard position** when its initial side is along the positive x-axis and its vertex is at the origin.

TRIGONOMETRIC FUNCTIONS (FIRST QUADRANT) 13.2

With the angle in standard position we may classify it according to the location of its terminal side. The *x*- and *y*-axes divide the plane into four regions called quadrants. Quadrants were introduced in Chapter 7. In Figure 13.18 the quadrants are marked and an angle in the second quadrant is drawn.

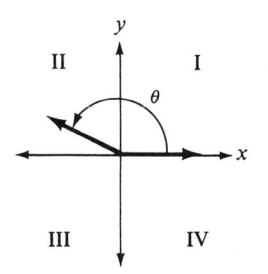

FIGURE 13.18

DEFINITION

QUADRANT OF AN ANGLE

An angle, placed in standard position, is said to be in the quadrant in which its terminal side is found.

EXAMPLE Sketch two angles with different signs that lie in the third quadrant.

SOLUTION Positive and negative angles in the third quadrant are shown in Figure 13.19.

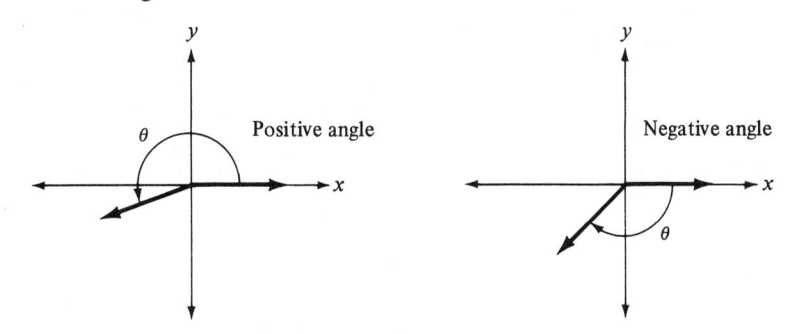

FIGURE 13.19

EXAMPLE Identify the range of values of the positive angles in each quadrant in degrees and radians.

SOLUTION The table of values gives the ranges.

Quadrant	I	II	III	IV
degree	$0° < \theta < 90°$	$90° < \theta < 180°$	$180° < \theta < 270°$	$270° < \theta < 360°$
radian	$0 < \theta < \frac{\pi}{2}$	$\frac{\pi}{2} < \theta < \pi$	$\pi < \theta < \frac{3\pi}{2}$	$\frac{3\pi}{2} < \theta < 2\pi$

If the terminal side of an angle lies along one of the axes, the angle is said to be a **quadrantal angle.** Quadrantal angles then have the values of 0° (0 rad), 90° $\left(\dfrac{\pi}{2} \text{ rad}\right)$, 270° $\left(\dfrac{3\pi}{2} \text{ rad}\right)$, 360° ($2\pi$ rad), etc. Negative quadrantal angles are also permitted.

DEFINITION

QUADRANTAL ANGLE

A **quadrantal angle** is one that has its terminal side along one of the axes when it is in standard position.

EXAMPLE Sketch the 270° quadrantal angle.

SOLUTION See Figure 13.20.

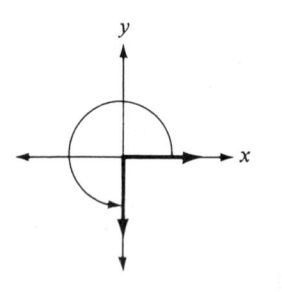

FIGURE 13.20

PERFORMANCE OBJECTIVE 5

You should now be able to identify an angle by quadrant or as a quadrantal angle.

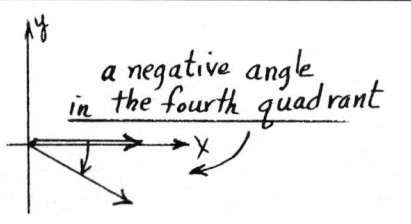

a negative angle in the fourth quadrant

In Chapter 8 we introduced three of the trigonometric functions of an angle. These functions were the sine, cosine, and tangent. They were defined in terms of the ratios of sides of a right triangle. We will now generalize on those definitions using the angle in the *x-y* coordinate system. Consider the angle in Figure 13.21(a).

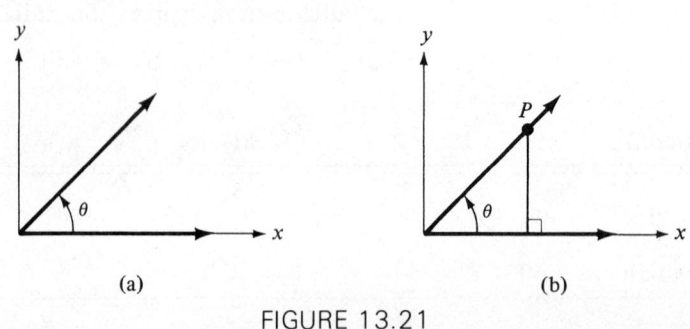

(a) (b)

FIGURE 13.21

Take any point along the terminal side of this angle, say point P, and drop a perpendicular to the x-axis as in Figure 13.21(b). The coordinates of point P can then be displayed on this diagram as in Figure 13.22.

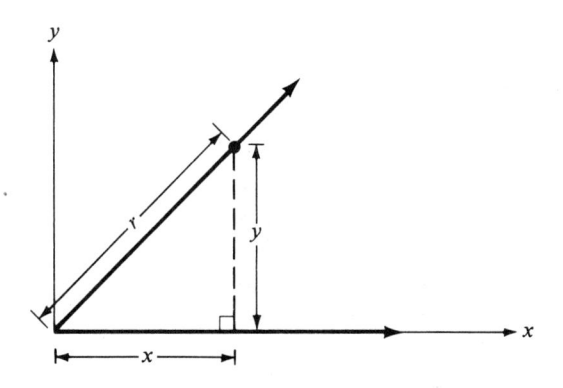

FIGURE 13.22

Let r be the straight-line distance from the origin to P. Using the right triangle formed by this construction, we can define three trigonometric functions of θ.

DEFINITION

TRIGONOMETRIC FUNCTIONS (USING COORDINATES)

$$\sin\theta = \frac{y}{r}, \quad \cos\theta = \frac{x}{r}, \quad \tan\theta = \frac{y}{x}$$

Notice that these definitions are identical with those made earlier in Chapter 8. As before, we will abbreviate the trigonometric functions to three letters (sin, cos, and tan). Finally, there are three additional trigonometric functions that are reciprocals of the first three.

1. The reciprocal of sine is cosecant (abbreviated csc).
   Since $\sin\theta = \frac{y}{r}$, we must have $\csc\theta = \frac{r}{y}$.
2. The reciprocal of cosine is secant (abbreviated sec).
   Since $\cos\theta = \frac{x}{r}$, we must have $\sec\theta = \frac{r}{x}$.
3. The reciprocal of tangent is cotangent (abbreviated cot).
   Since $\tan\theta = \frac{y}{x}$, we must have $\cot\theta = \frac{x}{y}$.

DEFINITION

TRIGONOMETRIC FUNCTIONS (USING COORDINATES)

$$\csc\theta = \frac{r}{y}, \quad \sec\theta = \frac{r}{x}, \quad \cot\theta = \frac{x}{y}$$

The following examples illustrate how the trigonometric functions may be determined from information about a point on the terminal side of the angle.

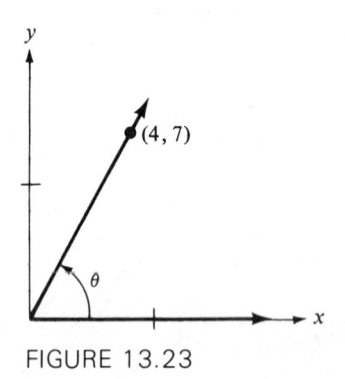

EXAMPLE In the angle in Figure 13.23 the point (4, 7) is on the terminal side. Evaluate the six trigonometric functions of θ.

SOLUTION To find all the trigonometric functions, we need x, y, and r. We can find r by using the Pythagorean Theorem. See Figure 13.24.

Then $r = \sqrt{4^2 + 7^2} = \sqrt{16 + 49} = \sqrt{65}$

$$\sin\theta = \frac{y}{r} = \frac{7}{\sqrt{65}} \qquad \csc\theta = \frac{r}{y} = \frac{\sqrt{65}}{7}$$

$$\cos\theta = \frac{x}{r} = \frac{4}{\sqrt{65}} \qquad \sec\theta = \frac{r}{x} = \frac{\sqrt{65}}{4}$$

$$\tan\theta = \frac{y}{x} = \frac{7}{4} \qquad \cot\theta = \frac{x}{y} = \frac{4}{7}$$

FIGURE 13.23

FIGURE 13.24

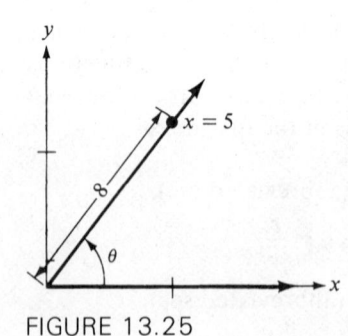

EXAMPLE A point on the terminal side of θ has an x-coordinate of 5 and lies 8 units from the origin. Find the trigonometric functions of θ. See Figure 13.25.

SOLUTION $y = \sqrt{r^2 - x^2} = \sqrt{8^2 - 5^2} = \sqrt{64 - 25} = \sqrt{39}$

$$\sin\theta = \frac{y}{r} = \frac{\sqrt{39}}{8} \qquad \csc\theta = \frac{r}{y} = \frac{8}{\sqrt{39}}$$

$$\cos\theta = \frac{x}{r} = \frac{5}{8} \qquad \sec\theta = \frac{r}{x} = \frac{8}{5}$$

$$\tan\theta = \frac{y}{x} = \frac{\sqrt{39}}{5} \qquad \cot\theta = \frac{x}{y} = \frac{5}{\sqrt{39}}$$

FIGURE 13.25

TRIGONOMETRIC FUNCTIONS (FIRST QUADRANT) 13.2

PERFORMANCE OBJECTIVE 6

You should now be able to evaluate trigonometric functions of an angle given a point on its terminal side (quadrant I).

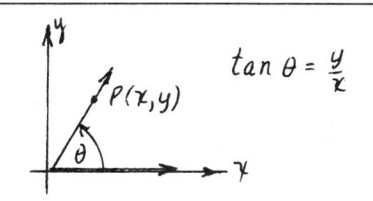

EXERCISE SET 13.2

1. Identify the quadrant of the following angles:

   a. 156° b. $\frac{\pi}{3}$ rad c. −110° d. $\frac{5\pi}{6}$ rad

   e. $\frac{-\pi}{4}$ rad f. −260° g. $\frac{5}{6}$ rotations

2. Identify the quadrant of the following angles:

   a. −79° b. $\frac{4\pi}{7}$ rad c. 266° d. $\frac{7}{12}$ rotations

   e. $\frac{3\pi}{2}$ rad f. $\frac{-7\pi}{8}$ rad g. $\frac{7\pi}{12}$ rad

3. Identify the range of values of the positive angles in each quadrant in terms of rotations.

4. Identify the range of values of the negative angles in each quadrant in terms of rotations.

5. Sketch the following quadrantal angles:

   a. 180° b. $\frac{-\pi}{2}$ c. −270°

6. Sketch the following quadrantal angles:

   a. −π b. 270° c. $\frac{3\pi}{2}$

7. Identify the axis (+ or − x- or y-axis) along which the terminal sides of the following angles are found:

   a. −180° b. $\frac{\pi}{2}$ rad c. $\frac{-3\pi}{2}$ rad

   d. $\frac{3}{4}$ rotation e. −π rad

8. Identify the axis (+ or − x- or y-axis) along which the terminal sides of the following angles are found:

   a. −π rad b. 270° c. $\frac{\pi}{2}$ rad

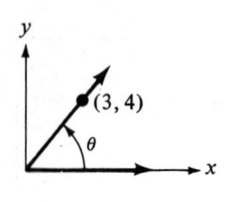

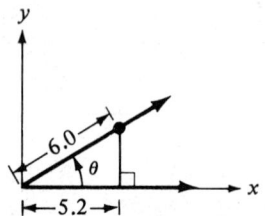

d. $-\dfrac{3}{4}$ rotation   e. $-90°$

9. Find $\tan\theta$, $\cot\theta$, $\cos\theta$, and $\sec\theta$ for each of the angles shown in Figure 13.26.

10. Find $\cos\theta$, $\sec\theta$, $\tan\theta$, and $\cot\theta$ for each of the angles shown in Figure 13.27.

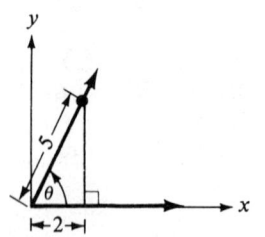

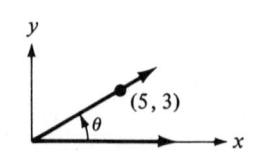

FIGURE 13.26

FIGURE 13.27

11. Find $\sin\theta$, $\csc\theta$, $\tan\theta$, and $\cot\theta$ for each of the angles shown in Figure 13.28.

12. Find $\sin\theta$, $\csc\theta$, $\tan\theta$, and $\cot\theta$ for each of the angles shown in Figure 13.29.

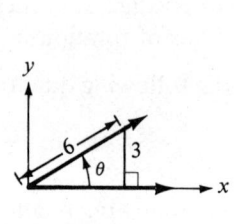

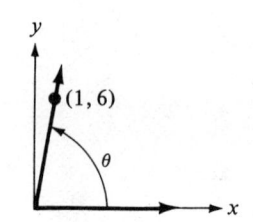

FIGURE 13.28

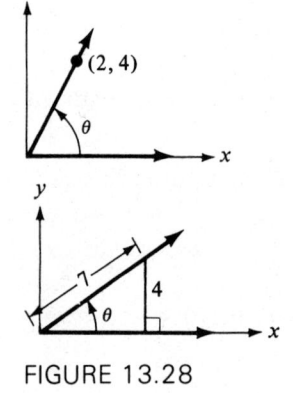

FIGURE 13.29

13. Find all six trigonometric functions of the angles shown in Figure 13.30.

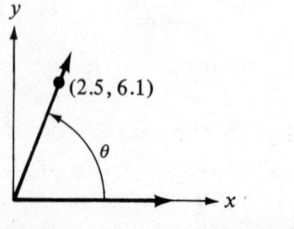

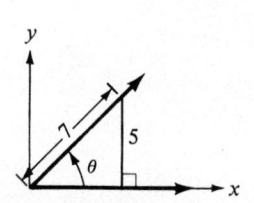

FIGURE 13.30

14. Find all six trigonometric functions of the angles shown in Figure 13.31.

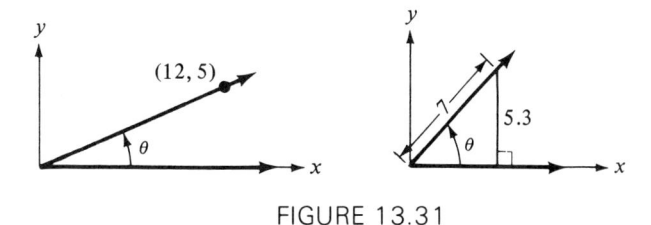

FIGURE 13.31

SECTION 13.3 TRIGONOMETRIC FUNCTIONS (ALL QUADRANTS)

In the last section we restricted our attention to angles less than 90° and greater than 0°. In such cases, the terminal side of the angle will always fall between the positive x- and y-axes; therefore, any point P on the terminal side will have coordinates (x, y), which are positive. The distance r from the origin to any point P is always positive. Thus, all trigonometric functions for angles in the first quadrant will be positive since they will be determined as the ratios of positive numbers.

However, if we allow the angle to be greater than 90°, a point on the terminal side will not have both x- and y-coordinates positive. At least one of the coordinates will be negative. We will exclude the special angles 0°, 90°, 180°, and 270° for a while because there are difficulties in defining all the trigonometric functions for these angles. The angle in Figure 13.32 is placed in standard position.

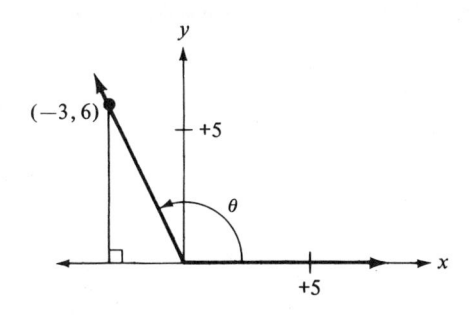

FIGURE 13.32

Choose some point P on the terminal side. Note that its x-

coordinate will be negative and its y-coordinate positive. Suppose the coordinates of P are (−3, 6). Then

$$r = \sqrt{x^2 + y^2} = \sqrt{(-3)^2 + (6)^2} = \sqrt{9 + 36} = \sqrt{45} = 3\sqrt{5}$$

The trigonometric functions of θ are found to be

$$\sin \theta = \frac{y}{r} = \frac{6}{3\sqrt{5}} = \frac{2}{\sqrt{5}}, \quad \csc \theta = \frac{r}{y} = \frac{3\sqrt{5}}{6} = \frac{\sqrt{5}}{2}$$

$$\cos \theta = \frac{x}{r} = \frac{-3}{3\sqrt{5}} = -\frac{1}{\sqrt{5}}, \quad \sec \theta = \frac{r}{x} = \frac{3\sqrt{5}}{-3} = -\sqrt{5}$$

$$\tan \theta = \frac{y}{x} = \frac{6}{-3} = -2, \quad \cot \theta = \frac{x}{y} = \frac{-3}{6} = -\frac{1}{2}$$

Note that four of the six trigonometric functions are negative.

EXAMPLE Determine the mathematical signs of the trigonometric functions of the angle in Figure 13.33. Notice that both the x- and y-coordinates of a point on its terminal side are negative.

SOLUTION

$$\sin \theta = \frac{y}{r} = \frac{-}{+} = -$$

$$\cos \theta = \frac{x}{r} = \frac{-}{+} = -$$

$$\tan \theta = \frac{y}{x} = \frac{-}{-} = +$$

$$\csc \theta = \frac{r}{y} = \frac{+}{-} = -$$

$$\sec \theta = \frac{r}{x} = \frac{+}{-} = -$$

$$\cot \theta = \frac{x}{y} = \frac{-}{-} = +$$

Again four of the six functions are negative.

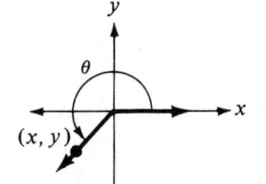

FIGURE 13.33

EXAMPLE Determine the value of the sine, cosine, and tangent of the angle shown in Figure 13.34.

SOLUTION Note that

$$r = \sqrt{x^2 + y^2} = \sqrt{(2)^2 + (-4)^2} = \sqrt{4 + 16} = \sqrt{20} = 2\sqrt{5}$$

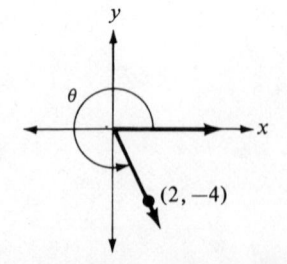

FIGURE 13.34

Then

$$\sin\theta = \frac{y}{r} = \frac{-4}{2\sqrt{5}} = -\frac{2}{\sqrt{5}}$$

$$\cos\theta = \frac{x}{r} = \frac{2}{2\sqrt{5}} = \frac{1}{\sqrt{5}}$$

$$\tan\theta = \frac{y}{x} = \frac{-4}{2} = -2$$

PERFORMANCE OBJECTIVE 7

You should now be able to evaluate the trigonometric functions of an angle given a point on its terminal side (all quadrants).

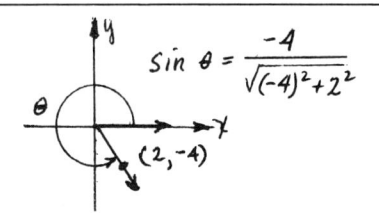

The use of trigonometric tables was introduced in Chapter 8. There we pointed out that a table could be abbreviated to 45° by using two sets of headings and by noting special relations between certain functions. Since the sin-cos, tan-cot, and sec-csc pairs have equal function values when the angles are complementary, we may limit the table to 45° for all six functions. A portion of such a table is shown here. The complete table is found in Appendix IV, Table A.

θ	sin	cos	tan	csc	sec	cot	θ
0	0.0000	1.0000	0.0000		1.000		90
1	0.0175	0.9998	0.0175	57.299	1.000	57.290	89
2	0.0349	0.9994	0.0349	28.654	1.001	28.636	88
3	0.0523	0.9986	0.0524	19.107	1.001	19.081	87
4	0.0698	0.9976	0.0699	14.336	1.002	14.301	86
5	0.0872	0.9962	0.0875	11.474	1.004	11.430	85
42	0.6691	0.7431	0.9004	1.494	1.346	1.111	48
43	0.6820	0.7314	0.9325	1.466	1.367	1.072	47
44	0.6947	0.7193	0.9657	1.440	1.390	1.036	46
45	0.7071	0.7071	1.0000	1.414	1.414	1.000	45
θ	cos	sin	cot	sec	csc	tan	θ

These tables give the function values for all angles from 0° to 90°. Angles greater than 90° present a slightly different problem. To see what happens for an angle, say of 120°, consider Figure 13.35.

$$\sin 120° = \frac{\sqrt{3}}{2}$$

$$\cos 120° = \frac{-1}{2}$$

$$\tan 120° = \frac{\sqrt{3}}{-1} = -\sqrt{3}$$

$$\csc 120° = \frac{2}{\sqrt{3}}$$

$$\sec 120° = \frac{-2}{1} = -2$$

$$\cot 120° = -\frac{1}{\sqrt{3}}$$

We have used $r = \sqrt{(-1)^2 + (\sqrt{3})^2} = \sqrt{1+3} = \sqrt{4} = 2$.

The preceding values are identical, except for sign, to the same function values for a 60° angle. But 60° is just the angle between the terminal side of the 120° angle and the negative x-axis. Note the 60° angle marked with the double arc in Figure 13.35.

It will always be true that the absolute value of the trigonometric function values for an angle greater than 90° will be identical to the function values for some angle less than 90°. The angle less than 90° is found by measuring the **smallest angle** between the terminal side of the angle and the x-axis—either its positive or negative extension. In Figure 13.35, the terminal side makes an angle of 60° with the negative x-axis.

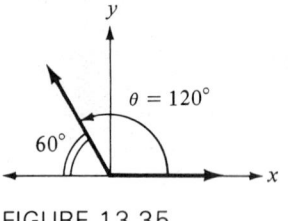

FIGURE 13.35

DEFINITION

SMALLEST ANGLE

The **smallest angle** is that angle (<90°) between the x-axis and the terminal side of an angle in standard position.

EXAMPLE Sketch the smallest angle for $\theta = 225°$.

SOLUTION See Figure 13.36.

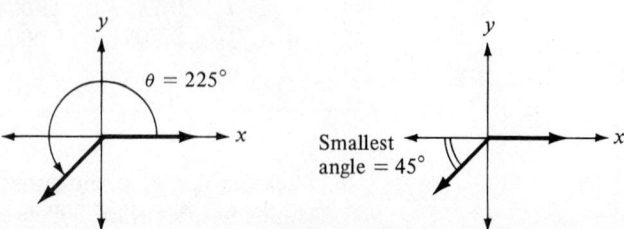

FIGURE 13.36

EXAMPLE Sketch the smallest angle for $\theta = 290°$.

SOLUTION See Figure 13.37.

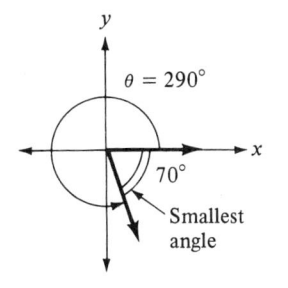

FIGURE 13.37

We now have a way to determine the absolute value of the trigonometric functions of angles larger than 90°. It only remains to determine the correct signs. The signs are determined by the definitions of the functions. For example, consider

$$\sin \theta = \frac{y}{r}$$

Whenever y is positive, $\sin \theta$ is positive (r is always positive). Whenever y is negative, then $\sin \theta$ is negative.

Recall that y is a coordinate of a point on the terminal side of the angle. Considering the sketches in Figure 13.38, we conclude:

$\sin \theta$ is positive when θ is between 0° and 180°.

$\sin \theta$ is negative when θ is between 180° and 360°.

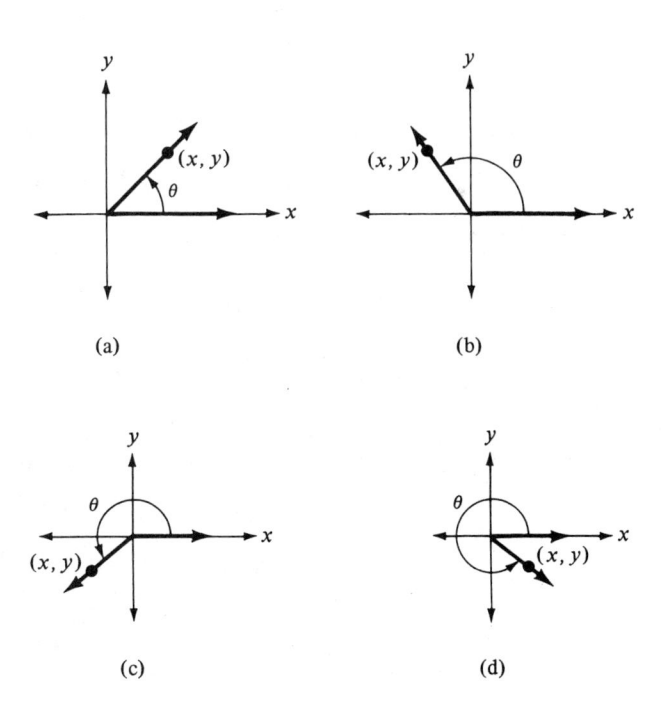

FIGURE 13.38

Similarly, since $\cos \theta = \frac{x}{r}$, $\cos \theta$ has the same sign as x. In the sketches in Figure 13.38, x is positive in (a) and (d) (quadrants I and

IV) and negative in (b) and (c) (quadrants II and III). We conclude

cos θ is positive when θ is between 0° and 90°.

cos θ is negative when θ is between 90° and 270°.

cos θ is positive when θ is between 270° and 360°.

Now consider tan $\theta = \dfrac{y}{x}$. Whenever y and x have the same sign, tan θ will be positive. Tan θ will be negative whenever they have different signs. We conclude:

tan θ is positive when θ is between 0° and 90°.

tan θ is negative when θ is between 90° and 180°.

tan θ is positive when θ is between 180° and 270°.

tan θ is negative when θ is between 270° and 360°.

We have identified the signs for the three basic trigonometric functions. Their reciprocals will show the same sign behavior. We work an example to illustrate the "smallest angle" method and sign determination.

EXAMPLE Find the sine, cosine, and tangent of 250°.

SOLUTION We sketch the angle in Figure 13.39.

The smallest angle between the terminal side of θ and the x-axis is 70°. So the absolute value of the trigonometric functions of 250° will equal the values of the functions of 70°. We find the following values in the tables:

$$\sin 70° = 0.9397$$
$$\cos 70° = 0.3420$$
$$\tan 70° = 2.747$$

Next we determine the sign by looking at the sketch and by observing that both the x- and y-coordinates of any point on the terminal side of the angle are negative. Therefore,

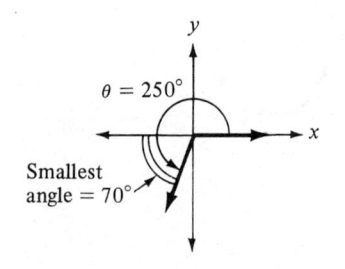

$$\sin \theta = \dfrac{-}{+} = -$$

$$\cos \theta = \dfrac{-}{+} = -$$

$$\tan \theta = \dfrac{-}{-} = +$$

FIGURE 13.39

Finally,

$$\sin 250° = -0.9397$$
$$\cos 250° = -0.3420$$
$$\tan 250° = +2.747$$

PERFORMANCE OBJECTIVE 8

You should now be able to evaluate the trigonometric functions of an angle using tables and the smallest angle (quadrants II to IV).

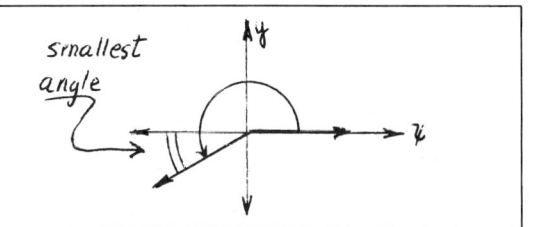

A scientific calculator simplifies finding the trigonometric function values since it provides correct signed values for all angles. The smallest angle need not be used. We list here the procedures and example problems for each function.

To find $\sin \theta$, enter θ, then press $\boxed{\text{SIN}}$.

EXAMPLE Find $\sin 121°$.

SOLUTION 121 $\boxed{\text{SIN}}$ and read 0.8572 to four decimal places.

To find $\cos \theta$, enter θ, then press $\boxed{\text{COS}}$.

EXAMPLE Find $\cos 263°$.

SOLUTION 263 $\boxed{\text{COS}}$ and read -0.1219 to four decimal places.

To find $\tan \theta$, enter θ, then press $\boxed{\text{TAN}}$.

EXAMPLE Find $\tan 282°$.

SOLUTION 282 $\boxed{\text{TAN}}$ and read -4.7046.

To find $\csc \theta$, enter θ, then press $\boxed{\text{SIN}}$ and $\boxed{1/x}$.

EXAMPLE Find $\csc 77°$.

SOLUTION 77 $\boxed{\text{SIN}}$ $\boxed{1/x}$ and read 1.0263.

To find sec θ, enter θ, then press $\boxed{\text{COS}}$ and $\boxed{1/x}$.

EXAMPLE Find sec 155°.

SOLUTION 155 $\boxed{\text{COS}}$ $\boxed{1/x}$ and read -1.1034.

To find cot θ, enter θ, then press $\boxed{\text{TAN}}$ and $\boxed{1/x}$.

EXAMPLE Find cot 143°.

SOLUTION 143 $\boxed{\text{TAN}}$ $\boxed{1/x}$ and read -1.3270.

PERFORMANCE OBJECTIVE 9

You should now be able to evaluate the trigonometric functions of an angle using a calculator (all quadrants).

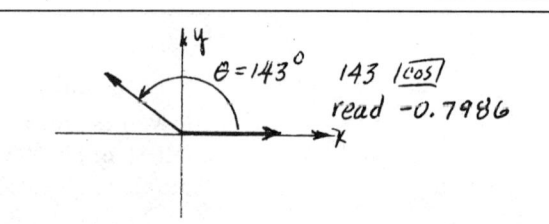

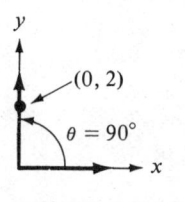

$\theta = 143°$ 143 /cos/ read -0.7986

FIGURE 13.40

The quadrantal angles were defined earlier in this chapter. Here we consider the special problems that arise in defining their trigonometric functions. Consider the 90° quadrantal angle shown in Figure 13.40. Take a point on the terminal side of that angle, say (0, 2).

From the definition of the trigonometric functions, we find

$$\sin 90° = \frac{y}{r} = \frac{2}{2} = 1$$

Note that $r = y$ since

$$r = \sqrt{x^2 + y^2} = \sqrt{0^2 + y^2} = \sqrt{y^2} = y$$

Continuing, we find

$$\cos 90° = \frac{x}{r} = \frac{0}{2} = 0 \quad \text{and} \quad \tan 90° = \frac{y}{x} = \frac{2}{0} = ?$$

Since division by zero is undefined, we have a problem. We cannot write down a value for tan 90°. Notice that the trigonometric tables do not give such a value. You might also try the following entry on your calculator:

90 $\boxed{\text{TAN}}$

An error message should result. Consider further.

$$\csc 90° = \frac{r}{y} = \frac{2}{2} = 1$$

$$\sec 90° = \frac{r}{x} = \frac{2}{0} = ? \quad \text{(undefined)}$$

$$\cot 90° = \frac{x}{y} = \frac{0}{2} = 0$$

The final result is that two of the six functions are not defined for the 90° angle. The function values are summarized in the following table:

$\theta =$	90°
$\sin \theta$	1
$\cos \theta$	0
$\tan \theta$	undefined
$\csc \theta$	1
$\sec \theta$	undefined
$\cot \theta$	0

Consider the quadrantal angle, $\theta = 180°$ shown in Figure 13.41. Take a point on the terminal side, say $(-2, 0)$. Then,

$$\sin \theta = \frac{y}{r} = \frac{0}{2} = 0$$

$$\cos \theta = \frac{x}{r} = \frac{-2}{2} = -1$$

$$\tan \theta = \frac{y}{x} = \frac{0}{-2} = 0$$

$$\csc \theta = \frac{r}{y} = \frac{2}{0} \quad \text{(undefined)}$$

$$\sec \theta = \frac{r}{x} = \frac{2}{-2} = -1$$

$$\cot \theta = \frac{x}{y} = \frac{-2}{0} \quad \text{(undefined)}$$

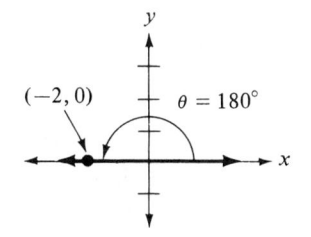

FIGURE 13.41

Again two functions are undefined. We expand the table started earlier.

$\theta =$	90°	180°
$\sin \theta$	1	0
$\cos \theta$	0	−1
$\tan \theta$	undefined	0
$\csc \theta$	1	undefined
$\sec \theta$	undefined	−1
$\cot \theta$	0	undefined

As we develop these function values, we should see a pattern emerging. That pattern is now summarized.

TRIGONOMETRIC FUNCTIONS OF QUADRANTAL ANGLES

Function values for quadrantal angles are

$$-1, \quad 0, \quad +1$$

or are undefined.

EXAMPLE Find the trigonometric functions of 180° using the calculator.

SOLUTION Use the following sequences:

180 [SIN] and read 0
180 [COS] and read −1
180 [TAN] and read 0
180 [SIN] [1/x] and read E (error)
180 [COS] [1/x] and read −1
180 [TAN] [1/x] and read E (error)

The results agree with the table developed earlier. The error messages result from an attempted division by zero and indicate that the function is undefined.

PERFORMANCE OBJECTIVE 10

You should now be able to evaluate the trigonometric functions of a quadrantal angle.

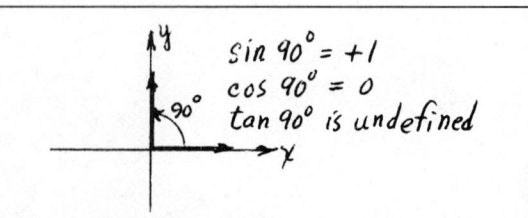

$\sin 90° = +1$
$\cos 90° = 0$
$\tan 90°$ is undefined

EXERCISE SET 13.3

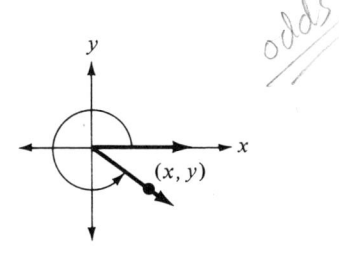

FIGURE 13.42

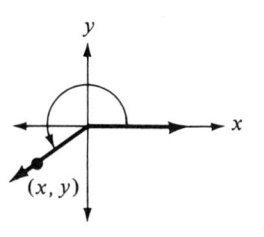

FIGURE 13.43

1. Determine the signs of all six trigonometric functions of an angle in the fourth quadrant. See Figure 13.42.

2. Determine the signs of all six trigonometric functions of an angle in the third quadrant. See Figure 13.43.

3. Determine numerical values (including the correct sign) for all six trigonometric functions of the angles shown.

   a. $(-12, 5)$

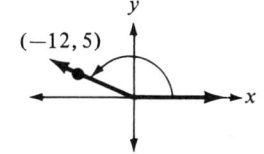

   b. P is 5 units from the origin with an x-coordinate of 3.4.

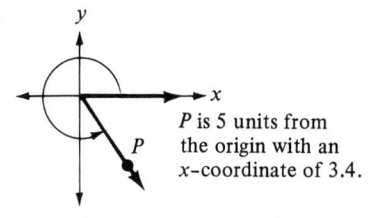

   c. $(-2, -2\sqrt{3})$

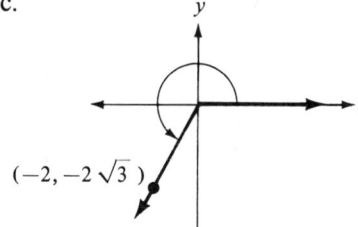

   d. $(-3, 3)$

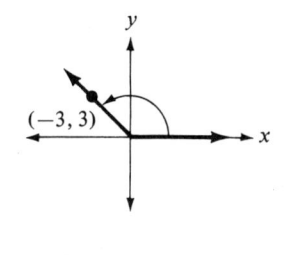

4. Determine numerical values (including the correct sign) for all six trigonometric functions of the angles shown.

   a. $(3, -5)$

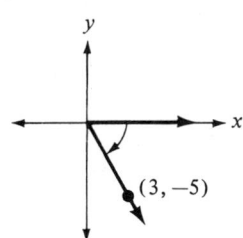

   b. $(-4, -3)$

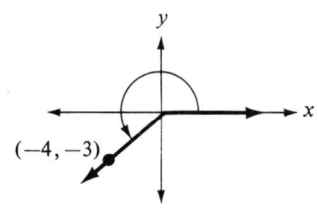

   c. $(-3, 8)$

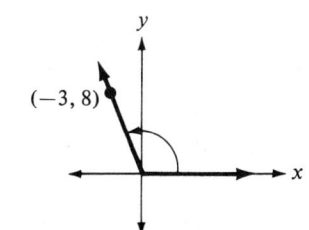

   d. P is 8 units from the origin and has a y-coordinate of -4.

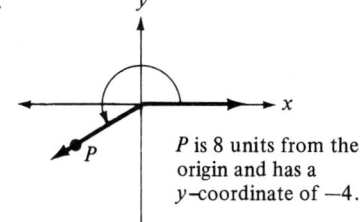

13 ADDITIONAL TOPICS IN TRIGONOMETRY

5. Determine the value of the smallest angle for each of the following angles in standard position:

   a. 119° b. 323° c. −149°

   d. $\frac{2\pi}{3}$ rad e. 229° f. π rad

   g. $\frac{2}{3}$ rotation h. 77°

6. Determine the value of the smallest angle for each of the following angles in standard position:

   a. −119° b. $\frac{\pi}{6}$ rad c. $\frac{4\pi}{5}$ rad

   d. $\frac{5}{7}$ rotation e. 224° f. $\frac{-3\pi}{2}$ rad

   g. 270° h. 280°

7. Complete the following table. An example entry is given. Report your answers rounded to three decimal places.

Trigonometric function	Angle	Smallest angle (SA)	Function value (SA)	Sign	Function value
sec	250°	70°	2.924	−	−2.924
cos	170°				
sin	100°				
tan	162°				
cot	300°				
cos	225°				
sin	345°				
cot	140°				
sec	245°				

TRIGONOMETRIC FUNCTIONS (ALL QUADRANTS) 13.3

8. Complete the following table. An example entry is given. Report your answers rounded to three decimal places.

Trigono- metric function	Angle	Smallest angle (SA)	Function value (SA)	Sign	Function value
sec	110°	70°	2.924	−	−2.924
cos	190°				
sin	−70°				
tan	143°				
cot	312°				
cos	262°				
sin	91°				
cot	209°	29°	1.8	+	+1.8
sec	354°				

Use the trigonometric table or your calculator to find the value of the functions in Exercises 9 and 10. Report your answer rounded to three decimal places.

9. a. sin 72°   b. cos 115°   c. tan 209°
   d. cot 132°   e. sec 290°   f. sin 245°
   g. cos 259°   h. csc 345°   i. tan 288°

10. a. sin 46°   b. cos 212°   c. tan 118°
    d. cot 99°   e. sec 255°   f. sin 219°
    g. cos 348°   h. csc 348°   i. tan 290°

Use the trigonometric table or your calculator to find the value of the following functions. Report your answer rounded to three decimal places. All angles are given in radians.

11. a. $\sin \dfrac{\pi}{2}$ b. $\cos 2\pi$ c. $\tan 1.4$
    d. $\cot 4.1$ e. $\sec \pi$ f. $\csc \dfrac{5\pi}{4}$

12. a. $\sin \dfrac{3\pi}{2}$ b. $\cos 3.5$ c. $\tan 2.1$
    d. $\cot 3.9$ e. $\cos \pi$ f. $\csc \dfrac{7\pi}{4}$

13. Complete the following table by using the definition of the trigonometric functions and a point on the terminal side of the quadrantal angle:

$\theta =$	270°
$\sin \theta$	
$\cos \theta$	
$\tan \theta$	
$\csc \theta$	
$\sec \theta$	
$\cot \theta$	

Use (0, −2) as the point on the terminal side.

14. Complete the following table by using the definition of the trigonometric functions and a point on the terminal side of the quadrantal angle:

$\theta =$	360°
$\sin \theta$	
$\cos \theta$	
$\tan \theta$	
$\csc \theta$	
$\sec \theta$	
$\cot \theta$	

Use (2, 0) as the point on the terminal side.

15. Complete the following table using a calculator. Interpret error messages as undefined values.

$\theta =$	0°	−90°
$\sin \theta$		
$\cos \theta$		
$\tan \theta$		

16. Complete the following table using a calculator. Interpret error messages as undefined values.

$\theta =$	180°	270°
$\sin \theta$		
$\cos \theta$		
$\tan \theta$		

SECTION 13.4 VECTOR RESOLUTION

In this section we leave angles and turn our attention to vectors. We develop the graphing of vectors in Appendix I-4. Here we analyze the representation of vectors in a rectangular coordinate system. We will use an arrow over a letter to designate a vector. Consider the vector $\vec{A}$ shown in Figure 13.44.

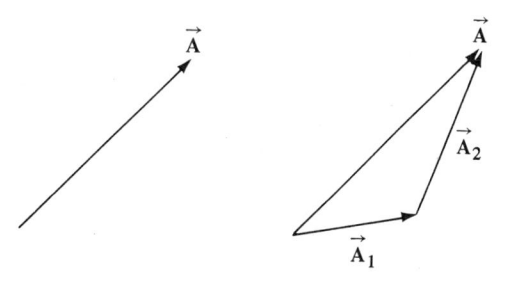

FIGURE 13.44

We will resolve this vector $\vec{A}$ into its components. A component of a vector is another vector, less than or equal to the vector, which can be considered a part (or component) of the vector. For example, in the sketch in Figure 13.44, $\vec{A}_1$ and $\vec{A}_2$ can be considered components of $\vec{A}$.

The rules for graphical addition of vectors require placing the vectors tail to tip. The sum is indicated by drawing a vector from the tail of the first to the tip of the last. Thus, in the sketch in Figure 13.44, $\vec{A} = \vec{A}_1 + \vec{A}_2$. When all the components are added, the original vector is obtained.

Place vector $\vec{A}$ on a set of x-y axes so that its tail is at the origin of coordinates. See Figure 13.45.

FIGURE 13.45

As components of $\vec{A}$ we choose two vectors, one parallel to the

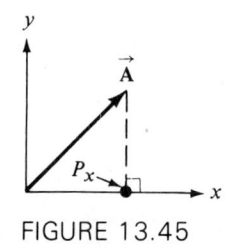

x-axis and one parallel to the y-axis, whose sum equals $\vec{A}$. To locate these vectors, drop a perpendicular from the tip of $\vec{A}$ to the x-axis. Call the point at which this perpendicular intersects the x-axis P_x. This is illustrated in Figure 13.46.

The components are the vectors shown in Figure 13.46. They are identified as $\vec{A}_x$ and $\vec{A}_y$.

It should be clear that these vectors satisfy the necessary condition of components; that is,

$$\vec{A}_x + \vec{A}_y = \vec{A}$$

From now on, we will refer to $\vec{A}_x$ as the **x-component** of $\vec{A}$ and $\vec{A}_y$ as the **y-component** of $\vec{A}$.

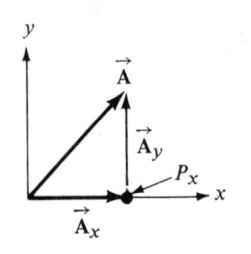

FIGURE 13.46

PERFORMANCE OBJECTIVE 11

You should now be able to sketch the x- and y-components of a vector.

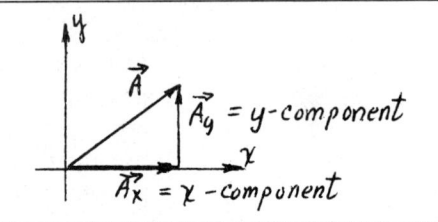

From the manner in which the x- and y-components have been formed, it is possible to derive algebraic relations between the magnitude of $\vec{A}$ and its components. In referring to the magnitude of a vector, we will use standard notaton. We will place the vector symbol between vertical bars, for example, magnitude of $\vec{A} = |\mathbf{A}|$.

Consider the right triangle formed by $\vec{A}_x$, $\vec{A}_y$, and $\vec{A}$. See Figure 13.47.

Let θ be the angle that vector $\vec{A}$ makes with the positive x-axis. From our definition of trigonometric functions, we can write

$$\cos \theta = \frac{A_x}{|\mathbf{A}|} \quad \text{and} \quad \sin \theta = \frac{A_y}{|\mathbf{A}|}$$

A_x and A_y are the signed lengths of $\vec{A}_x$ and $\vec{A}_y$ as defined previously. Rearranging these equations, we obtain the following formula:

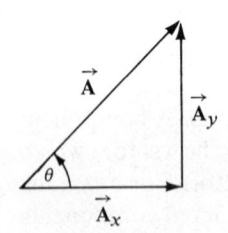

FIGURE 13.47

FORMULA

COMPONENTS OF VECTOR $\vec{A}$

$$A_x = |\mathbf{A}| \cos \theta, \quad A_y = |\mathbf{A}| \sin \theta$$

VECTOR RESOLUTION 13.4

Although we have worked out the components for a vector in the first quadrant only, the previous relations are true for a vector in any quadrant.

The calculator procedure for finding vector components is

Step 1. Enter magnitude.

Step 2. Press $\boxed{\times}$.

Step 3. Enter angle.

Step 4. Press $\boxed{\text{SIN}}$ or $\boxed{\text{COS}}$ as appropriate.

Step 5. Press $\boxed{=}$.

EXAMPLE Find the components of a vector 8.50 units long that makes an angle of 72.0° with the positive x-axis.

SOLUTION The x-component is obtained as follows:

8.5 $\boxed{\times}$ 72 $\boxed{\text{COS}}$ $\boxed{=}$ yields 2.63 to three significant digits

The y-component is obtained as follows:

8.5 $\boxed{\times}$ 72 $\boxed{\text{SIN}}$ $\boxed{=}$ yields 8.08 to three significant digits

EXAMPLE Find the components of a vector 17.3 units long that makes an angle of 145° with the positive x-axis.

SOLUTION

x-component: 17.3 $\boxed{\times}$ 145 $\boxed{\text{COS}}$ $\boxed{=}$ yields -14.2 to three significant digits

y-component: 17.3 $\boxed{\times}$ 145 $\boxed{\text{SIN}}$ $\boxed{=}$ yields 9.92

Note that the calculator produces the correct mathematical sign.

In practice, it is convenient to work with angles less than 90° since tables seldom go above 90°. This is possible by using a technique similar to that introduced in Section 13.3. There we determined that the trigonometric functions of an angle greater than 90° were always identical in absolute value to those of an angle less than 90°. We called that angle the smallest angle between the terminal side of the angle and the x-axis. In the vector case, it will be the smallest angle between the vector and the x-axis. Consider the sketch in Figure 13.48.

The trigonometric functions of the smallest angle (20°) are identical in magnitude to those of the 160° angle. We only have to determine signs. But the sign selection is obvious from the sketch. The

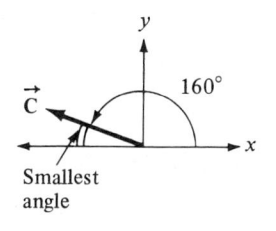

FIGURE 13.48

x-component must be negative and the y-component positive. So we can write

$$\mathbf{C}_x = -|\mathbf{C}|\cos 20°$$
$$\mathbf{C}_y = +|\mathbf{C}|\sin 20°$$

EXAMPLE Obtain numerical values for the x- and y-components of the vector shown in Figure 13.49 given that $\theta = 201°$ and $|\mathbf{A}| = 5.0$.

SOLUTION

$$\mathbf{A}_x = -|\mathbf{A}|\cos 21° = -(5.0)(0.9336) = -4.668 = -4.7$$
$$\mathbf{A}_y = -|\mathbf{A}|\sin 21° = -(5.0)(0.3584) = -1.792 = -1.8$$

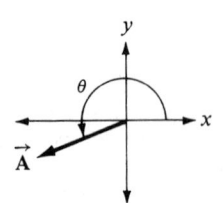

FIGURE 13.49

EXAMPLE Obtain numerical values for the x- and y-components of the vector shown in Figure 13.50 given that $\theta = 150°$ and $|\mathbf{B}| = 6.2$.

SOLUTION

$$\mathbf{B}_x = -|\mathbf{B}|\cos 30° = -(6.2)(0.866) = -5.369 = -5.4$$
$$\mathbf{B}_y = +|\mathbf{B}|\sin 30° = +(6.2)(0.5) = +3.1$$

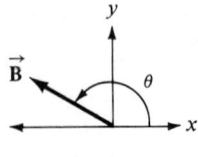

FIGURE 13.50

So far we have not discussed vectors along the coordinate axes. Finding components of such vectors will involve trigonometric functions of the quadrantal angles. Suppose the vector to be resolved lies along the x-axis as in Figure 13.51.

Then $\vec{\mathbf{A}}$ makes an angle of zero degrees with the positive x-axis. By the rule developed,

$$\mathbf{A}_x = |\mathbf{A}|\cos 0°$$
$$\mathbf{A}_y = |\mathbf{A}|\sin 0°$$

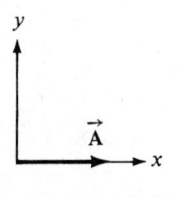

FIGURE 13.51

From the table, $\cos 0° = 1$ and $\sin 0° = 0$. Thus,

$$\mathbf{A}_x = |\mathbf{A}| \text{ and } \mathbf{A}_y = 0$$

The conclusion is that the vector's x-component equals the length of the vector itself and its y-component is zero. This makes sense because the vector is entirely in the x-direction.

EXAMPLE Find the x- and y-components of a vector along the negative y-axis.

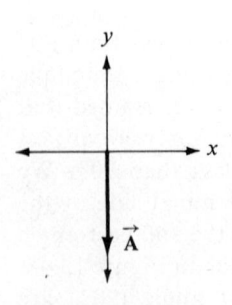

FIGURE 13.52

SOLUTION Consider the sketch in Figure 13.52. The smallest angle

VECTOR RESOLUTION 13.4

is 90°. Thus,

$$A_x = |\mathbf{A}| \cos 90° = |\mathbf{A}| \cdot 0 = 0$$

$$A_y = -|\mathbf{A}| \sin 90° = -|\mathbf{A}| \cdot 1 = -|\mathbf{A}|$$

It should be clear that we could have written down the answer to the preceding two examples by inspecting the diagrams. A vector along one of the axes always has only one nonzero component. That component is either the vector magnitude or the negative of that magnitude.

PERFORMANCE OBJECTIVE 12

You should now be able to evaluate the x- and y-components of a vector using tables or a calculator.

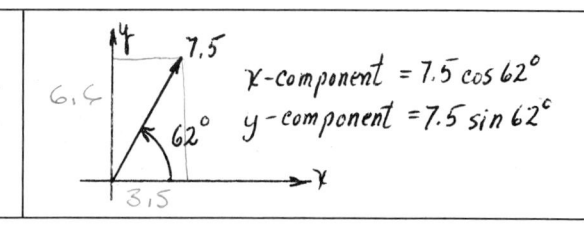

EXERCISE SET 13.4

1. Sketch the x- and y-components of the following vectors:

   a.   b.

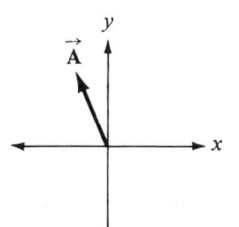

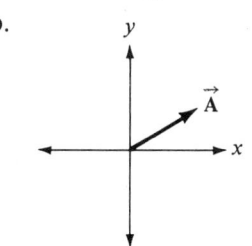

2. Sketch the x- and y-components of the following vectors:

   a.   b.

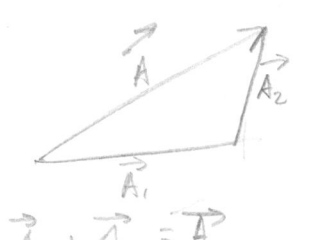

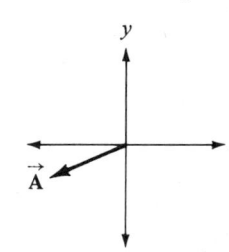

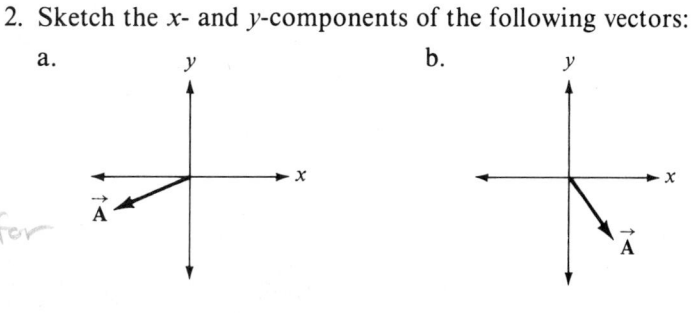

502   13 ADDITIONAL TOPICS IN TRIGONOMETRY

3. Determine the values of the x- and y-components of each vector $\vec{A}$ whose magnitude and direction are given. Directions are from the positive x-axis.

   a. $|A| = 8.50, \theta = 162°$
   b. $|A| = 4.01, \theta = 322°$
   c. $|A| = 7.52, \theta = 79°$
   d. $|A| = 15.6, \theta = 125°$
   e. $|A| = 6.62, \theta = -75°$
   f. $|A| = 9.23, \theta = \dfrac{2\pi}{3}$ rad
   g. $|A| = 19.1, \theta = \pi$ rad
   h. $|A| = 2.55, \theta = \dfrac{-3\pi}{4}$ rad

4. Determine the values of the x- and y-components of each vector $\vec{A}$ whose magnitude and direction are given. Directions are from the positive x-axis.

   a. $|A| = 7.90, \theta = 115°$
   b. $|A| = 5.12, \theta = 288°$
   c. $|A| = 8.35, \theta = 88°$
   d. $|A| = 12.1, \theta = -110°$
   e. $|A| = 4.91, \theta = 121°$
   f. $|A| = 8.51, \theta = \dfrac{3\pi}{4}$ rad
   g. $|A| = 15.6, \theta = \dfrac{-2\pi}{3}$ rad
   h. $|A| = 3.56, \theta = 1.2$ rad

5. List the smallest angle for each of the vectors in Problem 3.
6. List the smallest angle for each of the vectors in Problem 4.

SECTION 13.5 VECTOR COMPOSITION

In the preceding section we were given the magnitude and direction of a vector and asked to determine its components. Here we will be given the x- and y-components and asked to determine the vector's magnitude and direction.

Consider again vector $\vec{A}$ shown in Figure 13.53. Its x- and y-components are sketched.

We recall the relations between A_x, A_y and $|A|$. These are

$$A_x = |A| \cos \theta, \quad A_y = |A| \sin \theta$$

Rearranging these equations to solve for $|A|$, we obtain

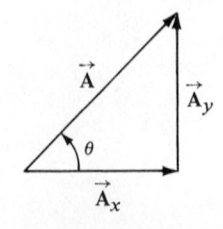

$$|A| = \dfrac{A_x}{\cos \theta} \quad \text{and} \quad |A| = \dfrac{A_y}{\sin \theta}$$

FIGURE 13.53

But these equations do not let us solve for $|A|$ yet because we do not

VECTOR COMPOSITION 13.5

know θ. To find θ, look again at the right triangle formed by $\vec{A}$ and its components. See Figure 13.54.

$$\text{Note that } \tan \theta = \frac{\text{opposite side}}{\text{adjacent side}} = \frac{\mathbf{A}_y}{\mathbf{A}_x}$$

Thus, θ is the angle whose tangent is given by $\frac{\mathbf{A}_y}{\mathbf{A}_x}$. We know the values of $\mathbf{A}_y$ and $\mathbf{A}_x$. Using the trigonometric tables or a calculator, we can find θ. In fact, knowing the numerical values of both $\mathbf{A}_y$ and $\mathbf{A}_x$ allows us to determine θ uniquely. (Knowing only the tangent value does not determine the angle uniquely). As an example, suppose $\mathbf{A}_y = -4.00$, $\mathbf{A}_x = 2.00$; then $\tan \theta = -\frac{4}{2} = -2$. Although the tangent is negative in both the second and fourth quadrants, each quadrant has a unique combination of signs for $\mathbf{A}_x$ and $\mathbf{A}_y$. That combination is shown in Figure 13.55.

In this example, since $\mathbf{A}_y = -4$ and $\mathbf{A}_x = 2$, the angle is in the fourth quadrant. From the tables, we find $\theta = 63°$ to the nearest degree. We can now sketch the direction of the vector. See Figure 13.56.

To find its magnitude, we may use the relation

$$|\mathbf{A}| = \frac{|\mathbf{A}_y|}{\sin \theta} = \frac{4}{\sin 63°} = \frac{4}{0.891} = 4.49$$

Note that the magnitude of $\mathbf{A}_y$ was used and, as a result, the smallest angle was obtained rather than the angle from the positive x-axis. This is a convenient shortcut that eliminates worrying about signs since $|\mathbf{A}|$ must be positive. We could have used

$$|\mathbf{A}| = \frac{|\mathbf{A}_x|}{\cos \theta} = \frac{2}{\cos 63°} = \frac{2}{0.454} = 4.41$$

The discrepancy in these answers is caused by the fact that the true angle is not 63° but 63.43°. This discrepancy results from a lack of precision in our tables and points to another advantage in using the calculator.

We now review the steps required to find the magnitude and direction of a vector.

RULES FOR FINDING THE MAGNITUDE AND DIRECTION OF A VECTOR (USING TABLES)

Step 1. Evaluate $\tan \theta = \frac{\mathbf{A}_y}{\mathbf{A}_x}$.

Step 2. Find θ from the tables.

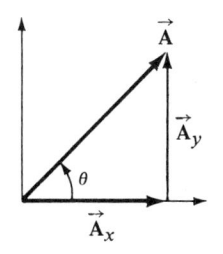

FIGURE 13.54

FIGURE 13.55

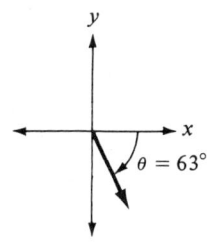

FIGURE 13.56

Step 3. Determine $|\mathbf{A}|$ using $|\mathbf{A}| = \dfrac{|\mathbf{A}_y|}{\sin \theta}$.

Step 4. Determine the quadrant of θ by looking at the signs of both $\mathbf{A}_x$ and $\mathbf{A}_y$.

Step 5. Plot the vector with correct direction and magnitude.

EXAMPLE Given $\mathbf{A}_x = -1.62$ and $\mathbf{A}_y = +5.65$, find the magnitude and direction of $\vec{\mathbf{A}}$ and plot $\vec{\mathbf{A}}$ on a set of coordinate axes.

SOLUTION

Step 1. $\tan \theta = \dfrac{5.65}{-1.62} = -3.49$.

Step 2. $\theta = 74°$ (tables).

Step 3. $|\mathbf{A}| = \dfrac{5.65}{\sin 74°} = \dfrac{5.65}{0.961} = 5.88$.

Step 4. θ is in the second quadrant.

Step 5. See Figure 13.57.

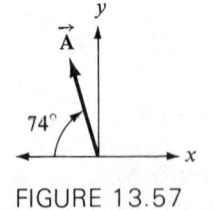

FIGURE 13.57

PERFORMANCE OBJECTIVE 13

You should now be able to determine the magnitude and direction of a vector given its components (using a table).

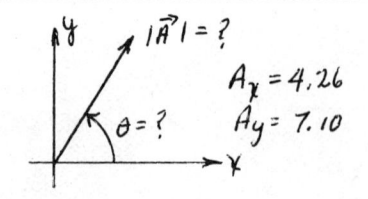

Finding the magnitude and direction of a vector when its components are known provides an excellent example of the power of a scientific calculator. This problem illustrates again the use of the **inverse** function capabilities of a calculator.

Steps 1 and 2 of the preceding procedure need to be combined to find an expression for θ. That is, we would like to solve the equation

$$\tan \theta = \frac{\mathbf{A}_y}{\mathbf{A}_x}$$

for θ. If asked to describe θ we might say "θ is the angle whose tangent equals the quotient $\dfrac{\mathbf{A}_y}{\mathbf{A}_x}$." In mathematical terms we are trying to solve the equation

$$\theta = \tan^{-1} \frac{\mathbf{A}_y}{\mathbf{A}_x}$$

VECTOR COMPOSITION 13.5

In Chapter 8 we discussed the use of the $\boxed{\text{INV}}$ key to find angles given the function values. We may find θ by a sequence of entries as follows:

$$\mathbf{A}_y \;\boxed{\div}\; \mathbf{A}_x \;\boxed{=}\; \boxed{\text{INV}} \;\boxed{\text{TAN}} \text{ and read } \theta$$

Step 3 stated previously requires the solution of the equation

$$|\mathbf{A}| = \frac{|\mathbf{A}_y|}{\sin \theta}$$

which we accomplish as follows:

$$\boxed{\text{SIN}} \;\boxed{\div}\; \mathbf{A}_y \;\boxed{=}\; \boxed{1/x} \text{ and read } \mathbf{A}$$

Disregard its sign if negative. This step assumes we have θ in the register.

RULES FOR FINDING THE MAGNITUDE AND DIRECTION OF A VECTOR (USING A CALCULATOR)

Step 1. Solve $\theta = \tan^{-1} \dfrac{\mathbf{A}_y}{\mathbf{A}_x}$

$$\mathbf{A}_y \;\boxed{\div}\; \mathbf{A}_x \;\boxed{=}\; \boxed{\text{INV}} \;\boxed{\text{TAN}}$$

Step 2. Solve

$$|\mathbf{A}| = \frac{|\mathbf{A}_y|}{\sin \theta}$$

$$\boxed{\text{SIN}} \;\boxed{\div}\; \mathbf{A}_y \;\boxed{=}\; \boxed{1/x}$$

Step 3. Determine the quadrant of θ and plot the vector.

You may wish to store θ before proceeding to Step 2 in case you forget to write down its value.

EXAMPLE Find the magnitude and direction of the vector with components $\mathbf{A}_x = -2.50$, $\mathbf{A}_y = 6.00$.

SOLUTION

Step 1. $6 \;\boxed{\div}\; 2.5 \;\boxed{+/-}\; \boxed{=}\; \boxed{\text{INV}} \;\boxed{\text{TAN}}$ and read -67.4 $\boxed{\text{STO}}$

Step 2. $\boxed{\text{SIN}} \;\boxed{\div}\; 6 \;\boxed{=}\; \boxed{1/x}$ and read 6.5 ignoring $-$ sign.

Step 3. See Figure 13.58.

FIGURE 13.58

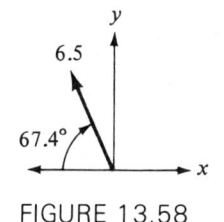

You may wonder why we seem to encounter some values with the wrong sign in solving these problems of vector composition from components. The problem arises because each function value does not correspond to a unique angle. For example, both 45° and 315° have the same cosine value of 0.7071. If we enter 0.7071 in the calculator and then press $\boxed{\text{INV}}$ and $\boxed{\text{COS}}$, we get 45° but the 315° answer doesn't appear. How has the decision been made to respond with 45° rather than 315°? The decision is somewhat arbitrary, but it does correspond to angles that are more likely to be used in calculations. We state here the range of values obtained when using the inverse functions.

Function	Range
inverse tangent	−90° to +90° (excluding −90° and +90°)
inverse cosine	0° to 180°
inverse sine	−90° to +90°

The easiest way to overcome this problem in the case of vector composition is to examine the sign of the angle at the end of Step 1. In the preceding example we obtained an angle of −67.38°. A look at the components (A_x was negative and A_y was positive) would indicate that the vector is in the second quadrant. The angle value with the positive x-axis is not −67.4° but is −67.4° + 180° = 112.6°. Using this value in Step 2 on the calculator will give the correct positive sign for the magnitude of the vector. This is illustrated in the revised solution procedure that follows:

Step 1. 6 $\boxed{\div}$ 2.5 $\boxed{+/-}$ $\boxed{=}$ $\boxed{\text{INV}}$ $\boxed{\text{TAN}}$ (read and check). $\boxed{+}$ 180 $\boxed{=}$ read 112.6 $\boxed{\text{STO}}$.

Step 2. $\boxed{\text{SIN}}$ $\boxed{\div}$ 6 $\boxed{=}$ $\boxed{1/x}$ and read 6.5 with correct sign.

EXAMPLE Find the magnitude and direction of the vector with components $A_x = -3.20$ and $A_y = -5.40$.

SOLUTION

Step 1. 5.4 $\boxed{+/-}$ $\boxed{\div}$ 3.2 $\boxed{+/-}$ $\boxed{=}$ $\boxed{\text{INV}}$ $\boxed{\text{TAN}}$ (read +59.3 and compare with components) $\boxed{+}$ 180 $\boxed{=}$ (read 239.3) $\boxed{\text{STO}}$.

Step 2. $\boxed{\text{SIN}}$ $\boxed{\div}$ 5.4 $\boxed{+/-}$ $\boxed{=}$ $\boxed{1/x}$ and read 6.28.

Step 3. See Figure 13.59.

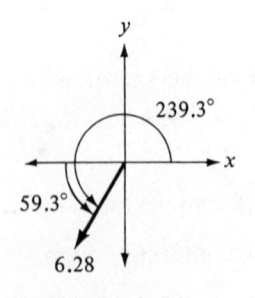

FIGURE 13.59

Note that the decision to add 180° is only appropriate if the vector components indicate that the vector is in the second or third

VECTOR COMPOSITION 13.5

quadrant. If the vector is in the first or fourth quadrant, an adjustment is not necessary.

EXAMPLE Find the magnitude and direction of the vector with components $A_x = 8.20$ and $A_y = -3.10$.

SOLUTION

Step 1. 3.1 $\boxed{+/-}$ $\boxed{\div}$ 8.2 $\boxed{=}$ $\boxed{INV}$ $\boxed{TAN}$ (read -20.7) $\boxed{STO}$.

Step 2. $\boxed{SIN}$ $\boxed{\div}$ 3.1 $\boxed{+/-}$ $\boxed{=}$ $\boxed{1/x}$ and read 8.77.

Step 3. See Figure 13.60.

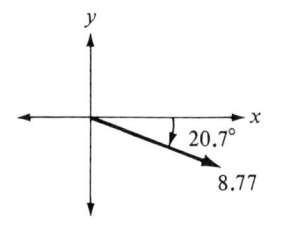

FIGURE 13.60

Determining the magnitude and direction of vectors that have only one nonzero component is easy. We should not try to follow the procedure outlined because a division by zero may result. A simple inspection of the nonzero component will indicate how to solve the problem.

EXAMPLE Find the magnitude and direction of the vector with components $A_x = 0$ and $A_y = -7.5$.

SOLUTION Since the vector has only one nonzero component, its magnitude must equal that component's absolute value. The vector must lie along the axis of the nonzero component and in the direction of its sign. Thus we have a vector of magnitude 7.5 along the negative y-axis.

PERFORMANCE OBJECTIVE 14	
You should now be able to determine the magnitude and direction of a vector given its components (using a calculator).	$A_x = -4.2, A_y = 3.5$ Procedure: 3.5 $\boxed{\div}$ 4.2 $\boxed{+/-}$ $\boxed{=}$ $\boxed{INV}$ $\boxed{TAN}$ $\boxed{+}$ 180 $\boxed{=}$ $\boxed{STO}$ $\boxed{SIN}$ $\boxed{\div}$ 3.5 $\boxed{=}$ $\boxed{1/x}$

EXERCISE SET 13.5

Find the magnitude and direction of the vectors having the following components. Plot each vector on a set of x-y-axes.

1. $A_x = -4.60, A_y = -3.47$
2. $A_x = 2.3, A_y = -3.4$
3. $A_x = -8.3, A_y = 6.5$
4. $A_x = -7.6, A_y = 0$
5. $A_x = 8.5, A_y = 9.78$
6. $A_x = 4.9, A_y = 7.1$

7. $A_x = -8.6$, $A_y = 3.2$

8. $A_x = -7.22$, $A_y = 4.91$

9. $A_x = 3.87$, $A_y = 0$

10. $A_x = -4.55$, $A_y = -10.6$

SECTION 13.6 VECTOR ADDITION

We are now ready to determine the resultant of two or more vectors. The term **resultant** is simply another way of saying sum. Resultant sometimes is taken to mean "net." For example, if a four pound force is acting due east and a three pound force is acting due west, we say the resultant or net force is a one pound force acting due east. The one pound force acting east is the sum of the two forces mentioned. Thus you may see the words *sum, resultant,* and *net* used interchangeably.

In Appendix I-4 we add vectors graphically. Such an addition is illustrated in Figure 13.61.

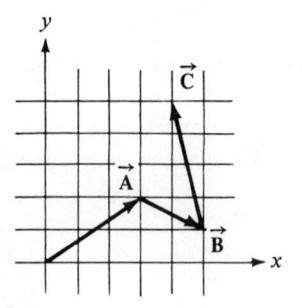

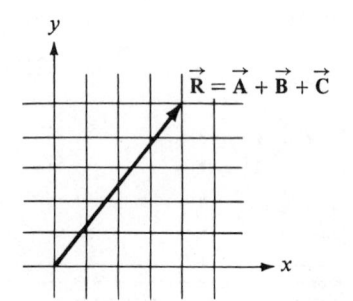

FIGURE 13.61

The resultant is obtained by drawing a vector from the tail of $\vec{A}$ to the tip of $\vec{C}$. You should verify that the order in which the vectors are placed on the diagram makes no difference in the resultant. That is, we could have drawn $\vec{C}$ first, then $\vec{A}$, and then $\vec{B}$. Also note the following breakdown of components of each vector:

Vector	x-component	y-component
$\vec{A}$	+3	+2
$\vec{B}$	+2	−1
$\vec{C}$	−1	+4
$\vec{R}$	+4	+5

VECTOR ADDITION 13.6

Looking at this table, we observe that

$$\mathbf{A}_x + \mathbf{B}_x + \mathbf{C}_x = \mathbf{R}_x$$

and that

$$\mathbf{A}_y + \mathbf{B}_y + \mathbf{C}_y = \mathbf{R}_y$$

In other words, the x-component of the resultant is obtained by summing the x-components of the vectors to be added. The y-component of the resultant is obtained similarly. This is the key to vector addition.

EXAMPLE Three vectors, $\vec{\mathbf{A}}_1$, $\vec{\mathbf{A}}_2$, and $\vec{\mathbf{A}}_3$, are given in component form. Find the magnitude and direction of their resultant.

Vector	x	y
$\vec{\mathbf{A}}_1$	5.1	6.7
$\vec{\mathbf{A}}_2$	−3.4	3.2
$\vec{\mathbf{A}}_3$	−6.5	−2.8

SOLUTION As a first step, we will find $\mathbf{R}_x$ and $\mathbf{R}_y$.

$$\mathbf{R}_x = 5.1 + (-3.4) + (-6.5) = -4.8$$
$$\mathbf{R}_y = 6.7 + 3.2 + (-2.8) = +7.1$$

Knowing the components of the resultant vector, we must now find its magnitude and direction. In the last section we solved such problems. Here are the steps.

Step 1. $\theta = \tan^{-1} \dfrac{7.1}{-4.8} = -55.9°$.

7.1 [÷] 4.8 [+/−] [=] [INV] [TAN] (read −55.9)
[+] 180 [=] and read 124.1.

Step 2. $|\mathbf{R}| = \dfrac{7.1}{\sin 124.1°} = 8.57$.

[SIN] [÷] 7.1 [=] [1/x].

Step 3. θ is in the second quadrant as shown in Figure 13.62.

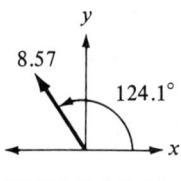

FIGURE 13.62

Unfortunately, we are not always given the vectors to be added in component form. Frequently, they are given as magnitudes and directions. When this happens we must first place each vector in component form. We have done this in Section 13.4.

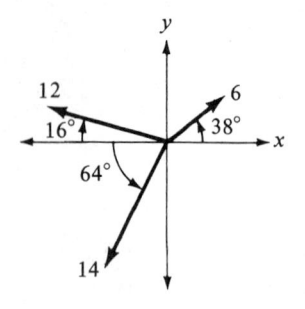

FIGURE 13.63

EXAMPLE Determine the resultant of the three vectors sketched in Figure 13.63. Each vector is labeled as to direction. Vector magnitude is indicated at the vector tip.

SOLUTION Establish a table in which to place the component values.

Vector	x	y
6	6 cos 38°	6 sin 38°
14	−14 cos 64°	−14 sin 64°
12	−12 cos 16°	+12 sin 16°

Notice that we have indicated the components in terms of sin and cos with correct mathematical signs. Each vector has been given the name corresponding to its length.

The table needs to be redrawn with numerical values. Looking up values and performing the multiplications indicated, we find

Vector	x	y
6	4.73	3.69
14	−6.14	−12.58
12	−11.54	3.31
R	−12.95	−5.58

The columns have been added to find $\mathbf{R}_x$ and $\mathbf{R}_y$. Then,

$$\tan \theta = \frac{\mathbf{R}_y}{\mathbf{R}_x} = \frac{-5.58}{-12.95} = 0.431$$

$$\theta = 23.3° + 180° = 203.3°$$

θ is in the third quadrant since both $\mathbf{R}_x$ and $\mathbf{R}_y$ are negative.

$$|\mathbf{R}| = \frac{\mathbf{R}_y}{\sin \theta} = \frac{-5.58}{\sin 203.3°} = \frac{-5.58}{-0.3955} = 14.1$$

The resultant is shown in Figure 13.64 to three significant figures.

This is the most difficult type problem we will encounter with vector resolution. You are encouraged to set up tables to solve these problems as illustrated in the example. Neatness and good bookkeeping help.

We illustrate a method of solving the vector addition problem using the calculator. Values should be entered in a table as they are determined on the calculator.

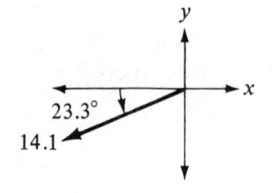

FIGURE 13.64

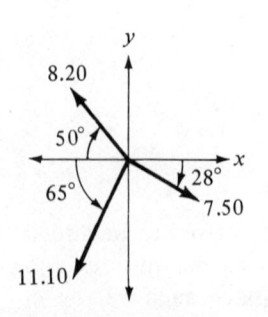

FIGURE 13.65

EXAMPLE Find the resultant of the vectors shown in Figure 13.65.

VECTOR ADDITION 13.6

SOLUTION Here is the table used to find resultant components.

Vector	x	y
7.50	6.622	−3.521
8.20	−5.271	6.282
11.10	−4.691	−10.060
R	−3.340	−7.299

These components were obtained as follows:

7.5 [×] 28 [COS] [=] and record 6.622
8.2 [×] 50 [COS] [=] and record −5.271
11.1 [×] 65 [COS] [=] and record −4.691
7.5 [×] 28 [SIN] [=] and record −3.521
8.2 [×] 50 [SIN] [=] and record 6.282
11.1 [×] 65 [SIN] [=] and record −10.060
6.622 [−] 5.271 [−] 4.691 [=] and record −3.340 [STO]
3.521 [+/−] [+] 6.282 [−] 10.060 [=] and record −7.299
[÷] [RCL] [=] [INV] [TAN] and record 65.41 [STO]
[SIN] [÷] 7.299 [=] [1/x] and record 8.027

The resultant is shown in Figure 13.66 (to three significant figures).

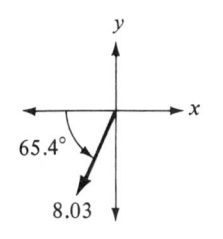

FIGURE 13.66

PERFORMANCE OBJECTIVE 15

You should now be able to determine the sum of two or more vectors using tables or a calculator.

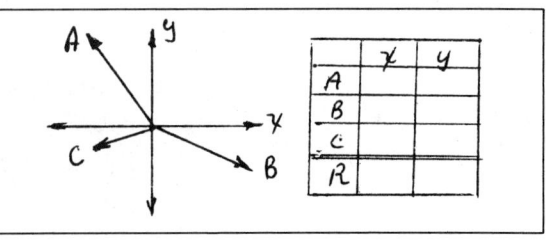

EXERCISE SET 13.6

1. Find the resultant of vectors $\vec{A}$ and $\vec{B}$. Give magnitude and direction and sketch the resultant vector.

Vector	x	y
$\vec{A}$	7.2	−1.9
$\vec{B}$	−3.6	−1.7

13 ADDITIONAL TOPICS IN TRIGONOMETRY

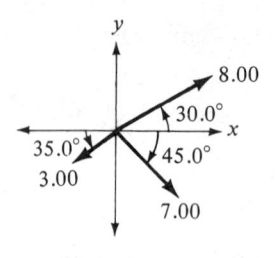

FIGURE 13.67

2. Find the resultant of the three vectors $\vec{A}$, $\vec{B}$, and $\vec{C}$.

Vector	x	y
$\vec{A}$	0	5.1
$\vec{B}$	4.7	−1.2
$\vec{C}$	−2.5	−2.6

3. Find the magnitude and direction of the resultant of the three vectors shown in Figure 13.67.

4. Find the magnitude and direction of the resultant of the three vectors shown in Figure 13.68.

5. Find the resultant of the following vectors. All angles are measured from the positive x-axis. Draw a sketch of the vectors and their resultant.

Vector	Magnitude	Direction
$\vec{A}$	15.2	−40.0
$\vec{B}$	7.10	+90.0
$\vec{C}$	10.2	+150.0

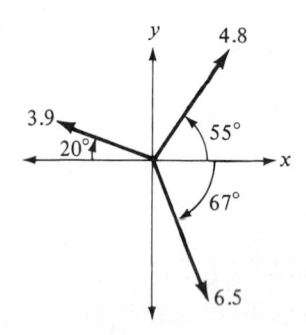

FIGURE 13.68

6. Find the resultant of the following vectors. All angles are measured from the positive x-axis. Draw a sketch of the vectors and their resultant.

Vector	Magnitude	Direction
$\vec{A}$	5.32	+180.0
$\vec{B}$	7.69	−71.0
$\vec{C}$	3.38	+135.0

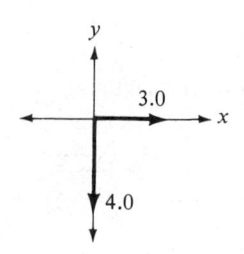

FIGURE 13.69

7. Find the components and the resultant of the vectors in Figure 13.69.

8. Find the components and the resultant of the vectors in Figure 13.70.

9. Find the components and the resultant of the vectors in Figure 13.71.

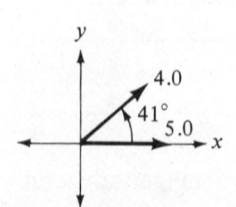

FIGURE 13.70

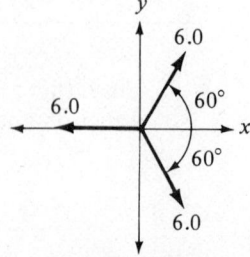

FIGURE 13.71

10. Find the components and the resultant of the vectors in Figure 13.72.

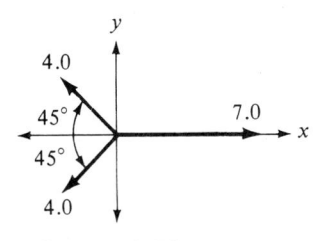

FIGURE 13.72

SECTION 13.7   STATICS PROBLEMS

Resolution of vectors is essential in analyzing the forces in a system at equilibrium. **Equilibrium** refers to a condition where the sum of the external forces on a system equals zero. When in equilibrium, the system is motionless (or moving with constant speed). We use this equilibrium condition to determine internal forces in a system.

FIGURE 13.73

EXAMPLE   A heavy weight **W** (1500 lb) is supported by cables attached to the ceiling. Find the tension in each of the cables. See Figure 13.73.

SOLUTION   The tension in each cable is the force being exerted along the length of the cable. We draw a vector diagram showing all forces acting at point P. See Figure 13.74.

We want the resultant of these forces to be zero since the system is in equilibrium.

Vector	x-component	y-component
1500	0	-1500
T_1	$-T_1 \cos 32°$	$T_1 \sin 32°$
T_2	$T_2 \cos 61°$	$T_2 \sin 61°$
R	0	0

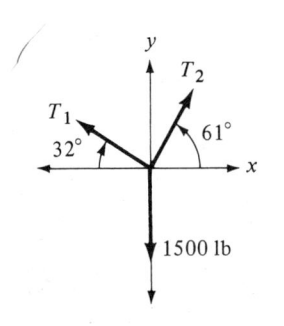

FIGURE 13.74

Note that the resultant is known. The problem is to find values for T_1 and T_2. We proceed as before by adding x- and y-components, but now we set them equal to zero.

(1) $\qquad 0 - T_1 \cos 32° + T_2 \cos 61° = 0$

(2) $\qquad -1500 + T_1 \sin 32° + T_2 \sin 61° = 0$

We now have two equations in two unknowns. We use the method of substitution to solve these equations. Solve equation (1) for T_2.

$$T_2 \cos 61° = T_1 \cos 32°$$

$$T_2 = \frac{T_1 \cos 32°}{\cos 61°}$$

Substitute T_2 into (2) to obtain

$$-1500 + T_1 \sin 32° + \left(\frac{T_1 \cos 32°}{\cos 61°} \sin 61°\right) = 0$$

$$T_1 \left(\sin 32° + \frac{\cos 32° \cdot \sin 61°}{\cos 61°}\right) = 1500$$

$$T_1 \left(0.53 + \frac{(0.848)(0.875)}{0.485}\right) = 1500$$

$$T_1 (2.06) = 1500$$

$$T_1 = \frac{1500}{2.06} = 728.2 \text{ lb}$$

Substitute back into $T_2 = \frac{T_1 \cos 32°}{\cos 61°}$ to find

$$T_2 = \frac{(728.2)(0.848)}{0.485} = 1273 \text{ lb}$$

PERFORMANCE OBJECTIVE 16

You should now be able to use vector addition to solve equilibrium problems.

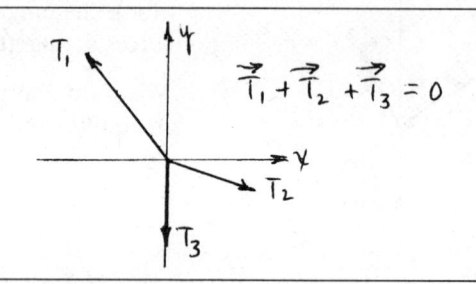

$$\vec{T_1} + \vec{T_2} + \vec{T_3} = 0$$

EXERCISE SET 13.7

1. Calculate the force necessary to hold the cable sideways as shown in Figure 13.75. Also calculate the increase in tension in the cable above the point P.

2. Calculate the force necessary to hold the cable sideways as shown in Figure 13.76. Also calculate the increase in tension in the cable above the point P.

SUMMARY 13.8

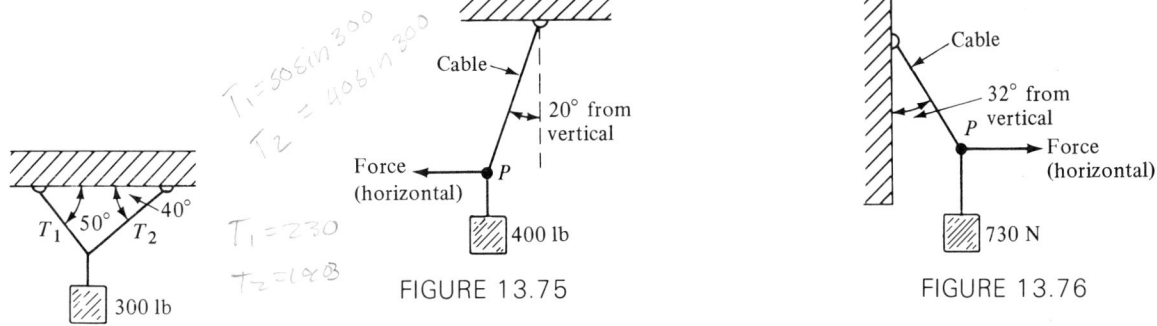

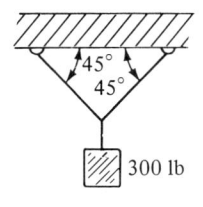

FIGURE 13.77 — hanging weight 300 lb, with T_1 at 50° and T_2 at 40° from ceiling.

Handwritten: $T_1 = 50 \sin 300$, $T_2 = 40 \sin 300$

FIGURE 13.75 — Cable at 20° from vertical, horizontal force at P, 400 lb weight.

Handwritten: $T_1 = 230$, $T_2 = 198$

FIGURE 13.76 — Cable at 32° from vertical, horizontal force at P, 730 N weight.

3. Calculate the tensions T_1 and T_2 in the system in Figure 13.77.
4. Calculate the tensions T_1 and T_2 in the system in Figure 13.78.
5. Calculate the weight of the suspended object and the tension in the upper part of the cable in Figure 13.79.
6. Calculate the maximum weight **W** if the tension in the upper part of the cable in Figure 13.80 is not to exceed 1200 N. Also determine the angle between the wall and the cable when this maximum weight is supported.

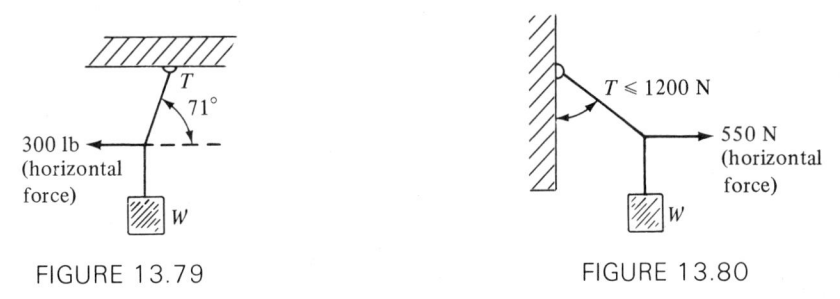

FIGURE 13.78 — 45°, 45°, 300 lb.

FIGURE 13.79 — T at 71°, 300 lb horizontal force, W weight.

FIGURE 13.80 — $T \leq 1200$ N, 550 N (horizontal force), W weight.

SECTION 13.8 SUMMARY

In this chapter we have redefined the angle and taken a more general look at the trigonometric functions in terms of that angle. We have applied the three basic trigonometric functions (sine, cosine, and tangent) to the problems of vector resolution and composition. Key terms and concepts are outlined here.

Systems of angular measurement: the three systems introduced are

degree measure: uses $\frac{1}{360}$ of a circle as the unit angle.

radian measure: uses the angle which subtends an arc equal to the arc radius as the unit angle.

rotation measure: uses the full circle as the unit angle.

Linear-angular relations: useful relations involving the angle in radian measure. They are

Arc length, radius, and angle: $s = r\theta$

Area, radius, and angle: $A = 0.5r^2\theta$

Linear speed, radius, and angular speed: $v = r\omega$

Standard position of an angle: a position in which the angle has its initial side along the positive x-axis and its vertex at the origin.

Quadrantal angle: an angle in standard position that has its terminal side along one of the coordinate axes.

Sine: a measure of an angle that equals the ratio of the y-coordinate of a point on its terminal side to the distance of that point from the origin.

Cosine: a measure of an angle that equals the ratio of the x-coordinate of a point on its terminal side to the distance of that point from the origin.

Tangent: a measure of an angle that equals the ratio of the y-coordinate of a point on its terminal side to the x-coordinate of that point.

Additional trigonometric functions: The three reciprocal functions cosecant, secant, and cotangent were defined. These functions can also be used to solve triangles and other practical problems.

Smallest angle: the smallest angle between the terminal side of an angle and the x-axis when the angle is in standard position. The smallest angle has trigonometric functions equal in magnitude to those of the original angle.

Vector components: the components of a vector are vectors less than or equal to the vector that add to form the vector. Components along the x- and y-axes are most commonly used in practical problems.

Vector resolution: the process of breaking a vector into its components.

Vector composition: the process of finding the magnitude and direction of a vector when its components are known.

Vector addition: the process of determining the sum of a set of vectors. In this chapter we have developed methods for finding the numerical values of the magnitude and direction of this vector sum.

Equilibrium the condition in which the vector sum of forces on a system is equal to zero.

CHAPTER 13 PRACTICE TEST

1. (PO-1) The wheel shown in Figure 13.81 rotates from position 1 to position 2 in a clockwise direction.

   Position 1   Position 2

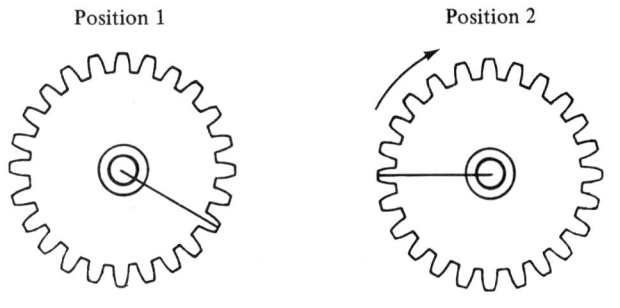

   FIGURE 13.81

   a. Draw an angle representing the rotation and label each part of the angle.
   b. Give the mathematical sign of the angle.

2. (PO-2) Express each of the following angles in the other two systems of measurement:
   a. 170°
   b. 1 radian
   c. 0.75 rotation
   d. 70°
   e. 5 radians

3. (PO-3) Consider that s, r, and θ represent arc length, radius and central angle of some circle. Complete the following table for the unknown quantities:

s	r	θ
3 meters	0.5 meters	
	2.00 feet	60.0°
16"		$\frac{\pi}{2}$ rad
	1.20 m	160°
2.5 f	5 f	

4. (PO-3, 4) A wheel of radius 13 cm is rotating at 30 revolutions per second.
   a. Calculate the velocity of a point on the edge (rim) of the wheel.
   b. Calculate the distance a point on the rim moves when the wheel turns through a quarter of a revolution.

5. (PO-5) Identify the quadrant of each of the angles in Problem 2. If the angle is quadrantal, so state and indicate the bordering quadrants.

6. (PO-6) Find all six trigonometric functions for the following angles:

   a.

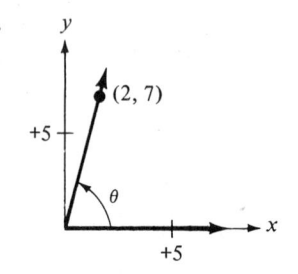

   b.

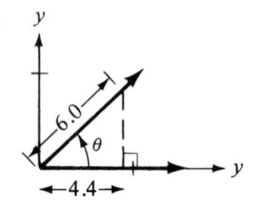

7. (PO-7) Find all six trigonometric functions for the following angles:

   a.

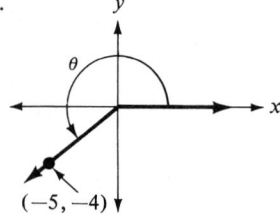

   b. P is 7 units from the origin with an x-coordinate of −3.

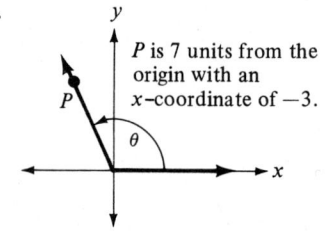

8. (PO-8) Evaluate all six trigonometric functions of the following angles using tables:

   a. $\theta = 132°$ b. $\theta = 240°$ c. $\theta = 295°$

9. (PO-9) Evaluate all six trigonometric functions of the following angles using the calculator:

   a. $\theta = 112°$ b. $\theta = 192°$ c. $\theta = 315°$

10. (PO-10) List all quadrantal values of the following functions by completing the table:

$\theta =$	90°	180°	270°	360°
$\cos \theta$				
$\sin \theta$				
$\tan \theta$				

CHAPTER 13 PRACTICE TEST

11. (PO-11, 12) Sketch and evaluate the x- and y-components of the following vectors:

    a. $\theta = 205°$, magnitude 8.0

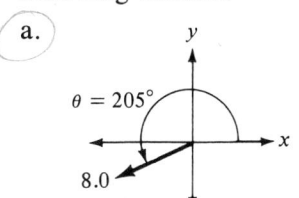

    b. $\theta = 113°$, magnitude 12.0

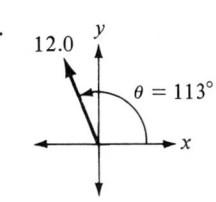

12. (PO-13, 14) Determine the magnitude and direction of the following vectors using either tables or calculator:

    a. $A_x = -3.50$
       $A_y = 5.50$

    b. $A_x = 7.00$
       $A_y = -4.20$

13. (PO-15) Determine the resultant of the following vectors:

    a. 7.00 along +y; 8.20 at 28.0° below +x

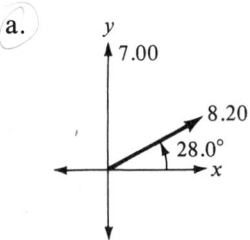

    b. 10.15 at 15.0° above −x; 6.00 at 20.0° above +x; 8.20 at 65.0° below +x (in the fourth-quadrant direction as shown)

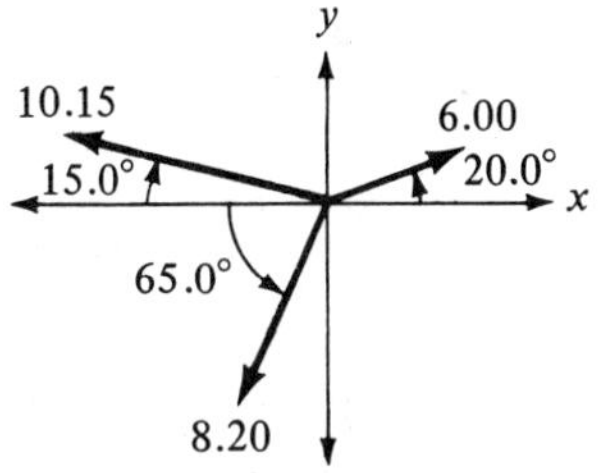

14. (PO-16) Calculate the indicated unknowns in the following statics problems:

    a. Rope at 58° from wall to a horizontal pulley; 500 lb over pulley supporting W. Evaluate T and W.

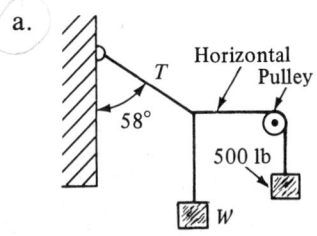

    b. Cable from ceiling at 41.0° and 39.0°; 820 N on one side, T and W on the other. Calculate T and W.

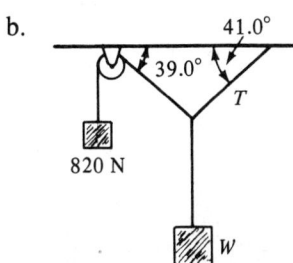

CHAPTER 14

GRAPHING THE TRIGONOMETRIC FUNCTIONS

14.1 Trigonometric Function Values by Quadrant
14.2 Graphs of the Sine, Cosine, and Tangent
14.3 The Sine and Cosine with Amplitude and Phase
14.4 Sine Curves and Rotating Vectors
14.5 Applications of Graphing (Electronics)
14.6 Summary

We have used the trigonometric functions (sine, cosine, and tangent) primarily to solve geometric type problems. However, these functions are interesting for their own sake. These functions have characteristic graphs just as we found for linear and quadratic functions. The graphs are often used to represent the behavior of real systems. For example, the position of a pendulum as a function of time can be represented by either a sine graph or cosine graph. Electronics students will recognize the sine wave as the usual output of an oscillator.

In this chapter we will concentrate on the graphical representation of the three trigonometric functions mentioned above. We will consider the graph of the sine function with both amplitude and phase as parameters. This can lead us to an understanding of interesting problems in electronics, mechanics, and basic physics.

SECTION 14.1 TRIGONOMETRIC FUNCTION VALUES BY QUADRANT

We have discussed the trigonometric functions as processors that accept as input an angle and return a number as output. In this section we will investigate the values that are acceptable as inputs to the processors. We will also determine the permitted output values.

Figure 14.1 displays the diagram from which the trigonometric functions were defined.

$$\sin \theta = \frac{y}{r}, \quad \cos \theta = \frac{x}{r}, \quad \tan \theta = \frac{y}{x}$$

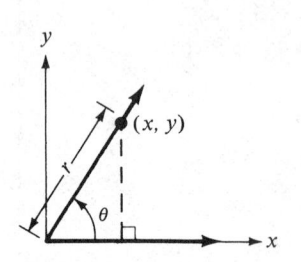

FIGURE 14.1

Consider first the equation $\sin \theta = \frac{y}{r}$. This expression establishes a relation between angle θ (the independent variable) and the ratio $\frac{y}{r}$ called $\sin \theta$ (the dependent variable). The ordered pairs that arise from this relation would be written $\left(\theta, \frac{y}{r}\right)$ or equivalently $(\theta, \sin \theta)$.

Think about the permitted values of θ. To this point we have considered positive and negative angles in the interval $-360°$ to $+360°$. That is, we have not considered angles greater than a single rotation in either direction. But there is no reason to exclude larger positive and negative values of θ. For example, consider an angle of $420°$ shown in Figure 14.2. The angle is formed by completing one rotation plus $60°$.

To evaluate the trigonometric functions of $420°$, we form certain ratios of x, y, and r, which are the sides of a triangle obtained when a perpendicular is dropped to the x-axis from the terminal side of the angle. But the same construction gives the trigonometric functions of $60°$. Thus, the trigonometric functions of $420°$ equal the corresponding functions of $60°$. It appears that θ (the independent variable) can take on any value whatsoever and the sine function will always be defined.

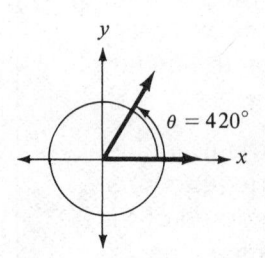

FIGURE 14.2

Next consider the permitted values of the trigonometric function (the dependent variable). That is, in the case of $(\theta, \sin \theta)$, what values can the ratio $\frac{y}{r}$ take on? Noting that y and r are respectively the side and hypotenuse of a right triangle and that the hypotenuse can never be smaller than a side, the ratio $\frac{y}{r}$ must be less than or equal to one in absolute value. We could also look at the values in trigonometric tables to find the limiting values of $\frac{y}{r}$.

TRIGONOMETRIC FUNCTION VALUES BY QUADRANT 14.1

PERMITTED VALUES OF THE PAIRS (θ, sin θ)

θ: all real values (also called domain)
sin θ: $-1 \leq \sin \theta \leq +1$ (also called range)

The pairs (θ, cos θ) have the same domain and range as the pairs (θ, sin θ).

Next we consider the permitted values of sin θ by quadrant for positive angles as the angle increases. This will help later in plotting (θ, sin θ). The diagrams in Figures 14.3(a), (b), (c), and (d) may help to visualize the change in sin θ as θ increases.

First quadrant

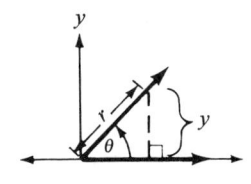

θ	sin θ
$0 \to \pi/2$	$0 \to +1$

Consider y/r as y increases from 0 to r.

(a)

Second quadrant

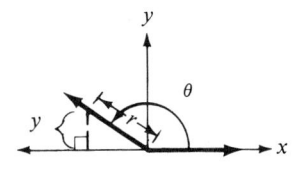

θ	sin θ
$\pi/2 \to \pi$	$+1 \to 0$

Consider y/r as y decreases from r to 0.

(b)

Third quadrant

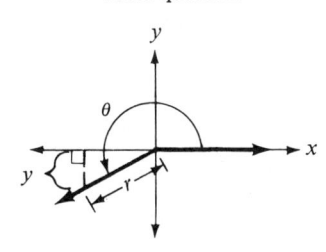

θ	sin θ
$\pi \to 3\pi/2$	$0 \to -1$

Consider y/r as y decreases from 0 to $-r$.

(c)

Fourth quadrant

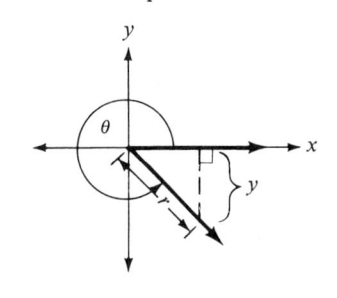

θ	sin θ
$3\pi/2 \to 2\pi$	$-1 \to 0$

Consider y/r as y increases from $-r$ to 0.

(d)

FIGURE 14.3

14 GRAPHING THE TRIGONOMETRIC FUNCTIONS

These results are summarized in Table 14.1 along with permitted values of $(\theta, \cos \theta)$. The values of $(\theta, \cos \theta)$ are obtained by considering the values of the ratio $\dfrac{x}{r}$ in each of the quadrants.

TABLE 14.1 $(\theta, \text{SIN } \theta)$ AND $(\theta, \text{COS } \theta)$ BY QUADRANT

θ	$\sin \theta$	$\cos \theta$
$0 \leq \theta < \frac{\pi}{2}$	$0 \leq \sin \theta < +1$	$+1 \geq \cos \theta > 0$
$\frac{\pi}{2} \leq \theta < \pi$	$+1 \geq \sin \theta > 0$	$0 \geq \cos \theta > -1$
$\pi \leq \theta < \frac{3\pi}{2}$	$0 \geq \sin \theta > -1$	$-1 \leq \cos \theta < 0$
$\frac{3\pi}{2} \leq \theta < 2\pi$	$-1 \leq \sin \theta < 0$	$0 \leq \cos \theta < +1$

Next we consider the permitted values of $\tan \theta$. Recall that $\tan \theta = \dfrac{y}{x}$. We introduce a notation that will be convenient in defining the permitted values of trigonometric functions. We can visualize the number line corresponding to the x- or y-axis as extending indefinitely to the left and right (x-axis) or up and down (y-axis). However, it will simplify notation to think of the number line terminating in an infinitely distant and undefined point called $-\infty$ (read "minus infinity") on the left and in a point called $+\infty$ on the right. Thus, the permitted values of θ in the relation $(\theta, \sin \theta)$ are $-\infty < \theta < +\infty$. This is another way of saying that all real values of θ are permitted. This, of course, will not be the case for all trigonometric functions since some of them are undefined at the quadrantal angles.

This new notation is used in describing values of $\tan \theta$ in Figures 14.4(a), (b), (c), and (d).

First quadrant

θ	$\tan \theta$
$0 \to \pi/2$	$0 \to +\infty$

Consider y/x as y is positive and increasing, x is positive and decreasing to zero.

(a)

Second quadrant

θ	$\tan \theta$
$\pi/2 \to \pi$	$-\infty \to 0$

Consider y/x as y is positive and decreasing to zero, x is negative and decreasing to $-r$.

(b)

FIGURE 14.4

TRIGONOMETRIC FUNCTION VALUES BY QUADRANT 14.1

Third quadrant

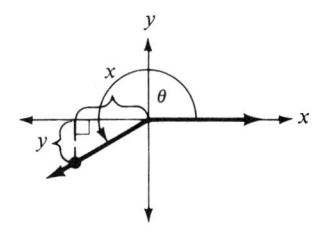

θ	$\tan \theta$
$\pi \to 3\pi/2$	$0 \to +\infty$

Consider y/x as y is negative and decreasing to $-r$, x is negative and decreasing to 0.

(c)

Fourth quadrant

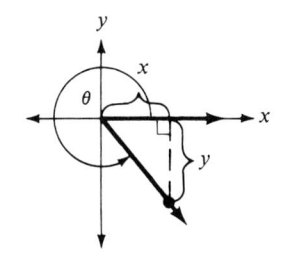

θ	$\tan \theta$
$3\pi/2 \to 2\pi$	$-\infty \to 0$

Consider y/x as y is negative and increasing to 0, x is positive and increasing from 0 to r.

(d)

FIGURE 14.4 (cont.)

We have previously observed that an expression of the form $\frac{a}{0}$ with $a \neq 0$ is undefined. However, for purposes of examining the permitted values of a function, it is often convenient to specify undefined numbers of this form as either $+\infty$ or $-\infty$. As noted earlier, infinity represents an undefined point on the end of the number line. Whether we have $+\infty$ or $-\infty$ is determined by whether the denominator in the undefined number approaches zero through the same or different signed values as the numerator. These results are summarized in Table 14.2.

TABLE 14.2 (θ, TAN θ) BY QUADRANT

θ	$\tan \theta$
$0 \leq \theta < \frac{\pi}{2}$	$0 \leq \tan \theta < +\infty$
$\frac{\pi}{2} < \theta < \pi$	$-\infty < \tan \theta < 0$
$\pi \leq \theta < \frac{3\pi}{2}$	$0 \leq \tan \theta < +\infty$
$\frac{3\pi}{2} < \theta < 2\pi$	$-\infty < \tan \theta < 0$

$\frac{\pi}{2}$ and $\frac{3\pi}{2}$ are not permitted values of θ in the relation (θ, $\tan \theta$) since the corresponding tangent values are undefined.

PERMITTED VALUES OF THE PAIRS (θ, TAN θ)

θ: all real θ except $\theta = \dfrac{n\pi}{2}$
(n an odd integer)

tan θ: all real numbers

PERFORMANCE OBJECTIVE 1

You should now be able to determine the range of values of the sine, cosine, and tangent by quadrant.

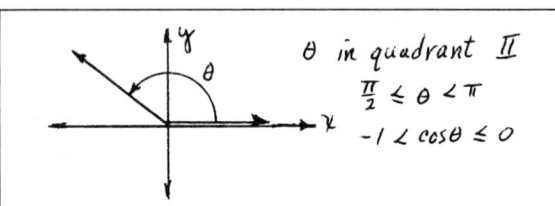

θ in quadrant II
$\dfrac{\pi}{2} \leq \theta \leq \pi$
$-1 \leq \cos\theta \leq 0$

EXERCISE SET 14.1

1. Without referring to the previous tables, complete the following statements:
   a. As θ increases from $\dfrac{\pi}{2}$ to π in quadrant II,
      sin θ (increases)(decreases) from _____ to _____.
      cos θ (increases)(decreases) from _____ to _____.
      tan θ (increases)(decreases) from _____ to _____.
   b. As θ increases from π to $\dfrac{3\pi}{2}$ in quadrant III,
      sin θ (increases)(decreases) from _____ to _____.
      cos θ (increases)(decreases) from _____ to _____.
      tan θ (increases)(decreases) from _____ to _____.

2. Without referring to the previous tables, complete the following statements:
   a. As θ increases from 0 to $\dfrac{\pi}{2}$ in quadrant I,
      sin θ (increases)(decreases) from _____ to _____.
      cos θ (increases)(decreases) from _____ to _____.
      tan θ (increases)(decreases) from _____ to _____.
   b. As θ increases from $\dfrac{3\pi}{2}$ to 2π in quadrant IV,
      sin θ (increases)(decreases) from _____ to _____.
      cos θ (increases)(decreases) from _____ to _____.
      tan θ (increases)(decreases) from _____ to _____.

3. Recalling that the cosecant function is the reciprocal of the sine function, develop a table of permitted values of the pairs $(\theta, \csc \theta)$ by quadrant.

4. Recalling that the secant function is the reciprocal of the cosine function, develop a table of permitted values of the pairs $(\theta, \sec \theta)$ by quadrant.

SECTION 14.2 GRAPHS OF THE SINE, COSINE, AND TANGENT

The material in the last section on permitted values of the trigonometric functions by quadrant was intended as preparation for graphing. In particular, if we understand Section 14.1, we will be able to sketch the general shape of the graphs of the trigonometric functions.

We will take a slightly different point of view in order to graph the trigonometric functions. Angle θ will no longer be displayed as an angle in standard position but will be thought of as a variable. Since θ is the first member of our ordered pairs, we will relabel it as x and use a rectangular x-y coordinate system in graphing. Thus, values of θ will be displayed along the x-axis and values of the trigonometric functions will be displayed along the y-axis. We will be plotting the following equations:

$$y = \sin x \text{ corresponding to } (x, \sin x)$$
$$y = \cos x \text{ corresponding to } (x, \cos x)$$
$$y = \tan x \text{ corresponding to } (x, \tan x)$$

Consider the coordinate system in Figure 14.5 on which x values corresponding to positive θ values in the four quadrants are indicated. The x-axis is labeled in radians.

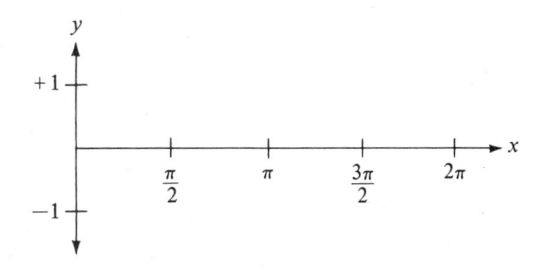

FIGURE 14.5

14 GRAPHING THE TRIGONOMETRIC FUNCTIONS

Suppose we want to graph the function $y = \sin x$ on this coordinate system. From the values of $\sin x$ at the quadrantal angles (which we should know) we can immediately locate five points on the graph. That is,

when $x = 0$, $\quad y = \sin 0 = 0 \quad \rightarrow (0, 0)$

when $x = \dfrac{\pi}{2}$, $\quad y = \sin \dfrac{\pi}{2} = 1 \quad \rightarrow \left(\dfrac{\pi}{2}, 1\right)$

when $x = \pi$, $\quad y = \sin \pi = 0 \quad \rightarrow (\pi, 0)$

when $x = \dfrac{3\pi}{2}$, $\quad y = \sin \dfrac{3\pi}{2} = -1 \quad \rightarrow \left(\dfrac{3\pi}{2}, -1\right)$

when $x = 2\pi$, $\quad y = \sin 2\pi = 0 \quad \rightarrow (2\pi, 0)$

These five ordered pairs are plotted in Figure 14.6.

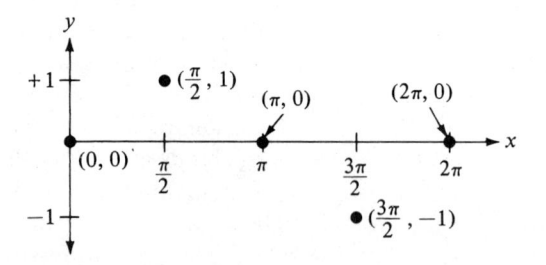

FIGURE 14.6

To identify further points on the graph of $\sin x$, we can go to the trigonometric tables, the calculator, or recall the sine values for the angles 30°, 45°, and 60°.

$$\sin \dfrac{\pi}{6} = \sin 30° = 0.5$$

$$\sin \dfrac{\pi}{4} = \sin 45° = 0.707$$

$$\sin \dfrac{\pi}{3} = \sin 60° = 0.866$$

These three values permit location of three additional points in each 90° interval along the x-axis. These additional points are plotted in Figure 14.7.

GRAPHS OF THE SINE, COSINE, AND TANGENT 14.2

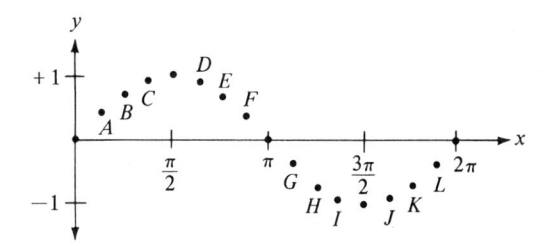

FIGURE 14.7

Their coordinates are

A $\left(\dfrac{\pi}{6}, 0.5\right)$ B $\left(\dfrac{\pi}{4}, 0.707\right)$ C $\left(\dfrac{\pi}{3}, 0.866\right)$

D $\left(\dfrac{2\pi}{3}, 0.866\right)$ E $\left(\dfrac{3\pi}{4}, 0.707\right)$ F $\left(\dfrac{5\pi}{6}, 0.5\right)$

G $\left(\dfrac{7\pi}{6}, -0.5\right)$ H $\left(\dfrac{5\pi}{4}, -0.707\right)$ I $\left(\dfrac{4\pi}{3}, -0.866\right)$

J $\left(\dfrac{5\pi}{3}, -0.866\right)$ K $\left(\dfrac{7\pi}{4}, -0.707\right)$ L $\left(\dfrac{11\pi}{6}, -0.5\right)$

With these additional points, we should be able to draw a smooth curve over the domain 0 to 2π of x. Our graph would then be as shown in Figure 14.8.

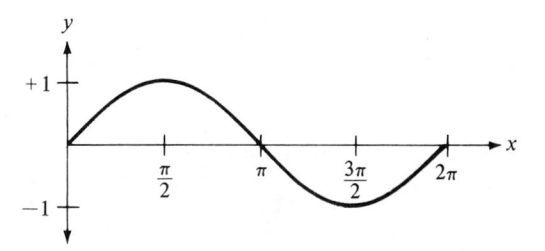

FIGURE 14.8

But what happens to this curve as x increases beyond 2π? We considered this in the last section when we discussed the fact that the angle 420° has trigonometric functions identical to the angle 60°. As a result the graph of sin x in the interval $x = 2\pi$ to 4π will be identical to the graph of sin x sketched in Figure 14.8 for $x = 0$ to 2π. This next interval is shown in Figure 14.9. Note that the scales on both the x- and y-axes have been reduced.

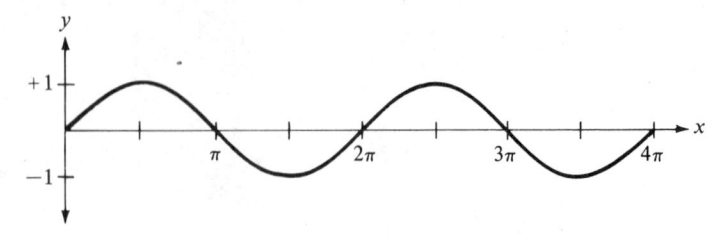

FIGURE 14.9

Functions that display this repetitive behavior are called **periodic.** The interval along the x-axis in which the function repeats itself is called the **period.** Then we say that $y = \sin x$ is a **periodic function** of x with a period of 2π radians. The graph in Figure 14.10 shows $y = \sin x$ extended to the left of the y-axis as well. Recall that the domain of $(x, \sin x)$ is $-\infty < x < +\infty$ and that the range is $-1 \leq \sin x \leq +1$.

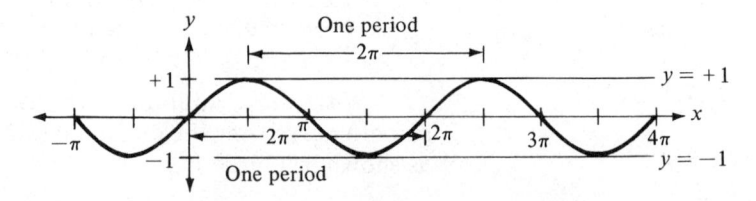

FIGURE 14.10

The horizontal lines $y = -1$ and $y = +1$, help to emphasize the limitations on the range of $(x, \sin x)$. Two different periods are shown in Figure 14.10. The "peak-to-peak" period from $x = \dfrac{\pi}{2}$ to $x = \dfrac{3\pi}{2}$ emphasizes the meaning of the period more clearly.

PERFORMANCE OBJECTIVE 2

You should now be able to graph the function $y = \sin x$ over any given values of x.

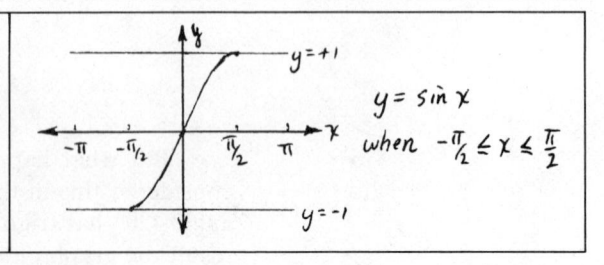

Next we consider the graph of $y = \cos x$. Values of $\cos x$ at well-known angles are first tabulated.

GRAPHS OF THE SINE, COSINE, AND TANGENT 14.2

$x =$	0	$\dfrac{\pi}{6}$	$\dfrac{\pi}{4}$	$\dfrac{\pi}{3}$	$\dfrac{\pi}{2}$	$\dfrac{2\pi}{3}$
cos x	1	0.866	0.707	0.5	0	-0.5

$x =$	$\dfrac{3\pi}{4}$	$\dfrac{5\pi}{6}$	π	$\dfrac{7\pi}{6}$	$\dfrac{5\pi}{4}$	$\dfrac{4\pi}{3}$
cos x	-0.707	-0.866	-1	-0.866	-0.707	-0.5

$x =$	$\dfrac{3\pi}{2}$	$\dfrac{5\pi}{3}$	$\dfrac{7\pi}{4}$	$\dfrac{11\pi}{6}$	2π
cos x	0	0.5	0.707	0.866	1

These points are plotted on a rectangular coordinate system. Connecting the points with a smooth curve we obtain the graph in Figure 14.11.

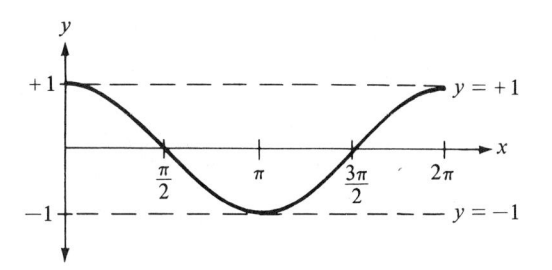

FIGURE 14.11

Notice that the $y = \cos x$ graph looks very much like the $y = \sin x$ graph. In fact, if we pushed it to the right by $\dfrac{\pi}{2}$ radians, it would look exactly like the $y = \sin x$ graph.

PERFORMANCE OBJECTIVE 3

You should now be able to graph the function $y = \cos x$ over any given values of x.

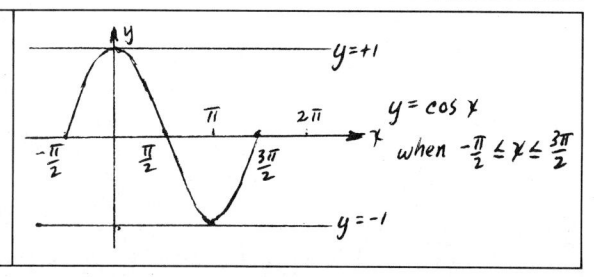

To graph the tangent we again start with a table of values.

$x =$	0	$\dfrac{\pi}{6}$	$\dfrac{\pi}{4}$	$\dfrac{\pi}{3}$	$\dfrac{\pi}{2}$	$\dfrac{2\pi}{3}$
$\tan x$	0	0.58	1	1.73	?	-1.73

$x =$	$\dfrac{3\pi}{4}$	$\dfrac{5\pi}{6}$	π	$\dfrac{7\pi}{6}$	$\dfrac{5\pi}{4}$	$\dfrac{4\pi}{3}$
$\tan x$	-1	-0.58	0	0.58	1	1.73

$x =$	$\dfrac{3\pi}{2}$	$\dfrac{5\pi}{3}$	$\dfrac{7\pi}{4}$	$\dfrac{11\pi}{6}$	2π
$\tan x$	?	-1.73	-1	-0.58	0

These values are plotted on a coordinate system in Figure 14.12.

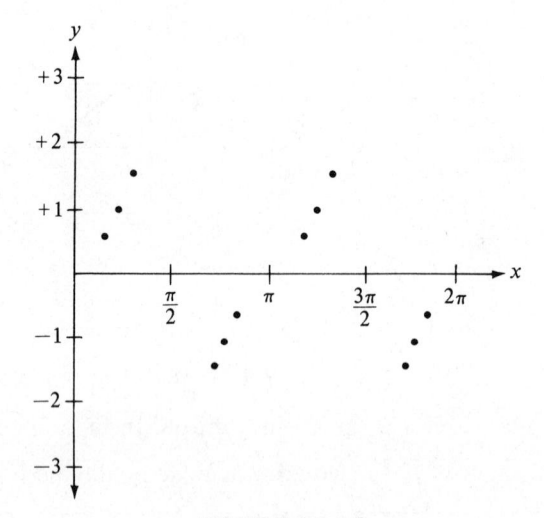

FIGURE 14.12

We encounter difficulty in plotting values of $\tan x$ for $x = \dfrac{\pi}{2}$ and $x = \dfrac{3\pi}{2}$. Recall that $\tan x$ is undefined at these values, and therefore $\dfrac{\pi}{2}$ and $\dfrac{3\pi}{2}$ are not permitted values of x. However, from the other points plotted and from the permitted values of $\tan x$

developed in the last section, we can sketch the behavior of the tan x graph around these values. Figure 14.13 illustrates this behavior.

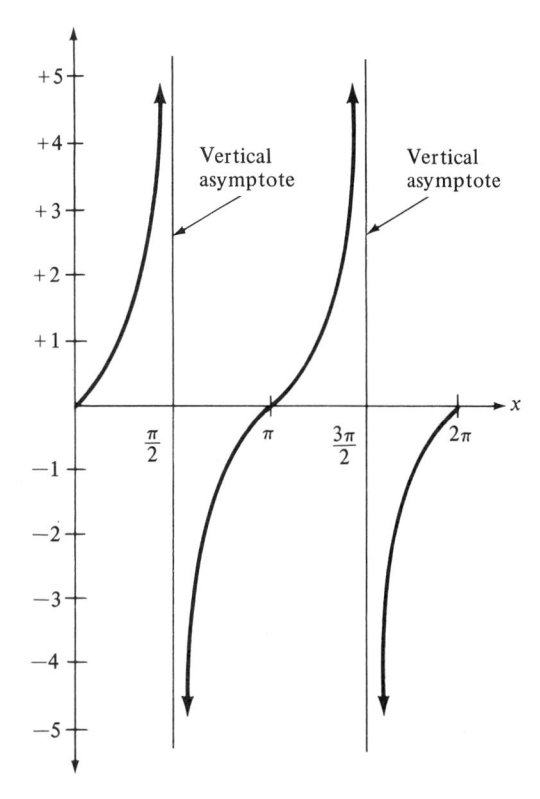

FIGURE 14.13

The vertical lines at $x = \dfrac{\pi}{2}$ and at $x = \dfrac{3\pi}{2}$ assist us in drawing the graph. As the x-values approach $\dfrac{\pi}{2}$ and $\dfrac{3\pi}{2}$, the graph moves off to $+$ or $-$ infinity along these vertical lines. Such lines are called **vertical asymptotes** for the graph.

From our definition of the range of $(x, \tan x)$ developed in the last section, we note that $\tan x$ approaches $+\infty$ as x approaches $\dfrac{\pi}{2}$ from lower values. Then $\tan x$ increases from $-\infty$ as x increases from $\dfrac{\pi}{2}$ to larger values. As x crosses over the value $\dfrac{\pi}{2}$, $\tan x$ takes a jump from very large positive values to very large negative values. This is referred to as a discontinuity in $\tan x$ at $x = \dfrac{\pi}{2}$. You will notice that the values of $\tan x$ (the range) are not limited as were the values of $\sin x$ and $\cos x$. All real values are in the range of $(x, \tan x)$.

14 GRAPHING THE TRIGONOMETRIC FUNCTIONS

There is another important difference between the tangent function and the sine and cosine functions. Although the tangent function is also periodic, the period is only π radians rather than 2π radians. Figure 14.14 shows the period of the tangent function.

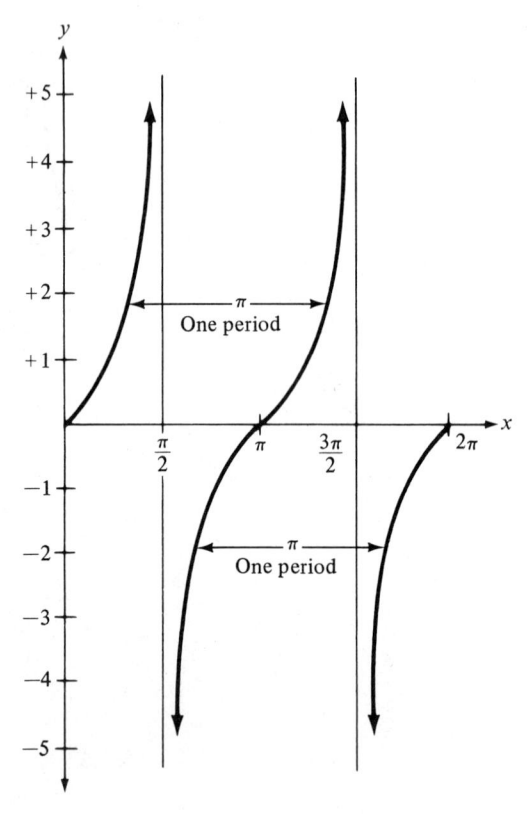

FIGURE 14.14

PERFORMANCE OBJECTIVE 4

You should now be able to graph the function $y = \tan x$ over any given values of x.

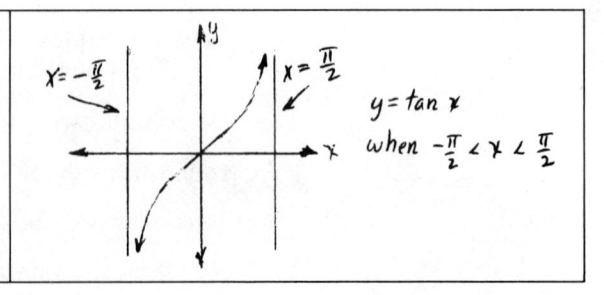

The characteristics of the sine, cosine, and tangent functions which we developed in this and the last section are summarized.

CHARACTERISTICS OF sin, cos AND tan

Function	Domain	Range	Period
$y = \sin x$	all real x	-1 to $+1$	2π rad
$y = \cos x$	all real x	-1 to $+1$	2π rad
$y = \tan x$	all real x*	$-\infty$ to $+\infty$	π rad

*(except $x = \dfrac{n\pi}{2}$, $n = \ldots -3, -1, +1, +3, +5, \ldots$ i.e., except $x = \dfrac{n\pi}{2}$, n any odd integer)

EXERCISE SET 14.2

In each of the following problems develop a table of x-y values at every 15° interval over the indicated values of x. Then sketch the graph by plotting these points and connecting them.

1. Graph $y = \sin x$ when x takes values from $-\pi$ to π.
2. Graph $y = \sin x$ when x takes values from $-\dfrac{\pi}{2}$ to $\dfrac{3\pi}{2}$.
3. Graph $y = \cos x$ when x takes values from π to 3π.
4. Graph $y = \cos x$ when x takes values from $-\pi$ to π.
5. Graph $y = \tan x$ when x takes values from $-\pi$ to π.
6. Graph $y = \tan x$ when x takes values from $-\dfrac{\pi}{2}$ to $\dfrac{3\pi}{2}$.

In each of the following problems sketch both graphs on the same set of axes.

7. Graph $y = \sin x$ and $y = \cos x$ when x takes values from $-\pi$ to 2π.
8. Graph $y = \sin x$ and $y = \cos x$ when x takes values from -2π to π.
9. Graph $y = \sin x$ and $y = \tan x$ when x takes values from $-\dfrac{\pi}{2}$ to $\dfrac{\pi}{2}$.
10. Graph $y = \cos x$ and $y = \tan x$ when x takes values from $-\pi$ to π.

SECTION 14.3 THE SINE AND COSINE WITH AMPLITUDE AND PHASE

In practical applications involving the sine and cosine functions, we frequently see a function such as $y = A \sin x$. We need to understand how the letter A changes the function. A is called the **amplitude** of the sine function. Suppose $A = 2$. Then the function is $y = 2 \sin x$. When we look for pairs of numbers in order to graph the function $y = 2 \sin x$, we first find the value of $\sin x$ and then multiply the value by 2. For example, if $x = \dfrac{\pi}{3}$, $\sin x = 0.866$ and $2 \sin x = 2 \cdot 0.866 = 1.732$.

Figure 14.15 shows graphs of $\sin x$ and $2 \sin x$ constructed on the same coordinate system.

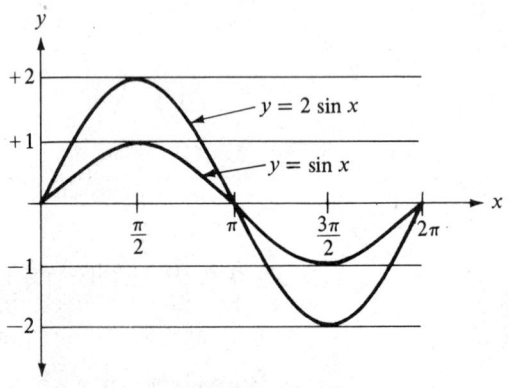

FIGURE 14.15

The presence of the multiplier 2 has changed the height of the curve but not its period or general behavior. Thus, the amplitude stretches or shrinks the curve. The equation $y = \sin x$ may be written as $y = 1 \sin x$ so that the sin curves plotted in the last section have amplitudes of one.

The cosine function with amplitude is written as $y = A \cos x$. As with the sine function, the value of A represents the maximum extent of the graph above and below the x-axis.

DEFINITION

AMPLITUDE OF THE SINE AND COSINE FUNCTIONS

The amplitude gives the maximum positive and negative values of these functions.

We can use the value of the amplitude to assist in graphing the sine or cosine function. In constructing the graph, draw horizontal lines at $y = +A$ and $y = -A$. These lines will indicate the vertical extent of the graph.

THE SINE AND COSINE WITH AMPLITUDE PHASE 14.3

EXAMPLE Graph the equation $y = 3 \sin x$ in the interval $0 \leq x \leq 2\pi$.

SOLUTION We first construct the coordinate system with the horizontal lines $y = +A$ and $y = -A$ plotted on them. See Figure 14.16.

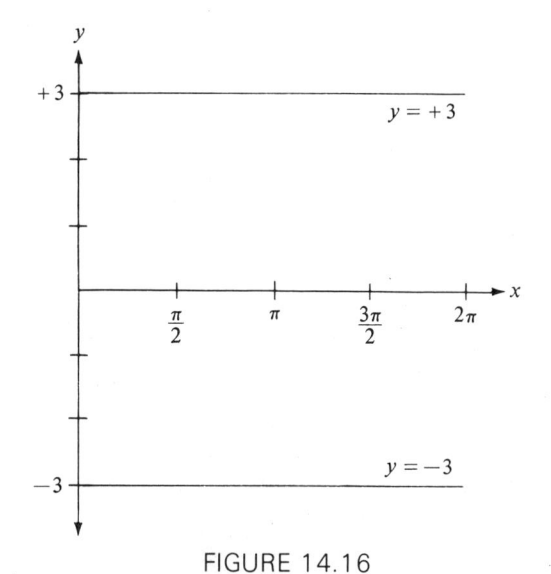

FIGURE 14.16

Next identify the values of the function at each 90° interval. The plot of these points is shown in Figure 14.17.

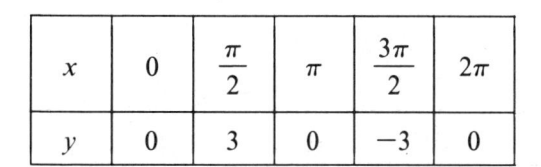

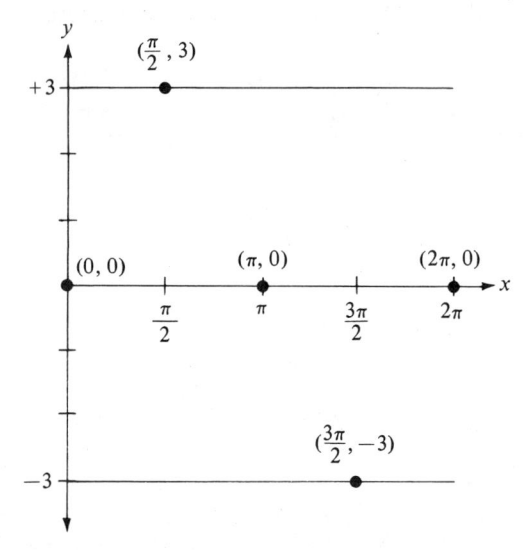

x	0	$\dfrac{\pi}{2}$	π	$\dfrac{3\pi}{2}$	2π
y	0	3	0	-3	0

FIGURE 14.17

538 14 GRAPHING THE TRIGONOMETRIC FUNCTIONS

Having plotted these points, we construct such intermediate points as are needed to produce a smooth line graph of the function. The final graph is shown in Figure 14.18.

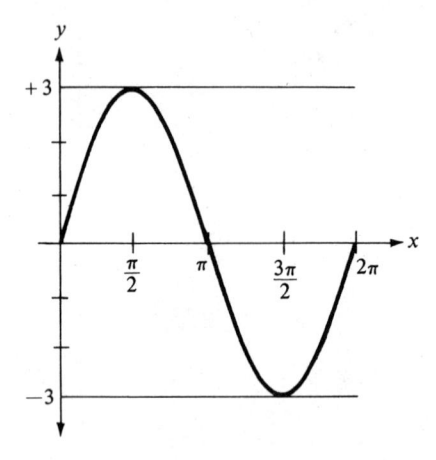

FIGURE 14.18

EXAMPLE Graph the equation $y = 2 \cos x$ in the interval $-\pi \leq x \leq +\pi$.

SOLUTION We first identify values of the function at 90° intervals.

x	$-\pi$	$-\dfrac{\pi}{2}$	0	$+\dfrac{\pi}{2}$	π
y	-2	0	$+2$	0	-2

The graph is shown in Figure 14.19.

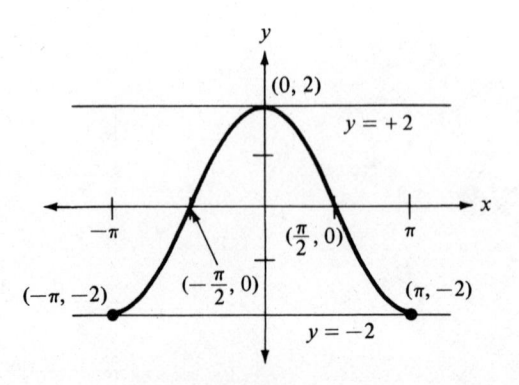

FIGURE 14.19

THE SINE AND COSINE WITH AMPLITUDE PHASE 14.3

PERFORMANCE OBJECTIVE 5

You should now be able to graph the sine and cosine functions with given amplitude.

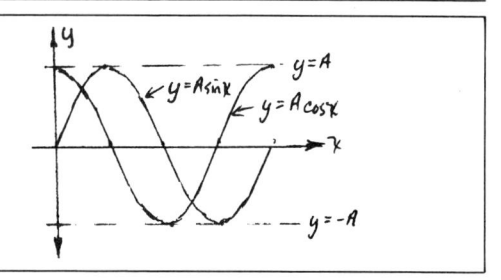

In dealing with practical problems involving wave motion (for example, sound waves, ac circuits, radio waves), we will often be concerned with the phase relation among these waves. The waves will most frequently be represented by sine functions. To illustrate, we draw two waves in Figure 14.20.

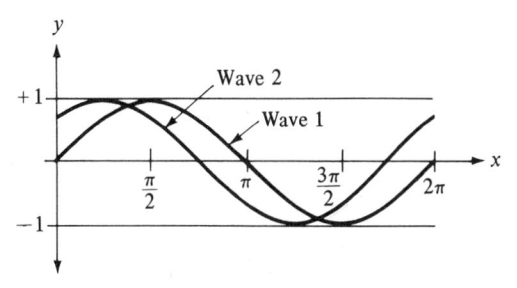

FIGURE 14.20

We represent wave 1 as a sine wave with amplitude 1. Wave 2 appears to have the form of a sine wave but it is shifted to the left by 45°. These waves are represented by the equations

$$y = \sin x \quad \text{(wave 1)}$$

$$y = \sin(x + 45°) \quad \text{(wave 2)}$$

x-y pairs for these equations are evaluated in the following table.

x	y (wave 1)	y (wave 2)
0°	$\sin 0° = 0$	$\sin(0° + 45°) = 0.707$
45°	$\sin 45° = 0.707$	$\sin(45° + 45°) = 1$
90°	$\sin 90° = 1$	$\sin(90° + 45°) = 0.707$
135°	$\sin 135° = 0.707$	$\sin(135° + 45°) = 0$
180°	$\sin 180° = 0$	$\sin(180° + 45°) = -0.707$
225°	$\sin 225° = -0.707$	$\sin(225° + 45°) = -1$
270°	$\sin 270° = -1$	$\sin(270° + 45°) = -0.707$
315°	$\sin 315° = -0.707$	$\sin(315° + 45°) = 0$

Examination of this table shows that wave 2 reaches the values of wave one 45° earlier than wave 1. We may generalize on this result as follows:

DEFINITION

> **PHASE DIFFERENCE**
>
> A wave defined by $y = \sin(x + \phi)$ differs in phase by ϕ from the wave $y = \sin x$.

If ϕ is positive, the wave leads the reference wave $y = \sin x$. If ϕ is negative, the wave lags the reference wave. The angle ϕ is called the **phase angle**.

EXAMPLE Determine the phase relation and amplitude of the waves

$y_1 = 2 \sin x$ and $y_2 = 3 \sin(x + 60°)$.

SOLUTION Wave 1 has an amplitude of 2 and a zero phase angle. Wave 2 has an amplitude of 3 and a phase angle of 60°. Wave 2 thus leads wave one by 60°.

EXAMPLE Graph the equations $y_1 = 2 \sin x$ and $y_2 = 3 \sin(x + 60°)$ over the domain $0° \leq x \leq 360°$.

SOLUTION A table of critical values should be generated.

x	y_1 (wave 1)	y_2 (wave 2)
0°	$2 \sin 0° = 0$	$3 \sin(0° + 60°) = 2.6$
30°	$2 \sin 30° = 1$	$3 \sin(30° + 60°) = 3$
60°	$2 \sin 60° = 1.73$	$3 \sin(60° + 60°) = 2.6$
90°	$2 \sin 90° = 2$	$3 \sin(90° + 60°) = 1.5$
120°	$2 \sin 120° = 1.73$	$3 \sin(120° + 60°) = 0$
etc.		

Note that wave 2 leads wave 1 by 60°. The equations are plotted in Figure 14.21.

Throughout this section statements made about sine functions apply equally to cosine functions.

EXAMPLE Graph the equations $y_1 = 2 \cos x$ and $y_2 = 2 \cos(x + 45°)$ over the domain $0° \leq x \leq 360°$.

SOLUTION The graphs are shown in Figure 14.22.

Note that y_2 leads y_1 by 45°.

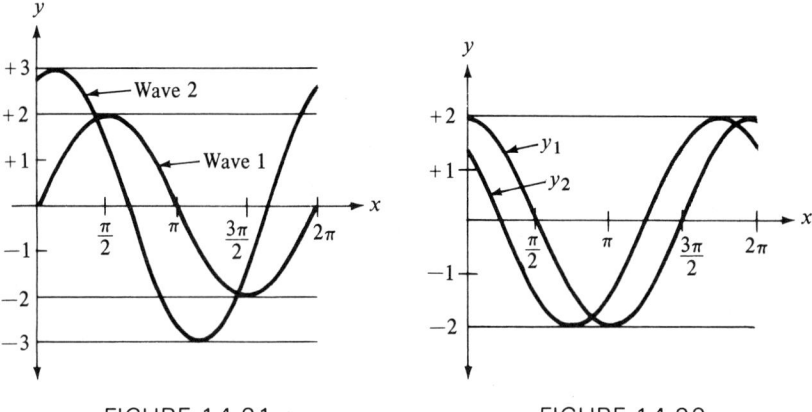

FIGURE 14.21

FIGURE 14.22

PERFORMANCE OBJECTIVE 6

You should now be able to graph the sine and cosine functions with given phase.

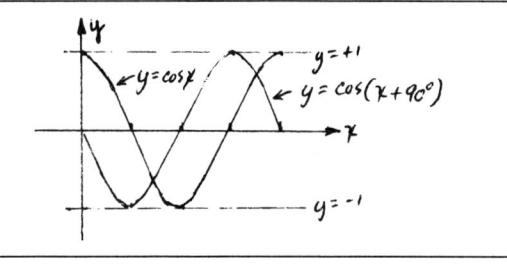

EXERCISE SET 14.3

1. Sketch the graph of $y = 4 \sin x$ over the interval from $x = 0$ to $x = \pi$ radians.

2. Sketch the graph of $y = 5 \sin x$ over the interval from $x = -\pi$ to $x = \dfrac{\pi}{2}$ radians.

3. Sketch the graph of $y = 3 \cos x$ over the interval from $x = 0$ to $x = 2\pi$ radians.

4. Sketch the graph of $y = 4 \cos x$ over the interval from $x = -2\pi$ to $x = 0$ radians.

5. Determine the amplitude of the following functions:
   a. $y = 0.5 \sin x$
   b. $y = 16 \cos x$
   c. $y = \sin x$
   d. $y = \sqrt{3} \sin x$

6. Determine the amplitude of the following functions:
   a. $y = 1.5 \cos x$
   b. $y = \sqrt{5} \sin x$
   c. $y = 12 \sin x$
   d. $y = \cos x$

7. Plot the following functions on the same set of coordinate axes:
   a. $y_1 = \sin x$, $y_2 = \sin(x - 60°)$, with $0° \leq x \leq 360°$
   b. $y_1 = 3 \sin(x + 30°)$, $y_2 = 2 \sin(x + 90°)$, with $0° \leq x \leq 360°$
   c. $y_1 = \cos x$, $y_2 = \cos(x + 45°)$, with $0° \leq x \leq 360°$

8. Plot the following functions on the same set of coordinate axes:
   a. $y_1 = \sin x$, $y_2 = \sin(x - 70°)$, with $0° \leq x \leq 360°$
   b. $y_1 = 4 \sin(x - 10°)$, $y_2 = 3 \sin(x + 80°)$, with $0° \leq x \leq 360°$
   c. $y_1 = \sin x$, $y_2 = 3 \sin(x - 30°)$, with $-180° \leq x \leq 180°$

SECTION 14.4 SINE CURVES AND ROTATING VECTORS

We illustrate a technique that will be particularly useful in analyzing electrical circuits that carry alternating current. This technique can also be helpful in understanding the interference of waves produced by several sources. Consider again the function $y = A \sin x$. The graph of this function is shown in Figure 14.23. Also shown is a vector of length A that makes an angle of x with the horizontal axis.

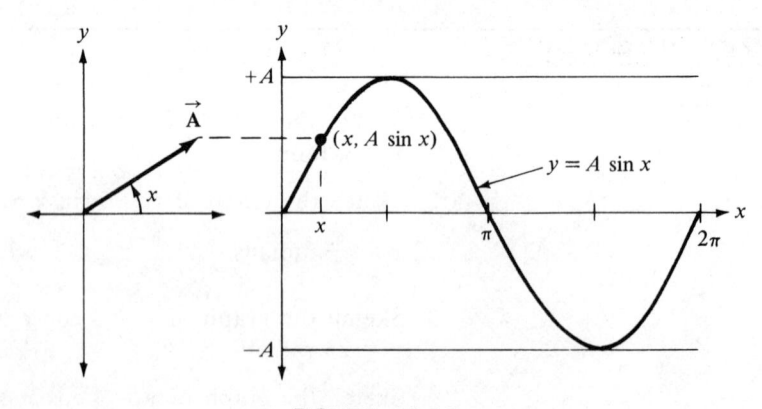

FIGURE 14.23

A horizontal line drawn through the tip of this vector intercepts the graph $y = A \sin x$ at the point $(x, A \sin x)$. As the vector **A** rotates about the origin in a counterclockwise direction it generates each point on the graph $y = A \sin x$. Other points are shown in Figure 14.24.

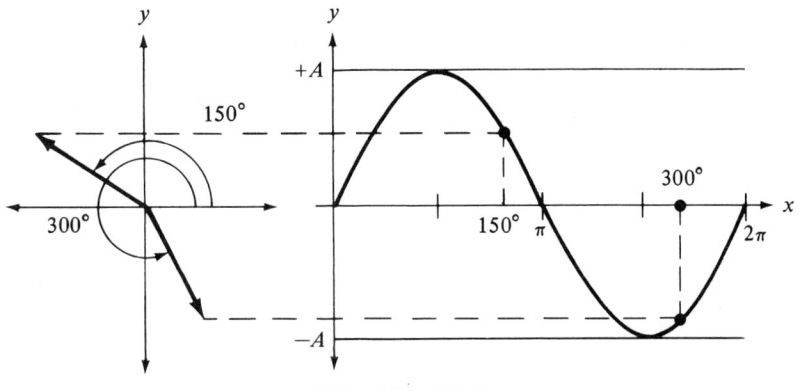

FIGURE 14.24

If the vector **A** were rotated at a constant angular speed, the equation for the curve would be $y = A \sin \omega t$ (x would equal ωt). The y-component of the vector would represent the instantaneous value (value at any time t) of an alternating quantity (current, voltage, electric field, etc.). The frequency recorded in cycles per second of that alternating quantity would be

$$f = \frac{\omega}{2\pi}$$

DEFINITION

TIME DEPENDENT WAVE

$y = A \sin \omega t$ represents a wave of amplitude A that oscillates in time with frequency $f = \dfrac{\omega}{2\pi}$. The angular frequency ω is expressed in radians/second; f is in cycles/second.

EXAMPLE Calculate the frequency of a wave generated by a vector that rotates at 1000 rad/sec.

SOLUTION $f = \dfrac{\omega}{2\pi} = \dfrac{1000}{2\pi} = 159.2$ cycles per second (hertz).

EXAMPLE Calculate the instantaneous amplitude of a wave generated by a vector 100 units long that has moved 30° from the reference position (at 0°).

SOLUTION Consider Figure 14.25.

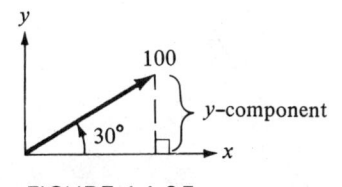

FIGURE 14.25

The instantaneous amplitude is the *y*-component of the vector. That component is

$$y = A \sin 30° = 100 \sin 30° = 100(0.5) = 50 \text{ units}$$

If the angular frequency or frequency of the wave is known, the instantaneous amplitude may be calculated at any given time. We assume that the angles are referred to the reference position of zero.

EXAMPLE Calculate the instantaneous amplitude of a wave generated by a vector of amplitude 110 V. Take the frequency to be 60 cycles per second, the time to be 0.01 sec, and the phase angle to be zero.

SOLUTION The angle value is obtained by multiplying angular frequency by time. Angular frequency is obtained from frequency by multiplying by 2π. We substitute into $y = A \sin \omega t$ with

$$\omega = 2\pi(60) = 377 \text{ rad/sec}$$
$$\omega t = (377)(0.01) = 3.77 \text{ rad}$$
$$A = 110 \text{ V}$$

Then $y = 110 \sin (3.77) = 110(-0.588) = -64.7$ V.

To generate the graph of an equation of the form

$$y = A \sin (x + \phi)$$

we would use a vector of length **A** units that was advanced from the reference position at time zero by an angle of ϕ. See Figure 14.26.

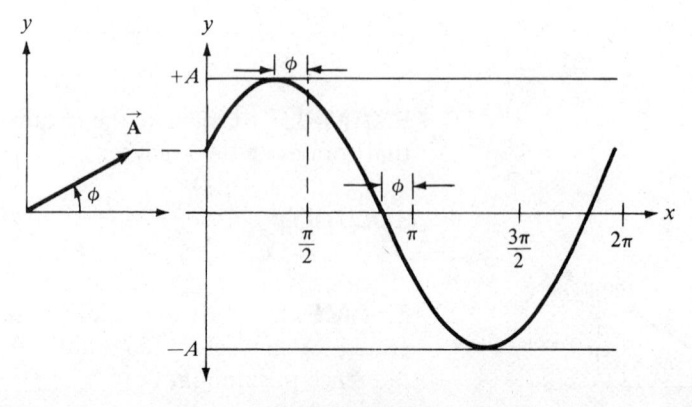

FIGURE 14.26

SINE CURVES AND ROTATING VECTORS 14.4

EXAMPLE Calculate the instantaneous amplitude of a wave at time 4×10^{-5} sec. The wave has an amplitude of 0.015 A and a frequency of 6000 cycles/sec. Its phase angle is $\frac{\pi}{3}$ radians.

SOLUTION We substitute into $y = A \sin(\omega t + \phi)$ with

$$A = 0.015 \text{ A}$$
$$\omega = 2\pi(6000) = 12{,}000\pi \text{ rad/sec}$$
$$\omega t = 12{,}000\pi(4 \times 10^{-5}) = 1.508 \text{ rad}$$
$$\phi = \frac{\pi}{3} \text{ rad} = 1.047 \text{ rad}$$
$$\omega t + \phi = 1.508 + 1.047 = 2.555$$

Then $y = 0.015 \sin 2.555 = 0.0083$ A $= 8.3$ mA.

PERFORMANCE OBJECTIVE 7

You should now be able to generate a sine wave of given amplitude and phase with a rotating vector.

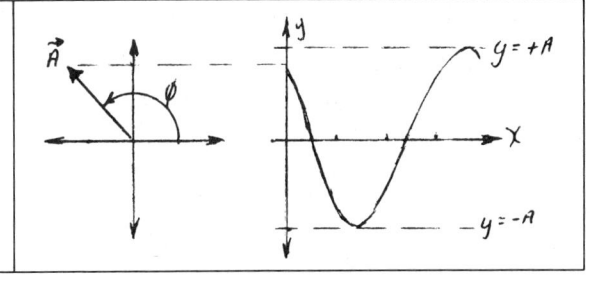

To add together the effects of two waves of the same frequency, we complete a vector addition with each vector in its starting position.

EXAMPLE Find the resulting voltage when the following two voltages are added:

Wave 1: 80.0 V at 110 cycles/sec

Wave 2: 70.0 V at 110 cycles/sec and leading wave 1 by 30.0°

FIGURE 14.27

SOLUTION Consider the diagram in Figure 14.27.

Vector	x	y
70	60.62	35
80	80	0
R	140.62	35

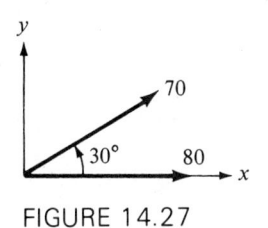

$$\theta = \tan^{-1} \frac{R_y}{R_x}$$
$$= \tan^{-1} \frac{35}{140.62}$$
$$= \tan^{-1} 0.2489$$
$$= 13.98° \quad \text{or} \quad 14.0°$$

$$R = \frac{R_y}{\sin \theta} = \frac{35}{\sin 13.98°} = 145 \text{ V}$$

The resultant vector has a magnitude of 145 V and starts with a phase angle of 14.0° (0.244 rad). See Figure 14.28.

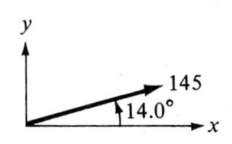

FIGURE 14.28

The equation of the resulting voltage is

$$V = 145 \sin (2\pi \cdot 110t + 0.244) \text{ V}$$

PERFORMANCE OBJECTIVE 8

You should now be able to add two sine waves of the same frequency using vector methods.

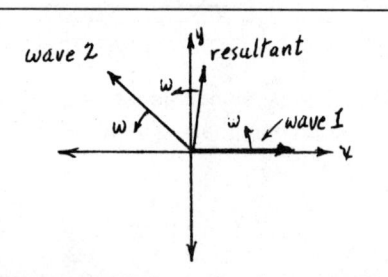

EXERCISE SET 14.4

1. Draw rotating vectors that will generate y-values in the following equations:
   a. $y = 15 \sin (x + 90°)$ when $x = 45°$
   b. $y = 10 \sin (x - 60°)$ when $x = 30°$

2. Draw rotating vectors that will generate y-values in the following equations:
   a. $y = 12 \sin (x + 45°)$ when $x = 60°$
   b. $y = 10 \cos (x - 30°)$ when $x = 60°$

3. Calculate the instantaneous amplitude of the following waves:
   a. $y = 15 \sin \left(10^6 t + \frac{\pi}{6}\right)$ at $t = 2 \times 10^{-5}$ sec
   b. $V = 60 \sin \left(2\pi \cdot 110t + \frac{\pi}{2}\right)$ at $t = 0.01$ sec

c. A wave with the following characteristics:

Amplitude = 5 A
Frequency = 110 cycles/sec
Phase angle = $\frac{\pi}{4}$ rad
At time $t = 0.25$ sec

d. A wave with the following characteristics:

Amplitude = 8×10^{-3} V
Frequency = 4×10^6 cycles/sec
Phase angle = $-\frac{\pi}{2}$ rad
At time $t = 4 \times 10^{-7}$ sec

4. Calculate the instantaneous amplitude of the following waves:

a. $y = 80 \sin\left(10^5 t + \frac{\pi}{3}\right)$ at $t = \pi \times 10^{-5}$ sec

b. $y = 110 \sin(2\pi \cdot 60t + \pi)$ at $t = 5 \times 10^{-3}$ sec

c. A wave with the following characteristics:

Amplitude = 3 A
Frequency = 60 cycles/sec
Phase angle = $\frac{\pi}{6}$ rad
At time $t = 0.4$ sec

d. A wave with the following characteristics:

Amplitude = 9×10^{-2} V
Frequency = 4×10^5 cycles/sec
Phase angle = $-\frac{\pi}{4}$ rad
At time $t = 0.25 \times 10^{-5}$ sec

5. Find an expression for the sum of the following equations using the rotating vector method. Show the starting position of the sum vector and determine its frequency.

a. $y = 10 \sin(10^3 \pi t)$

  $y = 15 \sin\left(10^3 \pi t + \frac{\pi}{6}\right)$

b. $y = 1.5 \sin(100\pi t)$

  $y = 0.8 \sin\left(100\pi t - \frac{\pi}{3}\right)$

6. Find an expression for the sum of the following equations using the rotating vector method. Show the starting position of the sum

vector and determine its frequency.

a. $y = 20 \sin(10^4 \pi t)$

$y = 15 \sin\left(10^4 \pi t + \dfrac{\pi}{3}\right)$

b. $y = 2.5 \sin(200 \pi t)$

$y = 1.8 \sin\left(200 \pi t - \dfrac{\pi}{2}\right)$

SECTION 14.5 APPLICATIONS OF GRAPHING (ELECTRONICS)

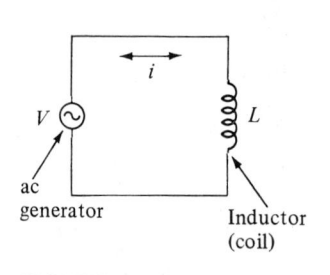

FIGURE 14.29

One of the common uses of the trigonometric functions is to represent electrical quantities in an alternating current (ac) circuit. We cite some properties of ac circuits in this section without proofs. Our intent is to illustrate the technical mathematics involved rather than the electrical principles. We hope you may be interested enough in these applications to consult a basic electricity text if you have questions concerning the principles.

A **pure inductive circuit** consists of an ac generator and an inductor that is assumed to have no resistance. A schematic of such a circuit is shown in Figure 14.29.

The generator develops a voltage amplitude V_m. Its angular frequency is ω. The inductor (or coil) has a constant inductance of L henries (H). The current in the circuit and the voltage across the inductor are given by

$$i = \dfrac{V_m}{\omega L} \sin \omega t$$

$$V_L = V_m \sin\left(\omega t + \dfrac{\pi}{2}\right)$$

EXAMPLE Graph the quantities i and V_L in the preceding equations. Allow t to run from zero to one period.

SOLUTION Note that V_L, the voltage applied across the inductor, leads the current by 90°. A sketch of these two sine curves is shown in Figure 14.30.

FIGURE 14.30

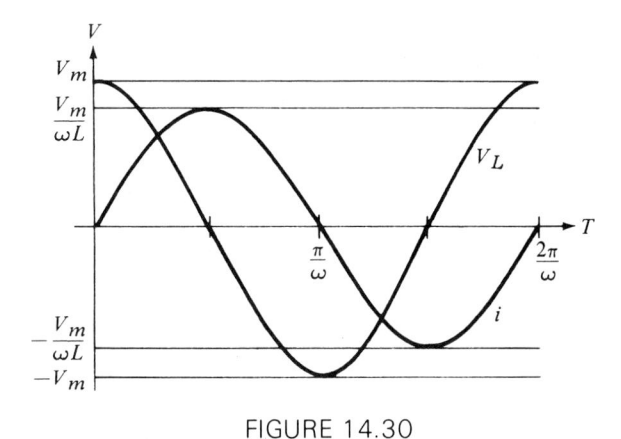

EXAMPLE Describe the applied voltage across the inductor and the current in the circuit in Figure 14.29 when $V_m = 200$ V, $\omega = 600\pi$ radians/sec and the inductance is 0.02 H.

SOLUTION Since $\omega = 600\pi$, the oscillator is operating at a frequency of $\dfrac{\omega}{2\pi} = \dfrac{600\pi}{2\pi} = 300$ cycles/sec. The current amplitude is

$$i_m = \frac{V_m}{\omega L} = \frac{200}{(600\pi)(0.02)} = 5.31 \text{ A}$$

Substituting these values into the equations in the last example, we find

$$i = 5.31 \sin 600\pi t \text{ A}$$
$$V_L = 200 \sin\left(600\pi t + \frac{\pi}{2}\right) \text{ V}$$

The current lags the applied voltage by 90°. Both quantities have a common frequency of 300 cycles per second.

An ac circuit containing a generator, an inductor, a resistor, and a capacitor is shown in Figure 14.31.

The generator develops a voltage amplitude of V_m. Its angular frequency is ω. The inductor has a constant inductance of L henries. The capacitor has a capacitance of C farads, and the resistor has a resistance of R ohms. The values of the quantities in the circuit are

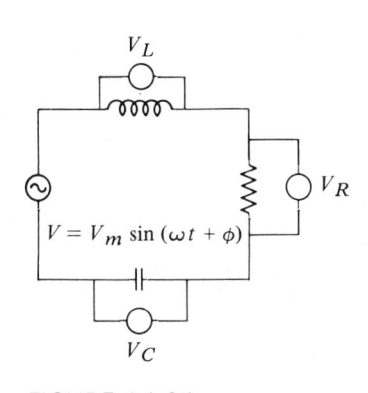

FIGURE 14.31

given by

$$V_R = i_m R \sin \omega t$$

$$V_L = i_m \omega L \sin\left(\omega t + \frac{\pi}{2}\right)$$

$$V_C = \frac{i_m}{\omega C} \sin\left(\omega t - \frac{\pi}{2}\right)$$

EXAMPLE Graph the preceding equations for V_R, V_L, and V_C over a one period interval.

SOLUTION See the graph in Figure 14.32.

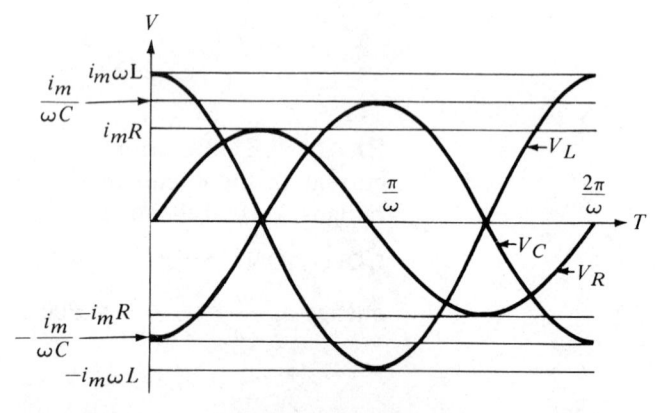

FIGURE 14.32

PERFORMANCE OBJECTIVE 9

You should now be able to use trigonometric functions to show the correct amplitude and phase of voltages and currents in an ac circuit.

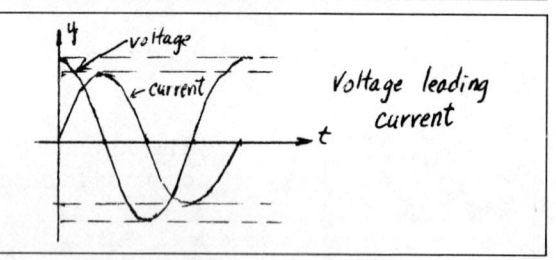

Next we consider the method of rotating vectors for the representation of voltages in an ac circuit.

EXAMPLE Display rotating vectors that would represent each of the quantities in the previous example.

SOLUTION See the sketch in Figure 14.33.

APPLICATIONS OF GRAPHING (ELECTRONICS) 14.5

The impedance **Z** of an ac circuit such as that sketched in Figure 14.33 is obtained by finding the resultant of the three vectors representing the effective resistance of each circuit component. These vectors have the same orientations as shown in Figure 14.33. The vectors are

X_L (called **inductive reactance**) $= \omega L$

X_C (called **capacitative reactance**) $= \dfrac{1}{\omega C}$

R (called **resistance**)

A sketch of these vectors is shown in Figure 14.34.

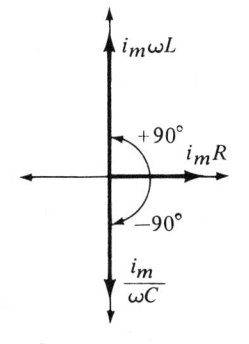

FIGURE 14.33

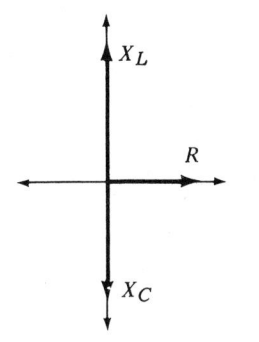

FIGURE 14.34

EXAMPLE For a circuit in which $\omega = 10^4$ rad/sec, $L = 2 \times 10^{-3}$ H, $C = 4 \times 10^{-6}$ F, and $R = 10\ \Omega$, calculate **Z**.

SOLUTION

$X_L = \omega L = (10^4)(2 \times 10^{-3}) = 20\ \Omega$

$X_C = \dfrac{1}{\omega C} = \dfrac{1}{(10^4)(4 \times 10^{-6})} = \dfrac{1}{0.04} = 25\ \Omega$

$R = 10\ \Omega$

The vector diagram is shown in Figure 14.35.

FIGURE 14.35

It is customary to combine X_L and X_C since they always oppose each other and then display **Z** graphically. See Figure 14.36.

$\tan \theta = \dfrac{X_L - X_C}{R} = \dfrac{-5}{10} = -\dfrac{1}{2}$

$\theta = -26.57°$

$Z = \dfrac{X_L - X_C}{\sin \theta} = \dfrac{-5}{-0.447} = 11.19\ \Omega$

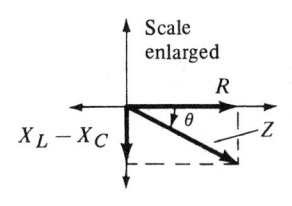

FIGURE 14.36

PERFORMANCE OBJECTIVE 10

You should now be able to use rotating vectors to represent phase relations of voltages in an ac circuit.

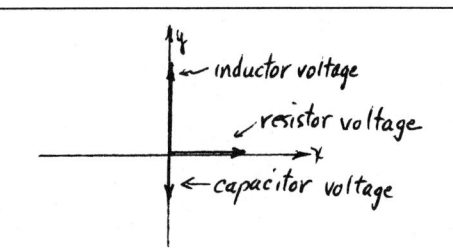

14 GRAPHING THE TRIGONOMETRIC FUNCTIONS

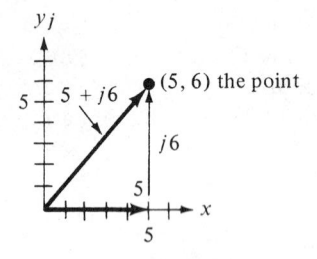

FIGURE 14.37

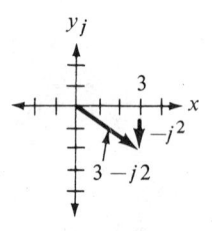

FIGURE 14.38

The evaluation of the impedance in an ac circuit is frequently done using a *j*-operator. In Chapter 10 we saw that a number such as $5 + 6j$ could be represented as a point in the complex plane. We plotted the point in the manner shown in Figure 14.37.

We make a vector interpretation of this plot. The sum of the vector 5 and the vector $j6$ is also shown in Figure 14.37.

As a second example, consider the vector $3 - j2$. This vector is the sum of a vector three units long to the right and two units long rotated 90° in the negative direction. See Figure 14.38.

This interpretation turns out to be convenient for ac circuits since phase differences between voltages across capacitors, resistors, and inductors always differ by $\pm 90°$. In the ac circuit of the previous example, we could write

$$\vec{Z} = \vec{R} + j(\vec{X_L} - \vec{X_C})$$

since the voltage across the inductor leads the voltage across the resistor and the voltage across the capacitor lags it by 90°. The driving voltage in the circuit is parallel to the impedance vector so that the lead or lag of the voltage with respect to the current in the circuit can be obtained from the relation

$$\cos \theta = \frac{R}{Z}$$

as illustrated in Figure 14.39.

$\vec{Z}$ is parallel to the driving voltage.

$\vec{R}$ is parallel to the current.

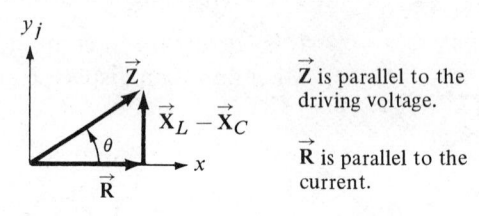

FIGURE 14.39

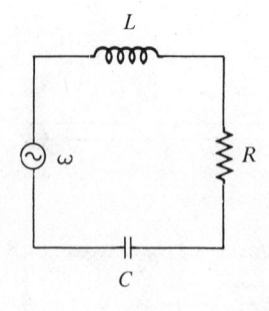

FIGURE 14.40

EXAMPLE Find the phase relation between the driving voltage and the current when the circuit in Figure 14.40 has the values

$$L = 2.50 \times 10^{-3} \text{ H}$$
$$R = 150 \text{ }\Omega$$
$$C = 1.40 \times 10^{-7} \text{ F}$$
$$\omega = 4.00 \times 10^{4} \text{ radians/sec}$$

APPLICATIONS OF GRAPHING (ELECTRONICS) 14.5

SOLUTION

$$X_L = \omega L = (4.00 \times 10^4)(2.50 \times 10^{-3}) = 100 \, \Omega$$

$$X_C = \frac{1}{\omega C} = \frac{1}{(4 \times 10^4)(1.4 \times 10^{-7})} = 179 \, \Omega$$

$$X_L - X_C = 100 - 179 = -79 \, \Omega$$

$$Z = 150 - j79 \, \Omega$$

$$\tan \theta = \frac{79}{150} = 0.527$$

So that $\theta = 27.8°$.

See the sketch in Figure 14.41.

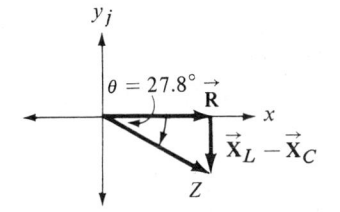

The driving voltage lags the current by a phase angle of $27.8°$.

FIGURE 14.41

NOTE: (to electronics students) We have used the maximum (peak) values of voltage and current in these examples. The RMS values can be substituted without changing the form of the equations. In dealing with power in the ac circuit you will want to use RMS values rather than the peak values.

PERFORMANCE OBJECTIVE 11	$\vec{Z} = \vec{R} + j(\vec{X_L} - \vec{X_c})$
You should now be able to use the j-operator to express the impedance in an ac circuit.	resistor contribution inductor contribution capacitor contribution

---

EXERCISE SET 14.5

1. Graph the following equations representing current and voltage in an ac circuit. Use the same set of axes but not necessarily the same scale on the vertical axis. Label the maximum values of

current and voltage in the circuit.

$$i = 0.9 \sin(800\pi t) \text{ A}$$

$$V = 50.0 \sin\left(800\pi t + \frac{\pi}{6}\right) \text{ V}$$

2. Graph the following equations representing current and voltage in an ac circuit. Use the same set of axes but not necessarily the same scale on the vertical axis. Label the maximum values of current and voltage in the circuit.

$$i = 1.1 \sin(400\pi t) \text{ A}$$

$$V = 35.0 \sin\left(400\pi t - \frac{\pi}{4}\right) \text{ V}$$

3. Graph the following equations representing voltages in an ac circuit. Use the same set of axes.

$$V_R = 15 \sin(120\pi t) \text{ V}$$

$$V_L = 30 \sin\left(120\pi t + \frac{\pi}{2}\right) \text{ V}$$

$$V_C = 20 \sin\left(120\pi t - \frac{\pi}{2}\right) \text{ V}$$

4. Graph the following equations representing voltages in an ac circuit. Use the same set of axes.

$$V_R = 20 \sin(240\pi t) \text{ V}$$

$$V_L = 15 \sin\left(240\pi t + \frac{\pi}{2}\right) \text{ V}$$

$$V_C = 35 \sin\left(240\pi t - \frac{\pi}{2}\right) \text{ V}$$

5. Sketch rotating vectors that would generate the sine curves in Problem 3.

6. Sketch rotating vectors that would generate the sine curves in Problem 4.

7. Evaluate the impedance in the following circuits:
   a. $R = 52 \; \Omega$
      $C = 2 \times 10^{-6}$ F
      $L = 12 \times 10^{-3}$ H
      $\omega = 2 \times 10^4$ radians/sec
   b. $R = 16.2 \; \Omega$
      $C = 0.4 \; \mu\text{F}$ (microfarads)
      $L = 0.3$ mH (millihenries)
      $\omega = 3 \times 10^5$ radians/sec

8. Evaluate the impedance in the following circuits:
   a. $R = 40\ \Omega$
      $C = 3 \times 10^{-6}\ F$
      $L = 2 \times 10^{-3}\ H$
      $\omega = 3 \times 10^4$ radians/sec
   b. $R = 30\ \Omega$
      $C = 1.2\ \mu F$ (microfarads)
      $L = 0.8$ mH (millihenries)
      $\omega = 2 \times 10^4$ radians/sec

9. Express each impedance in Problem 7 in vector form using the j-operator. Then find the phase relation between the driving voltage and the current.

10. Express each impedance in Problem 8 in vector form using the j-operator. Then find the phase relation between the driving voltage and the current.

SECTION 14.6 SUMMARY

The graphical representation of the sine and cosine functions with amplitude and phase is probably the most important part of this chapter. We have seen that such graphs can represent the behavior of real physical systems. The essential concepts and new definitions are reviewed here.

Infinity ($+\infty$ or $-\infty$): an undefined point on the end of the number line. Infinity is a useful concept in discussing the permitted values of trigonometric functions.

Periodic function: a function that repeats its values in certain regular intervals. The sine, cosine, and tangent functions are periodic functions.

Period: the interval (of the independent variable) over which a periodic function repeats itself.

Vertical asymptote: a vertical line that is approached by a graph as the function being graphed moves off to infinity.

Amplitude: a term used to describe the maximum positive and negative values of a sine or cosine function.

Phase difference: a term generally used to describe a property of two sine or cosine functions of the same frequency. The phase difference shows up graphically as a horizontal displacement between the graphs of the curves when they are plotted on the same set of axes.

Rotating vector: a vector that rotates about the origin of coordinates in such a fashion that a graph of its angle with the *x*-axis versus its *y*-component represents a sine curve.

Adding waves: finding the sum of two time-dependent waves of the same frequency. We have found the sum using vector methods.

Impedance: the effective resistance of an ac circuit containing some combination of resistance, capacitance, and inductance. The impedance may be determined by vector addition of the inductive reactance, the capacitative reactance, and the resistance considered as vectors in the complex plane.

CHAPTER 14 PRACTICE TEST

1. (PO-1, 2) Give the range of values of the following trigonometric functions in the quadrant indicated.

Function	Quadrant	Range
$\sin \theta$	II	
$\cos \theta$	IV	
$\tan \theta$	I	
$\cos \theta$	III	
$\tan \theta$	III	
$\tan \theta$	II	
$\sin \theta$	IV	

2. (PO-2, 3, 4) Graph the following functions over the *x*-interval indicated (*x* is indicated in radian measure):

    a. $y = \sin x$, interval: $-\dfrac{\pi}{2} \leq x \leq \dfrac{3\pi}{2}$

    b. $y = \tan x$, interval: $\dfrac{\pi}{2} < x < \dfrac{3\pi}{2}$

    c. $y = \cos x$, interval: $0 \leq x \leq 2\pi$

d. $y = \sin x$, interval: $-\pi \leq x \leq \pi$

e. $y = \tan x$, interval: $0 \leq x < \dfrac{\pi}{2}$

3. (PO-5, 6) Graph the following functions over the x-interval indicated:
   a. $y = 2 \sin (x - 45°)$, interval: $0° \leq x \leq 180°$
   b. $y = 1.5 \cos (x + 60°)$, interval: $0° \leq x \leq 360°$

4. (PO-7) Draw a rotating vector that will generate the following functions. Place the vector in the initial ($t = 0$) position.

   a. $y = 3 \sin \left(30\pi t - \dfrac{2\pi}{9}\right)$

   b. $y = 1.5 \sin \left(10^4 \pi t + \dfrac{7\pi}{8}\right)$

5. (PO-7) Calculate the instantaneous amplitude of the following time-dependent waves:

   a. $V = 15 \sin \left(220\pi t + \dfrac{\pi}{4}\right)$ V at $t = 2 \times 10^{-3}$ sec

   b. $E = 10^3 \sin \left(5 \times 10^4 \, t - \dfrac{\pi}{6}\right)$ V/m at $t = 6 \times 10^{-5}$ sec

6. (PO-8) Find an equation for the sum of the following time-dependent vectors. Show the starting position of the sum vector on a coordinate system. Evaluate the frequency of the rotating vector.

   $$y_1 = 8 \sin \left(220\pi t - \dfrac{\pi}{6}\right)$$

   $$y_2 = 5 \sin \left(220\pi t + \dfrac{\pi}{2}\right)$$

7. (PO-9, 10) Consider the following expressions for voltages in an ac circuit:

   $$V_R = 12 \sin (120\pi t)$$

   $$V_L = 36 \sin \left(120\pi t + \dfrac{\pi}{2}\right)$$

   $$V_C = 24 \sin \left(120\pi t - \dfrac{\pi}{2}\right)$$

   a. Graph these voltages on the same set of axes as t ranges from 0 to $\dfrac{1}{60}$th of a second.

   b. Sketch rotating vectors to represent the voltages.

8. (PO-11) Consider an ac circuit having the indicated characteristics:

$$R = 60 \ \Omega$$
$$C = 5 \times 10^{-6} \ F$$
$$L = 15 \times 10^{-3} \ H$$
$$\omega = 1500 \ \text{rad/sec}$$

a. Write an expression for the impedance of this circuit. Use the j-operator in the expression. Recall that $X_C = \dfrac{1}{\omega C}$ and that $X_L = \omega L$.

b. Determine the phase relation between the driving voltage and the current.

CHAPTER 15

LOGARITHMS AND EXPONENTIALS

15.1 The Exponential Function
15.2 The Logarithmic Function
15.3 The Laws of Logarithms
15.4 Common Logarithms
15.5 The Natural Logarithms
15.6 Computations with Logarithms
15.7 Applications of Logarithms and Exponentials
15.8 Summary

In Chapter 4 and again in Chapters 5 and 10, we worked with expressions involving exponents. In this chapter we will introduce exponential and logarithmic functions. These new functions involve exponents but in a different way; the variable appears in the exponent rather than as a base.

Equations with exponents and logarithms occur in practical problems involving electrical circuits, environmental studies, analysis of acidity, sound intensity, and others. We will emphasize the solution of such equations and include a brief discussion of the logarithm for computation.

SECTION 15.1 THE EXPONENTIAL FUNCTION

We begin by defining the exponential function.

DEFINITION

EXPONENTIAL FUNCTION

An **exponential function** has the form $y = b^x$ where $b > 0$ and $b \neq 1$.

The exponential function can be contrasted with expressions of the form $y = x^b$. In the exponential function, the base is constant and the exponent is the independent variable.

The restrictions on the base are necessary since, if $b = 1$, we would have the constant linear function $y = 1^x = 1$. If $b < 0$ and x were certain fractional values. we would have nonreal values for y, for example,

$$y = (-4)^{1/2} \quad \text{implies} \quad y = \sqrt{-4} = 2j$$

Some examples of exponential functions are

$$y = 2^x, \quad y = \left(\frac{1}{4}\right)^x, \quad y = \left(\frac{1}{2}\right)^x$$

$$y = 4^x, \quad y = 5^x, \quad y = \left(\frac{1}{5}\right)^x$$

Most scientific calculators are equipped to evaluate exponential functions for any base. The procedure is developed in example problems.

EXAMPLE Evaluate the exponential function $y = 2^x$ for the following values of x:

$$x = -2, \quad -1.5, \quad -1, \quad -0.5, \quad 0, \quad 0.5, \quad 1, \quad 1.5, \quad 2$$

SOLUTION To obtain 2^x we will use the special function key $\boxed{y^x}$. The sequence of operations is

Enter base (2), press $\boxed{y^x}$, enter exponent, press $\boxed{=}$, read result.

THE EXPONENTIAL FUNCTION 15.1

Here are typical results:

$2 \boxed{y^x} 2 \boxed{+/-} \boxed{=}$ yields 0.25

$2 \boxed{y^x} 1 \boxed{+/-} \boxed{=}$ yields 0.5

$2 \boxed{y^x} 0 \boxed{=}$ yields 1

$2 \boxed{y^x} 1 \boxed{=}$ yields 2

$2 \boxed{y^x} 2 \boxed{=}$ yields 4

PERFORMANCE OBJECTIVE 1	$y = 3^x$ for $-2 \le x \le 4$, x an integer
You should now be able to evaluate an exponential function for given values of the independent variable.	<table><tr><td>x</td><td>-2</td><td>-1</td><td>0</td><td>1</td><td>2</td><td>3</td></tr><tr><td>y</td><td>1/9</td><td>1/3</td><td>1</td><td>3</td><td>9</td><td>27</td></tr></table>

The results of the evaluation of $y = 2^x$ are displayed in a data table.

x	-2	-1.5	-1	-0.5	0
y	0.25	0.3536	0.5	0.7071	1

x	0.5	1	1.5	2
y	1.4142	2	2.8284	4

This data is graphed in Figure 15.1.

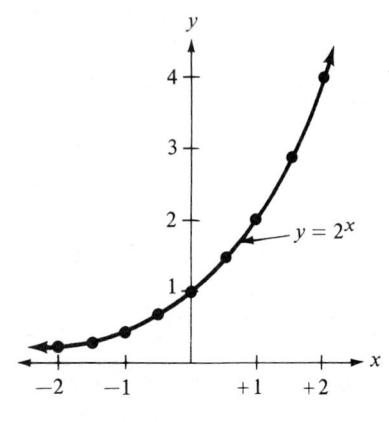

FIGURE 15.1

All exponential functions of the form $y = b^x$, $b > 1$, will have certain properties in common with the function in Figure 15.1. We illustrate these common properties in Figure 15.2. Three functions of the form $y = b^x$ are plotted on the same axes. Only the base changes from plot to plot.

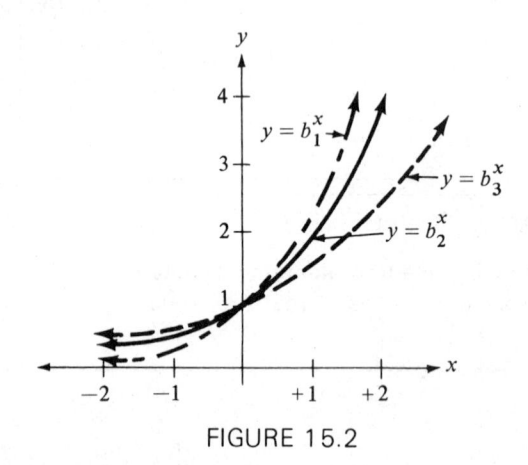

FIGURE 15.2

Note the following common properties of these graphs:

PROPERTIES OF GRAPHS OF EXPONENTIAL FUNCTIONS OF THE FORM $y = b^x$, $b > 1$

1. Each intersects the y-axis at (0, 1).
2. Each rises from left to right and lies above the x-axis.
3. Each approaches the x-axis as x gets larger and negative.

These graphs differ in their rate of increase. The larger the base the more quickly the function rises at the right of the y-axis and the more quickly it approaches zero on the left. In Figure 15.2, $b_1 > b_2 > b_3$.

PERFORMANCE OBJECTIVE 2	
You should now be able to graph an exponential function when the base (b) is greater than 1.	

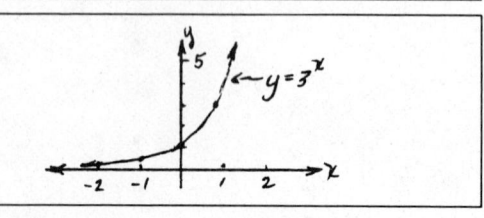

We generate a data table for the function

$$y = (0.5)^x$$

when $x = -2, -1, 0, +1, +2, +3$.

x	-2	-1	0	$+1$	$+2$	$+3$
y	4	2	1	0.5	0.25	0.125

The graph based on this table is shown in Figure 15.3.

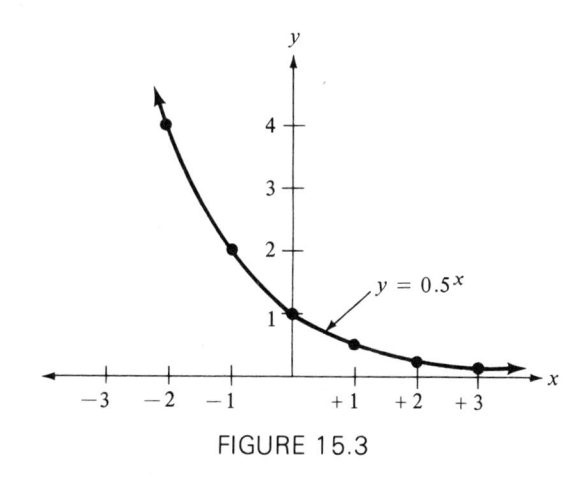

FIGURE 15.3

All exponential functions of the form $y = b^x$, $0 < b < 1$, will have certain properties in common with the function graphed in Figure 15.3. We state these common properties without proof.

PROPERTIES OF GRAPHS OF EXPONENTIAL FUNCTIONS OF THE FORM $y = b^x$, $0 < b < 1$

1. Each intersects the y-axis at (0, 1).
2. Each falls from left to right and lies above the x-axis.
3. Each approaches the x-axis as x gets larger and positive.

Graphs of these functions will differ in their rate of fall depending on the size of b. The smaller b is, the larger the rate of descent.

PERFORMANCE OBJECTIVE 3

You should now be able to graph an exponential function when the base (b) is in the interval $0 < b < 1$.

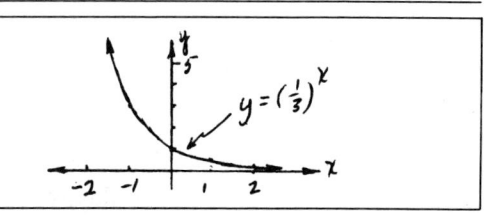

At this point we will introduce two exponential functions that have a slightly different mathematical form than those already discussed. They are, however, true exponential functions as we will illustrate. We refer to the functions $y = e^{kx}$ and $y = e^{-kx}$ where e is the base of the natural logarithm system. The base e is an irrational number approximately equal to 2.7183. The constant k is some number greater than zero (a positive constant).

Let us suppose that $k = 2$. We may then write the function $y = e^{kx}$ as $y = e^{2x} = (e^2)^x$. We evaluate e^2 on a calculator by squaring 2.7183 and find $e^2 = 7.3892$. Then we have the function $y = e^{2x} = 7.3892^x$, which is the first form of the exponential function ($b > 1$). We note that, for any value of k greater than zero, the base b (that is, e^k) is greater than one.

Again let $k = 2$. We may write the function $y = e^{-kx}$ as $y = e^{-2x} = (e^{-2})^x$. Evaluate e^{-2} on the calculator by squaring 2.7183 and taking the reciprocal of that result. We find $e^{-2} = 0.1353$. Then the function $y = e^{-2x}$ becomes $y = (0.1353)^x$, which is the second form of the exponential function ($0 < b < 1$).

SPECIAL EXPONENTIAL FUNCTIONS

$y = e^{kx}$ ($k > 0$) represents an exponential function that rises from left to right.

$y = e^{-kx}$ ($k > 0$) represents an exponential function that falls from left to right.

Exponential functions of the form $y = e^{kt}$ and $y = e^{-kt}$, where the independent variable t has been chosen to represent time, have very interesting characteristics. We graph one of each of these functions to illustrate their characteristics.

EXAMPLE   Graph $y = e^{0.4t}$ for integral values of t from -3 to $+3$ sec.

SOLUTION   We first develop a data table using the following procedure:

Step 1. Select t.

Step 2. Evaluate $0.4t$.

Step 3. Evaluate $e^{0.4t}$.

The procedure for evaluation on the calculator is given here.

$$t \; \boxed{\times} \; 0.4 \; \boxed{=} \; \boxed{e^x} \quad (\text{or } \boxed{\text{INV}} \; \boxed{\text{LN}})$$

yields the function value. If your calculator does not have an $\boxed{e^x}$

THE EXPONENTIAL FUNCTION 15.1

key, substitute [INV] [LN]. The reason for this particular substitution should be clear when natural logarithms are introduced later in this chapter.

t	-3	-2	-1	0	$+1$	$+2$	$+3$
$0.4t$	-1.2	-0.8	-0.4	0	0.4	0.8	1.2
$e^{0.4t}$	0.301	0.449	0.670	1	1.492	2.226	3.320

The graph based on this table is plotted in Figure 15.4.

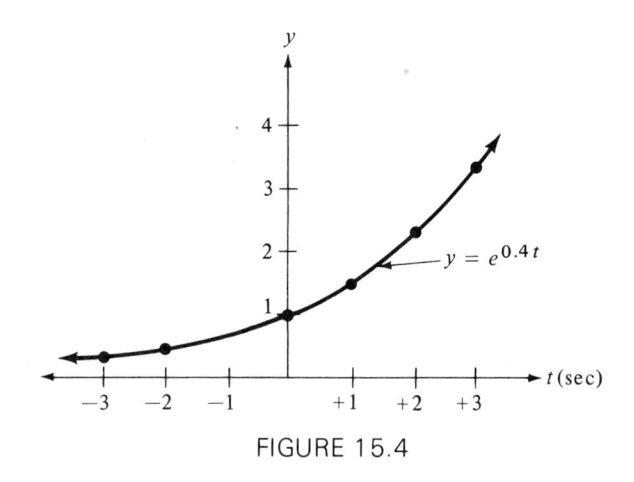

FIGURE 15.4

The graph in Figure 15.4 is said to represent **exponential growth** in time. It has the interesting property that in a time interval equal to approximately $\dfrac{0.693}{k}$ the function value doubles. For the function in the preceding example, that time interval is $\dfrac{0.693}{0.4} = 1.733$ sec. Two doubling times for this graph are shown in Figure 15.5. We use the symbol T_2 to represent **doubling time**.

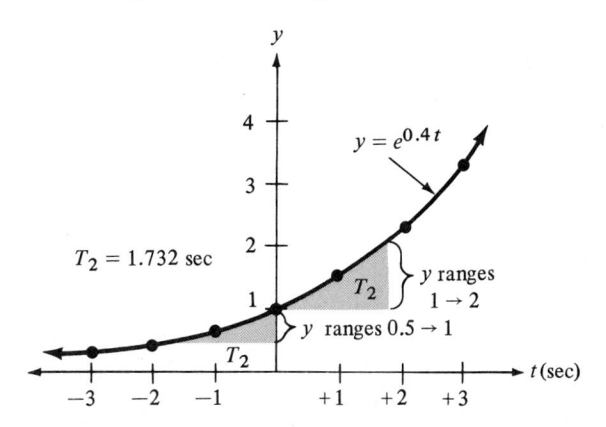

FIGURE 15.5

EXAMPLE Graph $y = e^{-0.5t}$ for $t = -4$ to $+4$ sec.

SOLUTION We first develop a data table using the procedure of the last example.

t	−4	−3	−2	−1	0	1	2	3	4
$-0.5t$	2	1.5	1	0.5	0	−0.5	−1	−1.5	−2
$e^{-0.5t}$	7.39	4.48	2.72	1.65	1	0.61	0.37	0.22	0.14

This data is graphed in Figure 15.6.

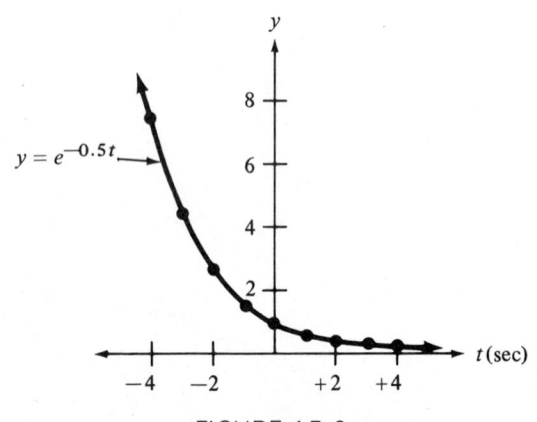

FIGURE 15.6

The graph in Figure 15.6 is said to represent exponential decay in time. **Exponential decay** has the property that the function value is halved in a time interval equal to approximately $\frac{0.693}{k}$. For the preceding function, that time interval is $T_{1/2} = \frac{0.693}{0.5} = 1.386$ sec. The time in which the function reduces its value by one-half is usually called the **half-life**. Three half-lives for the function $y = e^{-0.5t}$ are shown in Figure 15.7.

SPECIAL EXPONENTIAL FUNCTIONS OF TIME

$y = e^{kt}$ ($k > 0$) represents a function that doubles in value in each time interval $T_2 = \frac{0.693}{k}$.

$y = e^{-kt}$ ($k > 0$) represents a function that halves in value in each time interval $T_{1/2} = \frac{0.693}{k}$.

THE EXPONENTIAL FUNCTION 15.1

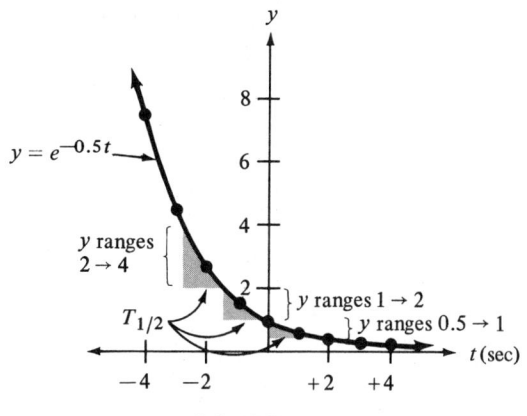

FIGURE 15.7

PERFORMANCE OBJECTIVE 4

You should now be able to graph exponential functions of the form $y = e^{kt}$ and $y = e^{-kt}$ ($k > 0$).

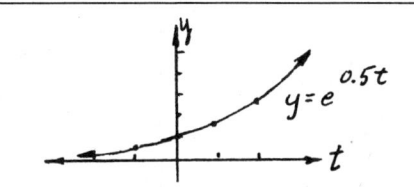

EXERCISE SET 15.1

1. Evaluate $y = b^x$ for the values given.
   a. $b = 3$, $x = -2, -1, 0, +1, +2$
   b. $b = 0.6$, $x = -3, -1, +1, +3$
   c. $b = 1.5$, $x = -4, -2, 0, +2, +4$

2. Evaluate $y = b^x$ for the values given.
   a. $b = 4$, $x = -2, -1, 0, +1, +2$
   b. $b = 0.2$, $x = -3, -1, +1, +3$
   c. $b = 1.3$, $x = -4, -2, 0, +2, +4$

3. Construct a graph of the data generated in Problem 1(a).

4. Construct a graph of the data generated in Problem 2(a).

5. Construct a graph of the data generated in Problem 1(b).

6. Construct a graph of the data generated in Problem 2(b).

7. Graph the exponential function $y = e^{-0.6t}$ for $t = -3$ to $+3$. Use integer values of t to develop ordered pairs.

8. Graph the exponential function $y = e^{-0.2t}$ for $t = -3$ to $+3$. Use integer values of t to develop ordered pairs.

9. Graph the exponential function $y = e^{+1.2t}$ for $t = -2$ to 2. Use integer values of t to develop ordered pairs.

10. Graph the exponential function $y = e^{+1.6t}$ for $t = -2$ to 2. Use integer values of t to develop ordered pairs.

11. Evaluate the doubling time for the function in Problem 9.

12. Evaluate the doubling time for the function in Problem 10.

13. Evaluate the half-life for the function in Problem 7.

14. Evaluate the half-life for the function in Problem 8.

SECTION 15.2 THE LOGARITHMIC FUNCTION

We have previously considered the function as a processor. That point of view will help us in understanding the relationship between the exponential and logarithmic functions.

Consider the general exponential function $y = b^x$. Let $b = 2$ and $x = 4$. Substituting we find

$$y = 2^4 = 2 \cdot 2 \cdot 2 \cdot 2 = 16$$

We know what to do with the 4 and 2 to generate the 16. Schematically, we have accomplished what is shown in Figure 15.8.

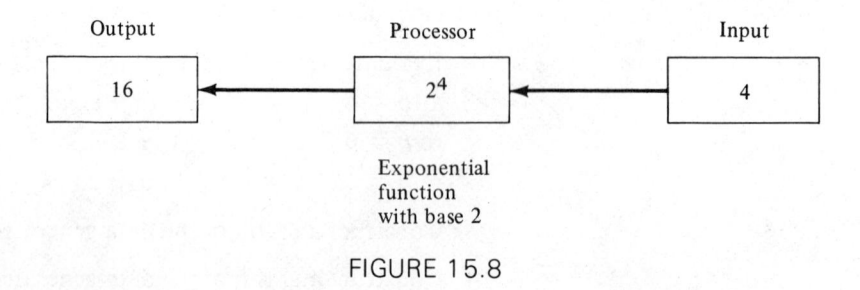

Output | Processor | Input
16 ← 2^4 ← 4

Exponential function with base 2

FIGURE 15.8

EXAMPLE Calculate the output values of the exponential function $y = 2^x$ for the following inputs:

$$x = -1, \quad 0, \quad +1, \quad +2, \quad +3$$

SOLUTION

Output	Processor	Input
0.5	2^{-1}	-1
1	2^0	0
2	2^{+1}	$+1$
4	2^{+2}	$+2$
8	2^{+3}	$+3$

We have thus generated the ordered pairs $(-1, 0.5)$, $(0, 1)$, $(1, 2)$, $(2, 4)$ and $(3, 8)$.

We see that some ordered pairs for our function are easily evaluated since we know how to raise numbers to integral powers. But suppose the problem were posed the other way. Consider again the general exponential function $y = b^x$ with $b = 2$. Now instead of x being given we assume y is known to be 16. Can we find x? That is, can we solve the equation

$$16 = 2^x$$

This is not so easily done because there is no single instruction that tells us what to do with 16 and 2 in order to find x. (We happen to know the answer, but, if we didn't, there would be no easy way to find it.) The first thing we should do is give x a name when it appears this way. To agree with common usage, x is called the logarithm of 16 to the base 2. In abbreviated notation,

$$x = \log_2 16$$

This equation is just another way of writing $16 = 2^x$.

However, in solving for x, we prefer the form in which x is isolated on the left. Writing the equation as $x = \log_2 16$ implies that there is some function (called logarithm) that will take 16 and process it to obtain the answer (which we know is 4). Such a function does exist, although there is no easy way to evaluate it. As a result, tables or calculators are normally used to find values of logarithms. A **logarithm** of a number is nothing more than an exponent of a base that yields that number. We show this inverse process in Figure 15.9.

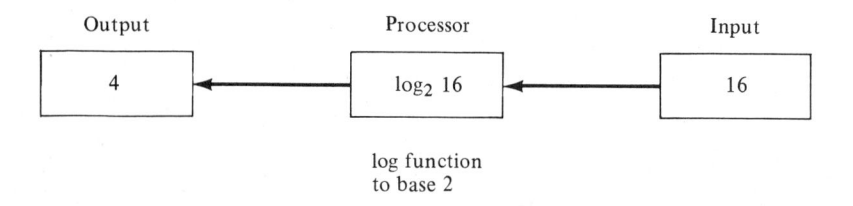

FIGURE 15.9

Since it is customary to let x be the independent variable and y be the dependent variable, we write the logarithmic function as

$$y = \log_b x$$

We say that the logarithmic function is the inverse of the exponential function with the same base. The exponential function is used to find the result of an exponentiation when the base and exponent are known. The logarithmic function is used to find the exponent when the base and the result of the exponentiation are known.

EXPRESSIONS INVOLVING EXPONENTS AND LOGARITHMS

$y = b^x$ and $x = \log_b y$ are different forms of the same equation.

$y = b^x$ and $y = \log_b x$ are inverse functions. If one processes 4 into 16, the other will process 16 into 4.

EXAMPLE Express the following in an equivalent logarithmic or exponential form:

a. $y = 3^x$ b. $y = \log_5 x$

SOLUTION Use the equivalent expressions $y = b^x$ and $x = \log_b y$.

a. $x = \log_3 y$ b. $x = 5^y$

EXAMPLE If $28 = b^3$, find $\log_b 28$.

SOLUTION Use the fact that the exponential and logarithmic functions are inverses. Note that the exponential function processes 3 into 28. The logarithmic function to the same base must process 28 into 3. Then the answer is

$$\log_b 28 = 3$$

PERFORMANCE OBJECTIVE 5

You should now be able to recognize the relationship between the exponential and logarithmic functions.

$y = 7^3 \iff 3 = \log_7 y$

$x = \log_2 16 \iff 16 = 2^x$

EXERCISE SET 15.2

In Exercises 1–8 write the equivalent logarithmic equations for the given exponential equations.

1. $y = 6^x$
2. $y = 7^x$
3. $y = 3^x$
4. $y = 2^x$
5. $y = 10^x$
6. $y = 9^x$
7. $y = \left(\dfrac{1}{8}\right)^x$
8. $y = \left(\dfrac{1}{4}\right)^x$

In Exercises 9–16 write the equivalent exponential equations for the given logarithmic equations.

9. $y = \log_6 x$
10. $y = \log_7 x$
11. $y = \log_{1/8} x$
12. $y = \log_{1/6} x$
13. $y = \log_{10} x$
14. $y = \log_9 x$
15. $y = \log_{1/7} x$
16. $y = \log_{1/4} x$

In Exercises 17–22 find the logarithms from your knowledge of powers of ten.

17. $\log_{10} 100$
18. $\log_{10} 10$
19. $\log_{10} 1000$
20. $\log_{10} 10{,}000$
21. $\log_{10} 0.001$
22. $\log_{10} 0.0001$

SECTION 15.3 THE LAWS OF LOGARITHMS

We mentioned in Section 15.1 that exponents have been studied previously. You should recall that exponents have a set of laws that they obey. Three of these Laws of Exponents can be used to develop the Laws of Logarithms.

The three Laws of Exponents to be used are

1. $a^m a^n = a^{m+n}$
2. $\dfrac{a^m}{a^n} = a^{m-n}$
3. $(a^m)^n = a^{mn}$

The Laws of Logarithms are

1. $\log_b(xy) = \log_b x + \log_b y$

This is read as: the logarithm of the product of two positive numbers x and y is equal to the sum of the logarithms of the numbers.

15 LOGARITHMS AND EXPONENTIALS

2. $\log_b \left(\dfrac{x}{y}\right) = \log_b x - \log_b y$

This is read as: the logarithm of the quotient of two positive numbers x and y is equal to the logarithm of the numerator minus the logarithm of the denominator.

3. $\log_b x^n = n \log_b x$

This is read as: the logarithm of the nth power of a positive number is equal to n times the logarithm of the number.

To establish the Laws of Logarithms, we use the Laws of Exponents. For instance, if we let $w = \log_b x$ and $v = \log_b y$, an equivalent statement is that $x = b^w$ and $y = b^v$. We then use these statements to write

$xy = b^w \cdot b^v = b^{w+v}$. An equivalent logarithmic statement is that

$$\log_b xy = w + v = \log_b x + \log_b y$$

$\left(\dfrac{x}{y}\right) = \dfrac{b^w}{b^v} = b^{w-v}$. An equivalent logarithmic statement is that

$$\log_b \left(\dfrac{x}{y}\right) = w - v = \log_b x - \log_b y$$

$x^n = (b^w)^n = b^{wn} = b^{nw}$. An equivalent logarithmic statement is that

$$\log_b (x^n) = nw = n \cdot \log_b x$$

In the following example problems, simplify by removing all products, quotients, and powers from the number whose logarithm is to be found. Use the Laws of Logarithms.

EXAMPLE Simplify

(a) $\log_b (4 \cdot 5)$ (b) $\log_b (2 \cdot 10)$

SOLUTION

(a) $\log_b (4 \cdot 5) = \log_b 4 + \log_b 5$
(b) $\log_b (2 \cdot 10) = \log_b 2 + \log_b 10$

EXAMPLE Simplify

(a) $\log_b \left(\dfrac{8}{3}\right)$ (b) $\log_b \left(\dfrac{1}{2}\right)$

SOLUTION

(a) $\log_b \left(\dfrac{8}{3}\right) = \log_b 8 - \log_b 3$

(b) $\log_b \left(\frac{1}{2}\right) = \log_b 1 - \log_b 2$

EXAMPLE Simplify
(a) $\log_b 2^3$ (b) $\log_b 5^4$ (c) $\log_b \sqrt[3]{16}$

SOLUTION

(a) $\log_b 2^3 = 3 \log_b 2$
(b) $\log_b 5^4 = 4 \log_b 5$
(c) $\log_b \sqrt[3]{16} = \frac{1}{3} \log_b 16$

EXAMPLE Simplify $\log_b \left[4 \cdot \left(\frac{2}{3}\right)^2 \right]$.

SOLUTION $\log_b 4 + \log_b \left(\frac{2}{3}\right)^2 = \log_b 4 + 2 \log_b \left(\frac{2}{3}\right)$
$= \log_b 4 + 2 (\log_b 2 - \log_b 3)$
$= \log_b 4 + 2 \log_b 2 - 2 \log_b 3$

PERFORMANCE OBJECTIVE 6	$\log [5 \cdot 2^6]$
You should now be able to use the Laws of Logarithms to simplify logarithmic expressions.	$= \log 5 + \log 2^6 = \log 5 + 6 \log 2$

EXERCISE SET 15.3

In the following exercises express each term as a sum, difference, or multiple of logarithms:

1. $\log_b (6 \cdot 3)$
2. $\log_b (4 \cdot 5)$
3. $\log_b \left(\frac{5}{4}\right)$
4. $\log_b \left(\frac{2}{3}\right)$
5. $\log_b \left(\frac{1}{5}\right)$
6. $\log_b \left(\frac{1}{8}\right)$
7. $\log_b (5^2)$
8. $\log_b (3^4)$
9. $\log_b (\sqrt{15})$
10. $\log_b (\sqrt{11})$
11. $\log_b \left(\frac{5}{7}\right)$
12. $\log_b \left(\frac{8}{9}\right)$

13. $\log_b (\sqrt[3]{25})$ 14. $\log_b (\sqrt[3]{23})$

15. $\log_b (8 \cdot 7)$ 16. $\log_b (3 \cdot 7)$

17. $\log_b \left[6 \cdot \left(\frac{4}{5}\right)^2 \right]$ 18. $\log_b \left[8 \cdot \left(\frac{5}{6}\right)^2 \right]$

19. $\log_b \left[\left(\frac{1}{2}\right)^4 \cdot 5 \right]$ 20. $\log_b \left[\left(\frac{1}{2}\right)^3 \cdot 7 \right]$

SECTION 15.4 COMMON LOGARITHMS

In Section 15.1 we noted that the b in the exponential function could be any real number greater than 0 but not equal to 1. This means that there are an infinite number of values for b and each of these values has its own logarithm system. However, only two logarithm systems are in common usage in mathematics. These two systems are the **common logarithms** and the **natural logarithms.** In this section we will introduce the common logarithm system.

DEFINITION

> **COMMON LOGARITHM**
>
> A logarithm to the base 10 is called a **common logarithm.**

As noted in the definition, common logarithms are logarithms that use 10 as a base, that is, $b = 10$. When we write common logarithms, we usually omit the subscript 10. The expression $\log_{10} N$, where N is any positive number, will be written as $\log N$. Common logarithms can be very useful for doing computational problems that involve multiplication, division, powers, and roots.

With common logarithms, the logarithms of powers of 10 have a very interesting pattern. For instance, the log $10 = 1$ since $10^1 = 10$ and the log $100 = 2$ since $10^2 = 100$. Common logarithms of some powers of 10 are listed here.

$$\log 10{,}000 = \log 10^4 = 4 \text{ since } 10^4 = 10{,}000$$
$$\log 100 = \log 10^2 = 2 \text{ since } 10^2 = 100$$
$$\log 10 = \log 10^1 = 1 \text{ since } 10^1 = 10$$
$$\log 1 = \log 10^0 = 0 \text{ since } 10^0 = 1$$
$$\log \frac{1}{10} = \log 10^{-1} = -1 \text{ since } 10^{-1} = \frac{1}{10}$$
$$\log \frac{1}{100} = \log 10^{-2} = -2 \text{ since } 10^{-2} = \frac{1}{100}$$
$$\log \frac{1}{10{,}000} = \log 10^{-4} = -4 \text{ since } 10^{-4} = \frac{1}{10{,}000}$$

Note that for powers of 10 the logarithm becomes the power on 10. This important property of logarithms of powers of 10 will be used in our further study of common logarithms. You should also note from the preceding list that $\log 1 = 0$ ($10^0 = 1$).

Before proceeding with the study of common logarithms, it would be helpful to review how to write a number in scientific notation. Scientific notation was introduced in Chapter 1.

We now consider the more general problem of finding the logarithm of any number to base 10. Table 15.1 is a partial table of common logarithms. A complete table can be found in Appendix IV.2. This table yields what is called the **mantissa** of the number entered. The relation of the mantissa to the common logarithm will be discussed later.

TABLE 15.1 PARTIAL TABLE OF COMMON LOGARITHMS

N	0	1	2	3	4	5	6	7	8	9
20	3010	3032	3054	3075	3096	3118	3139	3160	3181	3201
21	3222	3243	3263	3284	3304	3324	3345	3365	3385	3404
22	3424	3444	3464	3483	3502	3522	3541	3560	3579	3598
23	3617	3636	3655	3674	3692	3711	3729	3747	3766	3784
24	3802	3820	3838	3856	3874	3892	3909	3927	3945	3962
25	3979	3997	4014	4031	4048	4065	4082	4099	4116	4133
26	4150	4166	4183	4200	4216	4232	4249	4265	4281	4298
27	4314	4330	4346	4362	4378	4393	4409	4425	4440	4456
28	4472	4487	4502	4518	4533	4548	4564	4579	4594	4609
29	4624	4639	4654	4669	4683	4698	4713	4728	4742	4757
30	4771	4786	4800	4814	4829	4843	4857	4871	4886	4900
31	4914	4928	4942	4955	4969	4983	4997	5011	5024	5038
32	5052	5065	5079	5092	5105	5119	5132	5145	5159	5172
33	5185	5198	5211	5224	5237	5250	5263	5276	5289	5302
34	5315	5328	5340	5353	5366	5378	5391	5403	5416	5428

To use the partial logarithm table, locate the first two digits of the number in the column labeled N. The third digit is located across the top of the table. Find the intersection of these two numbers and we have the desired value. This value (the mantissa) is a four place decimal number although tables do not show the decimal points.

EXAMPLE Find the common logarithm mantissa of 296.

SOLUTION Locate the 29 under N and locate the 6 across the top. These two intersect at the number 4713. We record the number 0.4713. This number is the common logarithm mantissa of 296.

EXAMPLE Find mantissa values for each of the following numbers:

$$204, \quad 333, \quad 240$$

SOLUTION

204 yields 0.3096 from the table.

333 yields 0.5224 from the table.

240 yields 0.3802 from the table.

PERFORMANCE OBJECTIVE 7	
You should now be able to find the common logarithm mantissa of a number with less than 4 digits using a table.	344 yields 0.5366 as its mantissa. 278 yields 0.4440 as its mantissa.

If the number we are locating in the table has more than four nonzero digits, we must round the number to four nonzero digits. For instance, 23,481 becomes 23,480 while 56,446 becomes 56,450.

After doing this, we use the table as before to find the mantissa corresponding to the first three digits. To correct for the fourth digit we must use a table of proportional parts. Add the number found in the appropriate row (corresponding to the first two digits) under the fourth digit to the mantissa already determined. The result is a mantissa with four digits. Table 15.2 contains a partial table of logarithms with proportional parts added.

TABLE 15.2 PARTIAL TABLE OF COMMON LOGARITHMS WITH PROPORTIONAL PARTS

N	0		7	8	9	1	2	3	4	5	6	7	8	9
20	3010		3160	3181	3201	2	4	6	8	11	13	15	17	19
21	3222		3365	3385	3404	2	4	6	8	10	12	14	16	18
22	3424		3560	3579	3598	2	4	6	8	10	11	13	15	17
23	3617		3747	3766	3784	2	4	5	7	9	11	13	14	16
24	3802		3927	3945	3962	2	4	5	7	9	11	13	14	16
25	3979	3997		4116	4133	2	3	5	7	9	10	12	14	15
26	4150	4166		4281	4298	2	3	5	6	8	10	11	13	14
27	4314	4330		4440	4456	2	3	5	6	8	10	11	13	14
28	4472	4487		4594	4609	2	3	5	6	8	9	11	12	14
29	4624	4639		4742	4757	1	3	4	6	7	8	10	11	13

(Proportional parts columns 1–9)

EXAMPLE Find the common logarithm mantissa of 19,845.

SOLUTION Round 19,845 to four digits. It becomes 19,850. Enter the table with 198 to find a mantissa of 2967. Locate the 5 column in the proportional parts section and find the intersection with the row across from 19. This gives us 11. Now add the 2967 and the 11. The result is 2978, which we write as 0.2978.

EXAMPLE Find mantissa values for each of the following numbers:

$$29883, \quad 24754, \quad 21079$$

SOLUTION

29,883 rounds to 29,880. From the table we find 0.4753.

24,754 rounds to 24,750. From the table we find 0.3936.

21,079 rounds to 21,080. From the table we find 0.3238.

PERFORMANCE OBJECTIVE 8

You should now be able to find the common logarithm mantissa of a number with four or more digits using a table.

32187 rounds to 32190.
32190 yields 0.5077 as its mantissa.

So far we have only found the common logarithm mantissa. Its relation to the common logarithm will now be developed.

RULES FOR FINDING THE COMMON LOGARITHM OF N

Step 1. Write N in scientific notation.
$N = N_s \times 10^n$ where $1 < N_s < 10$ and n is an integer.

Step 2. Determine the characteristic (the exponent on 10).

Step 3. Determine the mantissa using a table of logarithms.

Step 4. Add the characteristic and mantissa to find log N.

The following examples will illustrate the technique for finding log N with a table of logarithms.

EXAMPLE   Find log 263.

SOLUTION

Step 1. $263 = 2.63 \times 10^2$

Step 2. Characteristic $= 2$

Step 3. Mantissa $= 0.4200$

Step 4. log $263 = 2 + 0.4200 = 2.4200$

EXAMPLE   Find log 5300.

SOLUTION

Step 1. $5300 = 5.30 \times 10^3$

Step 2. Characteristic $= 3$

Step 3. Mantissa $= 0.7243$

Step 4. log $5300 = 3 + 0.7243 = 3.7243$

EXAMPLE   Find log 4.239.

SOLUTION

Step 1. $4.239 = 4.239 \times 10^0$

Step 2. Characteristic $= 0$

Step 3. Mantissa $= 0.6272$

Step 4. log $4.239 = 0 + 0.6272 = 0.6272$

(Note that when N is between 1 and 10 the logarithm is equal to the mantissa.)

COMMON LOGARITHMS 15.4

EXAMPLE Find log 28567.

SOLUTION

Step 1. 28567 is rounded to 28570 so that $28570 = 2.857 \times 10^4$

Step 2. Characteristic = 4

Step 3. Mantissa = 0.4559

Step 4. log 28567 = 4.4559

When the characteristic is a negative number, there are two ways that we can write our answer in Step 4. For example, if we are finding log 0.0359, we would follow these steps.

Step 1. $0.0359 = 3.59 \times 10^{-2}$

Step 2. Characteristic = -2

Step 3. Mantissa = 0.5551

Step 4. We would write the characteristic -2 as a positive characteristic by adding and subtracting 10 so that $-2 = (10 - 2) - 10$. By doing this, Step 4 becomes

$$\begin{aligned}\log 0.0359 &= -2 + 0.5551 \\ &= (10 - 2) + 0.5551 - 10 \\ &= 8 + 0.5551 - 10 \\ &= 8.5551 - 10\end{aligned}$$

An alternative way of writing Step 4 is to add the characteristic -2 and the mantissa 0.5551 using the rules for signed numbers. Then Step 4 becomes

$$\log 0.0359 = -2 + 0.5551 = -1.4449$$

A note of caution is needed here since we can no longer assume that the characteristic is -1 and the mantissa is 0.4449. In fact, we know that the characteristic is -2 and the mantissa is 0.5551. The following examples will illustrate the techniques for finding log N when the characteristic is negative.

EXAMPLE Find log 0.00358.

SOLUTION

Step 1. $0.00358 = 3.58 \times 10^{-3}$

Step 2. Characteristic = -3

Step 3. Mantissa = 0.5539

Step 4. log 0.00358 = −3 + 0.5539
$$= (10 - 3) + 0.5539 - 10$$
$$= 7.5539 - 10$$

or

log 0.00358 = −3 + 0.5539 = −2.4461

EXAMPLE Find log 0.0759.

SOLUTION

Step 1. $0.0759 = 7.59 \times 10^{-2}$

Step 2. Characteristic = −2

Step 3. Mantissa = 0.8802

Step 4. log 0.0759 = −2 + 0.8802
$$= (10 - 2) + 0.8802 - 10$$
$$= 8.8802 - 10$$

or

log 0.0759 = −2 + 0.8802 = −1.1198

EXAMPLE Find log 0.0002142.

SOLUTION

Step 1. $0.0002142 = 2.142 \times 10^{-4}$

Step 2. Characteristic = −4

Step 3. Mantissa = 0.3308 (3304 + 4 = 3308)

Step 4. log 0.0002142 = −4 + 0.3308
$$= (10 - 4) + 0.3308 - 10$$
$$= 6.3308 - 10$$

or

log 0.0002142 = −4 + 0.3308 = −3.6692

PERFORMANCE OBJECTIVE 9	
You should now be able to find the common logarithm of a number using a table of logarithms.	log 0.002286 = log 2.286 × 10⁻³ characteristic = −3 ; mantissa = 0.3590 Then log 0.002286 = −3 + 0.3590 = 7.3590 − 10

COMMON LOGARITHMS 15.4   581

If you have a calculator with a $\boxed{\text{LOG}}$ key, you will find that the logarithm value will be given with the characteristic and the mantissa already added. Also, if you want the calculator answer to be equal (in most cases) to the answer from the four-place logarithm table, you need to round the calculator answer to four decimal places.

The procedure for finding log N is

Enter N, press $\boxed{\text{LOG}}$, read display

EXAMPLE   Find log 263.

SOLUTION   263 $\boxed{\text{LOG}}$ read 2.4199557

Rounding to 2.4200 gives the same value as an earlier example using the log table.

EXAMPLE   Find log 5300.

SOLUTION   5300 $\boxed{\text{LOG}}$ read 3.7242759

Rounding to 3.7243 agrees with the table value in an earlier example.

EXAMPLE   Find log 4.239.

SOLUTION   4.239 $\boxed{\text{LOG}}$ read 0.6272634

Rounding to 0.6273 gives a slightly different result than found in an earlier example.

EXAMPLE   Find log 28567.

SOLUTION   28567 $\boxed{\text{LOG}}$ read 4.4558646

Rounding to 4.4559 agrees with the result in an earlier example.

EXAMPLE   Find log 0.00358.

SOLUTION   0.00358 $\boxed{\text{LOG}}$ read -2.446117

Rounding to -2.4461 agrees with the table value in an earlier example.

EXAMPLE   Find log 0.0759.

SOLUTION   0.0759 $\boxed{\text{LOG}}$ read -1.1197582

Rounding to -1.1198 agrees with the table value in an earlier example.

EXAMPLE   Find log 0.0002142.

SOLUTION   0.0002142 $\boxed{\text{LOG}}$ read -3.6691805

Rounding to -3.6692 agrees with the table value in an earlier example.

PERFORMANCE OBJECTIVE 10	
You should now be able to find the common logarithm of a number using the calculator.	367 [log] gives 2.5647

We include some examples in which the logarithm must be evaluated in order to solve an equation.

EXAMPLE The difference in intensity in two sounds is typically measured on a decibel scale. The equation for this difference is

$$D = 10 \log\left(\frac{I_2}{I_1}\right) \text{ decibels}$$

Find D when $\frac{I_2}{I_1} = 50$.

SOLUTION When using logarithms in a problem such as this, it is best to find the logarithm and then substitute into the given formula.

Step 1. $50 = 5.0 \times 10^1$

Step 2. Characteristic $= 1$

Step 3. Mantissa $= 0.6990$

Step 4. $\log 50 = 1 + 0.6990 = 1.6990$

or 50 [LOG] read 1.69897

Substituting into the equation we find

$$D = 10(1.6990) = 16.99 \text{ or } 17 \text{ decibels (db)}$$

to two significant figures.

EXAMPLE It has been agreed that the threshold of the human ear is $\frac{10^{-12} \text{ W}}{m^2}$. Sounds below this level of intensity cannot be detected. If the average sound has an intensity level of $\frac{10^{-6} \text{ W}}{m^2}$, calculate the difference in these two sounds on the decibel scale.

SOLUTION Use the equation in the last example.

$$D = 10 \log\left(\frac{I_2}{I_1}\right).$$

Consider $I_2 = 10^{-6}$ and $I_1 = 10^{-12}$. On the calculator

1 [EE] 6 [+/−] [÷] 1 [EE] 12 [+/−] [=] [LOG] [×] 10 [=]
read 60

By hand

$$\frac{I_2}{I_1} = \frac{10^{-6}}{10^{-12}} = 10^6$$

But $\log 10^6 = 6$ so that $D = 6(10) = 60$.

The answer is that these two sounds differ in intensity by 60 decibels.

EXAMPLE An environmental technologist is testing a stream for acid mine drainage. A measure of acidity is pH (potency of hydrogen). We define pH as the negative value of the exponent of 10 used to express the molecular concentration of H_3O^+. In mathematical terms, the pH of a solution is the negative logarithm of the molar concentration of H_3O^+. That is,

$$pH = -\log (\text{concentration of } H_3O^+)$$

If the concentration of H_3O^+ in the stream water is 4.625×10^{-3} moles/ℓ, find the pH.

SOLUTION

$pH = -\log (4.625 \times 10^{-3})$
$pH = -(-3 + 0.6651) = -(-2.3349) = 2.335$ (to 4 significant digits)

On the calculator

4.625 [EE] 3 [+/−] [LOG] [+/−] read 2.3349

pH values less than 7 indicate acidic conditions.

PERFORMANCE OBJECTIVE 11	$D = 10 \log(I_2/I_1)$. If $I_2/I_1 = 40$
You should now be able to solve numerical problems involving common logarithms.	then $D = 10 \log 40 = 10 (1.6021)$ $= 16.021$ or 16 db

EXERCISE SET 15.4

Find the logarithm of each of the numbers in Exercises 1–20.

1. 790
2. 860
3. 31.60
4. 23.50
5. 0.6120
6. 0.7210
7. 0.04030
8. 0.03080
9. 900.5
10. 600.8
11. 0.4500
12. 0.6300
13. 0.9042
14. 0.5055
15. $(9065)^8$
16. $(7123)^8$
17. 0.8256
18. 0.9227
19. $\sqrt{641.6}$
20. $\sqrt{326.8}$

21. In the formula for sound intensity, if $\left(\dfrac{I_2}{I_1}\right) = 15$, find the value of D.

22. In the formula for sound intensity, if $\left(\dfrac{I_2}{I_1}\right) = 25$, find the value of D.

23. The intensity of a sound that will cause pain to the human ear is about $\dfrac{1 \text{W}}{m^2}$. Recall that the threshold for the human ear is about $\dfrac{10^{-12} \text{W}}{m^2}$. What is the difference in these two intensity levels on the decibel scale?

24. The maximum intensity sound response in a certain receiver is $\dfrac{75 \text{W}}{m^2}$. The maximum intensity response of the human ear is about $\dfrac{2 \text{W}}{m^2}$. What is the difference in these two intensity levels on the decibel scale?

25. Find the pH of a solution if the concentration of H_3O^+ is 0.0000036 moles/ℓ.

26. Find the pH of a solution if the concentration of H_3O^+ is 0.0000049 moles/ℓ.

SECTION 15.5 THE NATURAL LOGARITHMS

In Section 15.4 we mentioned that the second commonly used logarithm system is the **natural logarithm system.** In this section we introduce the natural logarithm. The natural logarithm system uses as a base the number e (the irrational number discussed previously, $e \cong 2.7183$). When we write natural logarithms, the subscript e is omitted and we write ln in place of log. That is, $\log_e N$ will be written as ln N where N is any positive number. Natural logarithms are so called because they occur in the description of certain 'natural' systems in science and engineering.

DEFINITION

> **NATURAL LOGARITHM**
>
> A logarithm to the base e is called a **natural logarithm.**

We can now see why we used the sequence $\boxed{\text{INV}}$ $\boxed{\text{LN}}$ previously to find powers of e.

Tables of ln N are written for numbers N where $1 < N < 10$. You must be careful when finding ln N for numbers not in this range since the natural logarithms of integral powers of 10 do not have the nice relationship that the integral powers of 10 had for the common logarithms (see Section 15.4). To find the ln N, we make use of the fact that ln $10 = 2.3026$ and that ln $10^n = n(2.3026)$. You should also note that ln $e = 1$ in the same way that log $10 = 1$.

In order to determine the natural logarithm of a number, use the following procedure.

RULES FOR FINDING THE NATURAL LOGARITHM OF N

Step 1. Round N to three nonzero digits if necessary.

Step 2. Write N in scientific notation. Suppose N is written as

$$N = N_s \times 10^n$$

where N_s is between 1 and 10.

Step 3. Find the value of ln N_s in the table.

Step 4. Find the value of ln 10^n by evaluating $2.3026n$.

Step 5. Add the value of ln N_s to the value of ln 10^n.

15 LOGARITHMS AND EXPONENTIALS

Table 15.3 contains a partial table of natural logarithms. This table will be used in the examples that follow. A complete table can be found in Appendix IV.3.

TABLE 15.3 PARTIAL TABLE OF NATURAL LOGARITHMS

N	0.00	0.01	0.02	0.03	0.04	0.05	0.06	0.07	0.08	0.09
2.0	0.6931	0.6981	0.7031	0.7080	0.7129	0.7178	0.7227	0.7275	0.7324	0.7372
2.1	0.7419	0.7467	0.7514	0.7561	0.7608	0.7655	0.7701	0.7747	0.7793	0.7839
2.2	0.7885	0.7930	0.7975	0.8020	0.8065	0.8109	0.8154	0.8198	0.8242	0.8286
2.3	0.8329	0.8372	0.8416	0.8459	0.8502	0.8544	0.8587	0.8629	0.8671	0.8713
2.4	0.8755	0.8796	0.8838	0.8879	0.8920	0.8961	0.9002	0.9042	0.9083	0.9123
2.5	0.9163	0.9203	0.9243	0.9282	0.9322	0.9361	0.9400	0.9439	0.9478	0.9517
2.6	0.9555	0.9594	0.9632	0.9670	0.9708	0.9746	0.9783	0.9821	0.9858	0.9895
2.7	0.9933	0.9969	1.0006	1.0043	1.0080	1.0116	1.0152	1.0188	1.0225	1.0260
2.8	1.0296	1.0332	1.0367	1.0403	1.0438	1.0473	1.0508	1.0543	1.0578	1.0600
2.9	1.0647	1.0682	1.0716	1.0750	1.0784	1.0818	1.0852	1.0886	1.0919	1.0953
3.0	1.0986	1.1019	1.1053	1.1086	1.1119	1.1151	1.1184	1.1217	1.1249	1.1282
3.1	1.1314	1.1346	1.1378	1.1410	1.1442	1.1474	1.1506	1.1537	1.1569	1.1600
3.2	1.1632	1.1663	1.1694	1.1725	1.1756	1.1787	1.1817	1.1848	1.1878	1.1909
3.3	1.1939	1.1969	1.2000	1.2030	1.2060	1.2090	1.2119	1.2149	1.2179	1.2208
3.4	1.2238	1.2267	1.2296	1.2326	1.2355	1.2384	1.2413	1.2442	1.2470	1.2499

EXAMPLE Find ln 2.46.

SOLUTION

Step 1. 2.46 has only 3 nonzero digits.

Step 2. $2.46 = 2.46 \times 10^0$

Step 3. ln 2.46 = 0.9002

Step 4. $\ln 10^0 = 0(2.3026) = 0$

Step 5. ln 2.46 = 0.9002 + 0 = 0.9002

EXAMPLE Find ln 34,526.

SOLUTION

Step 1. 34,526 is rounded to 34,500.

Step 2. $34,500 = 3.45 \times 10^4$

Step 3. ln 3.45 = 1.2384

Step 4. $\ln 10^4 = 4(2.3026) = 9.2104$

Step 5. ln 34,526 = 1.2384 + 9.2104 = 10.4488

THE NATURAL LOGARITHMS 15.5

EXAMPLE Find ln 0.00256.

SOLUTION

Step 1. 0.00256 only has 3 nonzero digits.

Step 2. $0.00256 = 2.56 \times 10^{-3}$

Step 3. $\ln 2.56 = 0.9400$

Step 4. $\ln 10^{-3} = -3(2.3026) = -6.9078$

Step 5. $\ln 0.00256 = 0.9400 + (-6.9078) = -5.9678$

PERFORMANCE OBJECTIVE 12	$\ln 2769$ rounds to $\ln 2770$
You should now be able to find the natural logarithm of a number using a table.	$2770 = 2.77 \times 10^3$
	$\ln 2.77 = 1.0188 \ ; \ \ln 10^3 = 3(2.3026)$
	$\ln 2769 = 1.0188 + 3(2.3026) = 7.9266$

If you have a calculator with a $\boxed{\text{LN}}$ key, you can generate values of the natural logarithm more precisely than by using the table. To compare calculator and table values, round the calculator values to four decimal places. In the next three examples we redo the preceding problems using a calculator and compare results.

EXAMPLE Find ln 2.46.

SOLUTION   2.46 $\boxed{\text{LN}}$ read 0.9001614

Round to 0.9002 and note the agreement with the table value.

EXAMPLE Find ln 34,526.

SOLUTION   34,526 $\boxed{\text{LN}}$ read 10.449468

Round to 10.4495 and note the difference from the table value.

EXAMPLE Find ln 0.00256.

SOLUTION   0.00256 $\boxed{\text{LN}}$ read -5.967748

Round to -5.9677 and note the difference from the table value.

PERFORMANCE OBJECTIVE 13	$\ln 2769$
You should now be able to use a calculator to find the natural logarithm of a number.	$2769 \boxed{\ln}$ gives 7.9262415 round to 7.9262

Numerical problems involving natural logarithms may require us to find values of the variable t in an equation such as

$$A = A_0 e^{-kt} \quad \text{or} \quad A = A_0 e^{+kt} \quad (k > 0)$$

To solve for t in these equations we write the equation in its equivalent logarithmic form. Recall that

$$y = b^x \quad \text{and} \quad x = \log_b y$$

are equivalent expressions. Taking the second equation $A = A_0 e^{+kt}$, we first divide by A_0.

$$\frac{A}{A_0} = e^{+kt}$$

The equivalent form is $kt = \ln\left(\dfrac{A}{A_0}\right)$. Then

$$t = \frac{1}{k} \ln\left(\frac{A}{A_0}\right)$$

EXAMPLE A laboratory technician is to investigate the bacteria count in a culture that is being treated. The bacteria present at any time t is assumed to be given by the equation

$$B(t) = 200{,}000 e^{-0.02t}$$

where t is measured in hours. Predict the times when the bacteria count will have decreased to 150,000, 100,000, and 50,000.

SOLUTION First let $B(t) = 150{,}000$. Then

$$150{,}000 = 200{,}000 e^{-0.02t}$$

$$\frac{150{,}000}{200{,}000} = e^{-0.02t}$$

The equivalent logarithmic form of this equation is

$$\ln\left(\frac{150{,}000}{200{,}000}\right) = -0.02t$$

Then t is given by

$$t = \frac{1}{-0.02} \ln\left(\frac{150{,}000}{200{,}000}\right) = -50(-0.2877) = 14.385$$
$$= 14 \text{ hr} \quad \text{(to the nearest hour)}$$

To evaluate the time corresponding to the other two counts substitute these counts where 150,000 occurred.

$$t = -50 \ln\left(\frac{100{,}000}{200{,}000}\right) = -50(-0.6931) = 34.655$$
$$= 35 \text{ hr} \quad \text{(to the nearest hour)}$$

$$t = -50 \ln\left(\frac{50{,}000}{200{,}000}\right) = -50(-1.3863) = 69.315 \text{ or } 69 \text{ hr}$$

Note that the second time corresponds to the half-life as defined in Section 15.1. You may want to verify that this time is equal to $\dfrac{0.693}{k}$.

PERFORMANCE OBJECTIVE 14

You should now be able to solve numerical problems involving natural logarithms.

$B(t) = 200{,}000\, e^{-0.02t}$

Find the time when $B(t) = 25{,}000$.

EXERCISE SET 15.5

Find the natural logarithm for Exercises 1–20.

1. 1.6253
2. 2.5633
3. 2646
4. 8969
5. 83.29
6. 23.46
7. 0.2145
8. 0.4682
9. 479
10. 863
11. 25
12. 49
13. 489.79
14. 286.39
15. 5614
16. 6829
17. 0.008765
18. 0.004965
19. 0.8843
20. 0.5526

21. Suppose that $D(t)$, the amount of a certain radioactive substance present at time t, is given by

$$D(t) = 2000 e^{-0.04t}$$

where t is measured in days and $D(t)$ in kilograms. Find the half-life of the radioactive substance. Find the amount present after 30 days.

22. Suppose that $D(t)$, the amount of a certain radioactive substance present at time t, is given by

$$D(t) = 4000e^{-0.06t}$$

where t is measured in days and $D(t)$ in kilograms. Find the half-life of the radioactive substance. Find the amount present after 20 days.

SECTION 15.6 COMPUTATIONS WITH LOGARITHMS

In this chapter we have evaluated logarithms for purposes of solving equations. However, logarithms were widely used for computations prior to the availability of low cost calculators. We only have to do a few computations using logarithm tables to be convinced of the value of our calculator. Why even bother to discuss logarithms for calculations? For those who like to understand their calculator or computer better, it may help to understand computations with logarithms. In many cases, electronic circuitry is now doing what we used to do by hand.

For example, if we ask the calculator to evaluate 2.5^5 (using the $\boxed{y^x}$ key, it will probably multiply 5 by ln 2.5 and then use the result as an exponent on e to obtain the answer. The calculator will not use 2.5 as a factor 5 times.

A brief experiment will indicate the degree of precision of your calculator in performing the $\boxed{y^x}$ function. Here is a result of an experiment on a calculator.

2.5 $\boxed{\times}$ 2.5 $\boxed{\times}$ 2.5 $\boxed{\times}$ 2.5 $\boxed{\times}$ 2.5 $\boxed{=}$ read 97.65625
(the exact answer)

2.5 $\boxed{y^x}$ 5 $\boxed{=}$ read 97.6562
(the answer using $\boxed{y^x}$ key)

The preceding results in finding 2.5^5 suggest that this particular calculator produces results to six significant figures in calculating powers.

If your calculator provides reasonable accuracy in the last decimal place in the register when you use the $\boxed{y^x}$ key, you may have difficulty detecting any difference between continued multiplications and the $\boxed{y^x}$ key result. Read the instruction manual provided to determine the accuracy you can expect when using $\boxed{y^x}$. The manual should tell you the maximum tolerance expected in the last digit reported.

We work three examples to illustrate multiplication, division and raising to a power using logarithm tables.

EXAMPLE Multiply 4.65 × 249.

SOLUTION First take the logarithm of the product and use the first law of logarithms.

$$\log (4.65 \times 249) = \log 4.65 + \log 249$$

Next find the logs.

$$\log 4.65 = 0.6675$$
$$249 = 2.49 \times 10^2$$
$$\log 249 = 2 + 0.3962 = 2.3962$$

Now add the logarithms to obtain

$$2.3962 + 0.6675 = 3.0637$$

Next find 0637 in the body of the table. We find 0607 and 0645, which bracket this value. Our value exceeds the lower value by 30. We find 30 (or the nearest number to it) in the proportional parts table under 8.

The four digit answer is 1.158.

The characteristic 3 tells us to multiply by 10^3.

The answer is 1158. We compare this number with the exact value of 1157.85.

EXAMPLE Divide 1575 by 9802.

SOLUTION First take the logarithm of the quotient and use the second law of logarithms.

$$\log \left(\frac{1575}{9802}\right) = \log 1575 - \log 9802$$

Next find the logs.

$$1575 = 1.575 \times 10^3; \log 1575 = 3.1973$$
$$9802 = 9.802 \times 10^3; \log 9802 = 3.9913$$
$$3.1973 - 3.9913 = -0.7940 = -1 + 0.2060$$

(mantissa must be positive)

The answer is $1.607 \times 10^{-1} = 0.1607$. We compare this with the exact value of 0.1606815.

EXAMPLE Evaluate $144^{3.5}$.

SOLUTION First take the logarithm of the power and use the third law of logarithms.

$$\log 144^{3.5} = 3.5 \log 144$$
$$144 = 1.44 \times 10^2; \log 144 = 2.1584$$
$$3.5 \times 2.1584 = 7.5544$$

The answer is 3.584×10^7.

This is compared with an exact value of 35,831,808. Note the error in the fourth digit of the table answer. You should compare the calculator result using the $\boxed{y^x}$ with this value. On the calculator you get at least six significant figures and perhaps even the exact answer.

PERFORMANCE OBJECTIVE 15	To find $249 \div 37$ evaluate
You should now be able to perform multiplications, divisions, and exponentiations using tables of logarithms.	$\log 249 - \log 37 = 2.3962 - 1.5682 = 0.828$ From the tables the quotient is 6.7298

EXERCISE SET 15.6

Perform the following calculations using a log table. In each case, compare the answer with a calculator value.

1. 365×1.59
2. 488×2.65
3. 8.67×1.98
4. 9.36×6.28
5. $8907 \div 254$
6. $3625 \div 300$
7. $735 \div 1.63$
8. $688 \div 2.56$
9. $3^{4.5}$
10. $6^{2.3}$
11. $\dfrac{1.67 \times 32.7}{2.56}$
12. $\dfrac{2.86 \times 58.9}{3.27}$

SECTION 15.7 APPLICATIONS OF LOGARITHMS AND EXPONENTIALS

Exponential curves are often used in the technologies to describe growth and decay processes. We illustrate with examples involving transient circuits and bacterial growth.

EXAMPLE Direct current is being driven through a circuit that has an effective inductance L and a resistance R. If the driving voltage is suddenly removed, leaving a current path including the inductor and resistor, the current varies with time as follows:

$$i = i_m e^{-Rt/L}$$

Note the schematic shown in Figure 15.10.

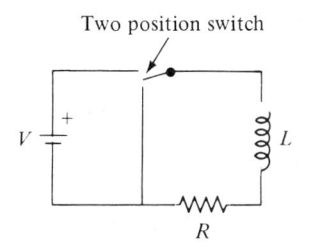

FIGURE 15.10

Plot current as a function of time at 10^{-4} sec intervals given the following values:

$$i_m = 2.50 \text{ A}, \quad R = 54.0 \text{ }\Omega, \quad L = 0.0150 \text{ H}$$

SOLUTION We rewrite the equation as follows:

$$i = 2.5e^{-54t/0.015} = 2.5e^{-3600t}$$

To develop a table of values, we follow the procedure

3600 $\boxed{\times}$ t $\boxed{=}$ $\boxed{+/-}$ $\boxed{\text{INV}}$ $\boxed{\text{LN}}$ $\boxed{\times}$ 2.5 $\boxed{=}$

entering $t = 0.0001, 0.0002, \ldots$.

We generate the following table:

$t \times 10^{-4}$	0	1	2	3	4
i	2.5	1.74	1.22	0.849	0.592

$t \times 10^{-4}$	5	6	7	8
i	0.413	0.288	0.201	0.140

From this table we produce the graph in Figure 15.11.

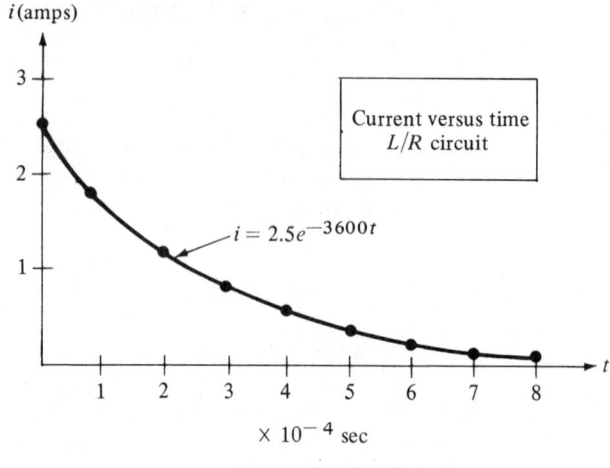

FIGURE 15.11

It is customary to describe the time it takes a **transient** (short-lived change in a circuit) to decay in terms of a **time constant.** The time constant measures how long it takes a quantity to come to within $\frac{1}{e}$ of its final value. We evaluate $\frac{1}{e}$ to find 0.3679 or about 37%. We can find this value for the previous circuit by setting

$$e^{-Rt/L} = e^{-1}$$

so that

$$\frac{Rt}{L} = 1; \quad t = \frac{L}{R}$$

The time constant (sometimes denoted by τ) for a circuit containing a resistance and an inductor is

$$\tau = \frac{L}{R}$$

For the circuit in Figure 15.11, we find

$$\tau = \frac{0.015}{54} = 0.000278 = 2.78 \times 10^{-4} \text{ sec}$$

The time constant is displayed on the graph in Figure 15.12.

Note that in two time constants we have come to within about $(0.37)(0.37) = 0.137$ or about 13.7% of the final value. In three time constants, we would reach to within about 5% of the final value.

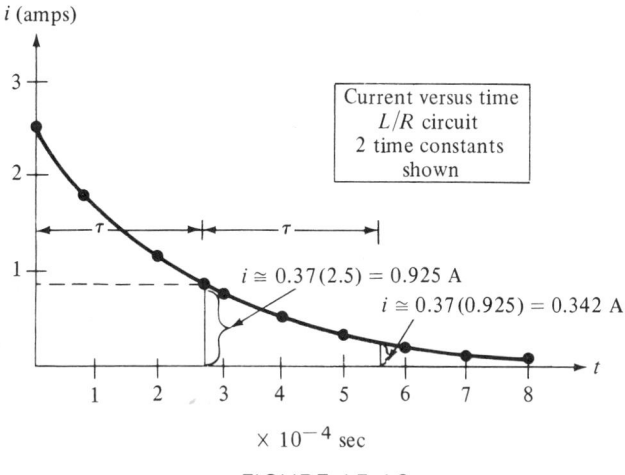

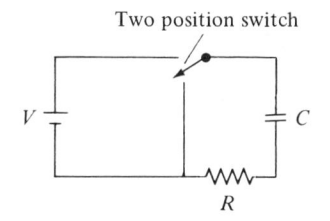

APPLICATIONS OF LOGARITHMS AND EXPONENTIALS 15.7

FIGURE 15.12 — Current versus time, L/R circuit, 2 time constants shown.

$i \cong 0.37(2.5) = 0.925$ A
$i \cong 0.37(0.925) = 0.342$ A

FIGURE 15.13

EXAMPLE Direct current is being driven through a circuit that has an effective capacitance C and resistance R. If the driving voltage is suddenly removed, leaving a current path including the capacitor and resistor, the current varies with time as follows:

$$i = i_m e^{-t/RC}$$

Note the schematic in Figure 15.13.

Plot current as a function of time at 10^{-3} sec intervals given the following values:

$$i_m = 3.60 \text{ A}, \quad R = 150 \text{ }\Omega, \quad C = 20.0 \text{ mF}$$

Calculate and display the time constant for this circuit.

SOLUTION Substitute into the equation to obtain

$$i = 3.6 e^{-t/(150)(20 \times 10^{-6})} = 3.6 e^{-333.3t}$$

Next, develop a table of values using the procedure

333.3 $\boxed{\times}$ t $\boxed{=}$ $\boxed{+/-}$ $\boxed{\text{INV}}$ $\boxed{\text{LN}}$ $\boxed{\times}$ 3.6 $\boxed{=}$

entering $t = 0, 0.001, 0.002, 0.003, \ldots$.

$t \times 10^{-3}$	0	1	2	3	4	5	6	7	8
i	3.6	2.58	1.85	1.32	0.95	0.68	0.49	0.35	0.25

From this table we produce the graph in Figure 15.14.

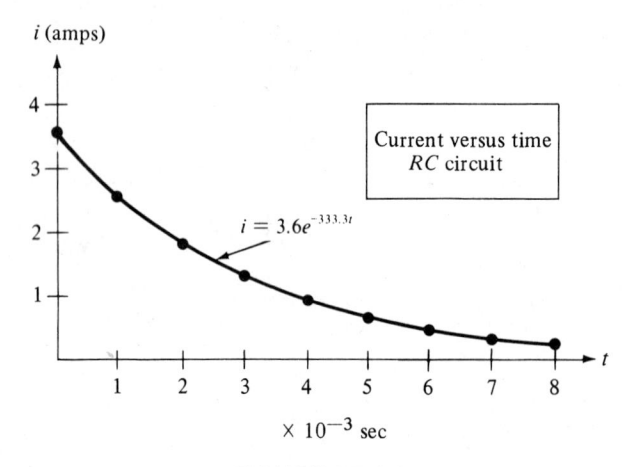

FIGURE 15.14

Current versus time RC circuit

$i = 3.6e^{-333.3t}$

The time constant for this circuit τ is obtained as in the previous example problem. Set

$$e^{-\tau/RC} = e^{-1}$$

This requires $\dfrac{\tau}{RC} = 1$, so that $\tau = RC$.

For this problem, $\tau = RC = (150)(20 \times 10^{-6}) = 3.0 \times 10^{-3}$ sec. The time constant is displayed in Figure 15.15.

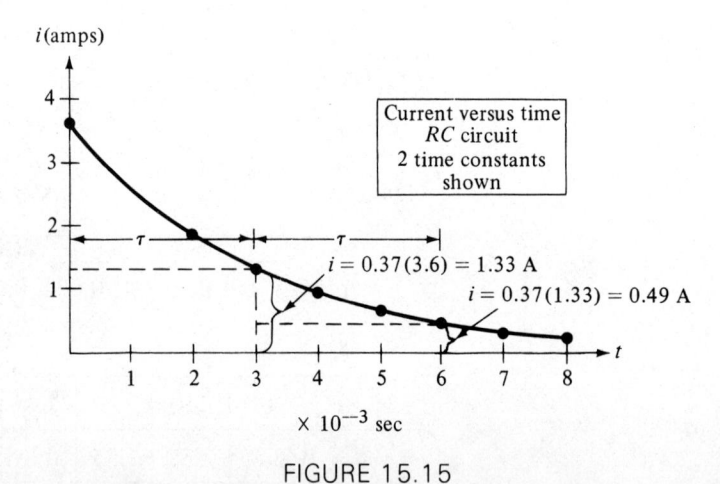

Current versus time RC circuit 2 time constants shown

$i = 0.37(3.6) = 1.33$ A
$i = 0.37(1.33) = 0.49$ A

FIGURE 15.15

PERFORMANCE OBJECTIVE 16

You should now be able to use exponential decay curves to describe physical processes.

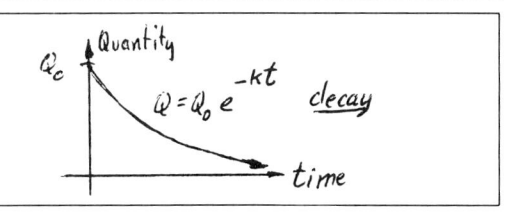

Exponential growth occurs when the rate at which a quantity increases is proportional to the quantity present at any instant. Exponential growth is then exactly equivalent to situations in finance where interest on money (rate of growth) is compounded continuously. Similarly, natural growth processes, such as population increases in the United States or bacterial growth, may often be described approximately by exponential growth. The advantage of using this description is that we may immediately apply features of the exponential curve. We can make predictions based on the doubling time characteristic of such curves.

EXAMPLE The bacteria count in a culture is found to be increasing at a rate of 5% each hour. Assuming a starting count of 10,000, graph the number of bacteria present at two-hour intervals over a period of 20 hr.

SOLUTION The growth will be described by the equation

$$N = N_0 e^{kt}$$

where $N_0 = 10,000$ and $k = 0.05/\text{hr}$. Then

$$N = 10,000 e^{0.05t}$$

We first construct a table using the previous equation and the following procedure:

$$0.05 \;\boxed{\times}\; t \;\boxed{=}\; \boxed{\text{INV}} \;\boxed{\text{LN}}\; \boxed{\times}\; 10,000 \;\boxed{=}$$

entering $t = 0, 2, 4, \ldots$.

t	0	2	4	6	8	10
N	10,000	11,052	12,214	13,499	14,918	16,487

t	12	14	16	18	20
N	18,221	20,138	22,255	24,596	27,183

From this table the graph in Figure 15.16 is constructed.

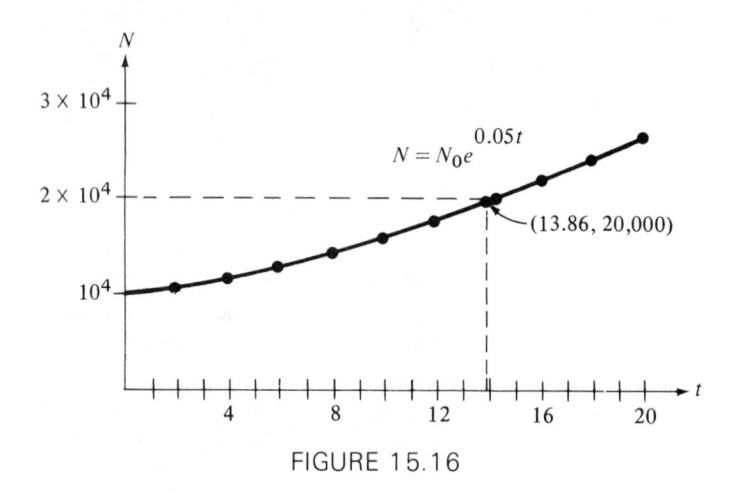

FIGURE 15.16

EXAMPLE Verify that the doubling time in the preceding example is given by

$$T_2 = \frac{0.693}{k}$$

where $k = 0.05$.

SOLUTION $T_2 = \dfrac{0.693}{0.05} = 13.86$ hr

Substitute into the equation $N = 10,000e^{0.05t}$ with $t = 13.86$ to find

$$N = 10,000e^{(0.05)(13.86)} = 19,997$$

This value is very nearly twice the initial amount of 10,000; so we conclude that the doubling time is given correctly by the expression

$$\frac{0.693}{k}$$

Note the point (13.86, 20,000) on the graph in Figure 15.16.

PERFORMANCE OBJECTIVE 17

You should now be able to use exponential growth curves to describe physical processes.

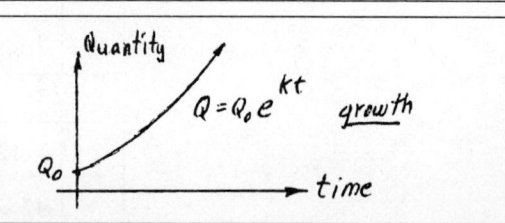

$Q = Q_0 e^{kt}$ growth

EXERCISE SET 15.7

1. Calculate the time constant for the following circuits:
   a. A resistor-capacitor circuit in which
   $$R = 2.5 \times 10^4 \ \Omega$$
   $$C = 3.2 \times 10^{-5} \ F$$
   b. A resistor-inductor circuit in which
   $$R = 1.8 \times 10^3 \ \Omega$$
   $$L = 6.5 \times 10^{-4} \ H$$

2. Calculate the time constant for the following circuits:
   a. A resistor-capacitor circuit in which
   $$R = 3.5 \times 10^2 \ \Omega$$
   $$C = 2.6 \times 10^{-6} \ F$$
   b. A resistor-inductor circuit in which
   $$R = 2.9 \times 10^4 \ \Omega$$
   $$L = 4.2 \times 10^{-2} \ H$$

3. Plot the current versus time in the following equations:
   a. $i = 2.5e^{-2500t}$ A
   (for $t = 0$ to $t = 1.2 \times 10^{-3}$ sec)
   b. $i = 3.8e^{-Rt/L}$ A when
   $$R = 7.0 \ \Omega$$
   $$L = 2.0 \times 10^{-3} \ H$$
   (for $t = 0$ to 10^{-3} sec)

4. Plot the current versus time in the following equations:
   a. $i = 3.6e^{-1500t}$ A
   (for $t = 0$ to $t = 1.2 \times 10^{-3}$ sec)
   b. $i = 2.4e^{-Rt/L}$ A when
   $$R = 7.0 \ \Omega$$
   $$L = 2.0 \times 10^{-4} \ H$$
   (for $t = 0$ to 10^{-4} sec)

5. The deoxygenation resulting from a source of pollution is given by
   $$D = 25e^{-0.30t} \text{ mg}/\ell \text{ where } t \text{ is measured in days}$$

Calculate the times for the value of D to reach 15, 10, and 5 mg/ℓ.

6. The deoxygenation resulting from a source of pollution is given by

$$D = 50e^{-0.10t} \text{ mg/ℓ where } t \text{ is measured in days}$$

Calculate the times for the value of D to reach 25, 19, and 5 mg/ℓ.

7. The growth of a bacteria culture is given by the equation

$$N = N_0 e^{kt}$$

k is the growth constant and N_0 is the amount present at time zero. Assuming arbitrary units for N and N_0, graph the values of N versus time when $k = 0.04/\text{day}$. Let t take on values from 1 to 20 days and let $N_0 = 1$.

8. The growth of a bacteria culture is given by the equation

$$N = N_0 e^{kt}$$

k is the growth constant and N_0 is the amount present at time zero. Assuming arbitrary units for N and N_0, graph the values of N versus time when $k = 0.08/\text{day}$. Let t take on values from 1 to 20 days and let $N_0 = 1$.

9. Solve Problem 7 with the following parameters:

$$k = 0.24/\text{min}, \quad N_0 = 2, \quad t = 0 \text{ to } 9 \text{ min}$$

10. Solve Problem 8 with the following parameters:

$$k = 0.48/\text{min}, \quad N_0 = 2, \quad t = 0 \text{ to } 9 \text{ min}$$

SECTION 15.8 SUMMARY

In this section we highlight the new concepts and definitions introduced in this chapter.

Exponential function: a function in which a constant base is raised to a linear variable power. The function has the form $y = b^x$ where $b > 0$ and $b \neq 1$.

Logarithmic function: a function that is the inverse of the exponential function. This function determines the exponent when the base and the result of an exponentiation of the base are known.

Common logarithm: a logarithmic function in which the base is 10.

Mantissa: the decimal part of the logarithm of any number.

Characteristic: the power on ten when a number is in scientific notation. The characteristic precedes the decimal when the number is written as a common logarithm.

Natural logarithm: a logarithmic function in which the base is e.

Transient: a time dependent quantity in a circuit that does not show periodic behavior.

Exponential growth: growth that occurs at a rate proportional to the quantity present. Under exponential growth a quantity doubles in regular intervals.

Exponential decay: decay that occurs at a rate proportional to the quantity present. Under exponential decay a quantity halves in regular intervals.

CHAPTER 15 PRACTICE TEST

1. (PO-1) Evaluate $y = b^x$ for the values given.
   a. $b = 4$, $x = -2, -1, 0, 1, 2$
   b. $b = 0.3$, $x = -3, -1, 0, 1, 3$
   c. $b = 1.2$, $x = -2, -1, 0, 1, 2$

2. (PO-2, 3, 4) Graph each exponential function.
   a. $y = 4^x$
   b. $y = \left(\dfrac{1}{5}\right)^x$
   c. $y = e^{1.5t}$ for $t = -2$ to 1

3. (PO-5) Write the equivalent logarithmic equations for the given exponential equations.
   a. $y = 7^x$
   b. $y = \left(\dfrac{1}{5}\right)^x$

4. (PO-5) Write the equivalent exponential equations for the given logarithmic equations.
   a. $y = \log_2 x$
   b. $y = \log_{16} x$

5. (PO-6) Express $\log\left[\left(\dfrac{1}{3}\right)^3 \cdot 2\right]$ as a sum, difference, or multiple of logarithms.

6. (PO-7, 8, 9, 10) Find the common logarithm of each of the following. In each case compare your answer with the calculator answer.
   a. log 45,013
   b. log 0.0264
   c. log 0.09869

7. (PO-11) Find the pH of a solution if the concentration of H_3O^+ is 0.000025 moles/ℓ.

8. (PO-12, 13) Find the natural logarithm of each of the following. In each case compare your answer with the calculator answer.
   a. 2367
   b. 0.5163
   c. 783.5

9. (PO-14) $H(t) = 5000e^{-0.08t}$ where t is measured in hours and $H(t)$ in ounces. Find the half-life of the substance. Find the amount present at $t = 10$ hr.

10. (PO-15) Perform the following calculation using a log table. Compare your answer with a calculator value.

$$\frac{37.8 \times 0.003}{16.5}$$

11. (PO-16, 17) Plot the following equations:
    a. $i = 5.0e^{-1800t}$ A for $t = 0$ to 1 msec
    (t is expressed in seconds in the equation)
    b. $N = N_0 e^{0.5t}$ for $t = 0$ to 3 days. Let $N_0 = 4.0$.
    (t is expressed in days in the equation)

12. (PO-16, 17) Calculate the decay and growth constants in Problem 11.

APPENDIX I

GRAPHING EXPERIMENTAL DATA

This appendix deals with graphical methods of presenting information. The following topics are discussed:

1. General techniques of displaying data
2. Semi-logarithmic plots
3. Pictorial graphs
4. Vector addition

SECTION I.1 GENERAL TECHNIQUES OF DISPLAYING DATA

We consider the preparation of a graph to display experimental results. Presenting experimental results in graphical form helps us to understand the relation between the independent and dependent variables. For example, we are often concerned with how a physical quantity changes with time. In such a case, the independent variable would be the time; the dependent variable would be the observed values of the measured physical quantity.

Experimental data sets are generally collected as a set of ordered pairs. These ordered pairs are used to produce a point plot on a rectangular coordinate system. Since we may be seeking to find a mathematical relation between the independent and dependent quantities, we may attempt to convert the point plot to a line graph representing this relation.

To illustrate the process we start with a set of measured data. To collect this data set we have used an experimental arrangement such as that shown in Figure A.1. Readings on the ammeter are recorded simultaneously with readings on the voltmeter.

The data collected is shown in Table A.1.

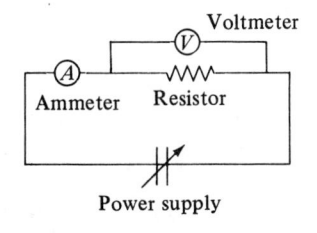

FIGURE A.1

TABLE A.1 CURRENT VERSUS VOLTAGE IN A RESISTOR

Volts	1.5	2.0	2.5	3.0	3.5	4.0	4.5	5.0	5.5
Amps	0.16	0.19	0.25	0.29	0.36	0.40	0.44	0.51	0.55

APPENDIX I GRAPHING EXPERIMENTAL DATA

To collect this data we adjusted the power supply to obtain readings on the voltmeter exactly one-half volt apart. For each such setting, we record the reading on the ammeter. In mathematical terms the voltage is treated as the independent variable; the current is the dependent variable (we set the voltage "independently" and observe the current's "dependence" on the voltage).

From Table A.1 it appears that current increases as voltage increases. To investigate this relation more closely we will graph the data. The first step in graphing the data is to determine a coordinate system. We will build a coordinate system on a piece of graph paper and develop the graph with a minimum of mathematical detail. We give a "prescription" for constructing a graph using the data in the preceding table.

Step 1. Select graph paper with a sufficiently fine grid. The grid is the system of horizontal and vertical lines. If there are too few lines, you may have to estimate points to be plotted. A section of the graph grid is shown in Figure A.2.

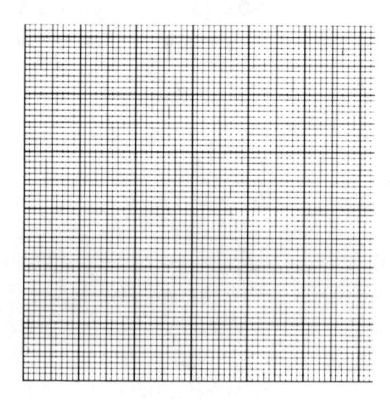

FIGURE A.2

Step 2. Draw horizontal and vertical lines to represent the axes for the independent and dependent variables. The horizontal axis will represent the independent variable. If we are plotting all positive data, the axes should intersect at the lower left of the graph. The axes are shown and labeled in Figure A.3.

Step 3. Enter a scale on each axis. The scale is developed so as to display the data over most of the graph. We find the range of both sets of data as follows:

Range = largest data value − smallest data value

Range (voltage) = 5.5 − 1.5 = 4.0 V

Range (current) = 0.55 − 0.16 = 0.39 A

GENERAL TECHNIQUES OF DISPLAYING DATA I.1

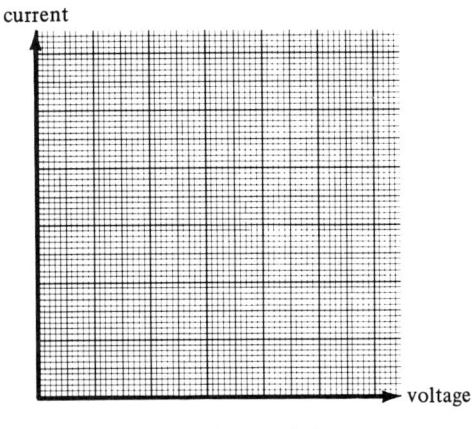

FIGURE A.3

We need at least four units on the voltage axis and 0.4 units on the current axis. If the smallest piece of data is not too far from zero, it is advisable to start each scale at zero. For our problem we will let voltage run from 0 to 6 volts and current run from 0 to 0.6 amps. Notice that we need not use the same size subdivisions on both axes. Always include units on the scale. The axes are scaled in Figure A.4.

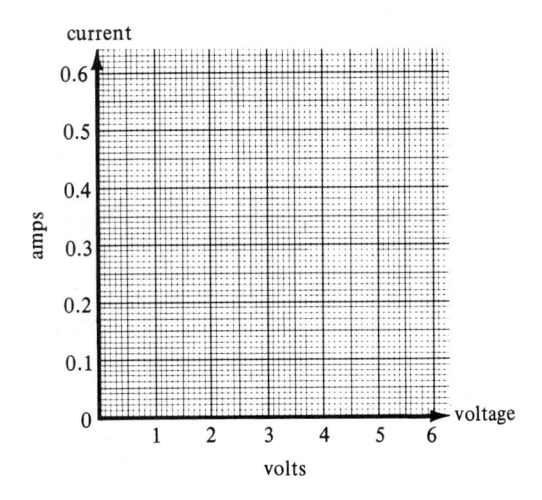

FIGURE A.4

Step 4. Plot each data point. When plotting we move horizontally to the voltage value and then up to the current value and place a point to mark the pair of values. The plot is shown in Figure A.5.

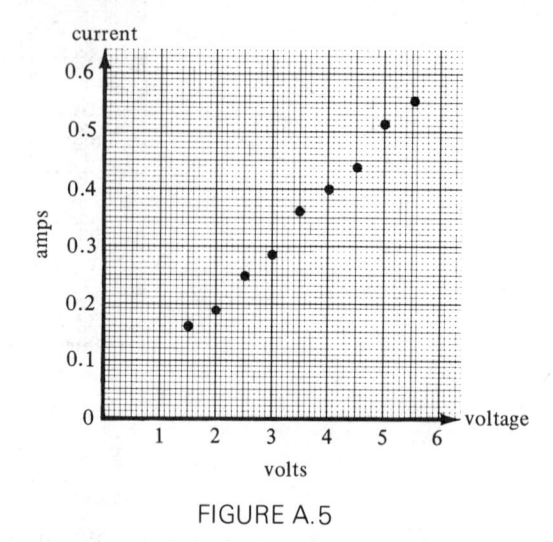

FIGURE A.5

Step 5. Construct lines, curves, etc. to complete the graphical display. In our problem we note that the data almost falls on a straight line. We may wish to draw a straight line to emphasize the trend of the data.

Step 6. Label the graph. Some identification of the graph, its purpose, and the source of data is important. The complete graph is displayed in Figure A.6.

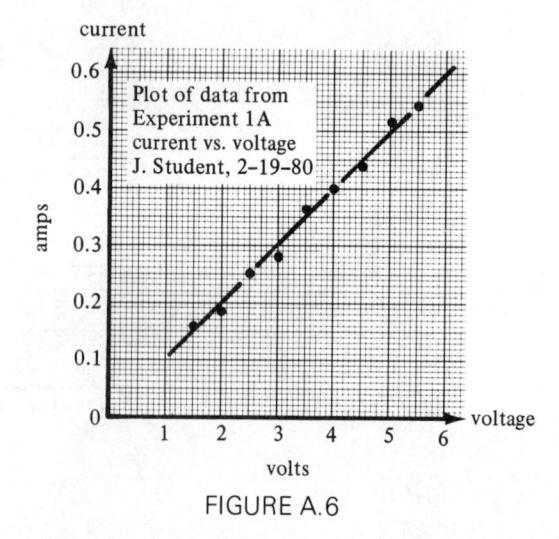

FIGURE A.6

The graph in Figure A.6 illustrates an approximately "linear" relationship between voltage and current in a resistor. The relation is called linear because the data falls generally on a straight line. In Appendix II

SEMI-LOGARITHMIC PLOTS I.2

we discuss such linear relations further. A "curvilinear" relation within a data set is illustrated in Table A.2.

TABLE A.2 POSITION VERSUS TIME FOR AN OBJECT IN FREE FALL

time-sec	0	0.05	0.10	0.15	0.20	0.25	0.30	0.35
position-cm	4.9	11.1	19.5	30.7	44.1	60.1	78.3	99.2

This data is plotted in Figure A.7.

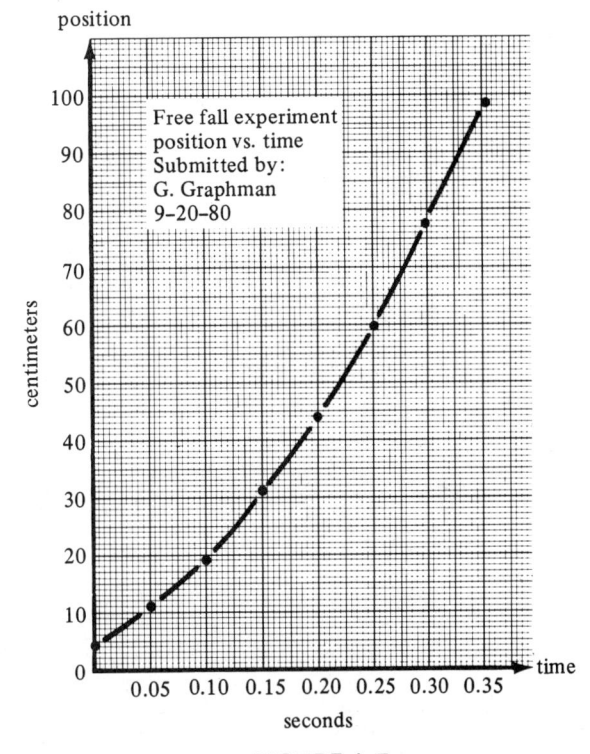

FIGURE A.7

SECTION I.2 SEMI-LOGARITHMIC PLOTS

The interest in semi-logarithmic plots arises because of the frequent occurrence of the processes of exponential growth or decay in real systems. Let us start by considering an example.

APPENDIX I GRAPHING EXPERIMENTAL DATA

A microprocessor reads and stores data from a digital voltmeter as a function of time. The circuit being analyzed is shown in Figure A.8.

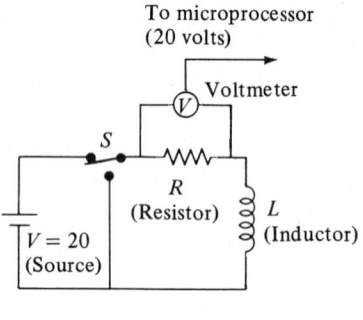

FIGURE A.8

After a steady voltage is developed across R, we find that the voltmeter reads 20.00 volts. The electronic switch (s) is then placed in the position shown in Figure A.9 so that the driving voltage (V) is removed from the circuit. The microprocessor records the data sent from the digital voltmeter at 0.4 millisecond intervals thereafter. The data collected is given in Table A.3.

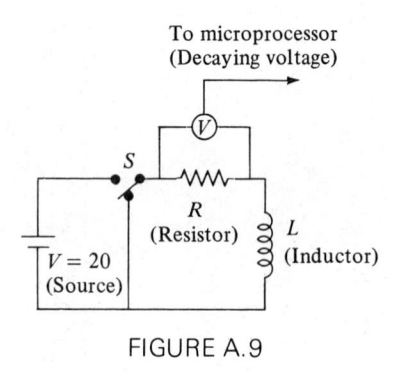

FIGURE A.9

TABLE A.3 VOLTAGE VERSUS TIME, RESISTOR-INDUCTOR CIRCUIT

Time-ms	0.0	0.4	0.8	1.2	1.6	2.0
Voltage-V	20.00	13.52	8.81	6.11	4.15	2.72
Time-ms	2.4	2.8	3.2	3.6	4.0	
Voltage-V	1.89	1.20	0.79	0.56	0.38	

SEMI-LOGARITHMIC PLOTS I.2

This data is plotted in Figure A.10. The plot displays a curvilinear relation in which the voltage decreases with time but at a decreasing rate. This is characteristic of exponential decay discussed in Chapter 15.

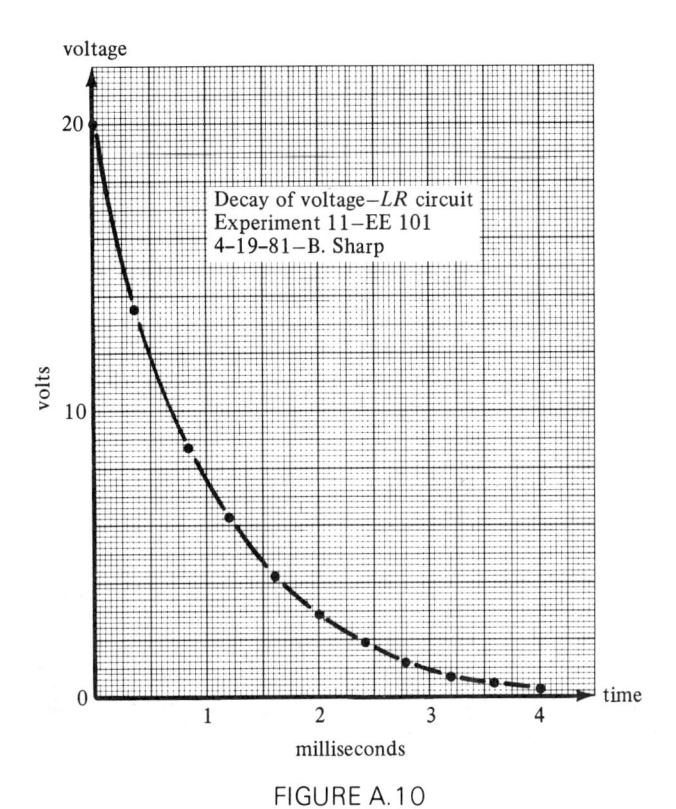

FIGURE A.10

A procedure commonly used to verify exponential decay or growth is to plot the data on semi-logarithmic graph paper. Semi-logarithmic graph paper has a logarithmic scale on one axis and a linear scale on the other. The logarithmic scale is divided into cycles. Each cycle differs from the next by a power of ten. We plot the data in Figure A.11.

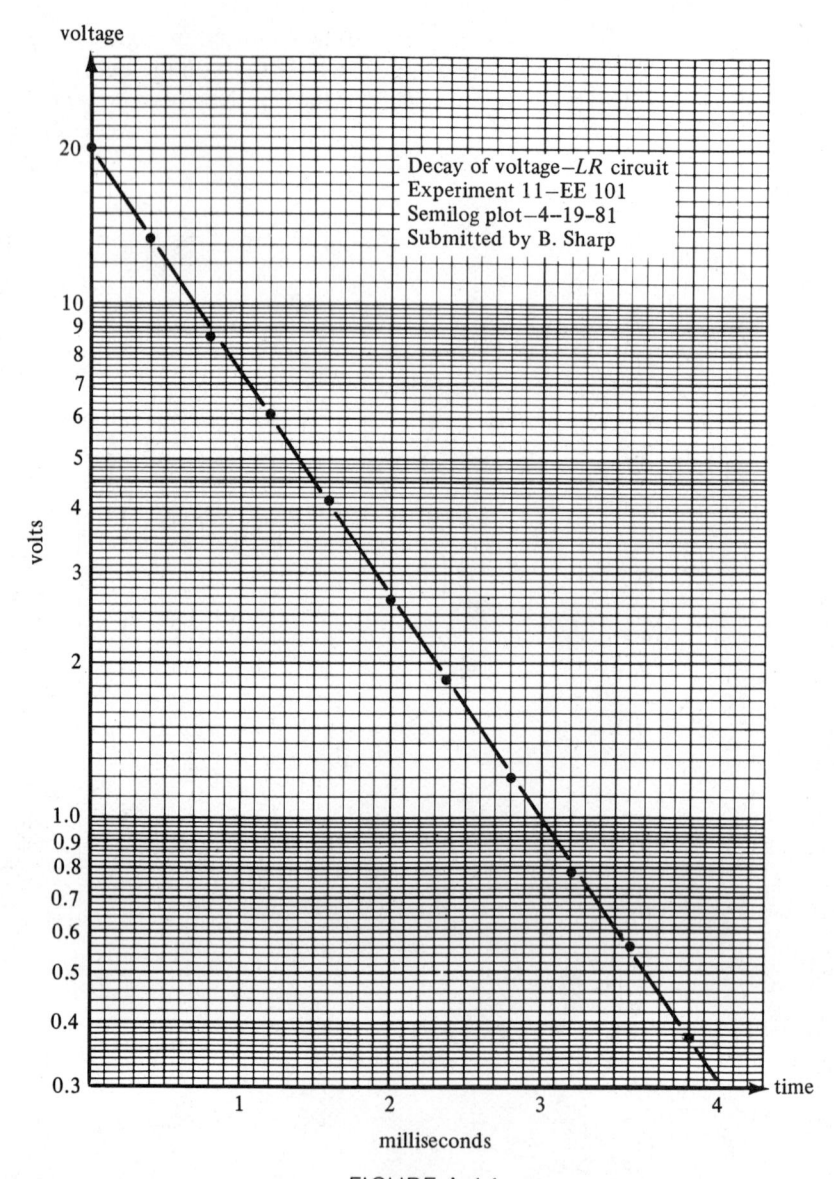

FIGURE A.11

Note that the data extends over parts of three cycles on the paper. It also appears to be linear. If the data plots as a straight line, then we know we are dealing with exponential growth or decay. The downward trend in this graph indicates exponential decay.

From this graph we may find the slope of the line using the methods of Chapter 7. We show this measurement in Figure A.12.

SEMI-LOGARITHMIC PLOTS I.2

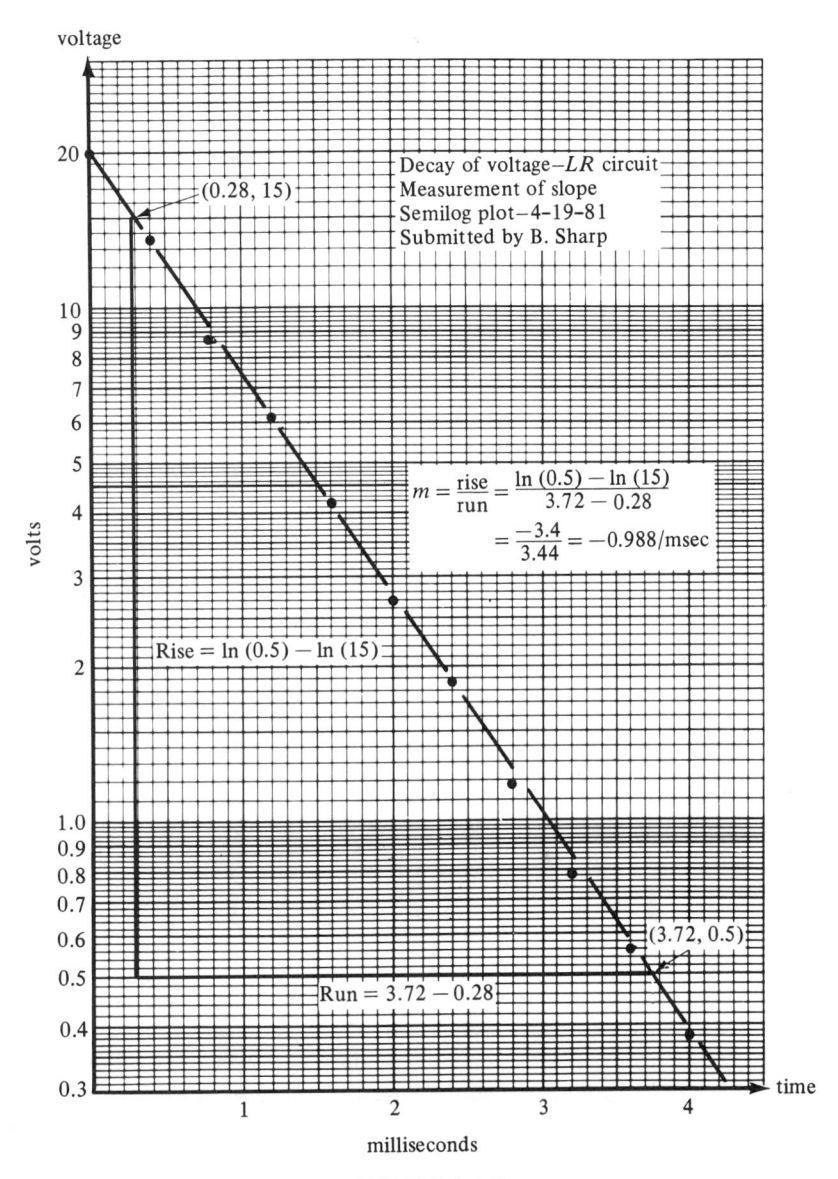

[Figure A.12: Semilog plot showing decay of voltage in an LR circuit. Submitted by B. Sharp, 4-19-81. Points marked at (0.28, 15) and (3.72, 0.5).

$$m = \frac{\text{rise}}{\text{run}} = \frac{\ln(0.5) - \ln(15)}{3.72 - 0.28}$$

$$= \frac{-3.4}{3.44} = -0.988/\text{msec}$$

Rise = $\ln(0.5) - \ln(15)$; Run = $3.72 - 0.28$. Axes: volts vs. time (milliseconds).]

FIGURE A.12

To find the slope (m) we took the difference in the natural logs of the voltage values divided by the difference in time. This slope is then the value of k in the equation

$$V = 20e^{kt}$$

Using the actual k value we write $V = 20e^{-0.988t}$. We compare the actual data with predicted data using the equation

$$V = 20e^{-0.988t}$$

Here t is measured in milliseconds. See Table A.4.

TABLE A.4 V (OBSERVED AND PREDICTED) VERSUS T

Time-ms	0.0	0.4	0.8	1.2	1.6	2.0
V (observed)	20.00	13.52	8.81	6.11	4.15	2.72
V (predicted)	20.00	13.47	9.07	6.11	4.12	2.77
Time-ms	2.4	2.8	3.2	3.6	4.0	
V (observed)	1.89	1.20	0.79	0.56	0.38	
V (predicted)	1.87	1.26	0.85	0.57	0.38	

Let us consider a case of exponential growth. Approximate figures for production of electricity in the United States are given in Table A.5.

TABLE A.5 ELECTRICAL ENERGY PRODUCTION (KW-HR $\times 10^9$) VERSUS TIME

Year	1935	1940	1945	1950	1955	1960	1965
Prod.	90	150	210	300	540	720	1100

This data is graphed in Figure A.13. A straight line is drawn to emphasize the linear trend of the data.

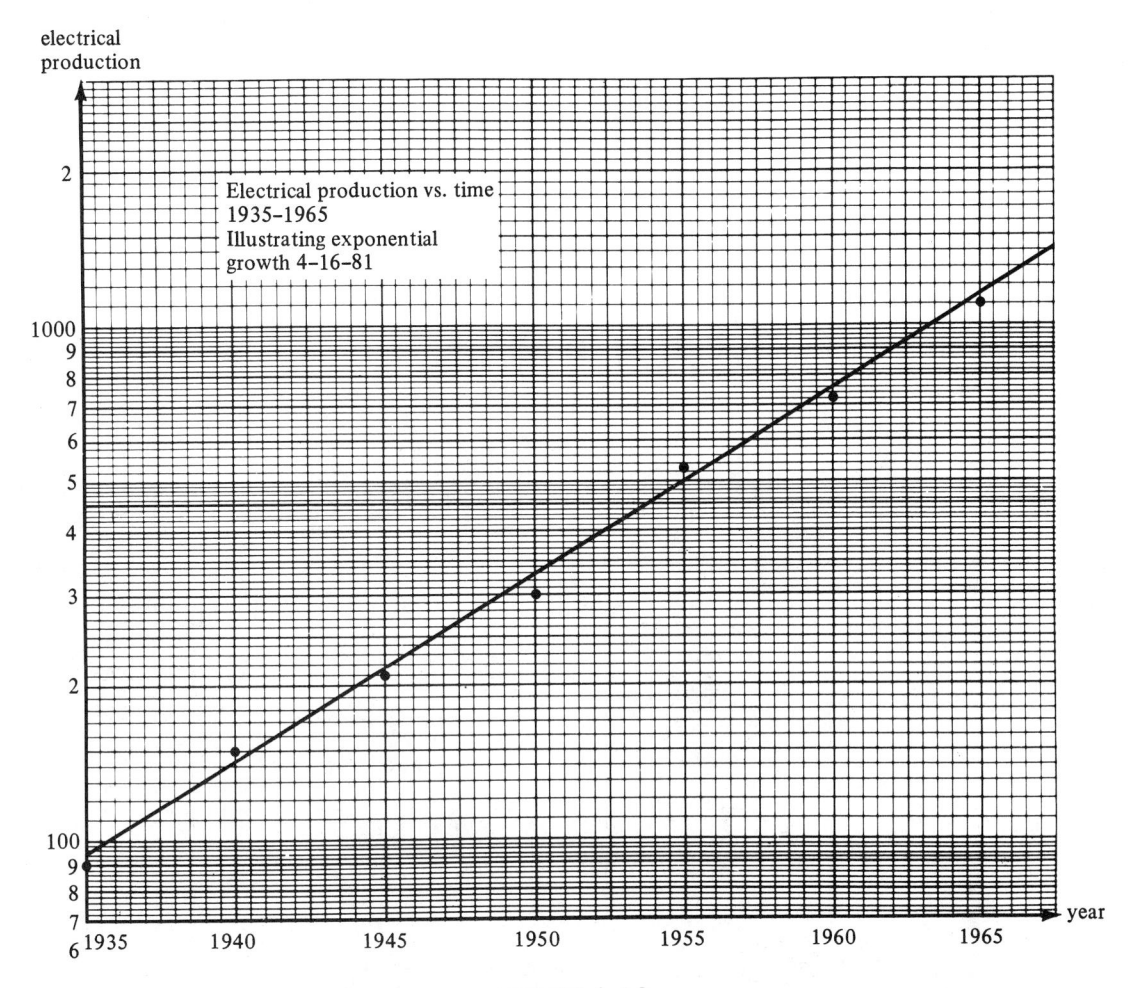

FIGURE A.13

In Figure A.14 we measure the slope of the line in Figure A.13 in order to estimate the growth rate.

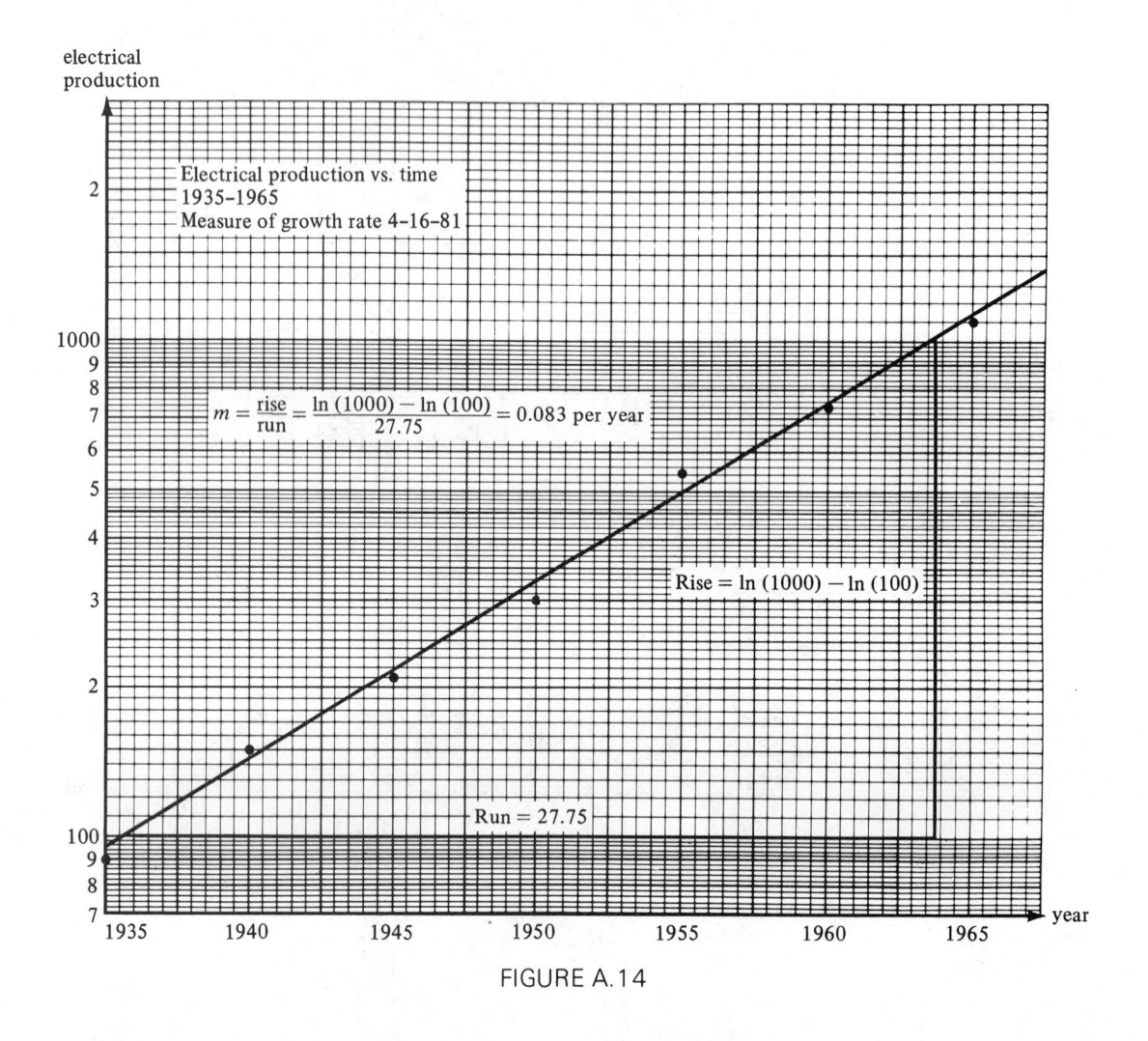

FIGURE A.14

Electrical production vs. time 1935–1965 Measure of growth rate 4-16-81

$$m = \frac{\text{rise}}{\text{run}} = \frac{\ln(1000) - \ln(100)}{27.75} = 0.083 \text{ per year}$$

Rise = $\ln(1000) - \ln(100)$

Run = 27.75

On the basis of this measurement we could write an equation for electrical energy production as follows:

$$y = 94e^{0.083t}$$

where t is time measured in years since 1935. We compare the actual data with predicted data using this equation. See Table A.6.

PICTORIAL GRAPHS I.3

TABLE A.6 ELECTRICAL ENERGY PRODUCTION VERSUS TIME

Year	1935	1940	1945	1950	1955	1960	1965
P (observed)	90	150	210	300	540	720	1100
P (predicted)	94	142	216	326	494	749	1134

SECTION I.3   PICTORIAL GRAPHS

Pictorial graphs are designed to give a quick and concise representation of data. These pictorial graphs are distinguished from the graphs and point plots discussed earlier. The pictorial graph, while forceful, often lacks detail. That is, we may lose some of the detailed information in the data, especially if the dependent variable has a large range of values.

We begin with a bar graph. Consider Table A.7, which displays production by day of the week.

TABLE A.7 AVERAGE PRODUCTION UNITS PER DAY OF WEEK AT SMALL MANUFACTURING CO.*

Day	Production
Monday	42.5
Tuesday	48.5
Wednesday	52.5
Thursday	45
Friday	40

* based on a four week sample

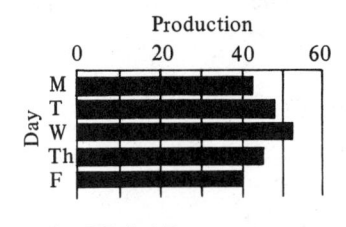

FIGURE A.15

This information is displayed in a bar graph in Figure A.15.

This bar graph illustrates one of the immediate advantages of the graph over the table. The graph shows the main features of the data at a glance.

Consider a more complex example as illustrated by the data in Table A.8.

TABLE A.8 SUPERCAR SALES BY REGION AND MODEL (FIRST QUARTER 1981)

Region → Model	NE	SE	MW	SW	W	Total
S	8,092	10,152	4,220	12,120	1,752	36,336
R	12,131	9,022	5,890	13,295	2,450	42,788
T	14,510	13,920	7,420	8,953	5,215	50,018
Total	34,733	33,094	17,530	34,368	9,417	129,142

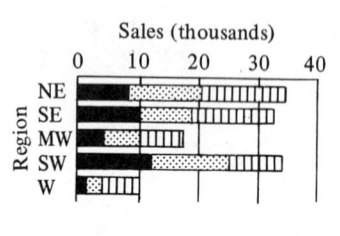

This table is more difficult to transform into a graph since there are two independent variables. We might try an approach as shown in Figure A.16.

A second type of pictorial graph commonly used is the pie chart. This chart is valuable as a means of conveying information about fractional parts or percentage parts of a whole.

Consider the operation of a small electronics business. Income for the year is distributed into the following categories.

Sales of major appliances	$89,200
Sales of components	$24,500
Repair services	$36,300
Contractual services	$5,800

We construct a pie chart in the following manner:

Step 1. Determine the total of all values of the dependent variable. This total is $155,800.

Step 2. Find the fraction and percentage of the total for each value. To find the fraction of the total divide each value by the total value. Multiply by 100 to find the percentage.

$$\frac{89,200}{155,800} = 0.573 = 57.3\%$$

$$\frac{24,500}{155,800} = 0.157 = 15.7\%$$

$$\frac{36,300}{155,800} = 0.233 = 23.3\%$$

$$\frac{5800}{155,800} = 0.037 = 3.7\%$$

FIGURE A.16

Supercar sales by region and model (first quarter 1981)

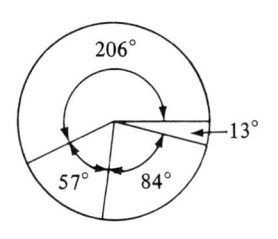

FIGURE A.17

Step 3. Convert the fraction into a degree reading by multiplying its decimal equivalent by 360°.

$$0.573 \times 360° = 206°$$
$$0.157 \times 360° = 57°$$
$$0.233 \times 360° = 84°$$
$$0.037 \times 360° = 13°$$

Step 4. Construct the pie chart using a protractor. See Figure A.17.

Step 5. Label the chart and each sector with the independent variable name and the percentage. See Figure A.18.

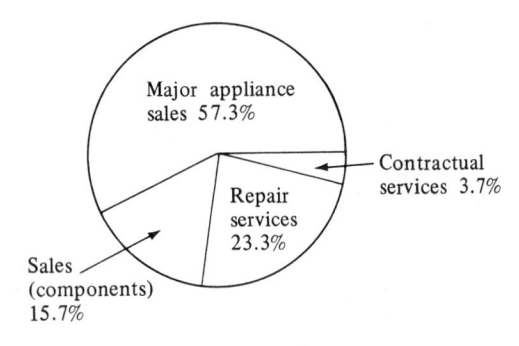

FIGURE A.18

We could develop other types of pictorial graphs. The draftsman will often be required to create complex pictorial graphics such as trilinear charts, surface charts, flow charts, etc. However, we will not consider these more technical tasks here.

SECTION I.4 VECTOR ADDITION

In Chapter 1 we noted that signed numbers could be added or subtracted by movement along a number line. We now view the movement as a vector and the vector as a representation of some physical quantity rather than as a signed number. We can represent the addition of vector quantities graphically using the same rules for adding signed numbers.

APPENDIX I GRAPHING EXPERIMENTAL DATA

To add the vectors $+3$ and -5 we draw either of these vectors along the number line. The vector $+3$ is shown in Figure A.19.

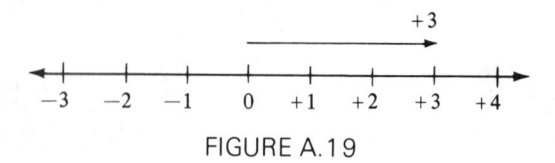

FIGURE A.19

Next draw -5 starting at the tip of $+3$ as shown in Figure A.20.

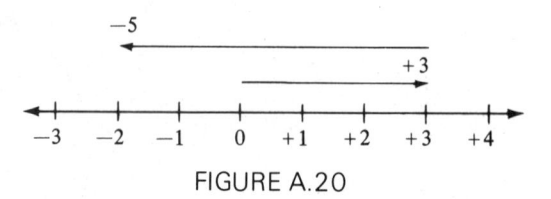

FIGURE A.20

Now form a new vector starting at the tail of vector $+3$ and ending at the tip of vector -5. This new vector represents the sum of the two vectors. See Figure A.21.

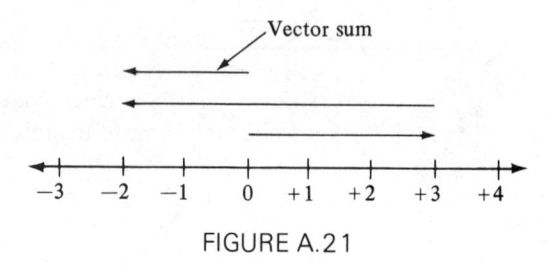

FIGURE A.21

Note that the sum vector is two units long and points left. We could make this addition without the number line present as shown in Figure A.22.

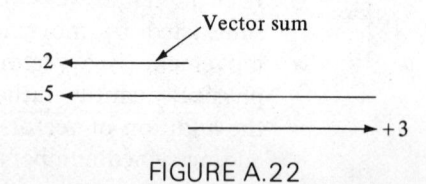

FIGURE A.22

VECTOR ADDITION 1.4   A17

As a second example, add the vectors −4, +2, and +5. The construction is shown in Figure A.23.

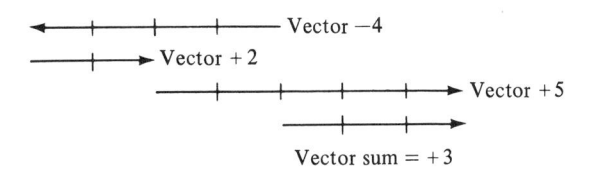

FIGURE A.23

Note that the vector sum starts at the tail of the first vector and ends at the tip of the last.

RULES FOR ADDING VECTORS GRAPHICALLY

Step 1. Place the vectors successively tail to tip.

Step 2. Form the sum by drawing a vector from the tail of the first to the tip of the last.

In the rectangular coordinate system, the vector is again specified by two quantities, a magnitude and a direction. We show a vector in Figure A.24.

Note that the + or − sign is no longer sufficient to specify the vector direction. We must give a particular angle value with respect to one of the axes. To add vectors graphically in this coordinate system, we use the procedure developed for addition.

Suppose we are to add the following vectors:

Vector 1 has a magnitude of 7 units and makes an angle of 65° with the positive x-axis.

Vector 2 has a magnitude of 4 units and makes an angle of 20° with the positive x-axis.

We plot the vectors with their correct magnitude and direction in Figure A.25.

To add them we slide either vector so that its tail touches the tip of the other. The vector orientation must not be changed in this process. The sum is obtained by starting at the tail of the first and drawing a straight line to the tip of the second (last). See Figure A.26.

The magnitude and direction of this sum vector could be obtained by measurement of its length and the angle it makes with the positive x-axis. Using a protractor and rule we find

$$\text{Magnitude of sum} = 10.2$$

$$\text{Angle with } x\text{-axis} = 49°$$

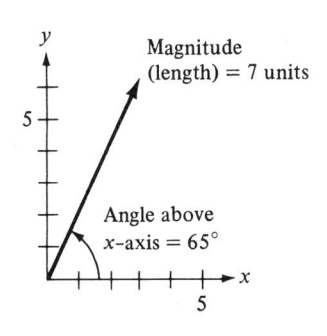

FIGURE A.24

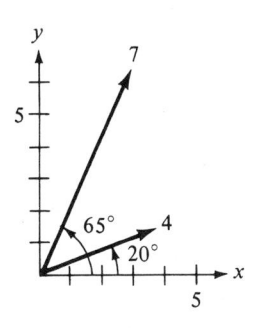

FIGURE A.25

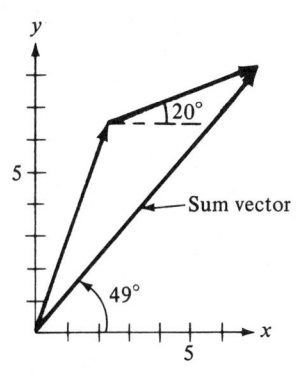

FIGURE A.26

As a second example consider the addition of the three vectors shown in Figure A.27.

FIGURE A.27

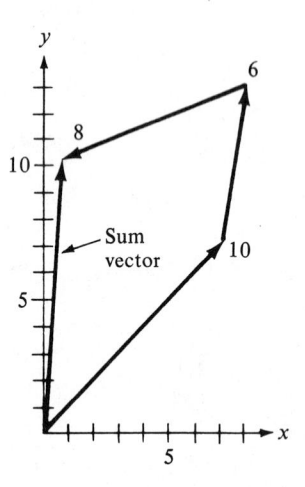

FIGURE A.28

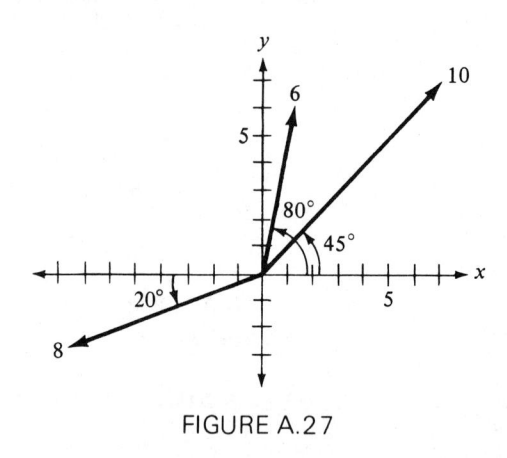

The vectors are added by joining them tip to tail. The sum vector is shown in Figure A.28. We measure its magnitude and direction to find

Magnitude = 10.3 units

Direction = 87° angle above positive x-axis

APPENDIX II

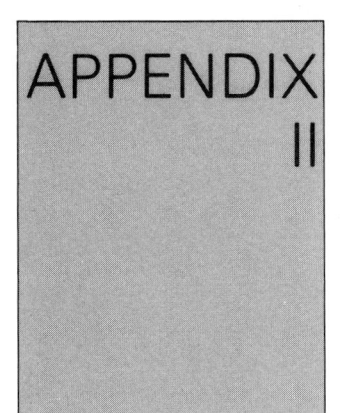

DATA ANALYSIS

In this appendix we will discuss certain mathematical techniques that allow us to interpret the meaning of a data set in a quantitative way. The techniques to be discussed are

1. Linear least squares fitting
2. Mean and standard deviation

SECTION II.1   LINEAR LEAST SQUARES FITTING

In Appendix I.1 we discussed graphs in which the data appeared to have a "linear trend." That is, the data points appeared to fall on or near a straight line. The purpose of this section is to take a more mathematical look at this linear trend. In so doing we will develop a value for the slope and y-intercept of the so-called "line of best fit" to the data. We will also calculate a number that will tell us whether or not the assumption of a linear trend is valid.

The equations used to calculate slope, y-intercept, etc. will not be discussed in this text. These equations are discussed in standard texts on statistics. We will simply outline the procedures to be followed in evaluating these equations. Consider the data set in Table A.9.

TABLE A.9   ELONGATION OF A SHOCK SPRING UNDER TENSION

Stretch (cm)	1.0	2.0	3.0	4.0	5.0	6.0	7.0	8.0
Force (N)	4.7	7.9	12.8	15.5	22.1	26.2	32.8	38.9

APPENDIX II DATA ANALYSIS

We graph this data set in Figure A.29.

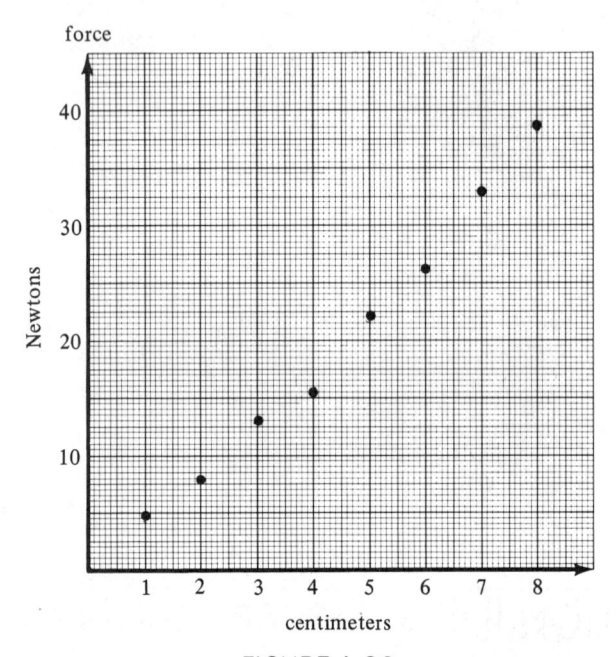

FIGURE A.29

We see that the data has a linear trend, yet it appears to turn upward a bit at the top right of the graph. Is it valid to consider this data as linear? We need some sort of mathematical measure to answer that question. The measure most commonly used is the **linear correlation coefficient** (usually designated by the small letter r). If the data can be considered linear, how do we select a straight line to best represent the data? To answer these questions we will set up a table involving the given data (see Table A.10). We call the quantity plotted on the horizontal axis x and that plotted on the vertical axis y.

The values needed for our calculations are shown in the bottom row. We must complete the entries in the last three columns. Column xy values are obtained by multiplying x and y together. Columns x^2-values and y^2-values are obtained by squaring the corresponding x- and y-values. The last row is a total of these values. Entries are completed in Table A.11.

LINEAR LEAST SQUARES FITTING II.1

TABLE A.10 TABLE FOR CALCULATING SLOPE, y-INTERCEPT, AND CORRELATION COEFFICIENT

Stretch (cm)—x	Force (N)—y	xy	x^2	y^2
1.0	4.7			
2.0	7.9			
3.0	12.8			
4.0	15.5			
5.0	22.1			
6.0	26.2			
7.0	32.8			
8.0	38.9			
$\Sigma x =$	$\Sigma y =$	$\Sigma xy =$	$\Sigma x^2 =$	$\Sigma y^2 =$

TABLE A.11 TABLE FOR CALCULATING SLOPE, y-INTERCEPT, AND CORRELATION COEFFICIENT

Stretch (cm)—x	Force (N)—y	xy	x^2	y^2
1.0	4.7	4.7	1	22.09
2.0	7.9	15.8	4	62.41
3.0	12.8	38.4	9	163.84
4.0	15.5	62.0	16	240.25
5.0	22.1	110.5	25	488.41
6.0	26.2	157.2	36	686.44
7.0	32.8	229.6	49	1075.84
8.0	38.9	311.2	64	1513.21
$\Sigma x =$ 36	$\Sigma y =$ 160.9	$\Sigma xy =$ 929.4	$\Sigma x^2 =$ 204	$\Sigma y^2 =$ 4252.49

So far we have only multiplied, squared, and added to complete the table. The expressions for the correlation coefficient, the slope, and the intercept that we are seeking are rather complicated if expressed in terms of the column totals given. We suggest the calculation of some intermediate quantities that we call

$\bar{x}$ (the mean value of the x-data) $= \dfrac{\Sigma x}{n}$

$\bar{y}$ (the mean value of the y-data) $= \dfrac{\Sigma y}{n}$

X_1 (related to the fluctuation in x-data) $= \Sigma x^2 - n(\bar{x})^2$
Y_1 (related to the fluctuation in y-data) $= \Sigma y^2 - n(\bar{y})^2$
Z_1 (related to the fluctuation in x and y) $= \Sigma xy - n\bar{x}\bar{y}$

We find

$$\bar{x} = \dfrac{\Sigma x}{n} = \dfrac{36}{8} = 4.5$$

$$\bar{y} = \dfrac{\Sigma y}{n} = \dfrac{160.9}{8} = 20.1125$$

$$X_1 = \Sigma x^2 - n(\bar{x})^2 = 204 - 8(4.5)^2 = 42$$
$$Y_1 = \Sigma y^2 - n(\bar{y})^2 = 4252.49 - 8(20.1125)^2 = 1016.4$$
$$Z_1 = \Sigma xy - n\bar{x}\bar{y} = 929.4 - 8(4.5)(20.1125) = 205.35$$

With these quantities we finally arrive at equations for r (the correlation coefficient), m (the slope), and b (the y-intercept). See Chapter 7 for the definition of m and b.

$$r = \dfrac{Z_1}{\sqrt{X_1 Y_1}} = \dfrac{205.35}{\sqrt{(42)(1016.4)}} = 0.9938887 = 0.99 \text{ with the correct number of significant figures.}$$

$$m = \dfrac{Z_1}{X_1} = \dfrac{205.35}{42} = 4.8892857 = 4.9$$

$$b = \bar{y} - m\bar{x} = 20.1125 - (4.8892857)(4.5) = -1.8892857 = -1.9$$

Let us first look at r, which has the value 0.99. The possible values of a linear correlation coefficient run from -1 to $+1$. We sketch those values on a line as shown in Figure A.30.

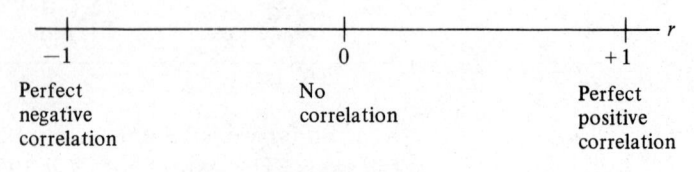

```
  +―――――――――――――+―――――――――――――+――― r
 -1             0            +1
Perfect         No         Perfect
negative    correlation    positive
correlation               correlation
```

FIGURE A.30

LINEAR LEAST SQUARES FITTING II.1

The endpoints and center of this line have the meanings displayed. If $r = +1$ then every point falls exactly on a straight line with positive slope. If $r = -1$ then every point falls on a straight line with negative slope. If $r = 0$ there is no linear trend at all. But what about intermediate values of r? We present a table that statisticians use to determine whether r is close enough to $+1$ or -1 to claim a linear relation. The table (see Table A.12) gives critical values of r. From this table we display the critical value of r for our problem. See Figure A.31.

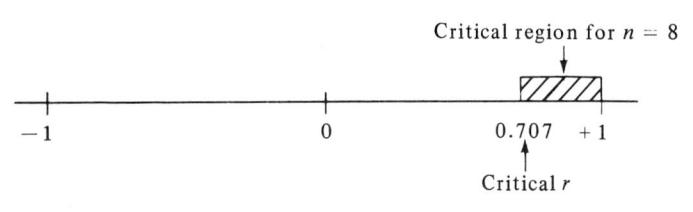

FIGURE A.31

If the calculated value of r lies between the critical value and the end point ($+1$ in this case), we can claim that a linear relationship exists (to be precise we can be 95% sure that it exists). The appropriate critical value is determined by entering Table A.12 with a value of n corresponding to the number of data points to be fitted.

Our value of $r = 0.99$ clearly meets the criterion for a linear relationship.

TABLE A.12 TABLE OF CRITICAL r's*

n	6	7	8	9	10	11
r	0.811	0.755	0.707	0.666	0.632	0.602
n	12	13	14	15	16	
r	0.576	0.553	0.532	0.514	0.497	

*Calculated from t-Distribution in Fadil H. Zuwaylif, *General Applied Statistics*, 3rd ed. (Reading: Addison-Wesley Publishing Co., Inc.), 1979, p. T-5.

Using m and b and the methods of Chapter 7, we can now plot the **line of best fit** along with the data points. See Figure A.32.

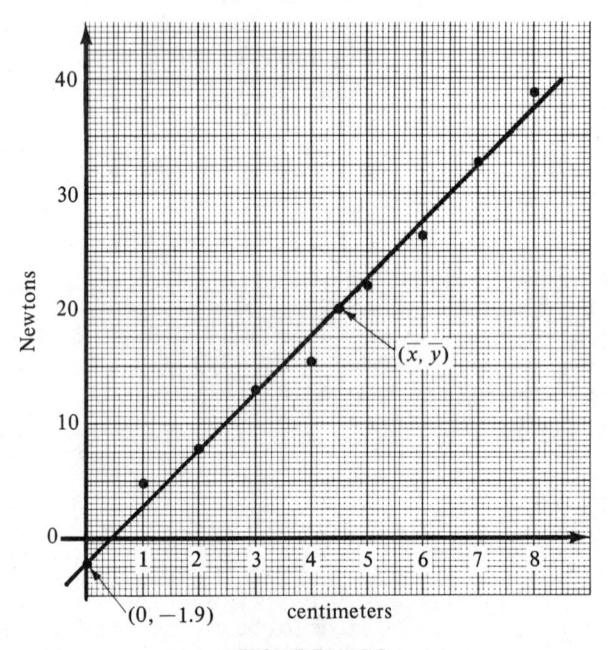

FIGURE A.32

However, the easiest and most accurate way to plot the line of best fit is by using $\bar{x}$ and $\bar{y}$. The point $(\bar{x}, \bar{y})$ must be on the line of best fit. As a result we know two points $(0, b)$ and $(\bar{x}, \bar{y})$ that determine the line.

This line of best fit serves as the most valid predictor of intermediate data values. For example, to find the best estimate of the force to extend the spring 5.5 cm we would move right to 5.5 on the horizontal scale and then measure the vertical distance to the line of best fit.

Alternatively, we could use the equation $y = mx + b$ to find an intermediate y. Using this technique we find

$$y = (4.9)(5.5) + (-1.9) = 25.05 \text{ N} \quad \text{or} \quad 25 \text{ N}$$

corresponding to 5.5 cm. This value should agree with the graphical result.

SECTION II.2 MEAN AND STANDARD DEVIATION

In Chapter 2 we emphasized that the result of a measurement was an approximate number. In dealing with measurements we also have to contend with fluctuations in our measuring apparatus. Two precise measurements of the diameter of a machined rod may not yield the same number because of fluctuations. There are also situations in which we measure quantities that themselves fluctuate. An example would be samples taken in a manufacturing process. Because of machine settings, operator error, etc., products that should show identical characteristics are different. We will use an example of quality control to introduce the mean and standard deviation.

Precision ball bearings are being produced with a diameter of 8 millimeters. To be usable the ball bearings must have no smaller diameter than 7.95 millimeters and no larger diameter than 8.05 millimeters. In other words, a tolerance of ±0.05 millimeters is permitted. A sample of ten bearings is selected randomly during each 30 minute interval to determine whether tolerances are being met. If the average value of the diameter of these bearings deviates by 0.02 millimeters from 8.00, the machinery is shut down and adjusted. Let us analyze a sample.

Bearing	Diameter (x)
#1	8.03
#2	7.96
#3	8.01
#4	8.02
#5	7.95
#6	7.97
#7	8.06
#8	7.98
#9	8.00
#10	7.96

$$\Sigma x = 79.94$$

To find the average of these values we sum them and divide by the number in the sample. The result is often called the **mean** of the data sample. The total of the diameters will be represented symbolically by Σx. (The Σ is shorthand for addition; that is, Σx means add together all values of x.) For our problem $\Sigma x = 79.94$. To find the mean, evaluate

$$\bar{x} = \frac{\Sigma x}{n}$$

where x is the symbol for the value of each piece of data and n is the number of pieces of data. For our problem, $n = 10$.

$$\text{Thus } \bar{x} = \frac{79.94}{10} = 7.994 \text{ mm} \quad \text{or} \quad 7.99 \text{ mm}$$

using the correct number of significant figures. The value is within the tolerance limit established for the mean value. However, we notice that one of the ball bearings has a diameter of 8.06 millimeters (over the tolerance limit) and another has a diameter of 7.95 millimeters (on the limit). This suggests that we need to be concerned with more than the average value. After all, five bearings at 7.8 millimeters and five at 8.2 millimeters will average to 8 millimeters yet none will be usable.

The manufacturer must have some standard for the size fluctuations as well as for the mean value. The measure of fluctuation most often used is the **standard deviation.** The standard deviation is calculated using the equation

$$s = \sqrt{\frac{n(\Sigma x^2) - (\Sigma x)^2}{n(n-1)}}$$

This equation needs some explanation. Table A.13 displays a method of evaluating the quantities needed in this equation.

TABLE A.13 STANDARD DEVIATION DATA

Bearing #	Diameter (x)	x^2
1	8.03	64.4809
2	7.96	63.3616
3	8.01	64.1601
4	8.02	64.3204
5	7.95	63.2025
6	7.97	63.5209
7	8.06	64.9636
8	7.98	63.6804
9	8.00	64.0
10	7.96	63.3616
$n = 10$	$\Sigma x = 79.94$	$\Sigma x^2 = 639.052$

The values of n, Σx, and Σx^2 are shown in the bottom row of Table A.13. With these values we may write

$$s = \sqrt{\frac{10(639.052) - (79.94)^2}{10(9)}}$$

MEAN AND STANDARD DEVIATION II.2

The procedure for solving this problem on the calculator is given here.

Step	Enter	Press	Display	
#1	10		10.	
#2		$\times$	10.	Evaluate and
#3	9		9.	store the
#4		=	90.	denominator.
#5		STO	90.	
#6	10		10.	
#7		$\times$	10.	
#8	639.052		639.052	
#9		−	6390.52	Evaluate the
#10	79.94		79.94	numerator.
#11		x^2	6390.4036	
#12		=	0.1164	
#13		÷	0.1164	
#14		RCL	90.	Complete the
#15		=	0.0012933	calculation.
#16		$\sqrt{x}$ or INV x^2	0.0359629	

If your calculator has a standard deviation key (denoted by $\boxed{SD}$, $\boxed{\sigma_n}$, or $\boxed{\sigma_{n-1}}$), you can compute the mean and standard deviation without establishing a table. Typical keystroke sequences are shown here.

8.03 $\boxed{\Sigma+}$ 7.96 $\boxed{\Sigma+}$ 8.01 $\boxed{\Sigma+}$ 8.02 $\boxed{\Sigma+}$ 7.95 $\boxed{\Sigma+}$ 7.97 $\boxed{\Sigma+}$ 8.06 $\boxed{\Sigma+}$ 7.98 $\boxed{\Sigma+}$ 8.00 $\boxed{\Sigma+}$ 7.96 $\boxed{\Sigma+}$

After you enter all the data, press the appropriate keys to obtain the mean and the standard deviation. For example, if the mean and standard deviation are second functions as they often are, use

$\boxed{2nd}$ $\boxed{\bar{x}}$ and find 7.994 and

$\boxed{2nd}$ ($\boxed{SD}$ or $\boxed{\sigma_{n-1}}$) to find 0.0359629

It appears that the standard deviation, rounded to the same precision as the measurements $\left(\frac{1}{100}\text{ of a millimeter}\right)$, is 0.04 millimeter. But what does this mean?

The standard deviation is a measure of fluctuation from the mean. The statistical interpretation of standard deviation is that about

68% of all the data will be within one standard deviation on either side of the mean (assuming a "normal distribution"). For this sample we could say that about 68% of the bearings will have dimensions between 7.99 − 0.04 millimeters and 7.99 + 0.04 millimeters (7.95 to 8.03 millimeters).

Looking at the sample data we see three bearings (30%) that are outside the standard deviation (before rounding). These are the bearings with diameters 8.03, 7.95, and 8.06 millimeters. If this does not meet quality control standards, the data will dictate adjustment of the machinery. In quality control sampling, the tolerance in both the mean and the standard deviation will probably be set in advance.

There are other measures of fluctuation that can be used. The range (largest measured value minus smallest measured value) is sometimes used. Another measure of fluctuation is the average deviation. However, the measure that is easiest to interpret mathematically is the standard deviation.

PROPERTIES OF THE STANDARD DEVIATION

About 68% of the sample will be within one standard deviation of the mean.

About 95% of the sample will be within two standard deviations of the mean. (These statements are based on a distribution of data called a normal distribution.)

If the manufacturer of ball bearings wants no more than 5% unusable bearings, a maximum standard deviation of 0.025 millimeter should be set for the samples. With this requirement the mean diameter will have to stay very near to 8.00 millimeters.

APPENDIX III

DIMENSIONAL ANALYSIS

We should perhaps say that dimensional analysis has nothing to do with determining the size of things. The term dimensional analysis refers to a method of determining the "dimensions" of combinations of physical quantities. We worked some dimensional analysis exercises in Chapter 2 (Section 2.2).

Suppose we are to find the dimensions of the quantity represented by the term $0.5at^2$ where a is acceleration and t is time. This term would be evaluated by multiplying 0.5 by the value of a and then multiplying by the value of t to the second power.

Acceleration is a derived quantity with the dimensions length/time^2 (see Chapter 2). The quantity time is a basic quantity and appears to the second power. We form an expression for the dimensions of $0.5at^2$ by replacing the variables with their dimensions. Thus $0.5at^2$ has dimensions of

$$\frac{\text{length}}{\text{time}^2} \cdot \text{time}^2$$

These dimensions can be simplified by cancellation just like a numerical or algebraic fraction. So we write

$$\frac{\text{length}}{\cancel{\text{time}^2}} \cdot \cancel{\text{time}^2} = [\text{length}]$$

Dimensions are frequently shown in square brackets. Thus the dimensions of the term $0.5at^2$ are length. Note that the pure number 0.5 did not enter into the determination of dimensions.

To simplify dimensional analysis we will abbreviate the basic quantities in the following manner:

$$M = \text{mass}$$
$$L = \text{length}$$
$$T = \text{time}$$
$$I = \text{current}$$

As a second example, we determine the dimensions of quantity mvr where m represents mass, v represents velocity, and r represents distance. The dimensions of velocity are

$$\left[\frac{L}{T}\right]$$

Thus the dimensions of *mvr* are

$$[M]\left[\frac{L}{T}\right][L] = \left[\frac{ML^2}{T}\right].$$

We determine the dimensions of two other expressions that occur frequently in physical problems.

Consider $\frac{1}{2}kx^2$ where $k = $ force/distance and $x = $ distance. We first note that force is a derived quantity with dimensions of $\left[\frac{ML}{T^2}\right]$. Then k has dimensions of

$$\left[\frac{\frac{ML}{T^2}}{L}\right] = \left[\frac{M}{T^2}\right]$$

and $\frac{1}{2}kx^2$ has dimensions of $\left[\frac{M}{T^2}\right][L^2] = \left[\frac{ML^2}{T^2}\right]$.

Consider $\frac{q}{m}$ where q is electrical charge and m is mass. Charge is the product of current and time. Thus the dimensions of q are $[IT]$. Then $\frac{q}{m}$ has dimensions

$$\frac{[IT]}{[M]} = \left[\frac{IT}{M}\right].$$

Dimensional analysis serves two very practical purposes.

1. First, dimensional analysis allows us to verify that a quantity being evaluated has the correct dimensions.

Suppose we encounter a statement such as "the energy of a flywheel is the product of its mass, radius, and angular velocity squared." We know the dimensions of these quantities to be

energy $\left[\frac{ML^2}{T^2}\right]$

mass $[M]$

radius $[L]$

angular velocity $\left[\frac{1}{T}\right]$

Thus the product of mass, radius, and angular velocity squared is $[M][L]\left[\frac{1}{T^2}\right] = \left[\frac{ML}{T^2}\right]$.

But energy has dimensions of $\left[\frac{ML^2}{T^2}\right]$. Thus the statement cannot possibly be correct. We could not determine the energy of a flywheel

APPENDIX III DIMENSIONAL ANALYSIS

by forming the product indicated. On the other hand, if we are told that the energy of a flywheel is obtained by forming the product of its mass, radius squared, and angular velocity squared, a dimensional analysis would indicate no contradictions. Whether the statement is true or not depends on the geometry of the flywheel. We can use dimensional analysis to detect errors in dimensional statements but not to prove the truth of an equality.

2. Second, dimensional analysis allows us to detect incorrect terms in an equation. Consider the following equation for flow along a streamline in a fluid:

$$p + \rho g h + \frac{1}{2}\rho v^2 = \text{constant}$$

where

p = pressure
ρ = mass density = mass/volume
g = acceleration of gravity
h = height above some reference level
v = fluid velocity

We can make a dimensional analysis of each term on the left side of this equation.

$$P \text{ (pressure)} = \frac{\text{force}}{\text{area}} = \left[\frac{ML}{T^2}\right]/[L^2] = \left[\frac{M}{LT^2}\right]$$

$$\rho g h = \left[\frac{M}{L^3}\right]\left[\frac{L}{T^2}\right][L] = \left[\frac{M}{LT^2}\right]$$

$$\rho v^2 = \left[\frac{M}{L^3}\right]\left[\frac{L}{T}\right]\left[\frac{L}{T}\right] = \left[\frac{M}{LT^2}\right]$$

The equation appears to be dimensionally correct.

Consider another example. An equation states that the displacement d of any uniformly accelerated object is given by

$$d = v_0 t + 0.5at$$

where

v_0 = starting velocity
a = acceleration
t = time

We make a dimensional analysis of each term as follows:

d (displacement) $= [L]$

$$v_0 t = \text{velocity} \cdot \text{time} = \left[\frac{L}{T}\right][T] = [L]$$

$$at = \text{acceleration} \cdot \text{time} = \left[\frac{L}{T^2}\right][T] = \left[\frac{L}{T}\right]$$

Dimensionally, the term $0.5at$ cannot be correct. It represents a velocity rather than a distance. The correct equation should be $d = v_0 t + 0.5at^2$. We verified earlier that $0.5at^2$ has the dimensions of length.

DIMENSIONAL ANALYSIS

1. Permits a check on consistency between a quantity name and its dimensions. That is, an energy must have dimensions $\left[\frac{ML^2}{T^2}\right]$.

2. Permits a check on the dimensional consistency of terms in an equation.

APPENDIX IV

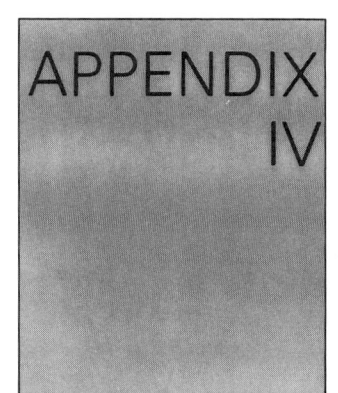

TABLES

This appendix contains the following tables:
- Table A Trigonometric Functions
- Table B Logarithms to Base 10
- Table C Natural Logarithms
- Table D Powers of Base *e*

TABLE A TRIGONOMETRIC FUNCTIONS

ANGLE	SIN	COS	TAN	CSC	SEC	COT	ANGLE
0	0.0000	1.0000	0.0000		1.000		90
1	0.0175	0.9998	0.0175	57.299	1.000	57.290	89
2	0.0349	0.9994	0.0349	28.654	1.001	28.636	88
3	0.0523	0.9986	0.0524	19.107	1.001	19.081	87
4	0.0698	0.9976	0.0699	14.336	1.002	14.301	86
5	0.0872	0.9962	0.0875	11.474	1.004	11.430	85
6	0.1045	0.9945	0.1051	9.567	1.006	9.514	84
7	0.1219	0.9925	0.1228	8.206	1.008	8.144	83
8	0.1392	0.9903	0.1405	7.185	1.010	7.115	82
9	0.1564	0.9877	0.1584	6.392	1.012	6.314	81
10	0.1736	0.9848	0.1763	5.759	1.015	5.671	80
11	0.1908	0.9816	0.1944	5.241	1.019	5.145	79
12	0.2079	0.9781	0.2126	4.810	1.022	4.705	78
13	0.2250	0.9744	0.2309	4.445	1.026	4.331	77
14	0.2419	0.9703	0.2493	4.134	1.031	4.011	76
15	0.2588	0.9659	0.2679	3.864	1.035	3.732	75
16	0.2756	0.9613	0.2867	3.628	1.040	3.487	74
17	0.2924	0.9563	0.3057	3.420	1.046	3.271	73
18	0.3090	0.9511	0.3249	3.236	1.051	3.078	72
19	0.3256	0.9455	0.3443	3.072	1.058	2.904	71
20	0.3420	0.9397	0.3640	2.924	1.064	2.747	70
21	0.3584	0.9336	0.3839	2.790	1.071	2.605	69
22	0.3746	0.9272	0.4040	2.669	1.079	2.475	68
23	0.3907	0.9205	0.4245	2.559	1.086	2.356	67
24	0.4067	0.9135	0.4452	2.459	1.095	2.246	66
25	0.4226	0.9063	0.4663	2.366	1.103	2.145	65
26	0.4384	0.8988	0.4877	2.281	1.113	2.050	64
27	0.4540	0.8910	0.5095	2.203	1.122	1.963	63
28	0.4695	0.8829	0.5317	2.130	1.133	1.881	62
29	0.4848	0.8746	0.5543	2.063	1.143	1.804	61
30	0.5000	0.8660	0.5774	2.000	1.155	1.732	60
31	0.5150	0.8572	0.6009	1.942	1.167	1.664	59
32	0.5299	0.8480	0.6249	1.887	1.179	1.600	58
33	0.5446	0.8387	0.6494	1.836	1.192	1.540	57
34	0.5592	0.8290	0.6745	1.788	1.206	1.483	56
35	0.5736	0.8192	0.7002	1.743	1.221	1.428	55
36	0.5878	0.8090	0.7265	1.701	1.236	1.376	54
37	0.6018	0.7986	0.7536	1.662	1.252	1.327	53
38	0.6157	0.7880	0.7813	1.624	1.269	1.280	52
39	0.6293	0.7771	0.8098	1.589	1.287	1.235	51
40	0.6428	0.7660	0.8391	1.556	1.305	1.192	50
41	0.6561	0.7547	0.8693	1.524	1.325	1.150	49
42	0.6691	0.7431	0.9004	1.494	1.346	1.111	48
43	0.6820	0.7314	0.9325	1.466	1.367	1.072	47
44	0.6947	0.7193	0.9657	1.440	1.390	1.036	46
45	0.7071	0.7071	1.0000	1.414	1.414	1.000	45
ANGLE	COS	SIN	COT	SEC	CSC	TAN	ANGLE

TABLE B LOGARITHMS TO BASE 10

Proportional Parts

N	0	1	2	3	4	5	6	7	8	9	1	2	3	4	5	6	7	8	9
10	0	43	86	128	170	212	253	294	334	374	4	8	13	17	21	25	29	34	38
11	414	453	492	531	569	607	645	682	719	755	4	8	11	15	19	23	27	30	34
12	792	828	864	899	934	969	1004	1038	1072	1106	4	7	11	14	18	21	25	28	32
13	1139	1173	1206	1239	1271	1303	1335	1367	1399	1430	3	6	10	13	16	19	22	26	29
14	1461	1492	1523	1553	1584	1614	1644	1673	1703	1732	3	6	9	12	16	19	22	25	28
15	1761	1790	1818	1847	1875	1903	1931	1959	1987	2014	3	6	8	11	14	17	20	22	25
16	2041	2068	2095	2122	2148	2175	2201	2227	2253	2279	3	5	8	10	13	16	18	21	23
17	2304	2330	2355	2380	2405	2430	2455	2480	2504	2529	3	5	8	10	13	15	18	20	23
18	2553	2577	2601	2625	2648	2672	2695	2718	2742	2765	2	5	7	9	12	14	16	18	21
19	2788	2810	2833	2856	2878	2900	2923	2945	2967	2989	2	4	7	9	11	13	15	18	20
20	3010	3032	3054	3075	3096	3118	3139	3160	3181	3201	2	4	6	8	11	13	15	17	19
21	3222	3243	3263	3284	3304	3324	3345	3365	3385	3404	2	4	6	8	10	12	14	16	18
22	3424	3444	3464	3483	3502	3522	3541	3560	3579	3598	2	4	6	8	10	11	13	15	17
23	3617	3636	3655	3674	3692	3711	3729	3747	3766	3784	2	4	5	7	9	11	13	14	16
24	3802	3820	3838	3856	3874	3892	3909	3927	3945	3962	2	4	5	7	9	11	13	14	16
25	3979	3997	4014	4031	4048	4065	4082	4099	4116	4133	2	3	5	7	9	10	12	14	15
26	4150	4166	4183	4200	4216	4232	4249	4265	4281	4298	2	3	5	6	8	10	11	13	14
27	4314	4330	4346	4362	4378	4393	4409	4425	4440	4456	2	3	5	6	8	10	11	13	14
28	4472	4487	4502	4518	4533	4548	4564	4579	4594	4609	2	3	5	6	8	9	11	12	14
29	4624	4639	4654	4669	4683	4698	4713	4728	4742	4757	1	3	4	6	7	8	10	11	13
30	4771	4786	4800	4814	4829	4843	4857	4871	4886	4900	2	3	5	6	8	9	11	12	14
31	4914	4928	4942	4955	4969	4983	4997	5011	5024	5038	1	3	4	6	7	8	10	11	13
32	5052	5065	5079	5092	5105	5119	5132	5145	5159	5172	1	3	4	5	7	8	9	10	12
33	5185	5198	5211	5224	5237	5250	5263	5276	5289	5302	1	3	4	5	7	8	9	10	12
34	5315	5328	5340	5353	5366	5378	5391	5403	5416	5428	1	3	4	5	7	8	9	10	12
35	5441	5453	5465	5478	5490	5502	5515	5527	5539	5551	1	2	4	5	6	7	8	10	11
36	5563	5575	5587	5599	5611	5623	5635	5647	5658	5670	1	2	4	5	6	7	8	10	11
37	5682	5694	5705	5717	5729	5740	5752	5763	5775	5786	1	2	4	5	6	7	8	10	11
38	5798	5809	5821	5832	5843	5855	5866	5877	5888	5899	1	2	3	4	6	7	8	9	10
39	5911	5922	5933	5944	5955	5966	5977	5988	5999	6010	1	2	3	4	6	7	8	9	10
40	6021	6031	6042	6053	6064	6075	6085	6096	6107	6117	1	2	3	4	6	7	8	9	10
41	6128	6138	6149	6160	6170	6180	6191	6201	6212	6222	1	2	3	4	5	6	7	8	9
42	6232	6243	6253	6263	6274	6284	6294	6304	6314	6325	1	2	3	4	6	7	8	9	10
43	6335	6345	6355	6365	6375	6385	6395	6405	6415	6425	1	2	3	4	5	6	7	8	9
44	6435	6444	6454	6464	6474	6484	6493	6503	6513	6522	1	2	3	4	5	6	7	8	9
45	6532	6542	6551	6561	6571	6580	6590	6599	6609	6618	1	2	3	4	5	6	7	8	9
46	6628	6637	6646	6656	6665	6675	6684	6693	6702	6712	**	2	3	4	5	5	6	7	8
47	6721	6730	6739	6749	6758	6767	6776	6785	6794	6803	**	2	3	4	5	5	6	7	8
48	6812	6821	6830	6839	6848	6857	6866	6875	6884	6893	**	2	3	4	5	5	6	7	8
49	6902	6911	6920	6928	6937	6946	6955	6964	6972	6981	**	2	3	4	5	5	6	7	8
50	6990	6998	7007	7016	7024	7033	7042	7050	7059	7067	**	2	2	3	4	5	6	6	7
51	7076	7084	7093	7101	7110	7118	7126	7135	7143	7152	**	2	3	4	5	5	6	7	8
52	7160	7168	7177	7185	7193	7202	7210	7218	7226	7235	**	2	2	3	4	5	6	6	7
53	7243	7251	7259	7267	7275	7284	7292	7300	7308	7316	**	2	2	3	4	5	6	6	7
54	7324	7332	7340	7348	7356	7364	7372	7380	7388	7396	**	2	2	3	4	5	6	6	7

Table B, cont'd

N	0	1	2	3	4	5	6	7	8	9			Proportional Parts						
											1	2	3	4	5	6	7	8	9
55	7404	7412	7419	7427	7435	7443	7451	7459	7466	7474	**	2	2	3	4	5	6	6	7
56	7482	7490	7497	7505	7513	7520	7528	7536	7543	7551	**	2	2	3	4	5	6	6	7
57	7559	7566	7574	7582	7589	7597	7604	7612	7619	7627	**	1	2	3	4	4	5	6	6
58	7634	7642	7649	7657	7664	7672	7679	7686	7694	7701	**	1	2	3	4	4	5	6	6
59	7709	7716	7723	7731	7738	7745	7752	7760	7767	7774	**	1	2	3	4	4	5	6	6
60	7782	7789	7796	7803	7810	7818	7825	7832	7839	7846	**	1	2	3	4	4	5	6	6
61	7853	7860	7868	7875	7882	7889	7896	7903	7910	7917	**	1	2	3	4	4	5	6	6
62	7924	7931	7938	7945	7952	7959	7966	7973	7980	7987	**	1	2	3	4	4	5	6	6
63	7993	8000	8007	8014	8021	8028	8035	8041	8048	8055	**	1	2	3	4	4	5	6	6
64	8062	8069	8075	8082	8089	8096	8102	8109	8116	8122	**	1	2	3	4	4	5	6	6
65	8129	8136	8142	8149	8156	8162	8169	8176	8182	8189	**	1	2	3	4	4	5	6	6
66	8195	8202	8209	8215	8222	8228	8235	8241	8248	8254	**	1	2	3	4	4	5	6	6
67	8261	8267	8274	8280	8287	8293	8299	8306	8312	8319	**	1	2	3	4	4	5	6	6
68	8325	8331	8338	8344	8351	8357	8363	8370	8376	8382	**	1	2	3	4	4	5	6	6
69	8388	8395	8401	8407	8414	8420	8426	8432	8439	8445	**	1	2	3	4	4	5	6	6
70	8451	8457	8463	8470	8476	8482	8488	8494	8500	8506	**	1	2	2	3	4	4	5	5
71	8513	8519	8525	8531	8537	8543	8549	8555	8561	8567	**	1	2	2	3	4	4	5	5
72	8573	8579	8585	8591	8597	8603	8609	8615	8621	8627	**	1	2	2	3	4	4	5	5
73	8633	8639	8645	8651	8657	8663	8669	8675	8681	8686	**	1	2	2	3	4	4	5	5
74	8692	8698	8704	8710	8716	8722	8727	8733	8739	8745	**	1	2	2	3	4	4	5	5
75	8751	8756	8762	8768	8774	8779	8785	8791	8797	8802	**	1	2	2	3	4	4	5	5
76	8808	8814	8820	8825	8831	8837	8842	8848	8854	8859	**	1	2	2	3	4	4	5	5
77	8865	8871	8876	8882	8887	8893	8899	8904	8910	8915	**	1	2	2	3	3	4	4	5
78	8921	8927	8932	8938	8943	8949	8954	8960	8965	8971	**	1	2	2	3	3	4	4	5
79	8976	8982	8987	8993	8998	9004	9009	9015	9020	9025	**	1	2	2	3	3	4	4	5
80	9031	9036	9042	9047	9053	9058	9063	9069	9074	9079	**	1	2	2	3	4	4	5	5
81	9085	9090	9096	9101	9106	9112	9117	9122	9128	9133	**	1	2	2	3	3	4	4	5
82	9138	9143	9149	9154	9159	9165	9170	9175	9180	9186	**	1	2	2	3	3	4	4	5
83	9191	9196	9201	9206	9212	9217	9222	9227	9232	9238	**	1	2	2	3	4	4	5	5
84	9243	9248	9253	9258	9263	9269	9274	9279	9284	9289	**	1	2	2	3	3	4	4	5
85	9294	9299	9304	9309	9315	9320	9325	9330	9335	9340	**	1	2	2	3	4	4	5	5
86	9345	9350	9355	9360	9365	9370	9375	9380	9385	9390	**	1	2	2	3	3	4	4	5
87	9395	9400	9405	9410	9415	9420	9425	9430	9435	9440	**	1	2	2	3	3	4	4	5
88	9445	9450	9455	9460	9465	9469	9474	9479	9484	9489	**	1	2	2	3	3	4	4	5
89	9494	9499	9504	9509	9513	9518	9523	9528	9533	9538	**	**	1	2	2	2	3	3	4
90	9542	9547	9552	9557	9562	9566	9571	9576	9581	9586	**	1	2	2	3	3	4	4	5
91	9590	9595	9600	9605	9609	9614	9619	9624	9628	9633	**	**	1	2	2	3	3	4	4
92	9638	9643	9647	9652	9657	9661	9666	9671	9675	9680	**	1	2	2	3	3	4	4	5
93	9685	9689	9694	9699	9703	9708	9713	9717	9722	9727	**	**	1	2	2	3	3	4	4
94	9731	9736	9741	9745	9750	9754	9759	9764	9768	9773	**	1	2	2	3	3	4	4	5
95	9777	9782	9786	9791	9795	9800	9805	9809	9814	9818	**	**	1	2	2	3	3	4	4
96	9823	9827	9832	9836	9841	9845	9850	9854	9859	9863	**	1	2	2	3	3	4	4	5
97	9868	9872	9877	9881	9886	9890	9894	9899	9903	9908	**	1	2	2	3	3	4	4	5
98	9912	9917	9921	9926	9930	9934	9939	9943	9948	9952	**	**	1	2	2	3	3	4	4
99	9956	9961	9965	9969	9974	9978	9983	9987	9991	9996	**	1	2	2	3	3	4	4	5

TABLE C NATURAL LOGARITHMS

N	0.00	0.01	0.02	0.03	0.04	0.05	0.06	0.07	0.08	0.09
1.0	0.0000	0.0100	0.0198	0.0296	0.0392	0.0488	0.0583	0.0677	0.0770	0.0862
1.1	0.0953	0.1044	0.1133	0.1222	0.1310	0.1398	0.1484	0.1570	0.1655	0.1740
1.2	0.1823	0.1906	0.1989	0.2070	0.2151	0.2231	0.2311	0.2390	0.2469	0.2546
1.3	0.2624	0.2700	0.2776	0.2852	0.2927	0.3001	0.3075	0.3148	0.3221	0.3293
1.4	0.3365	0.3436	0.3507	0.3577	0.3646	0.3716	0.3784	0.3853	0.3920	0.3988
1.5	0.4055	0.4121	0.4187	0.4253	0.4318	0.4383	0.4447	0.4511	0.4574	0.4637
1.6	0.4700	0.4762	0.4824	0.4886	0.4947	0.5008	0.5068	0.5128	0.5188	0.5247
1.7	0.5306	0.5365	0.5423	0.5481	0.5539	0.5596	0.5653	0.5710	0.5766	0.5822
1.8	0.5878	0.5933	0.5988	0.6043	0.6098	0.6152	0.6206	0.6259	0.6313	0.6366
1.9	0.6419	0.6471	0.6523	0.6575	0.6627	0.6678	0.6729	0.6780	0.6831	0.6881
2.0	0.6931	0.6981	0.7031	0.7080	0.7129	0.7178	0.7227	0.7275	0.7324	0.7372
2.1	0.7419	0.7467	0.7514	0.7561	0.7608	0.7655	0.7701	0.7747	0.7793	0.7839
2.2	0.7885	0.7930	0.7975	0.8020	0.8065	0.8109	0.8154	0.8198	0.8242	0.8286
2.3	0.8329	0.8372	0.8416	0.8459	0.8502	0.8544	0.8587	0.8629	0.8671	0.8713
2.4	0.8755	0.8796	0.8838	0.8879	0.8920	0.8961	0.9002	0.9042	0.9083	0.9123
2.5	0.9163	0.9203	0.9243	0.9282	0.9322	0.9361	0.9400	0.9439	0.9478	0.9517
2.6	0.9555	0.9594	0.9632	0.9670	0.9708	0.9746	0.9783	0.9821	0.9858	0.9895
2.7	0.9933	0.9969	1.0006	1.0043	1.0080	1.0116	1.0152	1.0188	1.0225	1.0260
2.8	1.0296	1.0332	1.0367	1.0403	1.0438	1.0473	1.0508	1.0543	1.0578	1.0613
2.9	1.0647	1.0682	1.0716	1.0750	1.0784	1.0818	1.0852	1.0886	1.0919	1.0953
3.0	1.0986	1.1019	1.1053	1.1086	1.1119	1.1151	1.1184	1.1217	1.1249	1.1282
3.1	1.1314	1.1346	1.1378	1.1410	1.1442	1.1474	1.1506	1.1537	1.1569	1.1600
3.2	1.1632	1.1663	1.1694	1.1725	1.1756	1.1787	1.1817	1.1848	1.1878	1.1909
3.3	1.1939	1.1969	1.2000	1.2030	1.2060	1.2090	1.2119	1.2149	1.2179	1.2208
3.4	1.2238	1.2267	1.2296	1.2326	1.2355	1.2384	1.2413	1.2442	1.2470	1.2499
3.5	1.2528	1.2556	1.2585	1.2613	1.2641	1.2669	1.2698	1.2726	1.2754	1.2782
3.6	1.2809	1.2837	1.2865	1.2892	1.2920	1.2947	1.2975	1.3002	1.3029	1.3056
3.7	1.3083	1.3110	1.3137	1.3164	1.3191	1.3218	1.3244	1.3271	1.3297	1.3324
3.8	1.3350	1.3376	1.3403	1.3429	1.3455	1.3481	1.3507	1.3533	1.3558	1.3584
3.9	1.3610	1.3635	1.3661	1.3686	1.3712	1.3737	1.3762	1.3788	1.3813	1.3838
4.0	1.3863	1.3888	1.3913	1.3938	1.3962	1.3987	1.4012	1.4036	1.4061	1.4085
4.1	1.4110	1.4134	1.4159	1.4183	1.4207	1.4231	1.4255	1.4279	1.4303	1.4327
4.2	1.4351	1.4375	1.4398	1.4422	1.4446	1.4469	1.4493	1.4516	1.4540	1.4563
4.3	1.4586	1.4609	1.4633	1.4656	1.4679	1.4702	1.4725	1.4748	1.4770	1.4793
4.4	1.4816	1.4839	1.4861	1.4884	1.4907	1.4929	1.4951	1.4974	1.4996	1.5019
4.5	1.5041	1.5063	1.5085	1.5107	1.5129	1.5151	1.5173	1.5195	1.5217	1.5239
4.6	1.5261	1.5282	1.5304	1.5326	1.5347	1.5369	1.5390	1.5412	1.5433	1.5454
4.7	1.5476	1.5497	1.5518	1.5539	1.5560	1.5581	1.5602	1.5623	1.5644	1.5665
4.8	1.5686	1.5707	1.5728	1.5748	1.5769	1.5790	1.5810	1.5831	1.5851	1.5872
4.9	1.5892	1.5913	1.5933	1.5953	1.5974	1.5994	1.6014	1.6034	1.6054	1.6074
5.0	1.6094	1.6114	1.6134	1.6154	1.6174	1.6194	1.6214	1.6233	1.6253	1.6273
5.1	1.6292	1.6312	1.6332	1.6351	1.6371	1.6390	1.6409	1.6429	1.6448	1.6467
5.2	1.6487	1.6506	1.6525	1.6544	1.6563	1.6582	1.6601	1.6620	1.6639	1.6658
5.3	1.6677	1.6696	1.6715	1.6734	1.6752	1.6771	1.6790	1.6808	1.6827	1.6845
5.4	1.6864	1.6882	1.6901	1.6919	1.6938	1.6956	1.6974	1.6993	1.7011	1.7029

Table C, cont'd

N	0.00	0.01	0.02	0.03	0.04	0.05	0.06	0.07	0.08	0.09
5.5	1.7047	1.7066	1.7084	1.7102	1.7120	1.7138	1.7156	1.7174	1.7192	1.7210
5.6	1.7228	1.7246	1.7263	1.7281	1.7299	1.7317	1.7334	1.7352	1.7370	1.7387
5.7	1.7405	1.7422	1.7440	1.7457	1.7475	1.7492	1.7509	1.7527	1.7544	1.7561
5.8	1.7579	1.7596	1.7613	1.7630	1.7647	1.7664	1.7681	1.7699	1.7716	1.7733
5.9	1.7750	1.7766	1.7783	1.7800	1.7817	1.7834	1.7851	1.7867	1.7884	1.7901
6.0	1.7918	1.7934	1.7951	1.7967	1.7984	1.8001	1.8017	1.8034	1.8050	1.8066
6.1	1.8083	1.8099	1.8116	1.8132	1.8148	1.8165	1.8181	1.8197	1.8213	1.8229
6.2	1.8245	1.8262	1.8278	1.8294	1.8310	1.8326	1.8342	1.8358	1.8374	1.8390
6.3	1.8405	1.8421	1.8437	1.8453	1.8469	1.8485	1.8500	1.8516	1.8532	1.8547
6.4	1.8563	1.8579	1.8594	1.8610	1.8625	1.8641	1.8656	1.8672	1.8687	1.8703
6.5	1.8718	1.8733	1.8749	1.8764	1.8779	1.8795	1.8810	1.8825	1.8840	1.8856
6.6	1.8871	1.8886	1.8901	1.8916	1.8931	1.8946	1.8961	1.8976	1.8991	1.9006
6.7	1.9021	1.9036	1.9051	1.9066	1.9081	1.9095	1.9110	1.9125	1.9140	1.9155
6.8	1.9169	1.9184	1.9199	1.9213	1.9228	1.9242	1.9257	1.9272	1.9286	1.9301
6.9	1.9315	1.9330	1.9344	1.9359	1.9373	1.9387	1.9402	1.9416	1.9430	1.9445
7.0	1.9459	1.9473	1.9488	1.9502	1.9516	1.9530	1.9544	1.9559	1.9573	1.9587
7.1	1.9601	1.9615	1.9629	1.9643	1.9657	1.9671	1.9685	1.9699	1.9713	1.9727
7.2	1.9741	1.9755	1.9769	1.9782	1.9796	1.9810	1.9824	1.9838	1.9851	1.9865
7.3	1.9879	1.9892	1.9906	1.9920	1.9933	1.9947	1.9961	1.9974	1.9988	2.0001
7.4	2.0015	2.0028	2.0042	2.0055	2.0069	2.0082	2.0096	2.0109	2.0122	2.0136
7.5	2.0149	2.0162	2.0176	2.0189	2.0202	2.0215	2.0229	2.0242	2.0255	2.0268
7.6	2.0281	2.0295	2.0308	2.0321	2.0334	2.0347	2.0360	2.0373	2.0386	2.0399
7.7	2.0412	2.0425	2.0438	2.0451	2.0464	2.0477	2.0490	2.0503	2.0516	2.0528
7.8	2.0541	2.0554	2.0567	2.0580	2.0592	2.0605	2.0618	2.0631	2.0643	2.0656
7.9	2.0669	2.0681	2.0694	2.0707	2.0719	2.0732	2.0744	2.0757	2.0769	2.0782
8.0	2.0794	2.0807	2.0819	2.0832	2.0844	2.0857	2.0869	2.0882	2.0894	2.0906
8.1	2.0919	2.0931	2.0943	2.0956	2.0968	2.0980	2.0992	2.1005	2.1017	2.1029
8.2	2.1041	2.1054	2.1066	2.1078	2.1090	2.1102	2.1114	2.1126	2.1138	2.1150
8.3	2.1163	2.1175	2.1187	2.1199	2.1211	2.1223	2.1235	2.1247	2.1258	2.1270
8.4	2.1282	2.1294	2.1306	2.1318	2.1330	2.1342	2.1353	2.1365	2.1377	2.1389
8.5	2.1401	2.1412	2.1424	2.1436	2.1448	2.1459	2.1471	2.1483	2.1494	2.1506
8.6	2.1518	2.1529	2.1541	2.1552	2.1564	2.1576	2.1587	2.1599	2.1610	2.1622
8.7	2.1633	2.1645	2.1656	2.1668	2.1679	2.1691	2.1702	2.1713	2.1725	2.1736
8.8	2.1748	2.1759	2.1770	2.1782	2.1793	2.1804	2.1815	2.1827	2.1838	2.1849
8.9	2.1861	2.1872	2.1883	2.1894	2.1905	2.1917	2.1928	2.1939	2.1950	2.1961
9.0	2.1972	2.1983	2.1994	2.2006	2.2017	2.2028	2.2039	2.2050	2.2061	2.2072
9.1	2.2083	2.2094	2.2105	2.2116	2.2127	2.2138	2.2148	2.2159	2.2170	2.2181
9.2	2.2192	2.2203	2.2214	2.2225	2.2235	2.2246	2.2257	2.2268	2.2279	2.2289
9.3	2.2300	2.2311	2.2322	2.2332	2.2343	2.2354	2.2364	2.2375	2.2386	2.2396
9.4	2.2407	2.2418	2.2428	2.2439	2.2450	2.2460	2.2471	2.2481	2.2492	2.2502
9.5	2.2513	2.2523	2.2534	2.2544	2.2555	2.2565	2.2576	2.2586	2.2597	2.2607
9.6	2.2618	2.2628	2.2638	2.2649	2.2659	2.2670	2.2680	2.2690	2.2701	2.2711
9.7	2.2721	2.2732	2.2742	2.2752	2.2762	2.2773	2.2783	2.2793	2.2803	2.2814
9.8	2.2824	2.2834	2.2844	2.2854	2.2865	2.2875	2.2885	2.2895	2.2905	2.2915
9.9	2.2925	2.2935	2.2946	2.2956	2.2966	2.2976	2.2986	2.2996	2.3006	2.3016

TABLE D POWERS OF BASE E

x	e^x	e^{-x}	x	e^x	e^{-x}
0.00	1.0000	1.0000	0.45	1.5683	0.6376
0.01	1.0101	0.9901	0.46	1.5841	0.6313
0.02	1.0202	0.9802	0.47	1.6000	0.6250
0.03	1.0305	0.9704	0.48	1.6161	0.6188
0.04	1.0408	0.9608	0.49	1.6323	0.6126
0.05	1.0513	0.9512	0.95	2.5857	0.3867
0.06	1.0618	0.9418	0.96	2.6117	0.3829
0.07	1.0725	0.9324	0.97	2.6379	0.3791
0.08	1.0833	0.9231	0.98	2.6645	0.3753
0.09	1.0942	0.9139	0.99	2.6912	0.3716
0.10	1.1052	0.9048	0.50	1.6487	0.6065
0.11	1.1163	0.8958	0.51	1.6653	0.6005
0.12	1.1275	0.8869	0.52	1.6820	0.5945
0.13	1.1388	0.8781	0.53	1.6989	0.5886
0.14	1.1503	0.8694	0.54	1.7160	0.5827
0.15	1.1618	0.8607	0.55	1.7333	0.5770
0.16	1.1735	0.8521	0.56	1.7507	0.5712
0.17	1.1853	0.8437	0.57	1.7683	0.5655
0.18	1.1972	0.8353	0.58	1.7860	0.5599
0.19	1.2093	0.8270	0.59	1.8040	0.5543
0.20	1.2214	0.8187	0.60	1.8221	0.5488
0.21	1.2337	0.8106	0.61	1.8404	0.5434
0.22	1.2461	0.8025	0.62	1.8589	0.5379
0.23	1.2586	0.7945	0.63	1.8776	0.5326
0.24	1.2713	0.7866	0.64	1.8965	0.5273
0.25	1.2840	0.7788	0.65	1.9155	0.5220
0.26	1.2969	0.7711	0.66	1.9348	0.5169
0.27	1.3100	0.7634	0.67	1.9542	0.5117
0.28	1.3231	0.7558	0.68	1.9739	0.5066
0.29	1.3364	0.7483	0.69	1.9937	0.5016
0.30	1.3499	0.7408	0.70	2.0138	0.4966
0.31	1.3634	0.7334	0.71	2.0340	0.4916
0.32	1.3771	0.7261	0.72	2.0544	0.4868
0.33	1.3910	0.7189	0.73	2.0751	0.4819
0.34	1.4050	0.7118	0.74	2.0959	0.4771
0.35	1.4191	0.7047	0.75	2.1170	0.4724
0.36	1.4333	0.6977	0.76	2.1383	0.4677
0.37	1.4477	0.6907	0.77	2.1598	0.4630
0.38	1.4623	0.6839	0.78	2.1815	0.4584
0.39	1.4770	0.6771	0.79	2.2034	0.4538
0.40	1.4918	0.6703	0.80	2.2255	0.4493
0.41	1.5068	0.6637	0.81	2.2479	0.4449
0.42	1.5220	0.6570	0.82	2.2705	0.4404
0.43	1.5373	0.6505	0.83	2.2933	0.4360
0.44	1.5527	0.6440	0.84	2.3164	0.4317

Table D, cont'd

x	e^x	e^{-x}	x	e^x	e^{-x}
0.85	2.3397	0.4274	2.55	12.8071	0.0781
0.86	2.3632	0.4232	2.60	13.4637	0.0743
0.87	2.3869	0.4190	2.65	14.1540	0.0707
0.88	2.4109	0.4148	2.70	14.8797	0.0672
0.89	2.4351	0.4107	2.75	15.6426	0.0639
0.90	2.4596	0.4066	2.80	16.4446	0.0608
0.91	2.4843	0.4025	2.85	17.2878	0.0578
0.92	2.5093	0.3985	2.90	18.1741	0.0550
0.93	2.5345	0.3946	2.95	19.1060	0.0523
0.94	2.5600	0.3906	3.00	20.0855	0.0498
1.00	2.7183	0.3679	3.05	21.1153	0.0474
1.05	2.8577	0.3499	3.10	22.1979	0.0450
1.10	3.0042	0.3329	3.15	23.3361	0.0429
1.15	3.1582	0.3166	3.20	24.5325	0.0408
1.20	3.3201	0.3012	3.25	25.7903	0.0388
1.25	3.4903	0.2865	3.30	27.1126	0.0369
1.30	3.6693	0.2725	3.35	28.5027	0.0351
1.35	3.8574	0.2592	3.40	29.9641	0.0334
1.40	4.0552	0.2466	3.45	31.5004	0.0317
1.45	4.2631	0.2346	3.50	33.1154	0.0302
1.50	4.4817	0.2231	3.55	34.8133	0.0287
1.55	4.7115	0.2122	3.60	36.5982	0.0273
1.60	4.9530	0.2019	3.65	38.4746	0.0260
1.65	5.2070	0.1921	3.70	40.4473	0.0247
1.70	5.4739	0.1827	3.75	42.5210	0.0235
1.75	5.7546	0.1738	3.80	44.7011	0.0224
1.80	6.0496	0.1653	3.85	46.9930	0.0213
1.85	6.3598	0.1572	3.90	49.4024	0.0202
1.90	6.6859	0.1496	3.95	51.9353	0.0193
1.95	7.0287	0.1423	4.00	54.5981	0.0183
2.00	7.3891	0.1353	4.05	57.3974	0.0174
2.05	7.7679	0.1287	4.10	60.3402	0.0166
2.10	8.1662	0.1225	4.15	63.4339	0.0158
2.15	8.5849	0.1165	4.20	66.6863	0.0150
2.20	9.0250	0.1108	4.25	70.1053	0.0143
2.25	9.4877	0.1054	4.30	73.6997	0.0136
2.30	9.9742	0.1003	4.35	77.4783	0.0129
2.35	10.4856	0.0954	4.40	81.4508	0.0123
2.40	11.0232	0.0907	4.45	85.6268	0.0117
2.45	11.5883	0.0863	4.50	90.0170	0.0111
2.50	12.1825	0.0821	4.55	94.6323	0.0106

ANSWERS TO ODD-NUMBERED EXERCISES

Exercise Set 1.1
(page 6)

1. a. True b. True c. False d. False   3. a. 0 b. 8 c. 0
5. a. $2 \cdot 3 \cdot 3 \cdot 3$ b. Prime c. $2 \cdot 2 \cdot 3 \cdot 3 \cdot 5$

Exercise Set 1.2
(page 12)

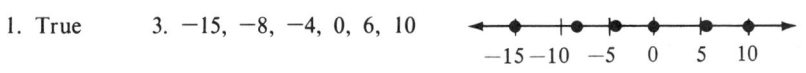

1. True   3. $-15, -8, -4, 0, 6, 10$

5. a. -12 b. 7 c. 0   7. a. $4, -4$ b. No such integers   9. a. -9 b. 0
11. a. The sum is -5.

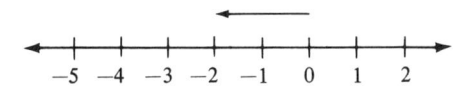

  b. The sum is -1.

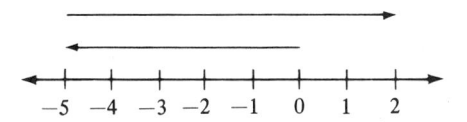

13. a. -5 b. 8 c. -10 d. 7 e. -88 f. 72 g. 4 h. -16

Exercise Set 1.3
(page 25)

1. True   3. $-\frac{1}{3}, -1, -0.8, 0, \frac{5}{6}, 1.5, \frac{11}{4}, 3$

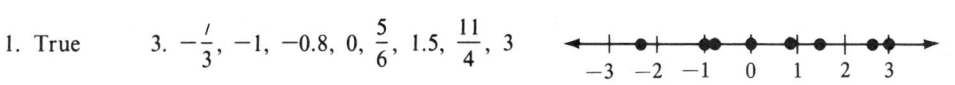

5. a. $\frac{3}{7}$ b. $-\frac{20}{21}$ 7. a. -36 b. 80 c. $\frac{14}{16}, \frac{21}{24}, \frac{28}{32}$ 9. a. $\frac{3}{4}$ b. $-\frac{1}{8}$ c. 1
11. a. $-\frac{7}{5}$ b. $\frac{1}{21}$ c. $-\frac{15}{56}$ d. 0 e. $\frac{5}{8}$ f. $-\frac{5}{12}$ g. $\frac{3}{2}$
13. a. -3.288 b. -3.431 c. -43.22016 d. 7.4
15. a. 2.375 b. $-0.181818\ldots$ c. 0.1875   17. a. $\frac{3}{40}$ b. $\frac{1}{2000}$ c. $\frac{2}{3}$

Exercise Set 1.4
(page 29)

1. a. Rational, real   b. Irrational, real   c. Rational, real   d. Irrational, real
   e. Rational, real   f. Irrational, real
3. a. 8   b. 10   c. $\frac{1}{2}$ d. $\frac{3}{4}$ e. 0.1   f. 0.7
5. a. Irrational   b. Rational   c. Rational   d. Irrational   e. Nonreal   f. Irrational

Exercise Set 1.5
(page 37)

1. a. 216   b. −32   c. $\frac{16}{81}$ d. 0.25   e. −256   f. 75   3. a. 10^5 b. 10^{-4}
5. a. 0.00001   b. 10,000   7. a. 3.2   b. −13.259   c. 0.020125   d. −5,400,000
9. a. 9.563×10^3 b. -4.95×10^1 c. 3.62×10^{-2} d. 8.91×10^5
   e. 1.8×10^{-2} f. 7.602×10^{-3}
11. a. 1.15×10^{-2} or 0.0115   b. -5.2×10^3 or −5200   c. 2.119×10^6 or 2,119,000
    d. 3.1125×10^{-1} or 0.31125   e. 8.9408×10^3 or 8940.8   f. 5×10^1 or 50
    g. 9.85×10^{-3} or 0.00985

Exercise Set 1.6
(page 46)

1. a. 0.0581   b. 3.49   c. 0.0075   3. a. $\frac{27}{50}$ b. $\frac{93}{1000}$ c. $\frac{33}{25}$ or $1\frac{8}{25}$
5. a. 7.6%   b. $33\frac{1}{3}$%   c. 162.5%   d. 0.05%   7. a. 139.2   b. 3.24
9. a. 59%   b. 250%   11. a. 120   b. 60   13. $\frac{2}{9}$ or 2:9   15. a. 24   b. 2.5
17. 5 ft   19. 182 mi

Exercise Set 2.1
(page 60)

1. Step 1: Find a measuring instrument, probably a ruler.
   Step 2: Place the ruler along the edge of the page and record the position of each end of the page. Subtract the smaller from the larger value to obtain the length.
   Step 3: Record the answer, for example, $10\frac{15}{16}$ in. (a number and a unit).

ANSWERS TO ODD-NUMBERED EXERCISES **A43**

3.

Quantity	Dimension	English	SI
time	time	second	second
acceleration	length/time^2	ft/s^2	m/s^2
energy	mass-length^2 / time^2	ft-lb	kg-m^2/s^2
mass	mass	slug	kg
momentum	mass-length / time	slug-ft/s	kg-m/s

5. kg/m^3 7. a. mass · length/time b. slug · ft/s
9. a. derived; length/time b. derived; mass/(length · time^2) c. basic; length d. derived; length^2
11. a. 3.6 kg b. 675 N c. 40 slugs; 1333 lb
13. a. 11.43 m b. 31.188 J c. 700.625 ft-lb/s d. 24.948 kg e. 56.7952 ft-lb

Exercise Set 2.2
(page 67)

1. a. 26.4 µm b. 3.81 mℓ c. 0.05 Mton d. 1.512 GHz e. 550 pF
   f. 22 × 10^3 pF (22,000 pF) g. 22.5 m^2
3. a. 3.1 µm b. 32 pF c. 1.6 km d. 42 mℓ
5. a. 340 × 10^{-3} m or 3.4 × 10^{-1} m b. 5.5 × 10^6 Hz c. 45 × 10^{-12} s or 4.5 × 10^{-11} s
   d. 0.32 × 10^{-2} m or 3.2 × 10^{-3} m

Exercise Set 2.3
(page 75)

1. a. 599.55 b. 21.6 c. 7.6 d. 261.625
3. a. 1.5497 −07 b. 1.52 09 c. 3.6 07 d. −2.14 −03 e. 1.521 06
   f. −2.136 −04 g. 1.59 09 h. 5.92 −11
5. a. 2. −10 b. 9.6 10 c. 3.15 −06 d. 4.5 −02 or 0.045 e. 6.5 −02 or 0.065
   f. 2.3 03 or 2300
7. a. 1.42 × 10^8 b. 4.843 × 10^{-20} c. −1.7582 × 10^{11} d. 1.246 × 10^{-10} e. 2.475 × 10^{-9}

Exercise Set 2.4
(page 82)

1. a. 11 b. 30<u>0</u>0 c. 7.3 × 10^{10} d. 2.27 × 10^{-7} e. 7.2 × 10^{-4}
3. a. 2.241 × 10^{-18} b. 6.700 × 10^{-27} c. 0.020 or 2.0 × 10^{-2} d. 2.10 e. 1.2 × 10^8
5. a. 2.52 × 10^{-3} b. 21<u>4</u>0 c. 9.2 × 10^{-31} d. 3.2 × 10^7 e. 7.220
7. a. 17<u>0</u>00 b. 4.56 c. 0.01700 d. 6.52 × 10^4 e. 34.0

Exercise Set 3.1
(page 94)

1. a. Ray $\vec{CD}$ b. Angle PQR ($\angle PQR$) or angle Q ($\angle Q$) c. Line segment $\overline{LM}$
3. a. Angle A and angle C are acute; angle B is a right angle.
   b. Angle E and angle F are acute angles; angle D is an obtuse angle.
5. a. ℓ_1 and ℓ_2 are parallel lines ($\ell_1 \parallel \ell_2$). b. $\angle ABC$ and $\angle CBD$ are supplementary angles.
   c. ℓ_1 and ℓ_2 are intersecting lines.
7. a. Triangle b. Rectangle c. Trapezoid
9. a. Right triangle b. Parallelogram c. Isosceles triangle 11. a. True b. False
13. Corresponding angles: angle A and angle D; angle B and angle E; angle C and angle F
    Corresponding sides: $\overline{AB}$ and $\overline{DE}$; $\overline{BC}$ and $\overline{EF}$; $\overline{CA}$ and $\overline{FD}$
15. No, they are not congruent. Yes, they are similar.

Exercise Set 3.2
(page 110)

1. 26.0 in. 3. 0.50 in. 5. 1350 ft^2 (1400 ft^2 to two significant digits)
7. 1984 ft^2 (2000 ft^2 to two significant digits) 9. 120° 11. 94.5 in. 13. 20 × 20 ft

Exercise Set 3.3
(page 116)

1. a. 40.2 ft b. 31 mm c. 3.2×10^4 m
   (*Note:* The addition rule was used in determining the number of significant digits.)
3. a. 99.2 in.^{2} b. 5.3×10^{-6} m^2 c. 710 ft^2
5. a. 1.3×10^{-2} m b. 59.173 ft c. 11 ft d. 4.0×10^6 m 7. 38 in. 9. 804 in.^{2}

Exercise Set 3.4
(page 126)

1. 8.56 in. 3. 17 ft^2 5. 37 gal 7. $28.30 9. 19.4 in.^{3}
11. Pay for a yard (at least 30 ft^3 in pile).

Exercise Set 4.1
(page 134)

1. a. 2 and 3 are constants; a, b, and c are variables. b. 6 is a constant; x and y are variables.
3. a. $12xy - 5z$ b. $2ac + 18b$ 5. a. $7a^5$ b. $x^2y^3zw^2$
7. a. Algebraic term b. Algebraic expression c. Algebraic expression d. Algebraic term

ANSWERS TO ODD-NUMBERED EXERCISES

Exercise Set 4.2
(page 139)

1. 1   3. −5   5. 21   7. 0   9. 47.5

Exercise Set 4.3
(page 142)

1. a. Binomial   b. Trinomial   c. Monomial   d. Trinomial
3. a. Fourth   b. Third   c. Eighth   d. Ninth

Exercise Set 4.4
(page 149)

1. True   b. True   c. False   d. True
3. a. $-30ab$ b. $\frac{1}{2}x$ c. $-3y + 1$ d. $10x - 7$
5. a. $21x - 2$ b. $9x - 5y - 3xy$ c. $18x^4 + 4x^2 + 25$ d. 0   e. $-4x^3 + 4x^2 + 7x + 2$
7. a. $-3x^2 + x - 3$ b. $-5x^4 + 9x^3$ c. $-2ax - 2by$ d. $x^2 + x - 4$ e. $-x^2 - x - 1$
9. $2x^3 - 3x^2 + 8x + 5$ 11. $6x^2 + x + 9$ 13. $4x^3 + 4x^2 + 5x + 3$ 15. $-3x^3 + 3x - 17$
17. $-a + 2b + 2c$

Exercise Set 4.5
(page 155)

1. a. 1   b. $-\frac{1}{216}$ c. x^7 d. $625x^4$ e. 243   f. y^8 g. 36   h. $\frac{a^4}{256}$
3. a. $15x^4y^5z$ b. $-4a^5b^9c^{13}$ c. $\frac{a^8}{6b^3}$ d. $-\frac{w^3}{3}$ e. $\frac{3x^4}{10y^4}$

Exercise Set 4.6
(page 162)

1. $15x^5 + 9x^4 + 12x^3$ 3. $10x^6y^2z^2 - 20xz^3$ 5. $-9x^4 + 6x^3 - 15x^2 + 12x$
7. $-8x^6 + 10x^5 + 12x^4$ 9. $2x^5y - 11x^3y + 4x^2y + 12xy - 16y$ 11. $2x^5 - 10x^3 + 10x^2 + 12x - 20$
13. $5x^6 - 13x^4 + 11x^2 - 10$ 15. $6x^4 - 23x^3 + 48x^2 - 49x + 30$ 17. $x^3 - 27$
19. $x^2 + 16x + 64$ 21. $x^2 - 256y^2$ 23. $x^3 - 12x^2 + 48x - 64$

Exercise Set 5.1
(page 172)

1. $-5x(5x - 2)$ 3. $3(x^2 - x + 3)$ 5. $7(x - 3)(x + 1)$ 7. $(x + 10)(x + 2)$

9. $(x-9)(x+2)$ 11. $(x-4)(x-4)$ or $(x-4)^2$ 13. $(2x+5)(3x+1)$ 15. $2(2x+3)(x-2)$
17. $(3x-2)(x+1)$ 19. Prime—cannot be factored

Exercise Set 5.2
(page 177)

1. $(x+5)(x^2-5x+25)$ 3. $(x+6)(x-6)$ 5. $(x-3)^2$ 7. $(x+1)^3$ 9. $(x-6)^3$
11. $(x+9)^2$ 13. $(x-4)(x^2+4x+16)$ 15. Prime—cannot be factored
17. $(x+7)(x^2-7x+49)$ 19. $(x+5)^2$

Exercise Set 5.3
(page 179)

1. a. Algebraic fraction   b. Polynomial   c. Algebraic fraction   d. Algebraic fraction
3. a. $P = 7x^2 + 1$; $Q = 3x$ b. $P = 3x^2 + 5x - 2$; $Q = 1$ c. $P = 1$; $Q = x^2 + 2$

Exercise Set 5.4
(page 183)

1. $\dfrac{x^2 - 3x + 9}{x - 1}$ 3. $\dfrac{3}{2x - 3}$ 5. -1 7. $\dfrac{x+2}{2}$ 9. $\dfrac{-2x - 14}{x^2 - 2x + 4}$

Exercise Set 5.5
(page 186)

1. $\dfrac{x-4}{4x}$ 3. -1 5. $\dfrac{x+3}{4}$ 7. $\dfrac{x^2 + x + 1}{x^2 - x + 1}$ 9. $\dfrac{x+1}{x}$

Exercise Set 5.6
(page 192)

1. $\dfrac{2(x^2+1)}{(x+1)(x-1)}$ 3. $\dfrac{2(x^2 - 4x - 2)}{(x-5)(x+5)(x-3)}$ 5. $\dfrac{2x^2 + 3x + 10}{(x-2)(x+3)(x+4)}$ 7. $\dfrac{-4(x+6)}{(x+3)(x+2)(x-3)}$
9. $\dfrac{-x+11}{(x+5)(x-3)(x+2)}$ 11. $\dfrac{x(28x^2 - 89x + 105)}{(3x+4)(3x-4)(x+3)(x-3)}$

Exercise Set 6.1
(page 198)

1. a. Yes   b. No   c. Yes   d. No   3. a. No   b. Yes   c. Yes

Exercise Set 6.2
(page 206)

1. $x = -14$
3. $x = 16$
5. $x = 0$
7. $x = -3$
9. $x = 8$
11. $x = -2$
13. $x = 5$
15. $x = 0.5$
17. $x = 3$
19. $x = 1.5$

Exercise Set 6.3
(page 217)

1. 13 and 22
3. 11 and 46
5. 9, 12, and 17
7. 40 ft; 80 ft; 60 ft
9. 15 ft × 60 ft
11. 49°; 49°; 82°
13. 108 oz
15. 200 lb of 25% alloy; 100 lb of 40% alloy
17. 18 lb of $2.75 nuts; 12 lb of $1.75 nuts
19. $\frac{2}{3}$ hr (40 min)
21. 40 mph; 55 mph
23. $\frac{2}{5}$ hr (24 min) in the car; $\frac{1}{5}$ hr (12 min) in the bus

Exercise Set 6.4
(page 222)

1. a. Yes   b. No   c. No   d. Yes   e. No
3. a. No   b. No   c. Yes

Exercise Set 6.5
(page 228)

1. $x \leq -14$ 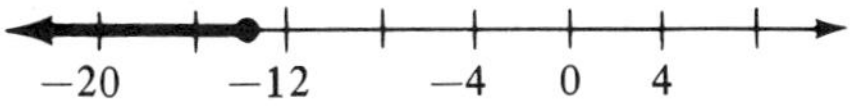
3. $x < 16$
5. $x < 0$
7. $x \geq -3$ 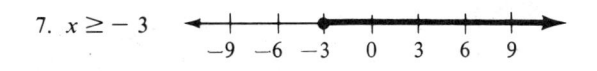
9. $x \geq 8$ 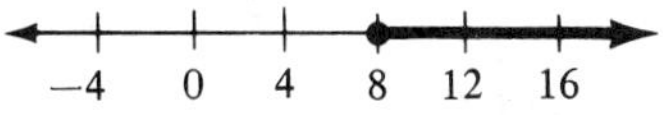
11. $x > -2$ 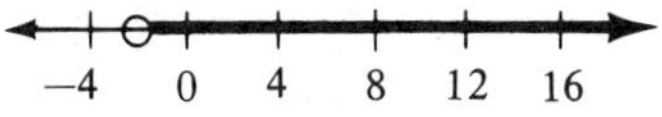
13. $x > 5$ 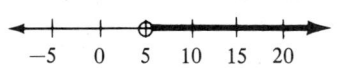
15. $x \leq 0.5$ 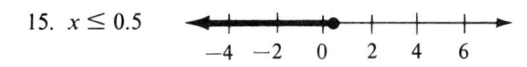
17. $x \geq 3$ 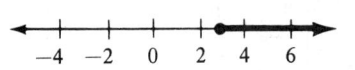
19. $x < 1.5$

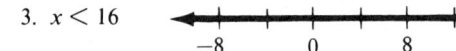

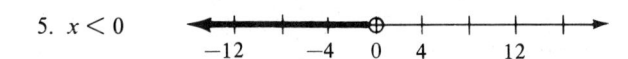

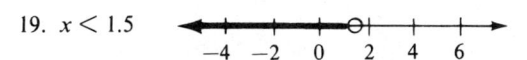

Exercise Set 7.1
(page 240)

1. a. $(-2, 8), (-1, 4), (0, 0), (1, -4), (2, -8)$ b. $(-5, 0), (-3, 2), (0, 5), (3, 8), (5, 10)$
   c. $(-2, -8), (-1, -1), (0, 0), (1, 1), (2, 8)$
3. a. Yes   b. Yes   c. Yes   d. No
5. a. $-8, -2, 4, 7, 13$ b. $11, -1, -5, -4, 4$ c. $5, 5, 5, 5, 5$ d. $-2, 0, 2, 3, 5$
7. a. $(-4, -8), (-2, -2), (0, 4), (1, 7), (3, 13)$ b. $(-4, 11), (-2, -1), (0, -5), (1, -4), (3, 4)$
   c. $(-4, 5), (-2, 5), (0, 5), (1, 5), (3, 5)$ d. $(-4, -2), (-2, 0), (0, 2), (1, 3), (3, 5)$
9. Yes

Exercise Set 7.2
(page 245)

1. a. Yes   b. No   c. Yes
3. a. $(-1, -12.5), (0, -10), (2, -5)$ b. $(-1, 5), (0, 2), (1, -1)$ c. $(-2, 16), (0, 12), (2, 8)$
   d. $(-1, 3), (0, 3), (1, 3)$
5. $y = 8 - 3x$, $(-2, 14), (0, 8), (1, 5), \left(\frac{8}{3}, 0\right)$ 7. $y = -2 + \frac{5}{4}x$, $(-4, -7), (0, -2), \left(\frac{8}{5}, 0\right), (4, 3)$
9. $y = 2 - \frac{2}{3}x$, $(-3, 4), (0, 2), \left(\frac{3}{2}, 1\right), (3, 0)$

Exercise Set 7.3
(page 248)

1. a.

x	-1	0	1	2	3
y	-1	1	3	5	7

b.

x	-3	-2	0	1	5
y	-3	-2	0	1	5

3.

x	0.00	0.05	0.10	0.15	0.20	0.25	0.30	0.35	0.40
y	0.0	0.4	0.8	1.2	1.6	2.0	2.4	2.8	3.2

5.

t	0	1	2	3	4	5	6
y	0	284	536	756	944	1100	1224
t	7	8	9	10	11	12	13
y	1316	1376	1404	1400	1364	1296	1196

7.

x	0	1	2	3	4	5	6
v	30	25	20	15	10	5	0
x	7	8	9	10	11	12	
v	−5	−10	−15	−20	−25	−30	

Exercise Set 7.4
(page 255)

1.

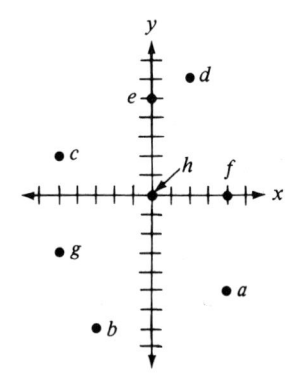

3. $A(2, 4)$, $B(-1, 1)$, $C(-4, 0)$, $D(-3, -2)$, $E(0, -2)$, $F(5, -3)$

5. a. b. 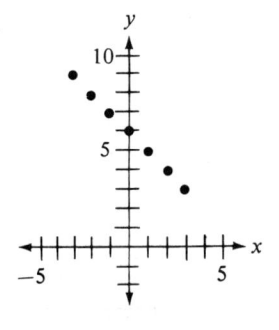 c.

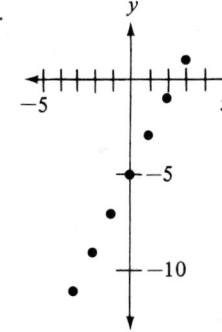

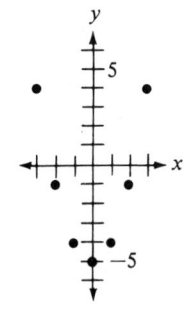

7. a. 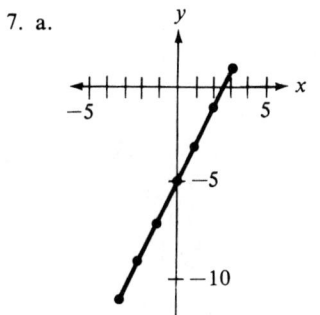 b. 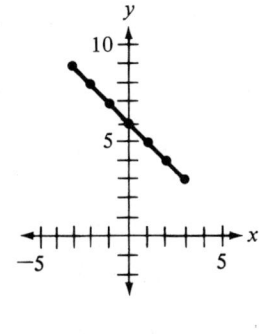 c.

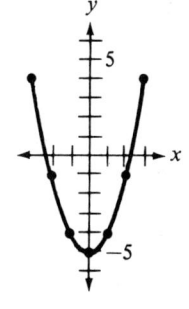

9. a. [graph] b. [graph] c. [graph]

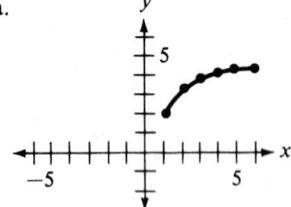

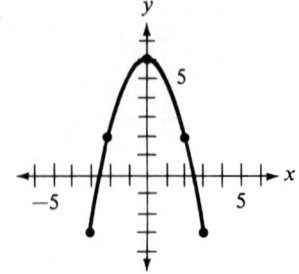

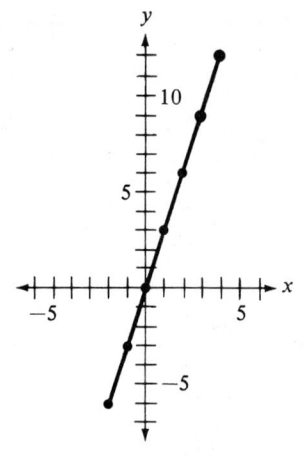

Exercise Set 7.5
(page 259)

1.

x	−3	0	3
y	0	3	6

[graph]

3.

x	−3	0	3
y	$-\frac{24}{5}$	−3	$-\frac{6}{5}$

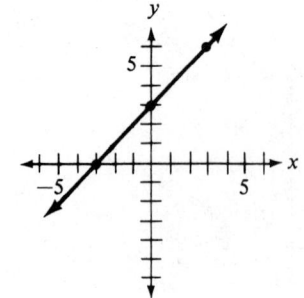

[graph]

5.

x	−3	0	3
y	$\frac{4}{3}$	$\frac{4}{3}$	$\frac{4}{3}$

[graph]

7.

x	−4	−4	−4
y	−3	0	3

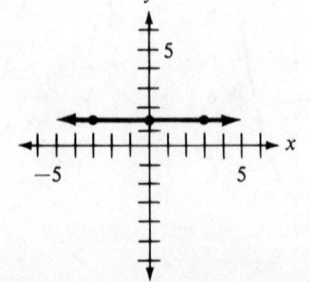

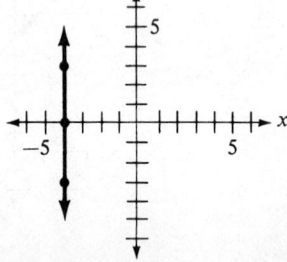

[graph]

ANSWERS TO ODD-NUMBERED EXERCISES **A51**

9.

x	−3	0	3
y	8	4	0

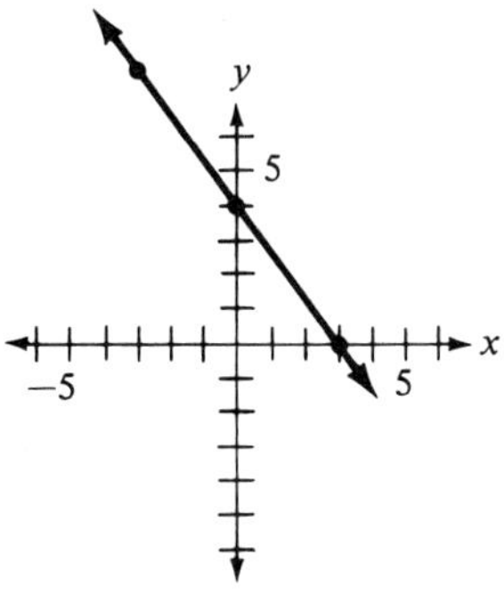

Exercise Set 7.6
(page 268)

1. a. Intercepts: (0, −4), (7, 0)　　b. Intercepts: $\left(\frac{1}{2}, 0\right)$, no y-intercept

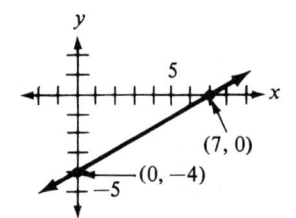

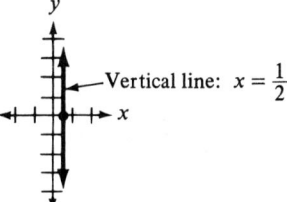

　Vertical line: $x = \frac{1}{2}$

　c. Intercepts: (0, 0) is the only intercept.
　　Use (0, 0) and (1, 3).

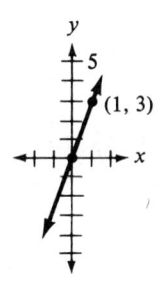

　d. Intercepts: (0, −3), no x-intercept

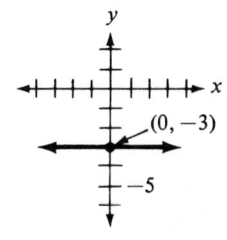

　e. Intercepts: (3, 0), $\left(0, \frac{4}{3}\right)$

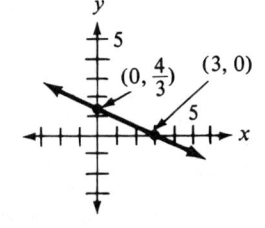

3. a. (6, 0), no y-intercept　　b. (0, −12), $\left(\frac{12}{5}, 0\right)$　　c. $\left(0, -\frac{9}{4}\right)$, no x-intercept　　d. (7, 0), (0, 2)
　e. (0, 0), x and y-intercepts are the same

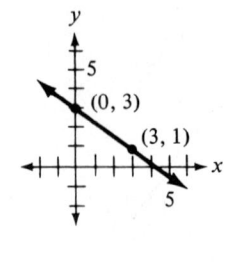

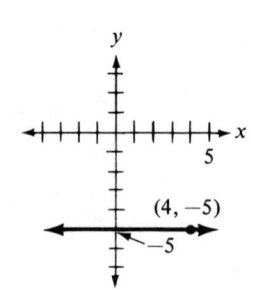

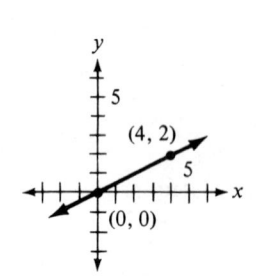

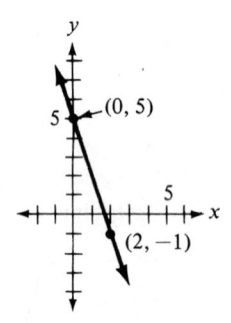

5. a. [graph: line through (−1, −2) and (0, 1)]
   b. [graph: line through (0, 3) and (3, 1)]
   c. [graph: horizontal line through (4, −5)]
   d. [graph: line through (0, 0) and (4, 2)]
   e. [graph: line through (0, 5) and (2, −1)]
   f. [graph: line through (−2, −3) and (2, 4)]

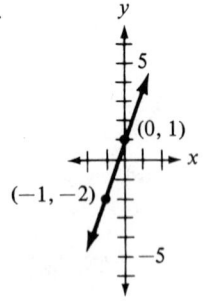

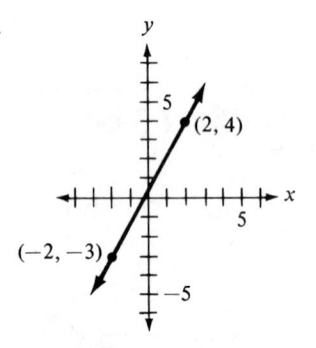

Exercise Set 7.7
(page 273)

1. a. $y = 3x + 6$ b. $y = 2$ c. $4x + 2y = 17$ d. $4x + 7y = -34$
3. a. $y = 3x + 6$, y-intercept is $(0, 6)$ b. $y = 0x + 2$, y-intercept is $(0, 2)$
   c. $y = -2x + \frac{17}{2}$, y-intercept is $\left(0, \frac{17}{2}\right)$ d. $y = -\frac{4}{7}x - \frac{34}{7}$, y-intercept is $\left(0, -\frac{34}{7}\right)$
5. a. $y = \frac{3}{7}x - 2$, slope $= \frac{3}{7}$; y-intercept $(0, -2)$ b. $y = -\frac{1}{2}x - 2$, slope $= -\frac{1}{2}$; y-intercept $(0, -2)$
   c. $y = -\frac{2}{7}x$, slope $= -\frac{2}{7}$; y-intercept $(0, 0)$ d. $y = 0x + \frac{8}{3}$, slope $= 0$; y-intercept $\left(0, \frac{8}{3}\right)$
   e. $y = \frac{4}{9}x + \frac{1}{3}$, slope $= \frac{4}{9}$; y-intercept $\left(0, \frac{1}{3}\right)$
7. a. $x + 2y = 10$ b. $x + 3y = 2$ c. $2x - 3y = 0$ d. $y = 2$ e. $x = 3$

Exercise Set 7.8
(page 280)

1. a. $\frac{1}{8}$ b. $\frac{3}{4}$ c. 26.7   3. a. 2.9   b. 18.9   c. 9.76×10^{-4}
5. a. z varies directly as the second power of x and inversely as the square root of y.
   b. A varies directly as the first power of r and directly as the first power of h.
   c. y varies inversely as the square of x.
7. a. $A = kbh$ b. $V = kr^2h$ c. $z = x^3/\sqrt{y}$

ANSWERS TO ODD-NUMBERED EXERCISES

Exercise Set 8.1
(page 288)

This exercise set requires constructions to scale. No solutions are given.

Exercise Set 8.2
(page 296)

(All results obtained using tables to the nearest degree)
1. a. 4.6, 9.5, 9.5, 10.6   b. y, x, y, z 3. a. $\frac{5.3}{6.7}, \frac{4.1}{6.7}, \frac{4.1}{5.3}$ b. $\frac{y}{z}, \frac{x}{z}, \frac{x}{y}$
5. a. 52°   b. 38°   c. 67°   d. 54°   7. a. 65°   b. 28°   c. 57°
9. a. 57°   b. 20°   c. 42°

Exercise Set 8.3
(page 301)

(All results obtained using tables to the nearest degree. Answers may vary slightly depending on the function used in the calculation.)
1. a. $A = 29°$, $B = 61°$, $c = 8.3$ b. $A = 57°$, $B = 33°$, $c = 24.9$
3. a. $A = 22°$, $B = 68°$, $b = 9.74$ b. $A = 71°$, $B = 19°$, $b = 2.90$
5. a. $c = 12.3$, $b = 10.2$, $A = 34°$ b. $c = 6.1$, $a = 3.8$, $B = 52°$ c. $a = 9.0$, $b = 1.9$, $B = 12°$
   d. $A = 27°$, $B = 63°$, $c = 4.0$ e. $A = 24°$, $B = 66°$, $b = 8.7$

Exercise Set 8.4
(page 307)

1. (Values reported to the nearest ten-thousandths)

Angle	79°	8°	42°	59°	5°	30°
sin	0.9816	0.1392	0.6691	0.8572	0.0872	0.5
cos	0.1908	0.9903	0.7431	0.5150	0.9962	0.8660
tan	5.1446	0.1405	0.9004	1.6643	0.0875	0.5774

3. (Values reported to the nearest tenth of a degree)
   60.9°, 24.5°, 56.1°, 81.4°, 45.0°, 13.3°, 24.2°
5. (Answers rounded to nearest tenths)
   a. $c = 5.7$, $b = 2.9$, $B = 31°$ b. $b = 5.1$, $c = 9.6$, $A = 58°$ c. $B = 25.9°$, $A = 64.1°$, $a = 8.6$
   d. $A = 32.1°$, $B = 57.9°$, $c = 6.0$

Exercise Set 8.5
(page 311)

1. 10.2 ft (rounded down)   3. 38.7°   5. 26.2 ft (rounded up)   7. 1384 m
9. 36.7 ft (rounded up)   11. 14.15 cm   13. 13.03 cm   15. 58 ft and 61 ft

Exercise Set 8.6
(page 319)

1. a. Law of cosines   b. Law of sines   c. Law of cosines   d. Law of sines
3. a. 11.75   b. $x = 10.53$, $y = 11.19$ c. $x = 6.77$ or $x = 10.84$ d. $x = 5.37$
5. a. $C = 26.7°$, $B = 57.7°$, $A = 95.6°$ b. $C = 74.8°$, $A = 60.2°$ c. $B = 86.1°$, $C = 59.9°$
   d. $B = 90°$, $A = 60°$
7. 9.90 mi and 14.61 mi   9. $x = 148$ ft, $y = 193$ ft

Exercise Set 8.7
(page 325)

1. (N 517.356, E 958.337)
3. Find $BD = 667.129$ m, bearing B to $D =$ N 53.88991 E, which gives $\Delta N = 393.165$ m, $\Delta E = 538.964$ m. These values added to the coordinates of B give the results in Problem 1 to within 1 mm.
5. Find $AC = 271.115$ m, bearing A to $C =$ S 43°50′ E, which gives $\Delta N = -195.571$, $\Delta E = 187.764$. Adding these to (N 297.672, E 49.835) yields the coordinates of C to within 1 mm of the previous result.

Exercise Set 9.1
(page 336)

1. a. Yes   b. No   3. a. No   b. Yes
5. a. $(3, -2)$ b. $\left(3, \frac{3}{2}\right)$ c. All (x, y) pairs satisfying either equation.

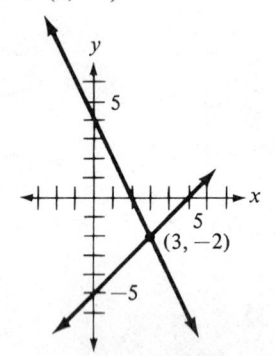

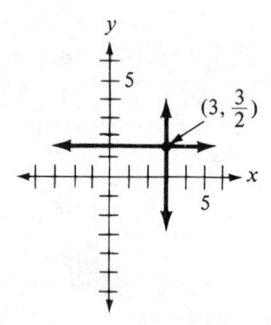

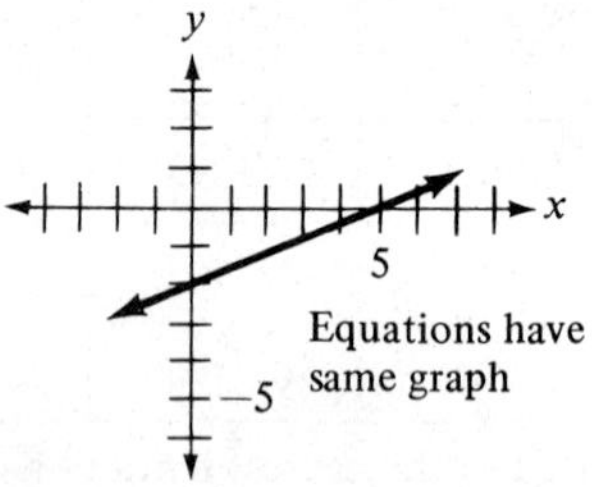

Exercise Set 9.2
(page 341)

1. a. $(2, 0)$ b. No solution   c. $(-3, 2)$ d. $(-2, 0)$
   e. All (x, y) pairs satisfying the equation $-4x + y = -9$
3. a. $(1, 1)$ b. All (x, y) pairs satisfying the equation $2y = x - 7$ c. $(1, -2)$ d. $(7, -2)$
   e. No solution

Exercise Set 9.3
(page 347)

1. $(-2, 0)$ 3. $(4, -2)$ 5. $(3, -2)$ 7. $(12, 6)$
9. All (x, y) pairs satisfying the equation $-4x + y = -9$ 11. No solution   13. $(2, 1)$
15. $\left(-\dfrac{9}{2}, -\dfrac{17}{5}\right)$ 17. All (x, y) pairs satisfying the equation $-4x + y = 2$ 19. $(7, -2)$

Exercise Set 9.4
(page 357)

1. 13, 22   3. 11, 46   5. 15 ft, 60 ft   7. 49°, 49°, 82°
9. 200 lb of 25% alloy, 100 lb of 40% alloy   11. 18 lb of $2.75 nuts, 12 lb of $1.75 nuts
13. 40 mph, 55 mph   15. $\dfrac{2}{5}$ hr (24 min) in the car, $\dfrac{1}{5}$ hr (12 min) in the bus

Exercise Set 9.5
(page 365)

1. a. $\begin{vmatrix} 4 & -3 \\ 1 & 0 \end{vmatrix}$ b. $\begin{vmatrix} 1 & -3 \\ 2 & 0 \end{vmatrix}$ c. $\begin{vmatrix} 1 & 4 \\ 2 & 1 \end{vmatrix}$ 3. -9
5. a. $x = \dfrac{1}{2}, y = 2, z = -1$ b. $x = -2, y = 2.8, z = -0.8$ 7. $i_1 = \dfrac{1}{5}, i_2 = 1, i_3 = \dfrac{4}{5}$

Exercise Set 10.1
(page 375)

1. 8   3. 17   5. 7   7. 13   9. $7\sqrt{2}$ 11. $8\sqrt{3}$ 13. $4\sqrt{2}$ 15. $12\sqrt{3}$
17. $9\sqrt{2}$ 19. $13\sqrt{3}$ 21. $3\sqrt[3]{2}$ 23. $y\sqrt{xy}$ 25. $\dfrac{3}{5}\sqrt{10}$ or $\dfrac{3\sqrt{10}}{5}$ 27. $\sqrt[6]{y^5}$
29. $\sqrt[4]{y^7}$ or $y\sqrt[4]{y^3}$ 31. $x^{5/2}$ 33. $y^{6/5}$ 35. $x^{5/3}y^{2/3}$

Exercise Set 10.2
(page 377)

1. $19\sqrt{3}$ 3. $-2\sqrt{7}$ 5. Cannot be combined   7. $31\sqrt{5}$ 9. $-4\sqrt{y}$ 11. $21\sqrt{17}$
13. $-14\sqrt{5}$ 15. $5\sqrt{x} - 5\sqrt{y}$ 17. 0   19. $-2x\sqrt{y}$

Exercise Set 10.3
(page 384)

1. $80\sqrt{21}$ 3. -180 5. $5\sqrt[3]{10} \cdot \sqrt{6}$ 7. $36xy\sqrt{x}$ 9. $30\sqrt{2} \cdot \sqrt[3]{2}$ 11. $\dfrac{2\sqrt{10} + \sqrt{5}}{15}$

13. $\dfrac{4\sqrt{5} - 3\sqrt{10}}{20}$ 15. $\dfrac{3\sqrt{x} + 6}{x - 4}$ 17. $\dfrac{2\sqrt{6} - \sqrt{3}}{3}$ 19. $2\sqrt{6}$

21. $\dfrac{3\sqrt{xy} + 3x - 2y\sqrt{x} - 2x\sqrt{y}}{y - x}$ 23. $\dfrac{13 + 5\sqrt{7}}{-2}$ or $\dfrac{-13 - 5\sqrt{7}}{2}$ 25. $\dfrac{\sqrt{x} - \sqrt{y} - x\sqrt{y} + y\sqrt{x}}{x - y}$

27. $0, 53.454, -13.114$ 29. $30x^{5/3} = 30x\sqrt[3]{x^2}$ 31. $3x^{11/12} = 3\sqrt[12]{x^{11}}$

33. $\dfrac{2x - x^{1/2}y^{1/2} - 3y}{y - x} = \dfrac{2x - \sqrt{xy} - 3y}{y - x}$

Exercise Set 10.4
(page 387)

1. a. -1 b. $8\sqrt{-1}$ or $8j$ c. 6 d. 1 e. -25 f. $-512\sqrt{-1}$ or $-512j$
3. a. $7j$ b. $3j$ c. $15j$ d. $3 - 5j$ e. $2 + 2j$ f. $3 + j$
5. a. $-3 + 4j$ b. $6 - 2j$ c. $-2 - 5j$ d. $10 - 7j$ e. $3 + 4j$ f. $-2 + 5j$

Exercise Set 10.5
(page 389)

1. $13 + 13j$ 3. $7 - j$ 5. $6 + 5j$ 7. $-7 + 15j$ 9. $14 + 16j$

Exercise Set 10.6
(page 391)

1. $36 + 8j$ 3. $-19 - 9j$ 5. 20 7. $8 - 56j$ 9. $50 - 10j$

Exercise Set 10.7
(page 392)

1. $-\dfrac{14}{13} - \dfrac{8j}{13}$ 3. $3 + 8j$ 5. $-1 + \dfrac{j}{3}$ 7. $\dfrac{11}{61} + \dfrac{60j}{61}$ 9. $\dfrac{89}{85} - \dfrac{18j}{85}$ 11. $\dfrac{128}{13} - \dfrac{16j}{13}$

Exercise Set 10.8
(page 395)

1. $|z| = \sqrt{61}$

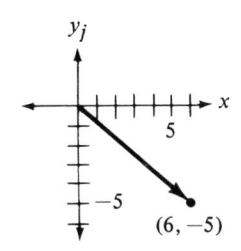

3. $|z| = 5\sqrt{5}$

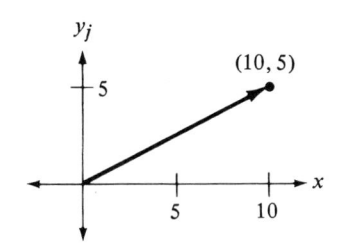

5. $|z| = 2\sqrt{5}$

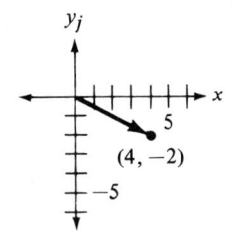

7. $|z| = 3\sqrt{2}$

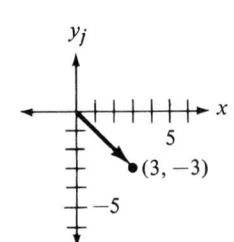

9. $|z| = \sqrt{41}$

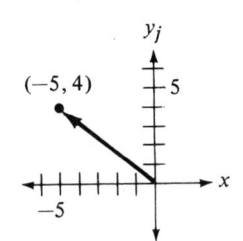

Exercise Set 10.9
(page 399)

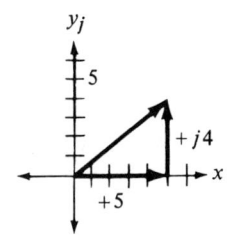

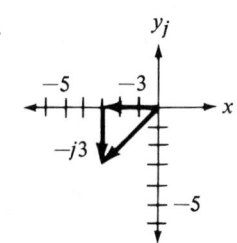

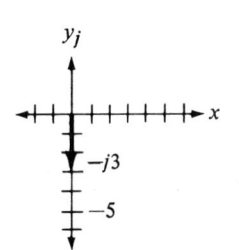

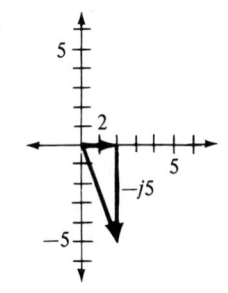

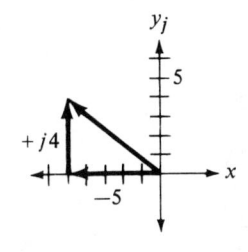

11. $5 - j2$

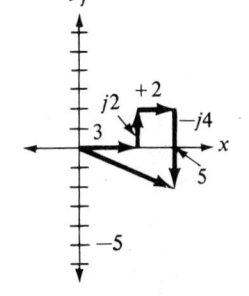

13. $-8 - j3$

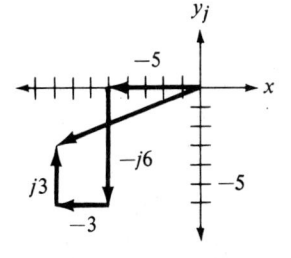

15. $j10$

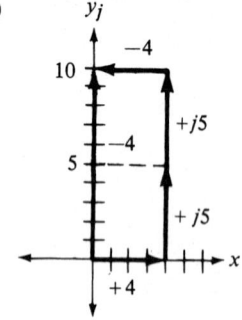

17. $3 - j4$

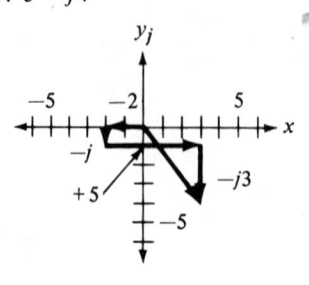

19. $3 + j7$

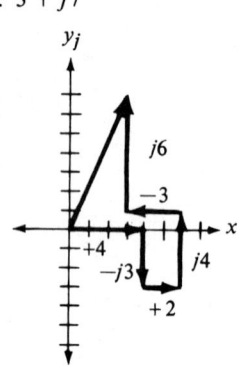

Exercise Set 11.1
(page 409)

1. $x = -11, x = 5$
3. $x = -11, x = 11$
5. $x = 0, x = 9$
7. $x = -6, x = 2$
9. $x = -7, x = 9$
11. $x = 2, x = \dfrac{5}{2}$
13. $x = -\dfrac{5}{2}, x = 1$
15. $x = \dfrac{4}{3}$
17. $x = \dfrac{1}{2}, x = \dfrac{3}{4}$
19. $x = -\dfrac{3}{2}, x = -\dfrac{5}{4}$

Exercise Set 11.2
(page 415)

1. $x = -\dfrac{3\sqrt{2}}{2}, x = \dfrac{3\sqrt{2}}{2}$
3. $x = -\dfrac{3}{2}, x = -\dfrac{5}{4}$
5. $x = \dfrac{5 + j\sqrt{5}}{3}, x = \dfrac{5 - j\sqrt{5}}{3}$
7. $x = -\dfrac{5}{2}, x = 1$
9. $x = \dfrac{5}{4}$

Exercise Set 11.3
(page 418)

1. $x = 1.25, x = -2$
3. $x = 7, x = -3$
5. $x = -1 + j\sqrt{3}, x = -1 - j\sqrt{3}$
7. $x = 0.2, x = -1.5$
9. $x = 1.25$

Exercise Set 12.1
(page 437)

1. Vertex: $\left(\frac{1}{2}, -\frac{25}{4}\right)$
   x-intercepts: $(3, 0), (-2, 0)$
   y-intercept: $(0, -6)$

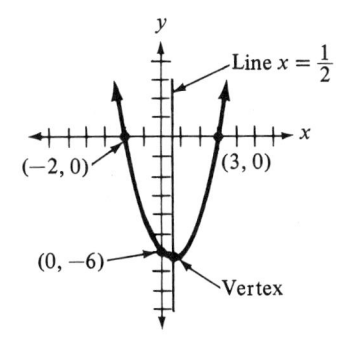

3. Vertex: $(-2, -16)$
   x-intercepts: $(-6, 0), (2, 0)$
   y-intercept: $(0, -12)$

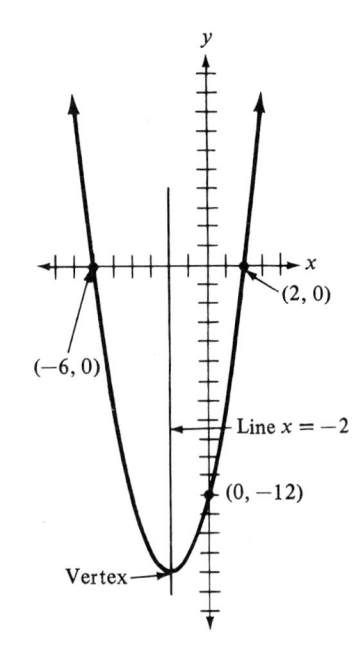

5. Vertex: $\left(-\frac{11}{4}, -\frac{49}{8}\right)$
   x-intercepts: $\left(-\frac{9}{2}, 0\right), (-1, 0)$
   y-intercept: $(0, 9)$

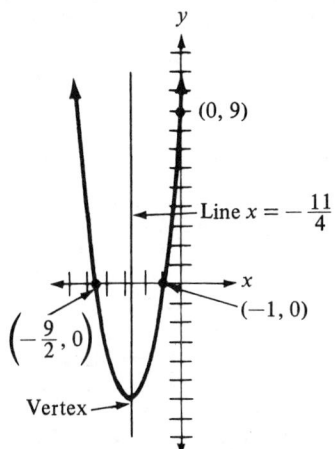

7. Vertex: $\left(\frac{1}{2}, \frac{49}{4}\right)$
   x-intercepts: $(-3, 0), (4, 0)$
   y-intercept: $(0, 12)$

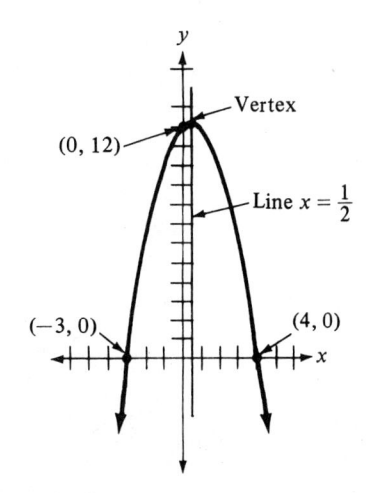

9. Vertex: $\left(-\dfrac{6}{5}, -\dfrac{16}{5}\right)$
   x-intercepts: $\left(-\dfrac{2}{5}, 0\right)$, $(-2, 0)$
   y-intercept: $(0, 4)$

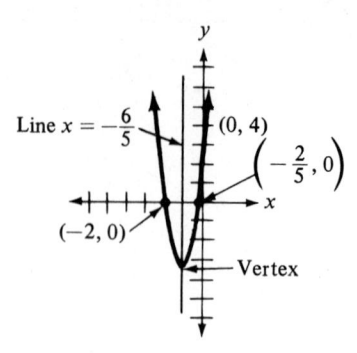

11. -4

Exercise Set 12.2
(page 444)

1. $\left(-\dfrac{4}{3}, -\dfrac{1}{9}\right)$, $(2, 1)$ 3. $(-1, 0)$, $\left(\dfrac{3}{2}, \dfrac{5}{4}\right)$ 5. $(1, -1)$

7. $\left(\dfrac{1 + \sqrt{5}}{2}, \dfrac{-5 + \sqrt{5}}{2}\right)$, $\left(\dfrac{1 - \sqrt{5}}{2}, \dfrac{-5 - \sqrt{5}}{2}\right)$ 9. $(\sqrt{7}, -11)$, $(-\sqrt{7}, -11)$ 11. $(1, -1)$, $(4, -1)$

13. $(0, 15)$ 15. No real solutions

Exercise Set 12.3
(page 457)

1. $x = 6.88$ in.
3. a. Times are 3.01 and 26.99. However, the latter corresponds to negative y. See graph.

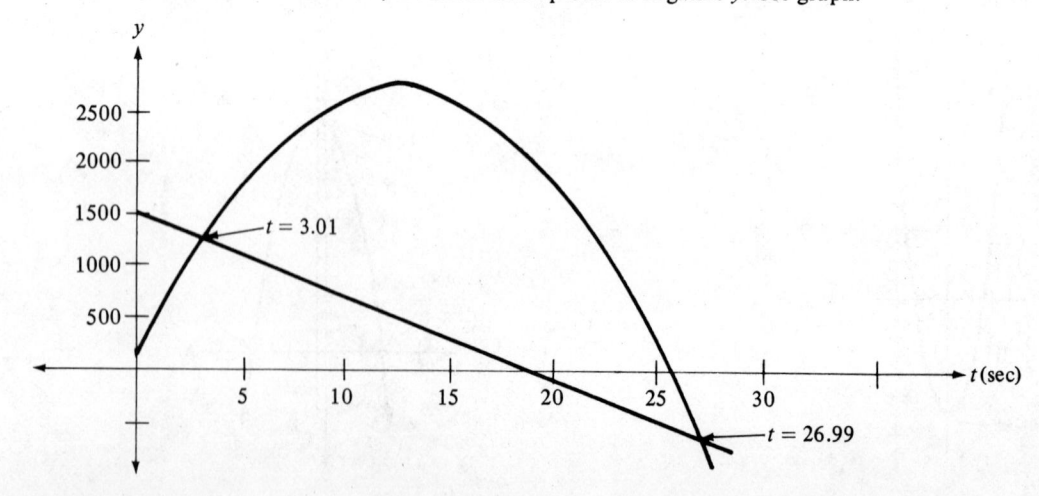

b. Times are 6.71 and 21.41. See graph.

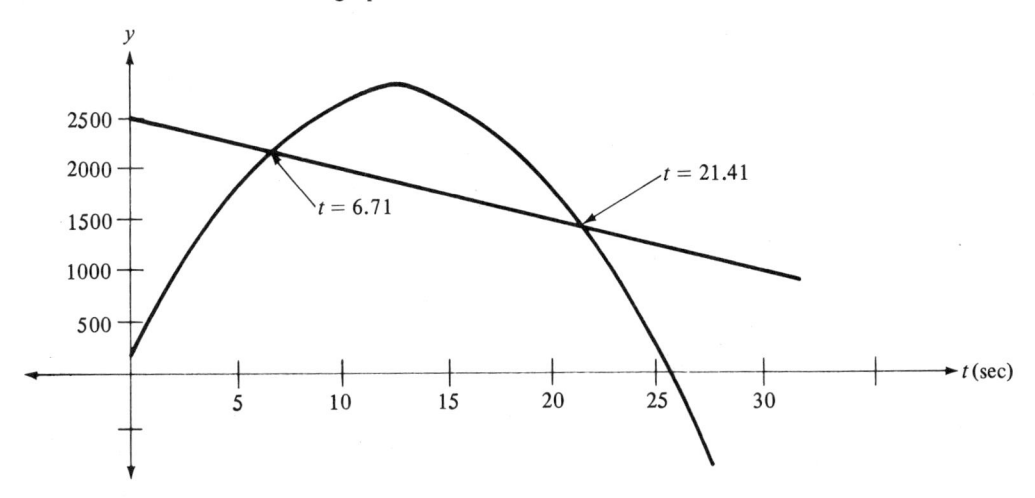

5. (13.65, 5.46). See graph.

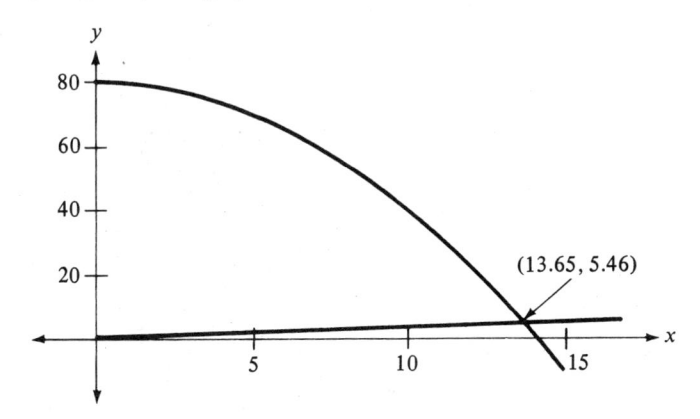

Exercise Set 13.1
(page 472)

1. a. [Initial side, Vertex, Terminal side]   b. [Vertex, Initial side, Terminal Side]   c. [Initial side, Vertex, Terminal side]

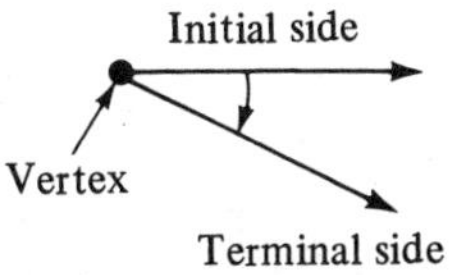

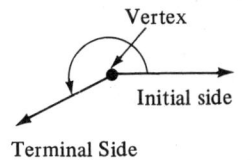

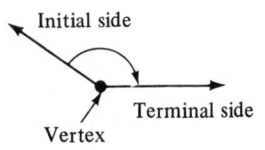

3. [Vertex, Initial Side, Terminal side — A negative angle]

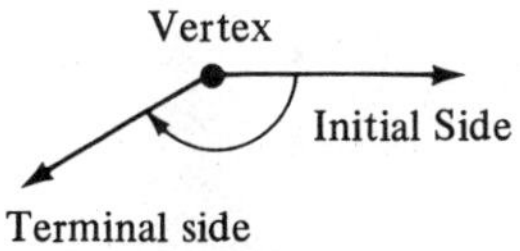

5. a. −   b. +   c. +

7. $-120°$, $-\dfrac{2\pi}{3}$ rad, $-\dfrac{1}{3}$ rotation

9.

Degree	Radian	Rotation
135°	$\frac{3\pi}{4}$	$\frac{3}{8}$
150°	$\frac{5\pi}{6}$	$\frac{5}{12}$
225°	$\frac{5\pi}{4}$	$\frac{5}{8}$
330°	$\frac{11\pi}{6}$	$\frac{11}{12}$
270°	$\frac{3\pi}{2}$	$\frac{3}{4}$
210°	$\frac{7\pi}{6}$	$\frac{7}{12}$
140°	$\frac{7\pi}{9}$	$\frac{7}{18}$

11.

s	r	θ
4 m	2 m	2 rad
2 m	$\frac{1}{\pi}$ m	2π rad
2π m	2 m	180°
$\frac{3\pi}{2}$ ft	3 ft	90°
π m	4 m	$\frac{\pi}{4}$ rad
π ft	6 ft	30°

13.

A	r	θ
$\frac{2\pi}{3}$ m²	2 m	60°
2.7 m²	2.12 m	1.2 rad
13.5 m²	3 m	3 rad
2π m²	4 m	45°

15. 21.2 lb   17. 6.28 ft

19.

v	r	ω
6 m/s	2 m	3 rad/s
4 m/s	5 m	0.8 rad/s
15π ft/s	1.5 ft	5 rev/s
π m/s	3 m	60°/sec
8 ft/s	1.5 ft	$\frac{16}{3}$ rad/s

21. 192π in./sec or 16π ft/sec

Exercise Set 13.2
(page 481)

1. a. II   b. I   c. III   d. II   e. IV   f. II   g. IV
3. Quadrant I : 0 to $\frac{1}{4}$ rotation

   Quadrant II : $\frac{1}{4}$ to $\frac{1}{2}$ rotation

   Quadrant III : $\frac{1}{2}$ to $\frac{3}{4}$ rotation

   Quadrant IV : $\frac{3}{4}$ to 1 rotation

5. a., 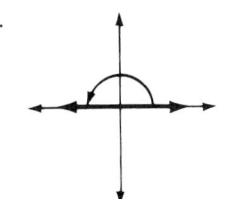 b., 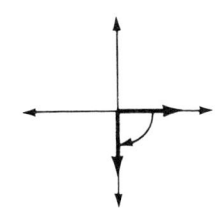 c.

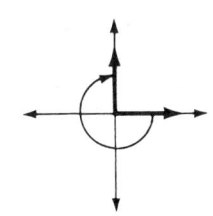

7. a. $-x$ b. $+y$ c. $+y$ d. $-y$ e. $-x$

9. $\tan \theta = \dfrac{4}{3} = 1.333$ $\tan \theta = \dfrac{2.993}{5.2} = 0.576$

$\cot \theta = \dfrac{3}{4} = 0.75$ $\cot \theta = \dfrac{5.2}{2.993} = 1.737$

$\cos \theta = \dfrac{3}{5} = 0.6$ $\cos \theta = \dfrac{5.2}{6} = 0.867$

$\sec \theta = \dfrac{5}{3} = 1.667$ $\sec \theta = \dfrac{6}{5.2} = 1.154$

11. $\sin \theta = \dfrac{4}{4.472} = 0.894$ $\sin \theta = \dfrac{4}{7} = 0.571$

$\csc \theta = \dfrac{4.472}{4} = 1.118$ $\csc \theta = \dfrac{7}{4} = 1.75$

$\tan \theta = \dfrac{4}{2} = 2$ $\tan \theta = \dfrac{4}{5.745} = 0.696$

$\cot \theta = \dfrac{2}{4} = 0.5$ $\cot \theta = \dfrac{5.745}{4} = 1.436$

13. $\sin \theta = \dfrac{6.1}{6.592} = 0.925$ $\sin \theta = \dfrac{5}{7} = 0.714$

$\csc \theta = \dfrac{6.592}{6.1} = 1.081$ $\csc \theta = \dfrac{7}{5} = 1.4$

$\cos \theta = \dfrac{2.5}{6.592} = 0.379$ $\cos \theta = \dfrac{4.899}{7} = 0.7$

$\sec \theta = \dfrac{6.592}{2.5} = 2.637$ $\sec \theta = \dfrac{7}{4.899} = 1.429$

$\tan \theta = \dfrac{6.1}{2.5} = 2.44$ $\tan \theta = \dfrac{5}{4.899} = 1.021$

$\cot \theta = \dfrac{2.5}{6.1} = 0.41$ $\cot \theta = \dfrac{4.899}{5} = 0.98$

Exercise Set 13.3
(page 493)

1. $\sin\,(-)$, $\cos\,(+)$, $\tan\,(-)$, $\csc\,(-)$, $\sec\,(+)$, $\cot\,(-)$
3. a. $\sin = 0.3846$, $\cos = -0.9231$, $\tan = -0.4167$, $\csc = 2.6$, $\sec = -1.0833$, $\cot = -2.4$
   b. $\sin = -0.7332$, $\cos = 0.68$, $\tan = -1.0783$, $\csc = -1.3639$, $\sec = 1.4706$, $\cot = -0.9274$
   c. $\sin = -0.8660$, $\cos = -0.5$, $\tan = 1.7321$, $\csc = -1.1547$, $\sec = -2$, $\cot = 0.5774$
   d. $\sin = 0.7071$, $\cos = -0.7071$, $\tan = -1$, $\csc = 1.4142$, $\sec = -1.4142$, $\cot = -1$
5. a. $61°$ b. $37°$ c. $31°$ d. $60°$ e. $49°$ f. $0°$ g. $60°$ h. $77°$

7.

Trigonometric function	Angle	Smallest angle (SA)	Function value (SA)	Sign	Function value
sec	250°	70°	2.924	−	−2.924
cos	170°	10°	0.985	−	−0.985
sin	100°	80°	0.985	+	0.985
tan	162°	18°	0.325	−	−0.325
cot	300°	60°	0.577	−	−0.577
cos	225°	45°	0.707	−	−0.707
sin	345°	15°	0.259	−	−0.259
cot	140°	40°	1.192	−	−1.192
sec	245°	65°	2.366	−	−2.366

9. a. 0.951 b. −0.423 c. 0.554 d. −0.900 e. 2.924 f. −0.906 g. −0.191
  h. −3.864 i. −3.078
11. a. 1.000 b. 1.000 c. 5.798 d. 0.702 e. −1.000 f. −1.414
13. $\sin 270° = \frac{-2}{2} = -1$, $\cos 270° = \frac{0}{2} = 0$

  $\tan 270° = \frac{-2}{0} = $ undefined, $\csc 270° = \frac{2}{-2} = -1$

  $\sec 270° = \frac{2}{0} = $ undefined, $\cot 270° = \frac{0}{-2} = 0$

15. $\sin 0° = 0$, $\sin -90° = -1$
  $\cos 0° = 1$, $\cos -90° = 0$
  $\tan 0° = 0$, $\tan -90° = $ undefined

Exercise Set 13.4
(page 501)

1. a.

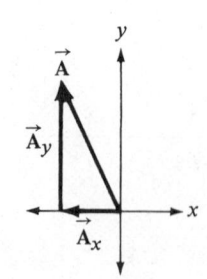

b.

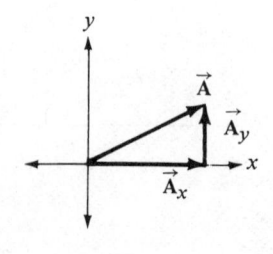

3. a. $A_x = -8.08$, $A_y = 2.63$ b. $A_x = 3.16$, $A_y = -2.47$ c. $A_x = 1.43$, $A_y = 7.38$
  d. $A_x = -8.95$, $A_y = 12.78$ e. $A_x = 1.71$, $A_y = -6.39$ f. $A_x = -4.62$, $A_y = 7.99$
  g. $A_x = -19.1$, $A_y = 0$ h. $A_x = -1.80$, $A_y = -1.80$
5. a. 18°   b. 38°   c. 79°   d. 55°   e. 75°   f. 60°   g. 0°   h. 45°

Exercise Set 13.5
(page 507)

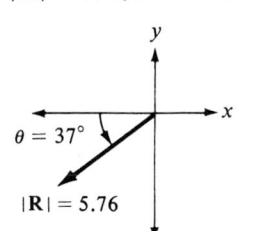

1. $|\mathbf{R}| = 5.76$, $\theta = 37°$ below $-x$-axis

3. $|\mathbf{R}| = 10.5$, $\theta = 38°$ above $-x$-axis

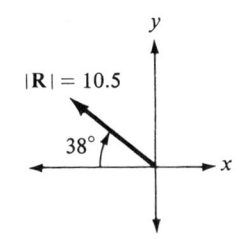

5. $|\mathbf{R}| = 13.0$, $\theta = 49°$ above $+x$-axis

7. $|\mathbf{R}| = 9.2$, $\theta = 20.4°$ above $-x$-axis

9. $|\mathbf{R}| = 3.87$, θ is along the x-axis

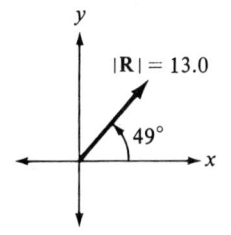

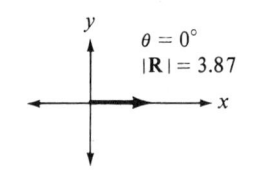

Exercise Set 13.6
(page 511)

1. $|\mathbf{R}| = 5.1$, $\theta = 45°$ below $+x$-axis

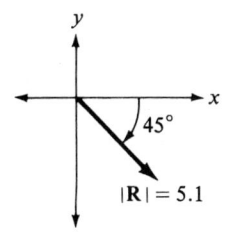

3. $|\mathbf{R}| = 9.79$, $\theta = 15.8°$ below $+x$-axis

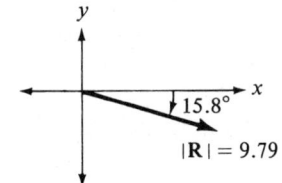

5. $|\mathbf{R}| = 3.7$, $\theta = 40.8°$ above $+x$-axis

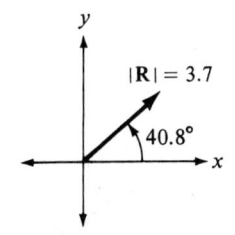

7.

Vector	x-component	y-component
3	3	0
4	0	−4
R	3	−4

$|\mathbf{R}| = 5$ at an angle of 53.1° below $+x$.

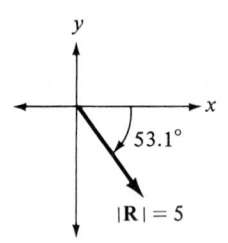

9.

Vector	x-component	y-component
Quad 1	3	+5.196
Neg x	−6	0
Quad 4	3	−5.196
R	0	0

$|\mathbf{R}| = 0$

Exercise Set 13.7
(page 514)

1. $F = 146$ lb, change in tension $= 26$ lb   3. $T_1 = 230$ lb, $T_2 = 193$ lb   5. $W = 871$ lb, $T = 921$ lb

Exercise Set 14.1
(page 526)

1. a. sin θ decreases from 1 to 0.
cos θ decreases from 0 to −1.
tan θ increases from −∞ to 0.
b. sin θ decreases from 0 to −1.
cos θ increases from −1 to 0.
tan θ increases from 0 to +∞.

3.

θ	csc θ
$0 < \theta < \dfrac{\pi}{2}$	$+\infty > \csc\theta > +1$
$\dfrac{\pi}{2} \leq \theta < \pi$	$1 \leq \csc\theta < +\infty$
$\pi < \theta < \dfrac{3\pi}{2}$	$-\infty < \csc\theta < -1$
$\dfrac{3\pi}{2} \leq \theta < 2\pi$	$-1 \geq \csc\theta > -\infty$

Exercise Set 14.2
(page 535)

1.

x	$-\pi$	$-\dfrac{11\pi}{12}$	$-\dfrac{5\pi}{6}$	$-\dfrac{3\pi}{4}$	$-\dfrac{2\pi}{3}$	$-\dfrac{7\pi}{12}$	$-\dfrac{\pi}{2}$
$\sin x$	0	-0.259	-0.5	-0.707	-0.866	-0.966	-1

x	$-\dfrac{5\pi}{12}$	$-\dfrac{\pi}{3}$	$-\dfrac{\pi}{4}$	$-\dfrac{\pi}{6}$	$-\dfrac{\pi}{12}$	0	$\dfrac{\pi}{12}$
$\sin x$	-0.966	-0.866	-0.707	-0.5	-0.259	0	0.259

x	$\dfrac{\pi}{6}$	$\dfrac{\pi}{4}$	$\dfrac{\pi}{3}$	$\dfrac{5\pi}{12}$	$\dfrac{\pi}{2}$	$\dfrac{7\pi}{12}$	$\dfrac{2\pi}{3}$
$\sin x$	0.5	0.707	0.866	0.966	1	0.966	0.866

x	$\dfrac{3\pi}{4}$	$\dfrac{5\pi}{6}$	$\dfrac{11\pi}{12}$	π
$\sin x$	0.707	0.5	0.259	0

3.

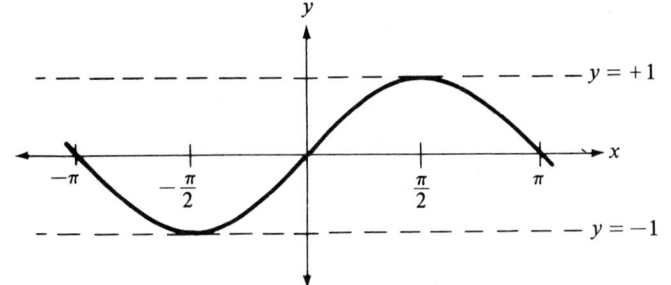

x	π	$\dfrac{13\pi}{12}$	$\dfrac{7\pi}{6}$	$\dfrac{5\pi}{4}$	$\dfrac{4\pi}{3}$	$\dfrac{17\pi}{12}$	$\dfrac{3\pi}{2}$
$\cos x$	-1	-0.966	-0.866	-0.707	-0.5	-0.259	0

x	$\dfrac{19\pi}{12}$	$\dfrac{5\pi}{3}$	$\dfrac{7\pi}{4}$	$\dfrac{11\pi}{6}$	$\dfrac{23\pi}{12}$	2π	$\dfrac{25\pi}{12}$
$\cos x$	0.259	0.5	0.707	0.866	0.966	1	0.966

x	$\dfrac{13\pi}{6}$	$\dfrac{9\pi}{4}$	$\dfrac{7\pi}{3}$	$\dfrac{29\pi}{12}$	$\dfrac{5\pi}{2}$	$\dfrac{31\pi}{12}$	$\dfrac{8\pi}{3}$
$\cos x$	0.866	0.707	0.5	0.259	0	-0.259	-0.5

x	$\dfrac{11\pi}{4}$	$\dfrac{17\pi}{6}$	$\dfrac{35\pi}{12}$	3π
$\cos x$	-0.707	-0.866	-0.966	-1

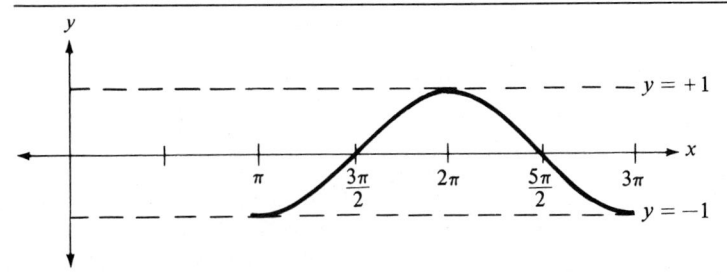

5.

x	$-\pi$	$-\dfrac{11\pi}{12}$	$-\dfrac{5\pi}{6}$	$-\dfrac{3\pi}{4}$	$-\dfrac{2\pi}{3}$	$-\dfrac{7\pi}{12}$	$-\dfrac{\pi}{2}$
tan x	0	0.268	0.577	1	1.732	3.732	undef.

x	$-\dfrac{5\pi}{12}$	$-\dfrac{\pi}{3}$	$-\dfrac{\pi}{4}$	$-\dfrac{\pi}{6}$	$-\dfrac{\pi}{12}$	0	$\dfrac{\pi}{12}$
tan x	−3.732	−1.732	−1	−0.577	−0.268	0	0.268

x	$\dfrac{\pi}{6}$	$\dfrac{\pi}{4}$	$\dfrac{\pi}{3}$	$\dfrac{5\pi}{12}$	$\dfrac{\pi}{2}$	$\dfrac{7\pi}{12}$	$\dfrac{2\pi}{3}$
tan x	0.577	1	1.732	3.732	undef.	−3.732	−1.732

x	$\dfrac{3\pi}{4}$	$\dfrac{5\pi}{6}$	$\dfrac{11\pi}{12}$	π
tan x	−1	−0.577	−0.268	0

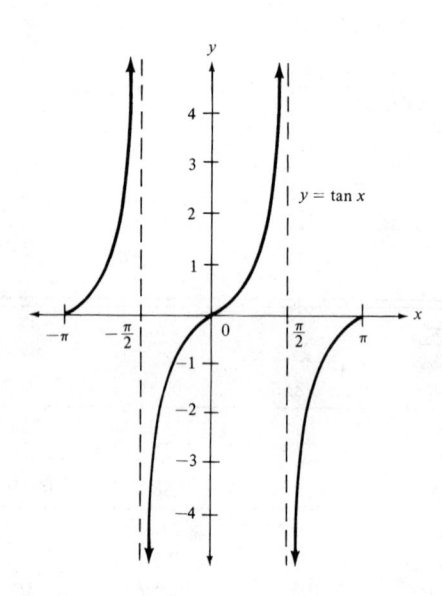

7. See graph.

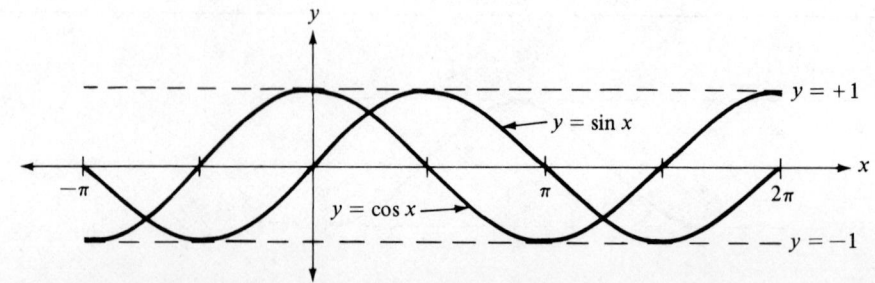

ANSWERS TO ODD-NUMBERED EXERCISES A69

9. See graph.

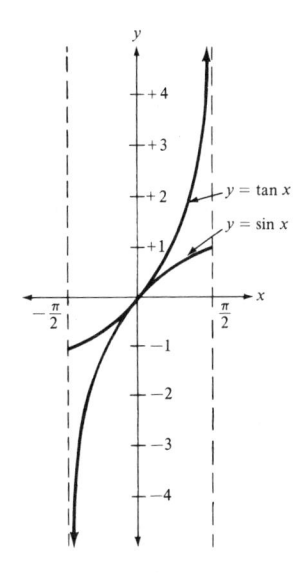

Exercise Set 14.3
(page 541)

1. See graph.

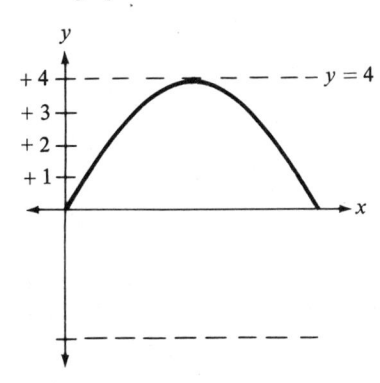

3. See graph.

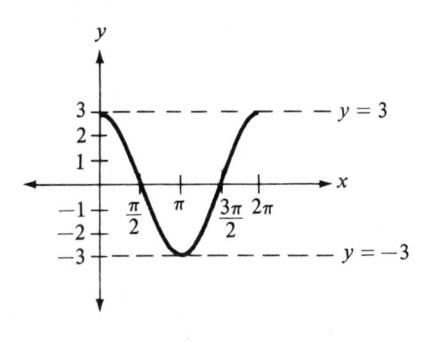

5. a. 0.5   b. 16   c. 1   d. $\sqrt{3}$

7. a. See graph.

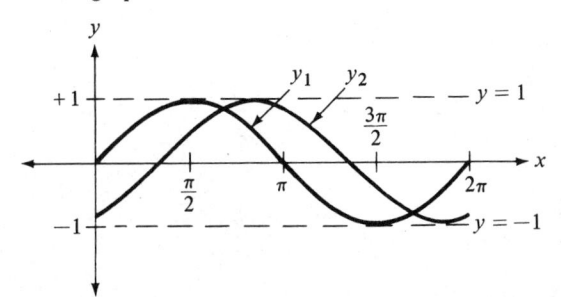

b. See graph.

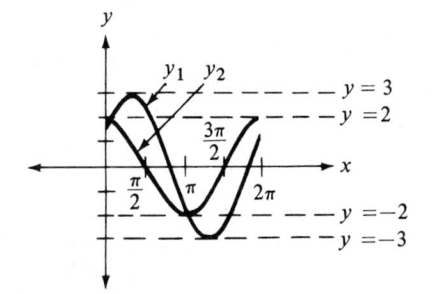

c. See graph.

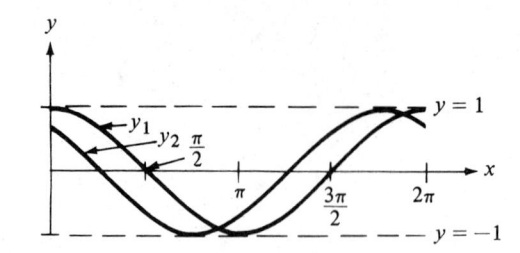

Exercise Set 14.4
(page 546)

1. a. See sketch.

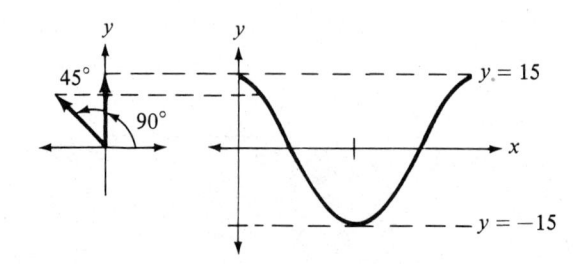

   b. See sketch.

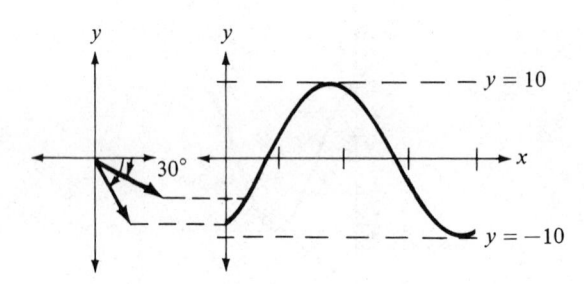

3. a. 14.92 b. 48.54 c. −3.54 A d. 0.00647 V
5. a. $y = 24.2 \sin(10^3 \pi t + 0.32)$. See sketch. b. $y = 2.02 \sin(100\pi t - 0.35)$. See sketch.

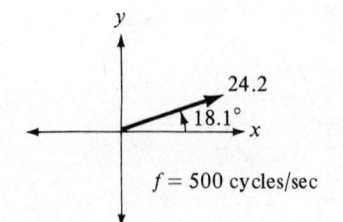

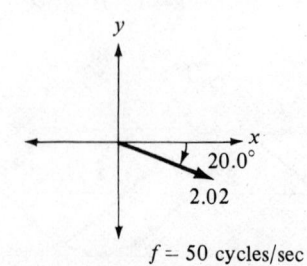

$f = 500$ cycles/sec

$f = 50$ cycles/sec

ANSWERS TO ODD-NUMBERED EXERCISES   A71

Exercise Set 14.5
(page 553)

1. See graph.

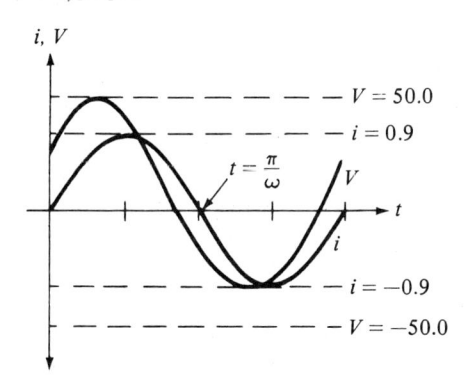

$\dfrac{\pi}{\omega} = \dfrac{1}{800}$

$\dfrac{2\pi}{\omega} = \dfrac{1}{400}$

3. See graph.

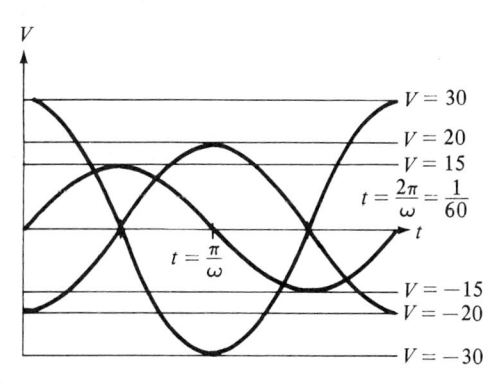

$t = \dfrac{2\pi}{\omega} = \dfrac{1}{60}$

5. See sketch.

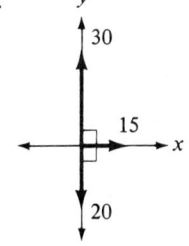

7. a. $\mathbf{Z} = 221.2\ \Omega$
   $\theta = 76.4°$

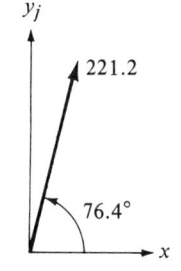

   b. $\mathbf{Z} = 83.3\ \Omega$
   $\theta = 78.8°$

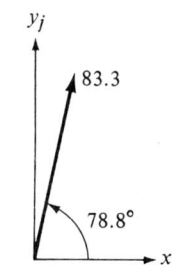

9. a. $\mathbf{Z} = 52 + j215$. The driving voltage leads the current by a phase angle of $76.4°$.
   b. $\mathbf{Z} = 16.2 + j81.7$. The driving voltage leads the current by a phase angle of $78.8°$.

Exercise Set 15.1
(page 567)

1. a. $y = 3^x$

x	y
-2	0.1111
-1	0.3333
0	1
1	3
2	9

b. $y = 0.6^x$

x	y
-3	4.6296
-1	1.6667
1	0.6
3	0.216

c. $y = 1.5^x$

x	y
-4	0.1975
-2	0.4444
0	1
2	2.25
4	5.0625

3. See graph.

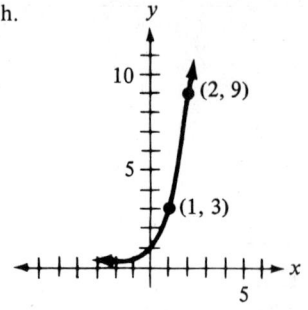

5. See graph.

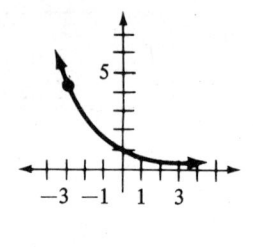

7. $y = e^{-0.6t}$

t	y
−3	6.050
−2	3.320
−1	1.822
0	1.000
1	0.549
2	0.301
3	0.007

9. $y = e^{+1.2t}$

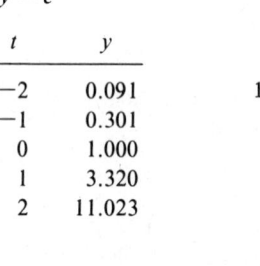

t	y
−2	0.091
−1	0.301
0	1.000
1	3.320
2	11.023

11. 0.5775 time units   13. 1.155 time units

Exercise Set 15.2
(page 571)

1. $x = \log_b y$ 3. $x = \log_3 y$ 5. $x = \log_{10} y$ 7. $x = \log_{1/8} y$ 9. $x = 6^y$
11. $x = \left(\frac{1}{8}\right)^y$ 13. $x = 10^y$ 15. $x = \left(\frac{1}{7}\right)^y$ 17. 2   19. 3   21. −3

Exercise Set 15.3
(page 573)

1. $\log_b 6 + \log_b 3$ 2. $\log_b 5 - \log_b 4$ 5. $\log_b 1 - \log_b 5$ or $-\log_b 5$ 7. $2 \log_b 5$
9. $\frac{1}{2} \log_b 15$ 11. $\log_b 5 - \log_b 7$ 13. $\frac{1}{3} \log_b 25$ 15. $\log_b 8 + \log_b 7$
17. $\log_b 6 + 2 \log_b 4 - 2 \log_b 5$ 19. $4 \log_b 1 - 4 \log_b 2 + \log_b 5$ or $-4 \log_b 2 + \log_b 5$

Exercise Set 15.4
(page 584)

(Answers found using a calculator)
1. 2.8976   3. 1.4997   5. −0.2132   7. −1.3947   9. 2.9545   11. −0.3466

13. −0.0437   15. 31.6589   17. −0.0832   19. 1.4036   21. $D = 12$ decibels (to two digits)
23. 120 decibels   25. 5.4

Exercise Set 15.5
(page 589)

(Answers found using a calculator)
1. 0.4857   3. 7.8808   5. 4.4223   7. −1.5394   9. 6.1717   11. 3.2189   13. 6.1940
15. 8.6330   17. −4.7370   19. −0.1230   21. 17.3 days, 602.4 kg

Exercise Set 15.6
(page 592)

Table	Calculator		Table	Calculator
1. 580.4	580.35	3.	17.17	17.1666
5. 35.07	35.0669	7.	450.9	450.920
9. 140.3	140.296	11.	21.33	21.3316

Exercise Set 15.7
(page 599)

1. a. 0.8 sec   b. 3.6×10^{-7} sec
3. a. See graph.   b. See graph.

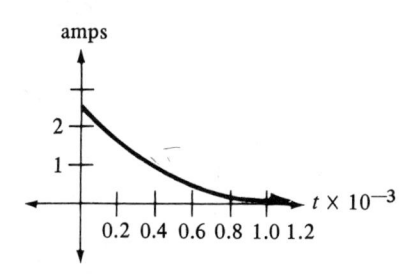

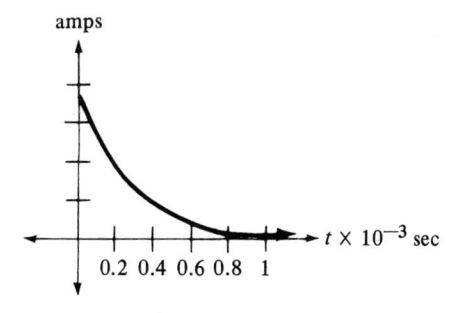

5. 1.7, 3.1, 5.4 days

7. See graph.

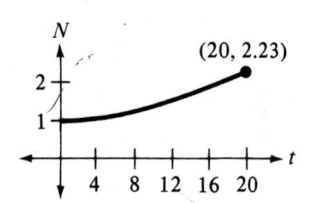

9. See graph.

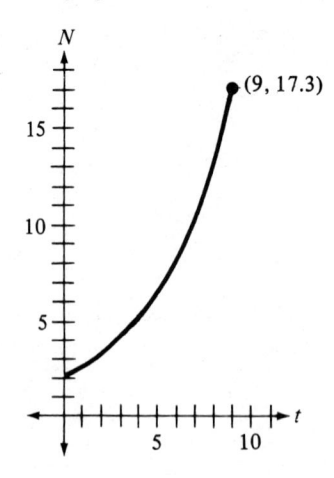

INDEX

Absolute value, 8
Addition method for solving a linear system, 340
Addition Principle
 for equations, 199
 for inequalities, 226
Adjacent side, in right triangle, 289–290
Algebraic expression
 definition, 133
 rational, 134
Algebraic fraction
 definition, 178
 lowest terms, 179
 rules for addition of, 188, 192
 rules for division of, 185, 186
 rules for finding LCD, 187
 rules for multiplication of, 183, 186
 rules for simplification of, 182
 rules for subtraction of, 192
 simplification of, 179, 181
Algebraic term
 additive inverse, 143
 definition, 133
 like, 142
 numerical coefficient of, 140
 rules for adding and subtracting like terms, 143–144
 rules for multiplying and dividing, 153–155
Amplitude, of sine and cosine, 536
Angle
 acute, 89
 complementary, 89
 defined by ratios, 289
 definition, 88
 measure of, 89
 obtuse, 89
 parts of, 462
 quadrant of, 477
 quadrantal, 478
 right, 89
 sides of, 88
 sign of, 463
 smallest, 436
 standard position, 476
 supplementary, 89
 systems of measurement, 463–464
 vertex of, 88

Angular-linear relations
 see linear-angular relations
Angular measure, conversion tables, 465
Applications of linear equations
 geometric, 209–211, 352–353
 mixture, 211–215, 353–356
 number, 208–209, 350–351
 rules for solving using linear equations, 207–208
 rules for solving using linear systems, 349
 uniform motion, 215–217, 220–222
Applications of logarithms and exponentials, 593–598
Associative Law
 for addition, 137
 for multiplication, 154
Asymptote, vertical, 533

Base, 31
Basic quantity, definition, 55
Binomial, 140

Calculator
 in arithmetic operations, 69–71
 exponential entry (power of ten), 73
 in expressions involving exponents, 153
 in expressions involving symbols of grouping, 104, 138
 in right triangle solution, 302–307
 in square root calculations, 102
 in vector composition, 505
 in vector resolution, 499
 key designations, 68
 pi key, 113, 115
 recall key, 101
 store key, 101
Circle
 area, 114
 circumference, 112–113
 definition and properties, 111
 inscribed, 112
Closed with respect to an operation, 3

Common logarithms
 applications of, 582–583
 calculator keystrokes, 581
 definition, 574
 mantissa, 574
 negative characteristic, 579
 positive characteristic, 578
 rules for finding, 578
Commutative Law
 for addition, 137
 for multiplication, 143
Components, of a vector, 497–499
Composite number, 5
Completing the square, 427
Complex form of a number, 385
Complex number
 conjugate of, 386
 definition, 386
 magnitude of, 394
 rules for addition of, 386
 rules for division of, 391
 rules for multiplication of, 389
 rules for subtraction of, 386
Complex z-plane, definition, 393
Constant, 131
Constant of proportionality, 275
Coordinates of a point, 250
Cosecant, 480
Cosine
 with amplitude and phase, 536–541
 coordinate definition, 479
 graph of, 530–531
 triangle definition, 292
Cotangent, 480
Cramer's Rule
 for solving a linear system in three variables, 363
 for solving a linear system in two variables, 345

Data table, for a graph, 246
Degree
 of a monomial, 141
 of a polynomial, 141
Dependent system, 332
Derived quantity, definition, 56
Dimension, definition, 56
Dimensional analysis, A29–A32

INDEX

Direct variation, 277
Distributive Law, 143, 168
Domain, 234
Doubling time, 565

English system of units
  basic quantities, 55
  conversion to SI, 59
Equilibrium, in statics, 513
Evaluation of an expression
  definition, 135
  involving symbols of grouping, 135–136
  order of operations, 136–137
Expansion by minors, 360
Exponential decay, 566
Exponential function
  calculator keystrokes, 560–561
  definition, 560
  special, 564
  special (of time), 566
Exponential growth, 565
Exponents
  laws of, 151
  natural numbers, 31
  negative integers and zero, 151
  use of the calculator with, 153
Expressions involving exponents and logarithms, 573

Factoring
  natural numbers, 5
  polynomials, 168
  second degree polynomials, 169–172
  special products, 173–177
Factorization (see factoring)
First degree equation in one variable, 196
Fractional exponents, 373–374
Function
  definition, 235
  inverse trigonometric, 305–307
  as a processor, 238
  trigonometric, 303–304

Geometry, 87
Graph of the general exponential function, 562–563
Graph of the general quadratic function
  characteristics of, 423
  critical points, 431

Graphical solution of a linear system, 333
Graphing experimental data
  pictorial graphs, A13–A15
  semi-logarithmic plots, A5–A13
  techniques, A1–A5
Graphing linear equations in two variables
  rules for, 257
  using a data table, 246
  using the line characteristics, 259–268
Greatest common divisor (GCD), 16

Half-life, 566
Horizontal method
  of addition, 144
  of multiplication, 156–157

Identity
  additive, 4
  multiplicative, 4
Imaginary number, definition, 386
Inconsistent system, 332
Independent system, 332
Index, of a radical, 370
Inequality
  definition, 219
  symbols, 219
Infinity, a point on the number line, 524
Integers
  definition, 7
  rules for performing operations on, 8–11
Intercepts, parabola, 423
Intercepts, straight line
  definition, 259
  rules for finding, 260
Inverse
  additive, 7
  multiplicative, 21
Inverse trigonometric functions, 305–307
Inverse variation, 278

j-operator, 395–398

Law of Cosines, 316
Law of Sines, 314
Least common denominator (LCD)
  definition, 18
  rules for finding, 18

Line
  parallel, 88
  perpendicular, 89
  segment, 88
  straight, 88
Line graph, 252–254
Line of symmetry, parabola, 422
Linear equation in one variable
  definition, 196
  solved, 199–206
Linear equation in two variables
  definition, 241
  graphed, 256–268
Linear inequality in one variable
  definition, 220
  solved, 223–228
Linear least squares fit, A19–A24
Linear-angular relations
  arc-angle, 467
  area-angle, 468
  speed-angular speed, 471
Literal number, 131
Logarithms
  common, 574
  computations with, 590
  definition, 569
  laws of, 571
  natural, 585

Mean and standard deviation, A25–A28
Measurement
  steps in, 52–53
  units of, 55
Minor, 360
Monomial, 140
Multiplication Principle
  for equations, 200
  for inequalities, 225

Natural logarithms
  applications of, 588–589
  calculator keystrokes, 587
  definition, 585
  rules for finding, 585
Number line
  construction of, 2, 7
  origin of, 2
Numbers
  integers, 7
  irrational, 27
  natural, 2
  nonreal, 29

INDEX

Numbers (*Cont.*)
  rational, 14
  real, 27
  whole, 2

Oblique triangles, solution, 313–318
Opposite of a number, 7
Opposite side, in right triangle, 289–290
Order of operations, 136–137
Ordered pairs, 234

Parabola, definition and characteristics, 422
Percent
  conversions involving, 39–40
  definition, 39
  rules for solving problems involving, 41
Period, 530
Periodic function, 530
Phase difference, 540
Pictorial graphs, A13–A15
Plane figure, 90
Point, 88
Point of equilibrium (supply and demand), 456
Point plot, 251–252
Point-slope form, straight line, 271
Polygon
  area formulas, 100–104
  circumscribed, 112
  congruent, 92
  definition, 90
  inscribed, 112
  names according to number of sides, 90
  perimeter, 97
  similar, 93
  special quadrilaterals, 91
  special triangles, 91
  sum of measures of angles, 105
  using similarity to solve, 106–109
Polynomial
  definition, 139
  degree, 141
  names according to number of terms, 140
  rules for adding and subtracting, 144–148
  rules for multiplying, 156–159

Power of tem
  definition, 32
  rules for multiplying and dividing, 36
  rules for multiplying by, 32–33
Prime factorization, 5
Prime number, 5
Proportion, definition and properties, 43
Pythagorean Theorem, 92

Quadrant, of an angle, 477
Quadrant, in a coordinate system, 249–250
Quadrantal angle, 478
Quadratic equation in one variable
  definition, 404
  rules for solving with the calculator, 415–417
  rules for solving by factoring, 405
  rules for solving with the quadratic formula, 411
Quadratic equation in two variables
  as a function, 436
  definition, 422
Quadratic formula, 410–411
Quadratic system
  applications of, 444–456
  definition, 437

Radian measure, 464
Radical
  calculator keystrokes, 382–383
  definition, 370
  fractional exponent form, 373–374
  rationalizing the denominator, 380
  rules for addition of, 376
  rules for division of, 380
  rules for multiplication of, 378
  rules for simplifying, 371
  rules for subtraction of, 377
Radical sign, 370
Radicand, 370
Range, 234
Ratio, 42
Rational numbers
  converting between fraction and decimal forms, 24
  decimal form, 14
  fraction form, 14
  improper fraction, 19
  lowest terms, 15

Rational numbers (*Cont.*)
  mixed number, 19
  raising to higher terms, 16
  repeating decimal, 15
  rules for performing operations on, 17–23
  terminating decimal, 14
Ray, 88
Reciprocal, 21
Rectangular coordinate system
  defined, 249–250
  origin of, 249
  quadrants, 249–250
Relation, 235
Resolution, of a vector, 497–501
Resultant, in vector addition, 508
Right triangle
  applied problems, 309–311
  hypotenuse, 92
  leg, 92
  Pythagorean Theorem, 92
  solution by calculator, 302–307
  solution by trigonometric ratios, 299–301
Rotating vectors, 542–546
Rounding, rules for, 81

Scientific notation
  definition, 33
  rules for performing operations on numbers in, 35–37
  rules for writing a number in, 33–34
Secant, 480
Second order determinant
  definition, 344
  evaluation of, 344
Semi-logarithmic plot, A5–A13
SI system of units
  advantages, 62, 66
  basic quantities, 55
  conversion to English, 59
Sine
  with amplitude and phase, 536–541
  coordinate definition, 479
  graph of, 527–530
  triangle definition, 291
Significant digits
  in addition and subtraction, 79
  definition, 77
  in multiplication and division, 78
  rules for indicating, 80

INDEX

Slope
  definition, 263
  observations about, 266
Slope-intercept form, straight line, 270
Smallest angle, 486
Solids
  circular cone, 118
  circular cylinder, 118
  definition, 117
  prism, 117
  sphere, 118
  surface area formulas, 119–122
  volume formulas, 123–125
Solution
  of an equation, 197
  of an inequality, 220–221
  of a linear system, 332
Special products
  factored, 173–176
  formed by multiplication, 159–162
Square root, 28
Standard position, angle, 476
Statics, 513–514
Statistical analysis
  linear least square fit, A19–A24
  mean and standard deviation, A25–A28
Straight line equation
  in point-slope form, 271
  in slope-intercept form, 270
Substitution method for solving a linear system, 337
Symbols of grouping, 135
System of linear equations
  in three variables, 360

System of linear equations (*Cont.*)
  in two variables, 332
  solution of, 332
  solved graphically, 333–336
  solved by addition method, 340–342
  solved by Cramer's Rule, 345–348, 363–366
  solved by substitution method, 337–340
Systems of units
  conversion tables, 59
  English, 54
  SI, 54

Tables
  trigonometric functions, A34
  logarithms to base 10, A35–A36
  natural logarithms, A37–A38
  powers of e, A39–A40
Tangent
  coordinate definition, 479
  graph of, 532–534
  triangle definition, 293
Third-order determinant
  definition, 360
  evaluation by expansion by minors, 360
Time constant, 594
Time dependent wave, 543
Transient, 594
Trigonometric functions
  of quadrantal angles, 303–304
  in a right triangle, 303–304
  using coordinates, 479
  values by quadrants, 522–526

Trigonometric ratio, 290
Trigonometry
  definition, 285, 326
  in surveying, 321–325
Trinomial, 140

Units (see systems of units)

Variable
  definition, 131
  dependent, 234
  domain, 234
  independent, 234
  notation for operations on, 132–133
  range, 234
Variation
  constant of proportionality in, 271
  direct, 277
  inverse, 278
Vector addition
  by components, 508–511
  graphically, A15–A18
Vector components, 497–499
Vector composition, 502–507
Vector magnitude 498, 503
Vector resolution, 497–501
Vertex of a parabola
  definition, 422
  rules for finding, 428
Vertical asymptote, 533
Vertical method
  of addition, 144
  of multiplication, 156–157